D1246700

TRANSITION METAL COMPLEXES OF PHOSPHORUS, ARSENIC AND ANTIMONY LIGANDS

ASPECTS OF INORGANIC CHEMISTRY

General Editor: C. A. McAuliffe

Transition Metal Complexes of Phosphorus, Arsenic and Antimony Ligands

edited by

C. A. McAULIFFE

University of Manchester Institute of Science and Technology

MACMILLAN

First published 1973 *by*
THE MACMILLAN PRESS LTD
London and Basingstoke
Associated companies in New York
Melbourne Dublin Johannesburg and Madras

SBN 333 13628 4

Printed in Great Britain by
William Clowes & Sons Limited
London, Colchester and Beccles

This volume is dedicated to
Frederick G. Mann, F.R.S.,
to whom the subject owes so much

FOREWORD

by

Professor J. Chatt
University of Sussex

I am very grateful to the Editor for his invitation to write a short introduction to
this book, which is dedicated to Frederick G. Mann, F.R.S. Dr Mann was my Research
Supervisor and mentor from 1937 to 1939. He introduced me to the fascinating
organic-type chemistry of complex compounds containing tertiary phosphines, arsines
and stibines at a time when remarkably little was known about them.

A few organic phosphines and arsines had been prepared during the latter half of
the last century. The chemists who prepared them had investigated their reactions
with various metal salts in analogy with amines, and had reported the formation of
addition compounds. Some were salts, such as the platinichlorides obtained from
acid solution, and others were substances now known as coordination compounds.
The *cis* and *trans* isomers of bis(triethylphosphine)dichloroplatinum(II) were isolated
in 1870 by Cahours and Gal, the yellow isomer being labelled α and the white one β.
With the advent in 1893 of Werner's theory of coordination the structures of complex
compounds and the nature of the isomerisation of the platinum(II) complexes
became understandable. Werner was able to assign the *cis* configuration to the brighter
yellow $[PtCl_2(NH_3)_2]$ isomer and the *trans* to the paler yellow. By analogy all the
other platinum(II) isomers were assigned configurations according to their colours.
The colourless β-$[PtCl_2(PEt_3)_2]$, now known to be *cis*, was thought until 1936 to
have a *trans* configuration. From the time of their discovery until about 1930 the
complex compounds of the tertiary phosphines, arsines and stibines received scant
investigation and the isomeric platinum(II) complexes, except for the ammines, were
given wrong configurations.

In the early part of this century many organo-arsenicals were found to have marked
therapeutic value, and others to have potential use as 'poison gases', vesicants or
sternutators, in war. Thus there was a great extension of organoarsenic chemistry and
by 1930 many more organoarsenicals were known than organic phosphorus, antimony and bismuth compounds added together. During the First World War the
Germans started to use shells containing organoarsenicals. Samples were sent to
Professor W. J. Pope, a member of the Chemical Warfare Committee, who started an
investigation of the compounds. He had two assistants, E. E. Turner and later F. G.
Mann, both of whom eventually became Fellows of The Royal Society. In this way
organoarsenic work came to Cambridge.

When the war was over Turner emigrated to Sydney in Australia and there he continued his study of organoarsenicals. Mann remained in Cambridge, becoming interested in the semi-polar double bond or coordinate link, whether it occurred in complex compounds or in organic compounds. There was still some doubt about the nature of the bond between the metal and ligand in complex compounds and the parachor, introduced by Sugden in 1924, was then a new and respectable method of determining bond type. It was thought to be an additive quantity, with each atom and bond having a definite parachor. Because its determination involved the measurement of the surface tension of a substance it was applicable only to liquids or easily meltable solids. By 1930 parachors of the common elements and of the common bond types in organic compounds had been estimated. Mann wished to apply the parachor to determine bond type in coordination compounds and for this purpose he needed readily fusible stable coordination compounds. Amongst those he chose were the tertiary arsine and tertiary phosphine complexes of palladium and mercury halides. His investigation served mainly to increase greatly the number of such complexes and to show that the parachor was useless as a method of determining bond type. However, on thermal decomposition some of the palladium compounds, $[PdCl_2(AsR_3)_2]$, they lost trialkylarsine, so being converted into halogen-bridged complexes, $[Pd_2Cl_4AsR_3)_2]$, for which he developed a better and more general synthetic method. Then, even the existence of halogen bridging was controversial, but the presence of such bridges was established by X-ray analysis and the basis of a whole new and interesting chemistry was opened up. In about a dozen papers in the 1930s, Mann and his coworkers described more new tertiary phosphine and arsine complexes than had been prepared in the whole previous history of complex chemistry.

A parallel line was developed in Copenhagen by K. A. Jensen, who prepared a great number of simple tertiary phosphine, tertiary arsine and dialkylsulphide complexes of platinum and determined their dipole moments, showing that all the isomers that were previously thought to be *trans* were *cis,* and *vice versa.* He also showed that tertiary phosphines would stabilise unusual oxidation states by preparing $[NiBr_3(PEt_3)_2]$. In the meantime another line stemming from Pope in Cambridge had slowly taken root in Australia. E. E. Turner handed on his techniques and interests to G. J. Burrows of Sydney before he returned to England in 1921 and Burrows turned to the study of arsine complexes. This interest was handed on to F. P. J. Dwyer and through Dwyer to R. S. Nyholm. Nyholm emigrated to London after the Second World War, where he developed a very flourishing school of arsine complex chemistry at University College, based mainly on the famous diarsine, $o\text{-}C_6H_4(AsMe_2)_2$, first prepared in Mann's laboratory in the late 1930s.

The early parachor work of F. G. Mann was done in collaboration with a young student named Donald Purdie. He was killed in the Second World War in Malaya, but not before he had prepared with Mann many series of complexes of various elements, palladium, mercury, cadmium, zinc etc., with tertiary phosphines and arsines and more than doubled the total number of such complexes known. I was brought into the study of Mann's bridged complexes in 1937, and realised the enormous value of tertiary phosphine and arsine ligands in bringing complex chemistry out of aqueous solution into organic solvents. In these ways the fusion of complex chemistry with organic chemistry started to grow, and after the war this fusion caught the imagination of young coordination chemists everywhere. These old, but nevertheless newly re-discovered, complex compounds were found to be fine material for the study of the behaviour of metal atoms in a non-aqueous environment. Out of this grew polymerisation, hydrogenation and carbonylation catalysts, the development of alkyl and aryl derivatives of the transition metals, the study of unusual stereochemistries and the magnetic properties of asymmetrically coordinated transition metal ions, and much of the vast new organometallic chemistry of the transition metals. This book is concerned with the chemistry that has led to these developments.

From the third quarter of the last century a small selection of tertiary phosphines and their complexes had been known, but their chemistry had failed to stir the imagination of chemists. It was F. G. Mann, K. A. Jensen and G. J. Burrows (mainly through his pupil Dwyer) who showed that they had a chemistry of great interest, and it was they who handed on this interest to their disciples and followers, until now we have the vast new knowledge that fills this volume.

J. CHATT
January 1973

PREFACE

There are few areas of inorganic chemistry in which research activity is as great as in the study of transition metal complexes containing Group VB ligands. There is a continuous increase in the number of papers which academic and industrial laboratories produce in this field.

Apart from a review of the whole area by Booth in 1964, and a review of metal complexes of diphosphines by Levason and McAuliffe in 1972, there have been no other specific contributions to the review literature of this field. The authors of this volume have thus reviewed transition metal complexes containing monodentate phosphines, monodentate arsines and stibines, bidentate arsines, and multidentate Group VB ligands. A section on the nature of the bond formed between transition metals and Group VB ligands has also been included, since this topic has been the subject of much discussion recently, and a reassessment of the nature of this interaction seems timely. In general, complexes containing organometallic ligands have been excluded, although carbonyls have been included. We have thus treated the general area of coordination chemistry and not of organometallic chemistry. The literature has been reviewed up to the end of June 1972.

As Editor I have attempted to produce uniformity in symbolism, nomenclature, etc., and any shortcomings in this respect are directly attributable to me.

The aim of the book is to provide a reference text for workers in this field, especially for workers new to the field who may find the literature already bewilderingly large. It is also hoped that undergraduates may find this book useful.

In putting this volume together I have been valuably assisted at all stages by my graduate student, William Levason. I wish also to thank my wife, Margaret, for her assistance, encouragement and forbearance.

C. A. McAULIFFE

CONTENTS

PART 3 · TRANSITION METAL COMPLEXES CONTAINING MONOTERTIARY ARSINES AND STIBINES

J. C. Cloyd, Jr and C. A. McAuliffe

PART 4 THE CHEMISTRY OF MULTIDENTATE LIGANDS CONTAINING HEAVY GROUP VB DONORS

B. Chiswell

Contents

PART 1
Group VB to Transition Metal Bonds

A. PIDCOCK

School of Molecular Sciences, University of Sussex, Brighton BN1 9QJ

1 INTRODUCTION

The structure of this introductory essay requires some explanation, because for much of the discussion the behaviour of platinum complexes of phosphorus-containing ligands is taken as the starting point. Considerable emphasis on phosphorus-containing ligands is unavoidable: parameters such as bond lengths, stretching frequencies, n.m.r. shifts and coupling constants and many more that bear reasonably directly on the question of the nature of the metal–ligand bonds are very much more plentiful for phosphorus ligands than for the ligands containing the heavier Group VB elements. Also, the study of platinum compounds has been very closely associated with the development of theories of metal–Group VB element bonding. During the 1940s some similarity had been noted between carbon monoxide, phosphorus donors and arsenic donors in their ability to form complexes with transition elements. In the late 1940s J. Chatt had the idea, based on Pauling's theory of metal–carbonyl double bonding, that ligands such as phosphines had vacant d-orbitals which could accept electrons from filled metal d-orbitals, thus forming a 'dative π-bond' in addition to the ligand–metal σ-bond. This transfer of electron density from the metal to ligand would be promoted by electronegative groups on phosphorus, so Chatt tested the theory by studying the interaction of phosphorus trifluoride with platinum(II) chloride and boron trifluoride. Platinum, which has suitable d-electrons, formed stable complexes, whereas boron, which has no electrons available for dative π-bond formation, did not have any significant interaction with phosphorus trifluoride. In 1950 Chatt[1] published these results with an account of the theory, and he also noted that the drift of electrons forming the π-bond would partly neutralise the inductive effect of the fluorine atoms on the lone pair, so that the availability of the lone pair would not be greatly impaired by electronegative groups on phosphorus. The tendency of one form of bond to neutralise the charges produced by the other form became an essential part of the theory and was later termed a 'synergic' effect. Subsequently Chatt and his associates examined platinum compounds by many of the physical techniques then available, and a variety of thermodynamic[2] and spectroscopic[3] parameters were determined. Craig *et al.* calculated overlap integrals for various types of metal–ligand π-interaction[4].

The experimental results and the calculations were consistent with the existence of strong dative π-bonding between platinum(II) and phosphine ligands. The theory became generally accepted and applied to many diverse phenomena such as the (a) and (b) classification of acceptors[5], and the ability of Group VB donors to stabilise transition-metal alkyls and aryls[6].

In this early period the experimental methods gave results that were determined largely by the overall characteristics of metal–ligand bonds rather than specifically the σ- or the π-components. The degree of participation of the d_π–d_π bonding could be inferred only after making assumptions about the σ-bonds. The position with regard to the σ-system at the time is perhaps best revealed by a quotation from a paper published in 1957[7]. In a discussion of the isomers of $[PtCl_2(R_3P)_2]$, it was assumed that in 'systems of *cis*- and *trans*-configurations in which all the ligands are the same, the (σ-electron) withholding due to the inductive effect should be nearly equal in both systems'. Thus, any differences in the metal–ligand bonds were attributed to d_π–d_π bonding and its synergic effect on the σ-system. More recently Graham[8], in an analysis of force constants in carbonyl complexes, has made the similar assumption that the σ-inductive effects of ligands are isotropic, that is, transmitted equally to *cis*- and *trans*-ligands.

Consideration of molecular orbitals or hybrid metal orbitals that may respond to the nature of the ligands makes the assumption of an isotropic inductive effect hard to justify, and there are now good reasons, based on both theory and experiment, for the belief that for some ligands the inductive effect is highly anisotropic. Symmetry certainly does not demand that *cis*- and *trans*-bonds should be equally affected, so an isotropic inductive effect may well be the exception rather than the rule. The existence of a strongly anisotropic σ-effect was first established by Chatt and Shaw for the hydride ligand in *trans*-$[PtHCl(Et_3P)_2]$, where the chloride ligand

TABLE 1   BOND LENGTHS (pm) FOR SOME PLATINUM(II) CHLORO-COMPLEXES

	l(Pt–Cl)	l(Pt–P)	Ref.
trans-$[PtHCl(EtPh_2P)_2]$	242.2(9)	226.7(8)	11
trans-$[PtCl_2(Et_3P)_2]$	229.4(9)	229.8(18)†	12
cis-$[PtCl_2(Me_3P)_2]$	236.4(8), 238.8(9)	225.6(8), 223.9(6)	3

† l(Pt–P) is 231.5(4) in *trans*-$[PtBr_2(Et_3P)_2)]$[12].

was found to be much more labile than when *trans* to chloride, and the high energy of hydrogen orbitals of π-symmetry precluded an explanation based on π-bonding[9]. X-ray crystallographic studies of the complexes *trans*-$[PtHBr(Et_3P)_2]$ [10], *trans*-$[PtHCl(EtPh_2P)_2]$ [11] and *trans*-$[PtCl_2(Et_3P)_2]$ [12] (see table 1) showed that the bonds *trans* to hydride were long, but the platinum–phosphorus bond lengths *cis* to hydride were similar to those in the dichloro-complex. Alkyl and aryl ligands were also shown to labilise specifically the *trans*-bond[6], so the early assumption of an isotropic σ-effect for phosphines became questionable. It is important to note that the early results of physical measurements on platinum compounds could be explained satisfactorily by making the assumption that phosphines weaken the binding of *trans* ligands by a mechanism that involves only the σ-electrons of the complex, without invoking platinum–phosphorus π-bonding. It is also possible that σ- and π-effects are each involved to a significant degree, and it is clear that great care is necessary with the interpretation of results. Both mechanisms are often expected to produce qualitatively similar effects, and only a small proportion of experimental methods differentiate between the σ- and π-effects.

Related to the question of the nature of metal–Group VB bonding is that of the mechanisms by which ligands influence neighbouring metal–ligand bonds. Apart from steric effects, ligands can also be considered to compete for metal σ-orbitals and π-electrons, but, in addition to this direct competition, any modification of the metal orbitals induced by a ligand will in general affect other bonds of all types formed by the metal. One of the most powerful general methods for the study of metal–Group VB bonds is to measure the changes they and other ligands induce in the bonds to or within neighbouring ligands. However, the derivation of information concerning the metal–Group VB bonds must then necessarily involve additional hypotheses for the mechanisms of transmission of electronic effects through the metal.

2 METALS IN POSITIVE OXIDATION STATES

2.1 BONDS TO PHOSPHINE LIGANDS

Alkyl and aryl phosphines form stable complexes with many non-transitional element acceptor atoms and with transition metals in a wide range of oxidation states. In comparison with phosphites and fluoro- and chloro-phosphines, they can be expected *a priori* to be the strongest σ-donors and the weakest π-acceptors. In this section we examine the results of physical measurements on phosphine and related complexes in order to assess the significance of each form of bonding and the way in which phosphines affect the bonding of other ligands.

Measurements of the equilibrium position in solution for the interconversion of *cis*- and *trans*-isomers of $[PtCl_2(R_3P)_2]$ showed that after allowance for solvation the *cis*-isomer was the more stable by about 40 kJ mole, clearly implying that the total metal–ligand bond energies are greater in the *cis*- than the *trans*-isomer[2]. In the original explanation of these results the extra bond energy of the *cis*-isomer was attributed to the platinum–phosphorus bonds, and explained solely in terms of platinum–phosphorus π-bonding and the consequential effects on the σ-bonds. In the *trans*-isomer the phosphines share platinum d_π-electrons in two orbitals whose plane of symmetry includes the P–Pt–P axis, whereas in the *cis*-isomer only the d_π-electrons in the plane of the complex are shared between the phosphorus atoms, and each of them competes favourably with chloride for a separate pair of electrons in a plane perpendicular to that of the complex. The π- and σ-platinum–phosphorus bonds should then be stronger in the *cis*-isomer. The platinum–phosphorus bond lengths (table 1) were found to be consistent with this explanation though the authors did not enter into a detailed discussion, because the *cis*-isomer was distorted from the square-planar structure[13]. However, these and other structure determinations have revealed that in transition-metal complexes bonds to chlorine are longer when *trans* to phosphines than when *trans* to chlorine, and that ligands in *cis*-relationship have a much smaller effect. In situations where steric effects are small, variation of metal–chlorine distances can be attributed to changes in the ionic character of the bonds and in the combination and energy of metal atomic orbitals involved in the covalent component of the σ-bond. The interaction between metal d_π-electrons and chloride ligand is probably very weak[14]. The relative weakness of platinum–chlorine bonds *trans* to phosphines is also apparent from platinum–chlorine stretching frequencies[15] and from the lability of the chloride ligands to nucleophilic substitution[16].

This analysis establishes that phosphines, like the hydride and alkyl ligands mentioned earlier, have a strong tendency to weaken metal–halogen σ-bonds in *trans* relationship, and it becomes highly probable that other σ-bonds *trans* to phosphines are similarly affected to some degree. Thus, metal–phosphorus bonds are expected to be longer when mutually *trans* than when *trans* to chloride. This being so, it is then not clear whether π-bonding has any appreciable effect on the bond lengths in, for example, *cis*- and *trans*-$[PtCl_2(R_3P)_2]$.

In 1966 on the basis of measurements of $^{195}Pt-^{31}P$ spin–spin coupling constants, the suggestion was made that in complexes of platinum(II) π-bonding between the metal and phosphine ligands may have a negligible influence on ground-state chemical properties[17]. Because p, d, f . . . atomic orbitals have zero density at the parent nucleus, the interaction between the nuclei is carried essentially by the s-orbital component of the platinum–phosphorus bond, and the coupling constants are therefore determined by the nature of the platinum–phosphorus σ-bond[18]. The s-component and the coupling constant are expected to be large for highly covalent platinum–phosphorus bonds, and also in bonds formed from hybrids of high s-character. As the s-character of the hybrids increases, the bonds are expected to become stronger[17-19], so for complexes of a given metal and phosphorus ligand, the coupling constants provide what is probably a reasonably reliable guide to the order of strengths of metal–phosphorus σ-bonds. Key results in the early work were obtained from the *cis*- and *trans*-bis(tributylphosphine) complexes of platinum in oxidation states II and IV. The ratio $^1J(cis$ complex$)/^1J(trans$ complex$)$ was found to be 1.47 for the platinum(II) complexes and 1.41 for the platinum(IV) complexes, and the very close similarity between these ratios suggested that the same bonding mechanisms are present in both oxidation states.

In more detail, there appear to be the following three possibilities: (a) that π-bonding has a significant synergic effect on the σ-bonds and is of fairly similar strength in the two oxidation states; (b) that back-donation by platinum(IV) is much weaker than in platinum(II), but the coupling constants are insensitive to the degree of π-bonding; (c) that π-bonding in both oxidation states is insignificant. Possibility (a) seems to be the least likely. The energy gap between the filled d_π-orbitals of platinum and the vacant $3d_\pi$-orbitals of phosphorus must be larger in the higher oxidation state, and significant back-donation is thereby less probable. For possibility (b) some judgement must be made of the sensitivity of the coupling constants to the strength of the π-bond. It has been shown that even for a single oxidation state of platinum with a single phosphorus ligand type, the coupling constants vary by over a factor of two, depending mainly on the nature of the ligand *trans* to phosphorus[18]. Few physical parameters of bonds vary so dramatically, so that even though the coupling constants would be affected only through the synergic effect of π-bonding, it seems unlikely that a significant degree of back-donation in platinum(II) would remain undetected. Nevertheless, the ratios of coupling constants are somewhat smaller in platinum(IV) than in platinum(II), and this difference persists if the more rigorous comparison is made of the coupling constants in $[PtCl(Et_3P)_3]^+$ and *mer*-$[PtCl_3(Et_3P)_3]^{+}$ [20]. Here the analogous ratios $^1J(trans$ to Cl$)/^1J(trans$ to P$)$ are 1.57 for platinum(II) and 1.50 for platinum(IV). However, even in the absence of π-bonding precise correspondence between platinum(II) and (IV) is hardly to be expected, as this would imply a precisely similar distribution of covalency and platinum 6s-orbital among the metal–ligand bonds in each oxidation state. These results therefore, are taken to suggest strongly that π-bonding does not have an appreciable synergic effect on platinum–trialkylphosphine σ-bonds. As the bond lengths can also be explained solely in terms of the strong σ-effects whose existence is now fully established, it is reasonable to adopt the working hypothesis that π-bonding does not contribute significantly to the binding of trialkylphosphines to platinum(II).

Before we examine this hypothesis in the light of other kinds of physical measurement, it is useful first to extend the comparison of the bonding in the situations P–M–Cl and P–M–P on the basis of nuclear resonance and X-ray studies to complexes of other transition metals. Suitable phosphorus–metal coupling constants have been measured for certain phosphine complexes of W(IV), Rh(I) and Rh(III) [20,23] and X-ray structures have been determined for the same or very similar complexes[22,25,26]. Results are given in table 2 for meridional tris(phosphine) complexes that have a chloride ligand *trans* to the unique phosphine ligand. It is evident that the ratio $^1J(P–M–Cl)/^1J(P–M–P)$ is substantially larger than unity for all cases, and the results suggest that the difference in covalency of P–M bonds *trans* to chloride compared with those *trans* to phosphine increases from tungsten in Group VI to platinum in Group VIII. Also included in table 2 are results for some mercury(II) phosphonate complexes, which have been shown to possess approximately linear P–Hg–Cl and P–Hg–P arrangements by X-ray structure determinations[21,24]. The results for these complexes are not precisely comparable with the others, but they appear to support the stated trend. Metal–phosphorus bond lengths for the mercury compounds, the *cis-* and *trans*-bis(phosphine)platinum(II) compounds and several meridional tris(phosphine) complexes are given in tables 1 and 3. The ratios of the metal–phosphorus bond lengths for phosphorus *trans* to chlorine and phosphorus *trans* to phosphorus are as follows

metal:	Mo(IV)	Re(III)	Os(III)	Ir(III)	Rh(I)	Pt(II)	Hg(II)
ratio:	1.018	1.024	1.025	1.038	1.046	1.027	1.028

For the meridional complexes of Mo, Re, Os, Ir and Rh, there is a gradual increase in the ratio in the order stated; the values for Pt and Hg do not follow this trend, but the bond lengths used in the ratios were not determined from single complexes. The results for the two methods appear to be compatible, and it seems probable that the trends in the results should be attributed to changes in the energies and distribution of the metal atomic orbitals involved in the metal–ligand σ-bonds. The results for molybdenum, tungsten and mercury complexes are particularly unlikely to be influenced by back-donation from the metals. For the molybdenum and tungsten complexes the oxidation state is high and it is more likely that the metals behave as acceptors of ligand π-electrons[27]. In mercury(II) it seems to be generally accepted that the d-electrons have become so tightly bound that back-donation is unlikely to be energetically advantageous[24].

The effects of both variation of *trans*-ligand and of central metal on the distribution of metal orbitals among the metal–ligand σ-bonds have been examined with theoretical models. Because of the complexity of the problem, none of the calculations can be regarded as being very satisfactory, but they are valuable in that they appear to reproduce σ-effects similar to those demonstrated by experiment, and they may serve to direct the attention of experimentalists towards the problem of measurement of σ-effects. Gray and Langford[28] considered the interaction of ligand σ-orbitals with the metal p_σ-orbital, and their results have been extended by

TABLE 2 PHOSPHORUS–METAL N.M.R. COUPLING CONSTANTS

Complex	*trans* ligands	^{1}J(P–M) (Hz)	$\dfrac{^1\text{J(P–M–Cl)}}{^1\text{J(P–M–P)}}$	Ref.
mer-[W(O)Cl$_2$(Me$_2$PhP)$_3$]	Cl	442	1.29	20
	Me$_2$PhP	342		
mer-[RhCl$_3$(Bu$_3$P)$_3$]	Cl	114	1.36	18
	Bu$_3$P	84		
[RhCl(Ph$_3$P)$_3$]	Cl	189	1.33	23
	Ph$_3$P	142		
mer-[PtCl$_3$(Et$_3$P)$_3$] (ClO$_4$)	Cl	2049	1.50	20
	Et$_3$P	1374		
[PtCl(Et$_3$P)$_3$] (ClO$_4$)	Cl	3499	1.57	20
	Et$_3$P	2233		
[HgCl{(EtO)$_2$PO}]	Cl	12 670		
			1.69	24
[Hg{(EtO)$_2$PO}$_2$]	(EtO)$_2$PO	7500		

TABLE 3 PHOSPHORUS–METAL BOND LENGTHS

Complex	*trans* ligands	*l*(P–M) (pm)	Ref.
mer-[Mo(O)Cl$_2$(Me$_2$PhP)$_3$]	Cl	250.3	25
	Me$_2$PhP	254.9	
mer-[ReCl$_3$(Me$_2$PhP)$_3$]	Cl	240.1	22
	Me$_2$PhP	245.8	
mer-[OsCl$_3$(Me$_2$PhP)$_3$]	Cl	235.0	22
	Me$_2$PhP	240.8	
mer-[IrCl$_3$(Me$_2$PhP)$_3$]	Cl	227.7	22
	Me$_2$PhP	236.3	
[RhCl(Ph$_3$P)$_3$]	Cl	221.8	26
	Ph$_3$P	232.1	
[HgCl{(EtO)$_2$PO}]	Cl	236.3	24
[Hg{(EtO)$_2$PO}$_2$]	(EtO)$_2$PO	241.1	21

McWeeny *et al.*[29]. Ligands having a large overlap integral with the metal p-orbital are found to correspond to ligands known to weaken *trans*-bonds, so the results certainly rationalise the directed σ-effect. However, although of the s, p, d bonding set of metal orbitals, the p-orbital is the only type which overlaps solely with two *trans*-placed ligands, consideration of the possible square-planar orthogonal hybrids in complexes of type *trans*-[MX$_2$YZ] shows that even a completely *trans*-directed σ-effect can involve the s-orbital and two d_σ-orbitals as well as the p_σ-orbital.

Zumdahl and Drago[19] showed that extended Hückel MO theory gave variations in the Pt-N bonding orbitals in *trans*-[PtCl$_2$(NH$_3$)L] in response to changes in the nature of the σ-orbital of L. Ligands such as H and PH$_3$ gave relatively weak *trans*-Pt—N bonds of low s-component, but the model appears to give results that conflict with experimental measurements of the effects on ligands in *cis* relationship. The model suggests that the effects on *cis* and *trans* ligands are qualitatively similar and not very different in magnitude, whereas the X-ray structures of the complexes [PtXCl(R$_3$P)$_2$] (X = H, Cl) (table 1) and many other results show that for several ligands and hydride in particular, the effect on the *cis*-ligands is small compared with the *trans* ligand and the changes in bond lengths may even be in opposite senses. The calculations gave some support to the interpretation of the coupling constants and they also indicated that platinum–phosphorus π-bonding to the model ligand PH$_3$ was likely to be insignificant.

A simpler model was adopted by Mason and Randaccio[30] for the analysis of the bond lengths in the meridional complexes of table 3 and for a series of complexes of type *trans*-[MCl$_4$P$_2$], where M is a third row transition element. The metal–ligand lengths were discussed in terms of centroids of overlap density which were derived from orbitals calculated on the basis of maximisation of the sum of overlap integrals for metal hybrid orbitals with σ-orbitals on the chloride and phosphine ligands. This approach gave qualitatively correct results: the calculated covalent radii of the third row metals in a metal–phosphorus bond vary much more rapidly than in a metal–chlorine bond, as indicated by the bond length measurements. However, the maximum overlap method gives equal weighting to overlap with the metal s-, p-, and d-orbitals, and the orbital energies are not directly included in the calculation. The method has not so far been applied to transition-metal problems involving other ligands, and it would be interesting to see whether it can give qualitatively correct results with, for example, hydride complexes.

To summarise, the *trans*-bond weakening effect or '*trans* influence'[17] of phosphine ligands compared with chloride can be loosely described as a rehybridisation of the metal orbitals in response to the form of the phosphorus σ-orbital, with the effect that the covalency and specifically the s-character of the *trans* bond are reduced, leaving the *cis* bonds relatively unaffected.

The phosphine–platinum bonds, as well as generating a strong *trans* influence, are themselves considerably affected by the nature of ligands in *trans* relationship. The results for phosphine metal chlorides are a strong indication that this is true also for complexes of other metals, but the combination of an available nuclear spin quantum number of $\frac{1}{2}$ with the ease of synthesis of complexes with a great variety of ligands has enabled the *trans* influence on phosphorus ligands to be investigated more thoroughly for platinum(II) complexes than for complexes of other metals. The following is an order of decreasing *trans* influence of ligands determined by measurements of ^{195}Pt–^{31}P coupling constants in platinum(II) complexes. More precisely it is the order of decreasing s-component of platinum–phosphorus bonds as inferred from the coupling constants. The connection between the coupling constants and *trans* influence can be made with most rigour when bonds within a single complex are compared or when coupling constants are determined for complexes of similar symmetry[18]. Other comparisons can involve assumptions about other variables in the MO expression for the coupling constants, but entries in the order are often supported by more than one measurement, so that reasonable reliability can be claimed provided that small differences are not given great weight. The order derived from currently available measurements is[18,31-4]

$$MePh_2Si > Ph \sim Me > H > R_3P \sim Me_2PhP \sim Ph_3P \sim (PhO)_3P,$$
$$CN > R_3As > NO_2 > amines, N_3, NCO, NCS > Cl, Br, I > ONO_2$$

Ligands in *cis*-relationship to phosphines in platinum(II) complexes have a much smaller effect on the parameters of the Pt–P bonds[18,32]. The interpretation of small differences in coupling constants may also be somewhat unreliable and it is probable that steric interactions between *cis* ligands can in some situations be comparable in magnitude with the electronic effects. Defining a *cis* influence as the tendency of a ligand to weaken a bond in *cis* relationship, the order of increasing *cis* influence of ligands on platinum–phosphorus bonds as assessed from platinum–phosphorus coupling constants is as follows[18,31-4]

$$CN > NCO, N_3 > Cl, R_3P, (RO)_3P > Ph_3Ge > NO_2, Ph_3Si > NO_3, Ph, Me, H$$

It must be emphasised that the range of coupling constants is relatively small, so that the *cis* influence may be difficult to study by bond length measurements. There is some indication, however,[11,12] that platinum–phosphorus bonds are shorter *cis* to hydride than when *cis* to bromide, but this trend could well have a steric origin.

Measurements of platinum–phosphorus bond lengths give the following order with respect to the donor atom of the *trans* ligand, and again there is good agreement between bond lengths and coupling constants[20]

$$Ge > C \sim P > S > N, Cl, O$$

Orders of *trans*-influence determined for other platinum–ligand bonds give fairly similar results, and for illustration, a selection are given as follows

$$l(Pt-Cl) \ (X\text{-ray})^{29}: Si > H > P > Cl, O$$

$$\nu(Pt-Cl) \ (i.r.)^{35}: Et_3P \sim Et_3As > (RO)_3P > RNC > CO$$

$$^1J(Pt-H) \ (n.m.r.)^{35,36}: Et_3P > (MeO)_3P > (PhO)_3P > Ph_3P > ArNC,$$
$$RNC > CO > Et_3As > py$$

$$^2J(Pt-H) \ (n.m.r.)^{37}: (PhO)_3P, PhMe_2P, Ph_3P, Ph_3Sb \gtrsim CO > Ph_3As > py$$
$$> RCN > ArCN$$

$$^2J(Pt-H) \ (n.m.r.)^{34}: CN > NO_2, I > Br, Cl, NCS, N_3, NCO > NO_3$$

$$\nu(Pt-C) \ (i.r.)^{34}: CN > NO_2, I > Br, Cl, NCS > NO_3$$

Group VB to Transition Metal Bonds

Although the interaction of a ligand σ-orbital with the metal and the transmission of the effect to the detector bond cannot be described simply, the *trans*-influence orders indicate that groups forming strong and covalent bonds with the metal have a powerful *trans* influence, whereas groups forming more ionic bonds have a weak *trans* influence. The *trans*-influence order may well be dependent on the metal, but it is clear for platinum(II) that the order is not always simply related to electronegativity (compare reference 29). The alkyl and iodide ligands are near opposite ends of the *trans*-influence scale, but have closely similar Pauling electronegativities.

The complexes *cis*- and *trans*-[PtCl$_2$(R$_3$P)$_2$] and related molecules have been examined by far i.r. spectroscopy[15,38,39], by chlorine nuclear quadrupole resonance spectroscopy[40] and by X-ray photoelectron spectroscopy[41]. The identification of metal–phosphorus stretching frequencies has not been without difficulty, and assignments for nickel and palladium complexes have been revised recently on the basis of measurements of metal isotope effects[42]. This method has not been used to check the assignment for platinum complexes, but in any case the conclusion drawn from a very extensive study of force constants in platinum

TABLE 4 BOND LENGTHS (pm) in *mer*-[RhCl$_3$(Bu$_3$P)$_2${(MeO)$_3$P}][49]

l(Rh−P)(MeO)$_3$P	219.9
l(Rh−P)Bu$_3$P	237.9, 240.0
l(Rh−Cl)[*trans* to Cl]	233.9, 234.9
l(Rh−Cl)[*trans* to (MeO)$_3$P]	241.8

complexes was that far i.r. spectroscopy is not a very suitable method with which to distinguish factors affecting metal–ligand bonds, except in the most closely related sets of molecules[43]. We note, however, that the platinum–phosphorus stretching frequencies[15,44] in the *cis*- and *trans*-bis(phosphine) complexes of the platinum(II) and platinum(IV) chlorides (table 4) are not inconsistent with our working hypothesis, and there is no indication that the bonding to phosphorus is greatly different in the two oxidation states.

Chlorine n.q.r.[40] and chlorine 2p binding energies[41] indicate a greater electron density on chlorine in the *cis* isomers of platinum(II) phosphine complexes and the phosphorus 2p binding energies are slightly greater in the *cis* than the *trans* isomer. The platinum–phosphorus bond lengths (table 1) suggest that electron drift from phosphorus to the metal is substantially greater in the *cis* isomer, so it would appear that the charge on phosphorus is largely buffered by electron drift from the organic groups on phosphorus[41].

Grim and his coworkers have studied the effect of variation of the organic groups on phosphorus on the phosphorus–metal coupling constants in dihalide complexes of platinum[46] and mercury[47]. For the platinum complexes the coupling constants in both *cis*- and *trans*-[PtCl$_2$(Bu$_{3−n}$Ph$_n$P)$_2$] (n = 0–3) increase with increase of n, whereas for [HgBr$_2$(Bu$_{3−n}$Ph$_n$P)] the coupling constants vary in the opposite sense.

This behaviour could be due to the participation of metal–phosphorus d_π–d_π-bonding in the platinum complexes alone. However, there are several difficulties with this interpretation[18]. Firstly, several terms in the MO expression for coupling constants could be affected by change of groups on phosphorus, and it is not clear whether the terms should vary in a similar manner in complexes of different stereochemistry. The trend in the platinum–phosphorus coupling constants is similar to that in proton–phosphorus coupling constants for the protonated phosphines[47], where the bond in question has no π-character. Also there is some doubt concerning the validity of the mercury–phosphorus coupling constants reported by Grim. Work in the author's laboratory[21] has demonstrated dissociation and phosphine exchange in mercury(II) bromide complexes of the methylphenylphosphines, and in complexes of type $[Hg_2Cl_4(phosphine)_2]$, where dissociation and exchange should be less extensive, the coupling constants vary in the same sense as for the platinum compounds. On the other hand, recently reported[48] results for analogous cadmium–phosphorus coupling constants determined at low temperatures are in the order found by Grim for the mercury compounds. It is probable that the effects of exchange in the cadmium system are negligible at the temperatures used, but whether this is so is not clear from the published description of the experiments.

2.2 BONDS TO PHOSPHITE LIGANDS

In comparison with the amount of information available for bonds to phosphines, relatively few parameters are known for bonds to phosphite ligands in complexes of metals in positive oxidation states. Several studies have been made of the effects of various Group VB donors on the stretching frequencies and coupling constants of neighbouring bonds, but we postpone discussion of most of these results to a later section.

Bond lengths have been determined[49] by X-ray crystallography for the complex *mer*-$[RhCl_3(Bu_3P)\{(MeO)_3P\}]$ (table 4). It is unfortunate that the structure of a meridional tris(phosphine) complex of rhodium(III) is not available for comparison, but bond lengths have been determined in the analogous iridium(III) compound[22]. If the length of a rhodium(III)–phosphine bond for a site *trans* to chlorine is estimated by taking the value for Rh(III)—P *trans* to phosphine and adding the difference between the Ir(III)—P lengths *trans* to phosphorus and *trans* to chloride, the result obtained is 228.5 pm. Even though this procedure is rather approximate, it is likely that bonds to phosphites are significantly shorter than those to phosphines in rhodium(III) complexes. A similar trend occurs in rhodium(I) complexes, where the Rh—P bond length is 214 pm in di-μ-chloro-bis(triphenylphosphite)(cyclo-octa-1,5-diene)dirhodium(I)[50], and is 221.8 pm for the phosphine *trans* to chlorine in the somewhat distorted complex chloro-tris(triphenylphosphine)rhodium(I)[26]. A further example is provided by the Ir(III)—P lengths in the meridional tris(phosphine) complex (table 3) and in the compound $[IrCl\{(PhO)_2(C_6H_4O)P\}_2\{(PhO)_3P\}]$[51]. The lengths of bonds *trans* to chlorine are 227.7(6) pm for the phosphine and 219.4 pm for the metallated phosphite ligands. The metal–chlorine distances in the rhodium(III) and iridium(III) complexes indicate that the *trans* influences of the phosphine and phosphite ligands are very similar. Metal–ligand stretching frequencies and coupling constants, in which small differences are more apparent than in bond lengths, indicate that the *trans* influence of a phosphite ligand is generally slightly lower than that of a phosphine in complexes of positive oxidation states[32,33,35]. The affinity of triethyl phosphite for the site *trans* to hydride in the ruthenium(II) complex [RuHCl(CO)(PhR$_2$P)$_2$L] was found on the basis of measured displacement equilibria to be about the same as dimethylphenylphosphine and somewhat greater than triethylphosphine[52].

These observations do not enable us to make a rigorous comparison between phosphites and phosphines, and if we take them as being a reliable indication of the behaviour of these ligands in positive oxidation states, they are rather difficult to explain on any reasonable basis. The implication is that the quite significantly shorter bond to phosphite ligands is not to be associated with a much greater bond strength, nor with a greater *trans* influence. If the shortness of the bond is the result of π-bonding, the consequentially short σ-bond must be relatively weak, as otherwise the bond would be stronger than the observations suggest. Such an arrangement may explain the comparable *trans* influences of phosphites and phosphines, but the synergic effect of the π- on the σ-bond must be virtually discounted, and the initial premise of significant back-donation from an oxidation state as high as rhodium(III) must be somewhat questionable.

On the other hand, if π-bonding is unimportant, all of the results must be explained on the basis of the effects of the oxygen atoms on the phosphorus donor orbital. There is little doubt that these effects are quite substantial[18], and the phosphite lone-pair is expected to have a greater s-character and to be somewhat contracted by the electronegative oxygen atoms. The tighter binding of the donor electrons need not lead to a weaker bond with the metal if closer metal–ligand approach is possible, and the rehybridisation of the metal orbitals in response to the changed donor orbital could produce in the *trans* position a site similar to that *trans* to the rather longer bond of phosphines. The two explanations are, of course, not mutually exclusive, but it would perhaps be unwise to speculate further until results are available that enable us to compare phosphines and phosphites in more closely related complexes.

2.3 SUMMARY

The foregoing examination of the results of physical measurements on phosphine complexes of transition-metal chlorides has revealed no instance of reliable evidence for metal–phosphorus π-bonding, and it is clear that in the early work π-bonding was invoked to explain phenomena now traced to the anisotropic character of the metal–ligand σ-bond interactions. However, phosphine complexes of metals in positive oxidation states are expected on basic principles to have a relatively low tendency towards d_π–d_π-bonding. The higher oxidation states have an undoubted tendency to accept π-electrons from suitable ligands[27], and even in complexes of nickel(II) there is evidence that n.m.r. contact shifts of the ligand resonances are caused by partial transfer of π-electron density from the ligand to the metal, and not in the reverse direction[53]. Also, as H. C. Clark *et al.* have noted very recently, ^{13}C n.m.r. shifts and the high value of ν_{C-O} (2098 cm^{-1}) in *trans*-[PtMe(CO)(Me$_3$As)$_2$]$^+$ indicate little π-backbonding even with carbon monoxide ligand, so the platinum d_π-electrons must be quite strongly bound to the metal[54]. Although back-donation to olefins in such compounds as Zeiss' salt is part of the normal description of the bonding, direct evidence is lacking, and it may be that the ability of such ligands as carbon monoxide and olefins to form reasonably strong bonds by interaction with transition-metal σ-orbitals may have been underestimated. Certainly, the properties of this class of compounds, and therefore probably also of analogous complexes of ligands containing heavier Group VB elements, can now be understood in terms of σ-bonding alone, and π-bonding may well be energetically insignificant.

Although not directly concerned with the question of the nature of metal-phosphorus bonding, mention should be made of the derivation by Parshall of an order of π-acceptor strengths of anionic ligands from ^{19}F shifts in the complexes *trans*-[PtXL(R$_3$P)$_2$], where X is an anionic ligand and L is *m*- or *p*-fluorophenyl[55]. In the *m*-fluorophenyl derivatives the shifts were considered to be affected only by the σ-characters of the Pt—X bonds, whereas the shifts in the *p*-fluorophenyl derivatives would be determined by the overall effects of the σ- and π-components of the Pt—X bonds. This interpretative framework has also been applied to gold(I) complexes[56]. However, the method involves assumptions about the origins of the shifts and the mechanisms by which the σ- and π-components of the Pt—X bond affect the electronic structure of the fluorophenyl groups. These assumptions cannot be fully justified[35] and much doubt is cast on the method by the work of Stone *et al.*[57] who showed that when perfluorophenyl compounds, in which the shifts of the *m*- and *p*-fluorine nuclei can be measured within a single compound, are used, the calculated π-acceptor character was greater for X = NO$_3$ than for X = CN.

We have concentrated in this chapter on an analysis of the results of physical measurements, in which the idea of metal–Group VB π-bonding has had perhaps the greatest impact for the interpretation of chemical properties. Mention may be made of its rôle in explanations of the 'soft' character of heavy Group VB donors[58], of the 'soft' or 'b' classification of certain transition-metal oxidation states, and of the stabilisation of transition-metal alkyls and aryls[6]. For the positive oxidation states of metals and ligands such as trialkylphosphines, the strength of metal–ligand bonding and the 'soft' classification can now be understood in terms of σ-bonding alone, but it must be emphasised that this is not to deny a possible rôle for metal–Group VB π-bonding in other situations. The very large number of stable complexes with alkyl and aryl ligands that are now known almost obviates the need to find an explanation for the stability of metal alkyls and aryls containing Group VB ligands. The transition-metal to carbon σ-bond is obviously not intrinsically as weak as was thought at one time, but transition-metal compounds containing no ligands other than alkyls and aryls would, however, generally be coordinately unsaturated. They would, therefore, be susceptible to attack by oxygen or other reagents or, for example, in certain alkyl complexes, by the hydrogen on the β-carbon atom of the alkyl ligands themselves. In so far as any special stabilising effect of Group VB ligands is real, it is probably to be attributed to their ability to form stable bonds (of essentially σ-character) with metal alkyls or aryls, thereby preventing preliminary coordination by the attacking reagent or blocking the site on the metal required for the β-elimination mechanism of decomposition. Thus, by use of the ligand Me_3SiCH_2-, which has no hydrogen atoms in β-position to the metal and which provides its own steric protection, two groups of workers have recently prepared a large number of stable compounds containing no extra 'stabilising' ligands[59,60].

3 LOW OXIDATION STATES

The tendency of metals to donate electrons into the d_π-orbitals of phosphines and other Group VB donors is expected to be greater in metal(0) complexes than in complexes of higher oxidation states. Such π-bonding is considered to be generally important for the stabilisation of low oxidation states[61], and particularly in compounds containing only Group VB ligands.

3.1 METAL CARBONYL COMPLEXES

The effects of phosphine ligands on the C—O frequencies and force constants of metal carbonyls have in the past been explained solely in terms of direct competition between carbon monoxide and the phosphines for metal d_π-electrons, but it now seems to be widely accepted that metal–phosphorus σ-bonding must also influence the carbonyl i.r. parameters. We may note at the outset to our brief discussion that the unambiguous recognition and delineation of the effects of π-bonding on the physical parameters of low oxidation states is severely hampered by uncertainties concerning the σ-bonds. The interactions between the σ-bonds can be loosely described as the response of metal orbitals to the form and energy of the ligand σ-orbitals, and these interactions cannot reasonably be assumed either to lead to an isotropic σ-effect or to take the same form as in higher oxidation states. These uncertainties will not easily be removed by experiment, because, unlike the higher oxidation states and platinum(II) in particular, it is not possible to synthesise complexes of a zero oxidation state metal that contain a wide range of suitably placed purely σ-bonding ligands. It is the behaviour of such ligands as alkyl, chloride and hydride that provides the basis for the understanding of platinum–phosphorus σ-bonding and its effects on neighbouring bonds.

The examination of Group VB–metal(0) bonding has been carried out principally through the analysis of C—O frequencies and force constants calculated by the Cotton-Kraihanzel method[62]. Although in this method the interaction constants are treated as if π-effects alone were in operation, the results obtained have been shown to be rather insensitive to the precise form of these assumptions[63], so the conclusions drawn from the derived parameters should not be appreciably prejudiced by the origins of the CK model. Several analyses of results of this kind have been presented, and this survey is not intended to be exhaustive (see reference 64).

If the assumptions are made that carbonyl parameters are determined by competition between ligands for metal d_π-electrons and that effects through the σ-bonds are insignificant in comparison, the order of π-acidities of phosphines and other ligands that can then be derived is a reasonable one, and is capable of providing a rationalisation of a great many results[65]. For example, on this basis, phosphites are shown to be better π-acceptors than phosphines, and Plastas *et al.*[66] have shown that this also leads to a satisfactory explanation of the differences in length of all corresponding bonds in $[Cr(CO)_5\{(PhO)_3P\}]$ and $[Cr(CO)_5(Ph_3P)]$ (table 5). As we have seen

TABLE 5 BOND LENGTHS IN $[Cr(CO)_5L]$ [66]

Bond	Bond length (pm)	
	$L = (PhO)_3P$	$L = Ph_3P$
Cr–P	230.9(1)	242.2
Cr–C (*trans*)[†]	186.1(4)	184.4(4)
C–O (*trans*)	113.6(6)	115.4(5)
Cr–C (*cis*, av.)	189.6(4)	188.0(4)
C–O (*cis*, av.)	113.1(6)	114.7(6)

† Cr–C in $Cr(CO)_6$ is 190.9(3) pm.

from the results for some quite high oxidation states, however, it is unlikely that the differences in the phosphorus σ-orbitals are without some effect on the parameters, especially those of the Cr–P bonds.

The correlation of CK force constants with inductive parameters for groups on phosphorus and with measured proton basicities of phosphines has led to the suggestion that metal–phosphorus bonds in carbonyl complexes are essentially of purely σ-character[68]. Tolman[69] has shown that the A_1 carbonyl stretching frequencies in the complexes $[Ni(CO)_3L]$, containing a wide range of substituents on the phosphorus donor L, can be fitted to the expression

$$\nu_{C-O}(A_1) = 2056.1 + \sum_{i=1}^{3} \chi_i \ (cm^{-1})$$

where χ_i is a parameter for substituent i. The values of χ_i and the proton basicities of phosphines correlate well with each other and with inductive parameters for substituents on phosphorus derived from the ionisation constants of the acids R_2POH in water[70]. These parameters give better correlations than the more usual Taft parameters which are determined by inductive effects through carbon. Although the correlation of carbonyl parameters with ligand proton basicity suggests that σ-donation by phosphorus raises the energy of the metal d_π-electrons and promotes π-bonding to carbonyl, Tolman noted that the effects of the substituents on the π-acidities of the phosphines would lead to a similar order of metal d_π-energies. There was therefore insufficient information to separate the effects of phosphorus–metal σ- and π-bonding on the carbonyl frequencies[69].

Graham[8] has analysed CK force constants for the series of complexes $[Mn(CO)_5X]$ and $[Mo(CO)_5L]$ and his method has also been applied to the analogous tungsten compounds $[W(CO)_5L]$[71]. Changes in the force constants *cis* and *trans* to X or L are used to determine two parameters, σ and π, defined as follows

$$\Delta k_{cis} = \sigma + \pi, \text{ and } \Delta k_{trans} = \sigma + 2\pi$$

The changes in the force constants are thus described in terms of an isotropic parameter σ and a parameter π whose numerical weighting is chosen to reflect the number of metal d_π-orbitals shared between X or L and the particular carbonyl groups. Graham[8] found that the values of π followed an intuitively reasonable order, but the values of σ did not correlate with the proton basicities of the ligands and they were felt to reflect the σ-donor properties of the ligands as they exist in the complex, that is, modified by the synergic effect[8,71]. As defined, the parameters σ and π refer to the effects of ligands on neighbouring C—O bonds and unless the interactions between the metal–ligand bonds take a particularly simple form, the parameters are not related to the correspondingly labelled components of the metal–ligand bonds. Clearly if variations in the metal–ligand σ-bond produce unequal effects on the C—O parameters of *cis* and *trans* ligands, this anisotropic σ-effect would be reflected in both parameters σ and π. This suggests that reliable information concerning the σ- and π-components of metal(0)-ligand bonds will not be obtained from the analysis of CK parameters alone[64].

Brown and Darensbourg[63,72] have attempted to utilise the additional information contained in the intensities of carbonyl stretching bands. Spectra of adsorbed carbon monoxide indicate that σ-donation by carbon monoxide *increases* ν(C—O) and reduces the intensity of the band, while occupation of carbon monoxide π^*-orbitals lowers ν(C—O) and the band intensity increases very markedly. Thus for carbon monoxide bound to a metal by σ- and π-interactions, although both components affect the band intensity, the effect of the π-component so outweighs that of the σ-component that a correlation is expected between the band intensities and the extent of π-bonding in metal carbonyl complexes. Application of this interpretation of intensities enabled Brown and Darensbourg[63] to distinguish between broad classes of bonding arrangements, but the limited accuracy of intensity measurements precludes a very detailed comparison of ligands. Amine substituents give lower values of ν(C—O) and greater band intensities than parent hexacarbonyls. Compared with carbon monoxide, amines have greater Lewis basicity and they transfer more charge to the metal. This reduces σ-donation by the carbonyl groups and promotes occupation of the π^*-orbitals of carbon monoxide by raising the energy of the metal d_π-electrons. Brown and Darensbourg suggested that the degree of electron transfer to metal from phosphorus could never be as high as that developed by nitrogen. Phosphines were regarded as weaker σ-donors than amines and whatever enhanced σ-donor properties they possess in carbonyls relative to the proton was ascribed to the synergic effect of the π-bonds, which reduce the charge on the metal. The intensity results for phosphine and arsine complexes were felt to demonstrate the existence of π-bonds between the metal and these ligands. The intensities give evidence that the same or less electron transfer from metal to carbonyl occurs in [Mo(CO)$_5$L] when L = phosphine, arsine as when L = CO. The lower C—O force constants in the Group VB ligand derivatives must then be due to the net decrease in the metal–carbon monoxide σ-bond strength that results from σ-donation by L. The increased π-bonding with CO that this produces in amine complexes is prevented by the stabilisation of the metal d_π-electrons as a result of their interaction with the dπ-orbitals of the heavier Group VB donor. This is a notable formulation of the bonding arrangements in metal carbonyl complexes and it will be interesting to see the theory tested further. The interpretation of i.r. intensities in simple molecules is a matter of some difficulty, so that the method outlined above for rather complex molecules cannot be accepted without some reservations[64]. The interpretation of the results for phosphine metal pentacarbonyls suggests that the carbon monoxide ligand *trans* to the phosphine should have a weaker M—C bond (similar π-component, reduced σ-donation) and a weaker C—O bond (owing to reduced σ-donation) than the parent hexacarbonyl. This contrasts with the more usual correlation of short M—C bonds[73] and is not supported by the Cr—C lengths given in table 5.

Dessy and Wieczorek[74] have measured the changes of CK force constants that result from the addition of electrons (by electrochemical reduction) to substituted metal carbonyls. Their interpretation of the results was based on the reasonable assumptions that the added electron occupies an orbital with a high degree of ligand π*-character and that the molecule does not severely alter its symmetry in the reduction process. Some support for the assumptions was obtained from electron spin resonance spectra of reduced species, and it was concluded that in the systems studied transmission of charge occurs primarily through anisotropic σ-inductive effects. Another indication of directed σ-effects is obtained from ^{181}W—^{31}P coupling constants[75], which appear to depend mainly on the nature of the ligand *trans* to the phosphorus atom. In the compounds studied such dependence is not expected from a model based on π-interactions, but thorough identification of the important variables in the MO expression for ^1J(W—P) has not yet been achieved. The variation of ^1J(W—P) with groups on phosphorus in [W(CO)$_5$L] was originally claimed to demonstrate the dominance of W—Lπ-bonding[76], but it was later shown that these coupling constants correlate well with values of ^1J(P—H) in protonated phosphines[47].

3.2 OTHER METAL(0) COMPLEXES

Intuition has long demanded an important rôle for π-bonding in complexes of phosphorus trifluoride[1,77]. Although the preparation of BH$_3$PF$_3$ [78] and AlCl$_3$PF$_3$ [79] shows that back-donation is not essential for the formation of complexes, these bonds are rather weak. Thus σ-bonding alone is unlikely to provide an explanation for the very stable transition metal complexes of PF$_3$. Some mechanism is almost certainly necessary for the removal of negative charge on the metal in the anionic complexes [M(PF$_3$)$_4$]$^-$ (M = Co, Rh, Ir)[77,80,81]. This formal transfer of electron density from metal dπ-orbitals into virtual orbitals of the phosphorus ligand can be a bonding interaction in its own right, or, since the transfer of charge promotes strong σ-bonding, the back-donation of d$_π$-electrons could conceivably leave their energy relatively unaffected. In the latter circumstances the metal–ligand bond energy would be mainly derived from the σ-interaction, but the π-interaction would be strongly evident in the electron density distribution.

Recently much insight into the nature of bonding in coordination compounds of phosphorus trifluoride has come from the study of molecular orbital energies by photoelectron spectroscopy[82-6] and *ab initio* SCF–MO calculations[82,84,87-90], and from the availability of some very accurate bond length measurements, mainly derived from electron diffraction and X-ray crystallography (table 6)[91-9]. The

TABLE 6 BOND LENGTHS IN PF_3 AND SOME OF ITS COMPLEXES

Molecule	l(M–P) (pm)	l(P–F) (pm)	Ref.
PF_3	–	157.0(1)	91
OPF_3	–	152.4(3)	92
[$Ni(PF_3)_4$]	211.6(10)	156.1(5)	93
[$Rh(PF_3)_3NO$]	224.6	156.0	†
[$Mo(PF_3)(CO)_5$]	236.9(10)	155.7(4)	96
[$CoH(PF_3)_4$]	205.2(5)	154.9(12)	97
[$Pt(PF_3)_4$]	223.0(10)	154.6(6)	93
BH_3PF_3	183.6(12)	153.8(8)	98
$B_2H_6(PF_3)_2$	184.8(28)	153.9(3)	99
$B_4F_6PF_3$	182.5(15)	151.0(1)	100

† Personal communication from J. F. Nixon.

calculations both with a near-minimal[82-5,88] and with extended[89,90] basis sets are in excellent agreement with the photoelectron results for PF_3, PF_3O [83] and BH_3PF_3 [84]. Calculations for the free donor molecules PH_3, PMe_3, and PF_3 [87,89,90] show that the highest occupied orbital is the phosphorus lone-pair, and for PH_3 and PMe_3 this has 60–70 per cent phosphorus 3p character, whereas in PF_3 the 3s and 3p contributions are about equal at about 35 per cent. The calculated energies of these orbitals -10.02 eV $(-1.618$ aJ) (PH_3), -8.49 eV $(-1.360$ aJ) (PMe_3) and -12.69 eV $(-2.033$ aJ) (PF_3) are in good agreement with the experimental ionisation potentials 9.93 eV (1.590 aJ) (PH_3), 8.58 eV (1.374 aJ) (PMe_3) and 12.31 eV (1.972 aJ) (PF_3). The energy separations of the lone-pair and the first virtual orbital are similar for all three molecules, so that PF_3 has the lowest lying virtual orbital. Upon formation of the oxide PF_3O the phosphorus lone-pair electrons are shown by photoelectron spectroscopy to be strongly stabilised (by about 3.2 eV (0.51 aJ))[83], and calculations indicate that the orbital is mainly of oxygen 2p character and the contribution from the phosphorus 3s orbital has decreased to about 5 per cent from the value of 34 per cent in the PF_3 molecule[82]. Because the phosphorus 3s-orbital is antibonding with $2p_\sigma$-orbitals of fluorine, the decrease of population of the phosphorus 3s orbital is accompanied by shortening of the P–F bonds in PF_3O compared with PF_3[82] (table 6). There is also an important contribution to P–O bonding from π-orbitals; the total contribution from phosphorus $3d_\pi$- and $3p_\pi$-orbitals is calculated to be 21 per cent; very significant back-donation of electrons occur in PF_3O therefore. The transfer of electrons to oxygen in the σ-bond is almost balanced by the delocalisation of oxygen

p_π-electrons over phosphorus and on oxidation of PF_3 the positive charge on phosphorus is calculated to increase only by about 0.1–0.2 e[88,89]. The energy of the fluorine non-bonding electrons appears to reflect the net migration of charge from phosphorus, and the photoelectron spectra show only a small increase in the ionisation potentials of these electrons from 15.8, 16.3, 17.5 and 18.6 eV (2.53, 2.61, 2.80 and 2.98 aJ) in PF_3 to 17.1, 17.7, 18.8 and 19.5 eV (2.74, 2.84, 3.01 and 3.12 aJ) in PF_3O[82,83]. Calculations for PH_3O [89] give a similar P—O total overlap population to that for PF_3O, but in PH_3O the σ-contribution is larger and the π-contribution smaller than in PF_3O.

For BH_3PF_3 [84,88] compared with PF_3O, the phosphorus 'lone-pair' orbital is stabilised to a rather smaller extent (about 0.3–0.8 eV (0.05–0.13 aJ)), and the highest filled molecular orbital, the one involved in back-bonding, is 95 per cent localised on BH_3. The delocalisation of the 'lone-pair' in BH_3PF_3 reduces the phosphorus 3s-contribution to 14 per cent and P—F bond shortening is again evident (table 6). The ionisation potentials of the non-bonding fluorine electrons (12.6, 16.8 and 17.6 eV (2.02, 2.69 and 2.82 aJ)[84]) are only slightly larger than in PF_3.

Calculations for the transition-metal complexes $[Ni(PF_3)_4]$ and $[Pt(PF_3)_4]$ have not yet been possible, but photoelectron spectra have been recorded[85,86], and accurate bond lengths are available (table 6). The ionisation potential of the phosphorus 'lone-pair' electrons is 13.2 eV (2.11 aJ) for the nickel complex and 14.5 eV (2.32 aJ) for the platinum complex, compared with 12.3 eV (1.97 aJ) for the free ligand. The greater degree of σ-donation in the platinum complex implied by these results is also reflected in the P—F bond lengths (table 6). The ionisation potentials of the non-bonding fluorine electrons are very similar in the two complexes: 15.97, 17.48 and 19.42 eV (2.558, 2.800 and 3.112 aJ) for $[Ni(PF_3)_4]$, and 15.87, 17.53 and 19.4 eV (2.541, 2.808 and 3.11 aJ) in $[Pt(PF_3)_4]$, so it would appear that charge on phosphorus developed by greater σ-donation in the platinum complex is compensated by greater π-back-bonding.

The application of photoelectron spectroscopy and interpretation based on the SCF calculations is eagerly awaited for other molecules in table 6. The similarity of P—F lengths in PF_3 and $[Ni(PF_3)_4]$ suggests that back-donation may cause some P—F bond weakening[93], and the result for $[Rh(PF_3)_3NO]$ seems to support this view (table 6). In this molecule the Rh—N—O axis is linear, indicating an NO^+ and Rh(−I) formulation[100], from which extensive back-bonding is expected. The P—F bond length in $[Mo(PF_3)(CO)_5]$ (table 6) indicates substantial π-bonding to PF_3 on this basis of interpretation, and this agrees with some evidence that PF_3 is a better π-acceptor than CO (for example see reference 86).

Other evidence that platinum(0) forms stronger bonds than nickel(0) with phosphorus ligands has been obtained from measurements of phosphorus–metal stretching frequencies for the compounds $[M\{(MeO)_3P\}_4]$ [101]. All three complexes have two bands at 340–290 cm^{-1} and one band at 220–190 cm^{-1}, and force constant calculations for tetrahedral MR_4 molecules give values in the order Ni < Pd < Pt. However, the frequencies observed for these molecules differ from those in $[M\{CH_3C(CH_2O)_3P\}_4]$ [102] by an unexpectedly large amount, so it was concluded that it may be necessary to include all the ligand atoms in the force constant calculations[101].

Study of the ligand substitution reactions of $[M(PF_3)_4]$ (M = Ni, Pt) [103] and $[M'\{(EtO)_3P\}_4]$ (M' = Ni, Pd, Pt) [104] with cyclohexyl isocyanide led to the suggestion that the order of phosphorus–metal bond dissociation energies was Ni > Pt > Pd for the trifluorophosphine complexes and Pt ~ Ni > Pd for the phosphite complexes. These orders, which are derived from the enthalpies of activation for S_N1 dissociation of the complexes, can be explained following Nyholm's suggestion[105] that the σ-bond strengths should follow the order Ni < Pd < Pt, and that the π-bonding abilities should correlate with the $d^{10} \rightarrow d^9$ ionisation energies and lie in the sequence Ni > Pd > Pt. The S_N1 activation enthalpies for the $[NiL_4]$ complexes follow the order with respect to ligand: $PF_3 \sim (EtO)_3P > CO$, suggesting a similar order of bond energies, although, as the authors point out[103], the dissociation processes do lead here to different product residues. Measurements of appearance potentials in the mass spectra of $[HCo(CO)_n(PF_3)_{4-n}]$ [106] and $[Ni(PF_3)_4]$ [107] indicate that the metal–ligand bond energies for carbon monoxide and phosphorus trifluoride are rather similar[106].

Bond lengths have been determined for a number of triphenylphosphine complexes of platinum in which the metal has formal oxidation state (0) (table 7)[108–15]. It is usual to compare the platinum–phosphorus bond lengths in the platinum(0) complexes with those in bis(phosphine) complexes of platinum(II) in which the

TABLE 7 TRIPHENYLPHOSPHINE COMPLEXES OF PLATINUM(0)

	l(Pt–P) (pm)	l(Pt–C)(pm)	Ref.
$[Pt(Ph_3P)_3]$	226–227		108
$[Pt(Ph_3P)_3(CO)]$	233.3(8)		
	233.5(8)	186(3)	109
	235.2(8)		
	235.7(8)		
	236.9(10)	184(2)	110
	235.3(10)		
$[Pt\{(NC)_2CC(CN)_2\}(Ph_3P)_2]$ †	229.1(9)	212(3)	111
	228.8(8)	210(3)	
$[Pt(Cl_2CCF_2)(Ph_3P)_2]$	230.3(6)	205(3)	112
	231.4(5)		
$[Pt(Cl_2CCCl_2)(Ph_3P)_2]$ ‡	229.2(7)	203(3)	112
	227.8(8)		
$[Pt\{Cl_2CC(CN)_2\}(Ph_3P)_2]$	233.9(6) ('*trans*' to CCl)	200(2) (CCl)	113
	226.0(6) ('*trans*' to CCN)	210(2) (CCN)	
$[Pt(H_2CCH_2)(Ph_3P)_2]$	227.0	211(3)	114
	226.5		
$[Pt(PhCCPh)(Ph_3P)_2]$ §	227	206	115
	228	201	

Dihedral angles between PtP_2 and PtC_2 planes:
† 8.3°; ‡ 12.3(1.5)°; § 14°.

phosphines are mutually *trans* (table 1)[109]. As the bond lengths in the two oxidation states are then similar, it is suggested that the weaker σ-donation expected in the lower oxidation state is compensated by the formation of stronger d_π-d_π bonds. However, in the *trans*-platinum(II) compounds, the platinum–phosphorus bonds are subject to the relatively high *trans*-influence of phosphorus, and few of the bonds in the platinum(0) compounds appear to approach the strength implied by the bond lengths in *cis*-[$PtCl_2(Me_3P)_2$] (table 1). The comparison is further complicated by the possibility of different effective radii for phosphorus in the alkyl and phenyl phosphines, and the rather different σ-bonding orbitals used by platinum in the two oxidation states.

Analysis of the bond lengths in the various platinum(0) compounds has some similar complications. The bond lengthening that occurs on addition of carbon monoxide to [$Pt(Ph_3P)_3$] is readily rationalised by the removal of d_π-electrons by CO [109], but again the stereochemistry change may affect the platinum–phosphorus σ-bonds, and the steric crowding of the phosphine ligands is greater in the tetrahedral complex. Similar difficulties attach to the comparison of lengths in [$Rh(Ph_3P)_3Cl$] and *trans*-[$RhCl(Ph_3P)_2(C_2F_4)$] reported by Mason *et al.*[26]. The rhodium–phosphorus bond lengthening on replacement of Ph_3P by C_2F_4 in combination with constancy of the Rh—Cl length is certainly strongly suggestive of significant changes in Rh—P π-bond order[26]. However, direct π-interaction between the olefinic and phosphine ligands can occur at all only because there is some distortion from square-planar geometry about rhodium and, because the distortion is smaller in the olefin complex, the phosphines are nearer to being genuinely mutually *trans*.

In the platinum(0) complexes with the halido- and cyanoethylenes (table 7),[111–13], the central carbon atoms of the ethylenes lie in or close to the PtP_2 plane, so that the stereochemistry of platinum is approximately trigonal as in [$Pt(Ph_3P)_3$] or it may be regarded as being pseudo-square-planar. Considerable electron transfer to the ethylenes is indicated by long bonds between the central carbon atoms, and much of the Pt—C bond length data is in accord with a σ–π model of bonding. It is suggested that the cyano-groups reduce σ-donation to a very low level, but with $C(CN)_2$ groups the π-interaction is so strong as to reach a saturation level[113]. Chlorine substituents lead to a shorter Pt—C bond due to a better overall σ- and π-interaction with the metal. It is interesting that the Pt—P lengths in the complex of $Cl_2CC(CN)_2$ appear to be determined mainly by σ-interactions with the approximately *trans*-placed carbon atoms[113].

On the basis of ESCA measurements, it appears that certain 'neutral' ligands are capable of removing considerable electron density from platinum[116], and for (formal) bis(triphenylphosphine) complexes the positive charge on the metal increases as a function of other ligands present in the order

$$2Ph_3P < PhC_2Ph < C_2H_4 < CS_2 < O_2 < 2Cl$$

The separation of the many factors determining this order is not yet possible[116].

4 STERIC EFFECTS

There can be little doubt that in certain circumstances steric interactions involving Group VB ligands are of great importance, and recently several systematic studies have been initiated. The structural consequences of steric interactions can be expected generally to be most evident in bond angles, especially in complexes of d^{10} metals where the absence of ligand field stabilisation gives low resistance to deformation. However, even quite small steric effects can have important consequences if there are alternative structures with similar energies. Thus, it appears that the mode of bonding of the thiocyanate ligand to metals can sometimes be influenced by steric factors because of the different spatial characteristics of the linear M—NCS grouping and the bent M—SCN group. Farona and Wojcicki[117] have shown that in the complexes $[Mn(CO)_3L_2(SCN)]$ (L = Ph_3As, Ph_3Sb) the linear MnNCS arrangement would give close contacts with the phenyl rings, whereas the S-bonded isomer, which is found, can be accommodated more easily. Basolo et al.[118] have suggested that in trans-$[Pd(SCN)_2L_2]$ (L = Ph_3P, Ph_3Sb) the longer bonds in the antimony complex give more space near the metal and permit S-bonding whereas the phosphine forces the adoption of the N-bonded structure. Meek et al.[119] have studied linkage isomerism in thiocyanate complexes of palladium with Group VB chelate ligands. In these cases the rôle of steric interactions in the formation of complexes containing both S- and N-bonded forms is supported by a crystal structure determination; one group is forced into a non-linear arrangement by the phenyl groups of a cis-placed Ph_2P— group. The study also confirms that in the immediate vicinity of the metal Ph_2P— groups leave less room than Ph_2As— groups.

A systematic examination of steric effects on bond angles has been made in the complexes $[CuXL_2]$ (X = anionic ligand, L = phosphorus donor), in which copper has the d^{10} configuration[120]. The PĈuP angle is large when X is planar and has low steric requirements, and the angle decreases when X occupies more space. Also the PĈuP angle was found to increase with the steric requirements of the phosphorus donor when X = NO_3. The angle was found to be $131(1)°$ for L = Ph_3P and $140(1)°$ for L = $(C_6H_{11})_3P$. As the PĈuP angle increases, the s-character of the copper orbitals in the Cu—P bonds increases and there is a definite trend towards shorter Cu—P distances.

In general the steric requirements of phenyl and cyclohexyl groups on phosphorus are larger than those of the normally used alkyl groups, but Shaw and his associates[121-5] have shown that t-butyl groups on phosphines are so bulky that the coordination numbers[121,122] and reactivities[123,124] of complexes can be very clearly affected and it is possible to observe rotational conformers of trans-$[RhCl(CO)(RBu^t_2P)_2]$ (R = Me, Et, Pr) at $-60°C$ by ^1H n.m.r.[125]. Shaw et al. have shown that t-butyl groups on phosphorus in square-planar complexes not only inhibit oxidative addition reactions by reagents as active as chlorine[123], but the complexes tend to release steric strain by expulsion of chloride ligand with formation of an hydrido-complex or a metal–carbon bond with an alkyl or aryl group also attached to phosphorus[124].

Tolman[126] has recently reported that the positions of equilibria set up on addition of free ligands to solutions of $[NiL_4]$ (L = phosphorus donor) are dominated by the steric interactions between ligands in the complexes. The absence of a correlation between the semiquantitative binding abilities of ligands and the frequency of the A_1-mode carbonyl stretching frequency in $[Ni(CO)_3L]$ suggests that electronic effects are relatively unimportant, and Tolman was able to show that the binding abilities correlated reasonably well with the apex angle formed at the centre of the nickel atom by the minimum cone containing the ligand.

5 COMPARISON OF GROUP VB DONORS

The nature of the bonding to phosphorus ligands appears to be reasonably clear at two extremes. Metals in oxidation state (II) or higher form essentially pure σ-bonds with alkyl and probably phenyl phosphines. The extensive participation of σ- and π-bonding appears to be established in complexes of metal(0) and ($-$I) that contain mainly fluorophosphine ligands. For intermediate situations, either suitable results are not yet available, or there is lack of agreement concerning their interpretation. For metal carbonyl complexes the 'σ-only' viewpoint seems improbable if applied to phosphorus trifluoride, and the 'π-only' viewpoint has no theoretical basis and is doubtful in the light of σ-interactions found in complexes of phosphines with higher oxidation states. Useful though the correlations suggested by Graham[8] may be, the fact that his σ- and π-parameters are both significant for heavier Group VB donors does not allow definite conclusions to be reached concerning the σ- and π-components of the metal–Group VB ligand bonds.

For ligands containing the heavier Group VB atoms the results are vastly fewer in number and systematic studies of, for example, bond lengths in complexes of a single oxidation state of a metal have not yet been made. The arsines appear to be fairly similar to the phosphines in their ligand behaviour and it may be reasonable to assume that they also form pure σ-bonds with higher oxidation states of metals. An opposite view has been taken by Pauling et al.[127] who note that the Co(III)—As bond in trans-$[CoCl_2(diars)_2] ClO_4$ is about 10 pm shorter than the estimated single bond distance and on a similar basis the bonds in $[\{Mn(\pi\text{-}MeC_5H_4)(CO)_2\}_2(diars)]$ appear to be shortened by about 19 pm. However, metal–phosphorus bonds show analogous 'shortening' even for metals in quite high oxidation states[21], and recent results suggest that metal radii are quite significantly dependent on the nature of the ligands[22]. This method of analysis is therefore probably subject to large errors.

The effects of arsenic and antimony donors on some physical parameters of platinum(II) complexes have been indicated in section 2.1, and Mössbauer parameters for iron(II) and iron(III) complexes have also been recorded[128,129]. Partial centre shifts indicate that the $\sigma + \pi$ interaction with low-spin iron(II) increases in the order $Ph_3As < Ph_3Sb < Ph_3P$ [128] and the study of pentacyano-iron(II) and (III) derivatives led to the suggestion that the π-affinities of Ph_3As and Ph_3Sb are similar and lower than that of Ph_3P [129]. Effects on carbonyl stretching frequencies have also been recorded[8,71].

Equilibrium constants for the displacement of amines from the complexes $[W(CO)_5(amine)]$ increases in the order

$$Ph_3Bi < (PhO)_3P \sim Ph_3Sb < Ph_3As < Ph_3P < (BuO)_3P < CH_3C(CH_2S)_3P$$
$$< EtC(CH_2O)_3P \sim (C_6H_{11})_3P \sim Bu_3P^{130}$$

and various correlations with ligand basicity suggested that σ-effects were dominant, but that the cyclohexylphosphine experienced some steric hindrance. A similar study by Douglas and Shaw for the site *trans* to hydride in the complexes $[RuHCl(CO)(R_2PhP)_2L]$ gave relative affinities in the order

$$Et_2PhAs < Me_2PhAs < Bu_2PhP \sim Pr_2PhP \sim Et_2PhP < Et_3P < (EtO)_3P \sim Me_2PhP$$
$$< (MeO)_2PhP^{52}$$

Acknowledgment

Thanks are due to Dr G. G. Mather for assistance with this article.

REFERENCES

1. Chatt, J., *Nature Lond.*, **165** (1950), 637; Chatt, J., and Williams, A. A., *J. chem. Soc.* (1951), 3061.
2. Chatt, J., and Wilkins, R. G., *J. chem. Soc.* (1952), 273; *J. chem. Soc.* (1952), 4300; *J. chem. Soc.* (1956), 525; Leden, I., and Chatt, J., *J. chem. Soc.* (1955), 2936.
3. Chatt, J., Duncanson, L. A., and Venanzi, L. M., *J. chem. Soc.* (1955), 4461.
4. Craig, D. P., Macoll, A., Nyholm, R. S., Orgel, L. E., and Sutton, L. E., *J. chem. Soc.* (1954), 332.
5. Ahrland, S., Chatt, J., Davies, N. R., and Williams, A. A., *Nature, Lond.*, **179** (1957), 1187; Ahrland, S., Chatt, J., and Davies, N. R., *Q. Rev. chem. Soc.*, **12** (1958), 265.
6. Chatt, J., and Shaw, B. L., *J. chem. Soc.* (1959), 705.
7. Ahrland, S., and Chatt, J., *J. chem. Soc.* (1957), 1379.
8. Graham, W. A. G., *Inorg. Chem.*, 7 (1968), 315.
9. Chatt, J., and Shaw, B. L., *J. chem. Soc.* (1962), 5075.
10. Owston, P. G., Partridge, J. M., and Rowe, J. M., *Acta crystallogr.*, **13** (1960), 246.
11. Eisberg, R., and Ibers, J. A., *Inorg. Chem.*, **4** (1965), 773.
12. Messmer, G. G., and Amma, E. L., *Inorg. Chem.*, **5** (1966), 1775.
13. Messmer, G. G., Amma, E. L., and Ibers, J. A., *Inorg. Chem.*, **6** (1967), 725.
14. Chatt, J., Gamlen, G. A., and Orgel, L. E., *J. chem. Soc.* (1958), 486.
15. Goggin, P. L., and Goodfellow, R. J., *J. chem. Soc. (A)* (1966), 1462.
16. Basolo, F., Chatt, J., Gray, H. B., Pearson, R. G., and Shaw, B. L., *J. chem. Soc.* (1961), 2207.
17. Pidcock, A., Richards, R. E., and Venanzi, L. M., *J. chem. Soc. (A)* (1966), 1707.
18. Pidcock, A., and Nixon, J. F., *Ann. Rev. n.m.r. Spectroscopy*, **2** (1969), 346.
19. Zumdahl, S. S., and Drago, R. S., *J. Am. chem. Soc.*, **90** (1968), 6669.
20. Pidcock, A., and Rapsey, G. J. N., unpublished results.
21. Pidcock, A., and Mather, G. G., unpublished results. Mather, G. G., D. Phil. Thesis (Sussex), 1971.
22. Aslanov, L., Mason, R., Wheeler, A. G., and Whimp, P. O., *Chem. Commun.* (1970), 30.
23. Brown, T. H., and Green, P. J., *J. Am. chem. Soc.*, **92** (1970), 2359.
24. Bennett, J., Pidcock, A., Waterhouse, C. R., Coggon, P., and McPhail, A. T., *J. chem. Soc. (A)* (1970), 2094.
25. Chatt, J., Manojlović-Muir, Lj., and Muir, K. W., *Chem. Commun.* (1971), 655.
26. Hitchcock, P. B., McPartlin, M., and Mason, R., *Chem. Commun.* (1969), 1367.
27. Brown, T. L., McDugle, W. G., and Kent, L. G., *J. Am. chem. Soc.*, **92** (1970), 3645; Cresswell, P. J., Fergusson, J. E., Penfold, B. R., and Scaife, D. E., *J. chem. Soc. (Dalton)*, 1972, 254.
28. Gray, H. B., and Langford, C. H., *Ligand Substitution Processes*, Benjamin, New York, 1966.
29. McWeeny, R., Mason, R., and Towl, A. D. C., *Discuss. Faraday Soc.*, **47** (1969), 20.
30. Mason, R., and Randaccio, L., *J. chem. Soc. (A)* (1971), 1150.
31. Heaton, B. T., and Pidcock, A., *J. organometal. Chem.*, **14** (1968), 235.
32. Allen, F. H., and Sze, S. N., *J. chem. Soc. (A)* (1971), 2054.
33. Allen, F. H., Pidcock, A., and Waterhouse, C. R., *J. chem. Soc. (A)* (1970), 2087.
34. Allen, F. H., and Pidcock, A., *J. chem. Soc. (A)* (1968), 2700.
35. Church, M. J., and Mays, M. J., *J. chem. Soc. (A)* (1968), 3074.
36. Church, M. J., and Mays, M. J., *J. chem. Soc. (A)* (1970), 1938.
37. Clark, H. C., and Ruddick, J. D., *Inorg. Chem.*, 9 (1970), 1226.
38. Adams, D. M., Chatt, J., Gerratt, J., Westland, A. D., *J. chem. Soc.* (1964), 734.
39. Chatt, J., Leigh, C. J., and Mingos, D. M. P., *J. chem. Soc. (A)* (1969), 2972.
40. Fryer, C. W., and Smith, J. A. S., *J. chem. Soc. (A)* (1970), 1029.
41. Clark, D. T., Adams, D. B., and Briggs, D. B., *Chem. Commun.* (1971), 602.

42. Shobatake, K., and Nakamoto, K., *J. Am. chem. Soc.,* **92** (1970), 3332.
43. Duddell, D. A., Goggin, P. L., Goodfellow, R. J., Norton, M. G., and Smith, J. G., *J. chem. Soc. (A)* (1970), 545. See also Murrell, J. N., and Nikolskii, A. B., *J. chem. Soc. (A)* (1970), 1363.
44. Adams, D. M., and Chandler, P. J., *J. chem. Soc. (A)* (1967), 1009.
45. Grim, S. O., Keiter, R. L., and McFarlane, W., *Inorg. Chem.,* **6** (1967), 1133.
46. Keiter, R. L., and Grim, S. O., *Chem. Commun.* (1968), 521.
47. McFarlane, W., and White, R. F. M., *Chem. Commun.* (1969), 744.
48. Mann, B. E., *Inorg. nucl. Chem. Lett.,* 7 (1971), 595.
49. Allen, F. H., Chang, G., Cheung, K. K., Lai, T. F., Lee, L. M., and Pidcock, A., *Chem. Commun.* (1970), 1297.
50. Coetzer, J., and Gagner, G., *Acta crystallogr.,* **B26** (1970), 985.
51. Guss, J. M., and Mason, R., *Chem. Commun.* (1971), 58.
52. Douglas, P. G., and Shaw, B. L., *J. chem. Soc. (A)* (1970), 1556.
53. Eaton, D. R., and Zaw, K., *Coord. Chem. Revs.,* 7 (1971), 197.
54. Chisholm, M. H., Clark, H. C., Manzer, L. E., and Stothers, J. B., *Chem. Commun.* (1971), 1627. See also Dobson, G. R., *Inorg. Chem.,* 4 (1965), 1673.
55. Parshall, G. W., *J. Am. chem. Soc.,* 88 (1966), 704.
56. Nicholls, D. I., *J. chem. Soc. (A)* (1970), 1216.
57. Hopton, F. J., Rest, A. J., Rosevear, D. T., and Stone, F. G. A., *J. chem. Soc. (A)* (1966), 1326.
58. Pearson, R. G., *Chem. Br.,* 3 (1967), 103.
59. Collier, M. R., Lappert, M. F., and Truelock, M. M., *J. organometal. Chem.,* **25** (1970), C36.
60. Yagupsky, G., Mowat, W., Shortland, A., and Wilkinson, G., *Chem. Commun.* (1970), 1369.
61. Nyholm, R. S., and Tobe, M. L., *Adv. inorg. Chem. Radiochem.,* **5** (1963), 1.
62. Cotton, F. A., and Kraihanzel, C. S., *J. Am. chem. Soc.,* 84 (1962), 4432.
63. Brown, T. L., and Darensbourg, D. J., *Inorg. Chem.,* 7 (1968), 959.
64. Haines, L. M., and Stiddard, M. H. B., *Adv. inorg. Chem. Radiochem.,* **12** (1969), 53.
65. Cotton, F. A., *Inorg. Chem.,* **3** (1964), 702; Kraihanzel, C. S., and Cotton, F. A., *Inorg. Chem.,* **2** (1963), 533; Horrocks, W. D., and Taylor, R. C., *Inorg. Chem.,* **2** (1963), 723.
66. Plastas, H. J., Stewart, J. M., and Grim, S. O., *J. Am. chem. Soc.,* **91** (1969), 4326.
67. Angelici, R. J., and Malone, M. D., *Inorg. Chem.,* **6** (1967), 1731; Angelici, R. J., *J. inorg. nucl. Chem.,* **28** (1966), 2627.
68. Bigorgne, M., *J. inorg. nucl. Chem.,* **26** (1964), 107.
69. Tolman, C. A., *J. Am. chem. Soc.,* **92** (1970), 2953.
70. Kabachnik, M. I., *Dokl. Akad. Nauk SSSR,* **110** (1956), 393; *Proc. Acad. Sci. USSR, Chem. Sect.,* **110** (1956), 577.
71. Stewart, R. P., and Treichel, P. M., *Inorg. Chem.,* 7 (1968), 1942.
72. Brown, T. L., and Darensbourg, D. J., *Inorg. Chem.,* 6 (1967), 971.
73. Sim, G. A., *A. Rev. phys. Chem.,* 18 (1967), 57.
74. Dessy, R. E., and Wieczorek, L., *J. Am. chem. Soc.,* 91 (1969), 4963.
75. Mather, G. G., and Pidcock, A., *J. chem. Soc. (A)* (1970), 1226.
76. Grim, S. O., Wheatland, D. A., and McFarlane, W., *J. Am. chem. Soc.,* 89 (1967), 5573; Grim, S. O., and Wheatland, D. A., *Inorg. nucl. Chem. Lett.,* 4 (1968), 187.
77. Kruck, T., *Angew. Chem. (Int. Ed. Engl.),* **6** (1967), 53; Nixon, J. F., *Adv. inorg. Chem. Radiochem.,* **13** (1970), 363.
78. Parry, R. W., and Bissot, T. C., *J. Am. chem. Soc.,* 78 (1956), 1524.
79. Alton, E. R., *Diss. Abstr.,* 21 (1961), 3620.
80. Kruck, T., Lang, W., and Derner, N., *Z. Naturf.,* **20B** (1965), 705.
81. Kruck, T., and Lang, W., *Z. anorg. allg. Chem.,* 343 (1966), 181.
82. Bassett, P. J., Lloyd, D. R., Hillier, I. H., and Saunders, V. R., *Chem. Phys. Lett.* **6** (1970), 253.
83. Bassett, P. J., and Lloyd, D. R., *J. chem. Soc. (Dalton)* (1972), 248.
84. Hillier, I. H., Marriott, J. C., Saunders, V. R., Ware, M. J., Lloyd, D. R., and Lynaugh, N., *Chem. Commun.* (1970), 1586.

85. Hillier, I. H., Saunders, V. R., Ware, M. J., Bassett, P. J., Lloyd, D. R., and Lynaugh, N., *Chem. Commun.* (1970), 1316.
86. Green, J. C., King, D. I., and Eland, J. D. H., *Chem. Commun.* (1970), 1121.
87. Hillier, I. H., and Saunders, V. R., *Chem. Commun.* (1970), 316.
88. Hillier, I. H., and Saunders, V. R., *J. chem. Soc. (A)* (1971), 664.
89. Serafini, A., Labarre, J.-F., Veillard, A., and Vinot, G., *Chem. Commun.* (1971), 996.
90. Guest, M. F., Hillier, I. H., and Saunders, V. R., *J. Chem. Soc. Faraday II,* **68** (1972), 114.
91. Morino, Y., Kuchitsu, K., and Moritani, T., *Inorg. Chem.,* **8** (1969), 867.
92. Moritani, T., Kuchitsu, K., and Morino, Y., *Inorg. Chem.,* **10** (1971), 344.
93. Marriott, J. C., Salthouse, J. A., Ware, M. J., and Freeman, J. M., *Chem. Commun.* (1970), 595.
94. Almenningen, A., Anderson, B., and Astrup, E. E., *Acta chem. scand.,* **24** (1970), 1579.
95. Bridges, D. M., Holywell, G. C., Rankin, D. W. H., and Freeman, J. M., *J. organometal. Chem.,* **32** (1971), 87.
96. Frenz, B. A., and Ibers, J. A., *Inorg. Chem.,* **9** (1970), 2403.
97. Kuczkowski, R. L., and Lide, D. R., *J. chem. Phys.,* **46** (1967), 857.
98. Lory, E. R., Porter, R. F., and Bauer, S. H., *Inorg. Chem.,* **10** (1971), 1072.
99. De Boer, B. G., Zalkin, A., and Templeton, D. H., *Inorg. Chem.,* **8** (1969), 836.
100. Mingos, D. M. P., Robinson, W. T., and Ibers, J. A., *Inorg. Chem.,* **10** (1971), 1043.
101. Myers, V. G., Basolo, F., and Nakamoto, F., *Inorg. Chem.,* **8** (1969), 1204.
102. Vandenbrouke, A. C., Hendricker, D. G., McCarley, R. E., and Verkade, J. G., *Inorg. Chem.,* **7** (1968), 1825.
103. Johnston, R. D., Basolo, F., and Pearson, R. G., *Inorg. Chem.,* **10** (1971), 247.
104. Meier, M., Basolo, F., and Pearson, R. G., *Inorg. Chem.,* **8** (1969), 795.
105. Nyholm, R. S., *Proc. chem. Soc.* (1961), 273.
106. Saalfeld, F. E., McDowell, M. V., Surinder, K. G., MacDiarmid, A. G., *J. Am. chem. Soc.,* **90** (1968), 3684.
107. Kiser, R. W., Krassoi, M. A., and Clark, R. J., *J. Am. Chem. Soc.,* **89** (1967), 3654.
108. Albano, V., Bellon, P. L., and Scatturin, V., *Chem. Commun.* (1966), 507.
109. Albano, V., Ricci, G. M. B., and Bellon, P. L., *Inorg. Chem.,* **8** (1969), 2109.
110. Albano, V. G., Bellon, P. L., and Sansoni, M., *Chem. Commun.* (1969), 899.
111. Bombieri, G., Forsellini, E., Panattoni, C., Graziani, R., and Bandoli, G., *J. chem. Soc. (A)* (1970), 1313.
112. Francis, J. N., McAdam, A., and Ibers, J. A., *J. organometal. Chem.,* **29** (1971), 131.
113. McAdam, A., Francis, J. N., and Ibers, J. A., *J. organometal. Chem.,* **29** (1971), 149.
114. Nyburg, S. C., personal communication reported in ref. [112].
115. Glanville, J. O., Stewart, J. M., and Grim, S. O., *J. organometal. Chem.,* **7** (1967), 9.
116. Cook, C. D., Wan, K. Y., Gelius, U., Hamrin, K., Johansson, G., Olson, E., Siegbahn, H., Nordling, C., and Siegbahn, K., *J. Am. chem. Soc.,* **93** (1971), 1904.
117. Farona, M. F., and Wojcicki, A., *Inorg. Chem.,* **4** (1965), 1402.
118. Basolo, F., Baddley, W. H., and Burmeister, J. L., *Inorg. Chem.,* **3** (1964), 1202.
119. Meek, D. W., Nipcon, P. E., and Meek, V. I., *J. Am. chem. Soc.,* **92** (1970), 5351.
120. Lippard, S. J., and Palenik, G. J., *Inorg. Chem.,* **10** (1971), 1322.
121. Masters, C., McDonald, W. S., Roper, C. I., and Shaw, B. L., *Inorg. Chem.,* **3** (1964), 1202. 210.
122. Masters, C., Shaw, B. L., and Stainbank, R. E., *Chem. Commun.* (1970), 209.
123. Shaw, B. L., and Stainbank, R. E., *J. chem. Soc. (Dalton)* (1972), 223.
124. Cheyney, A. J., Mann, B. E., Shaw, B. L., and Slade, R. M., *Chem. Commun.* (1970), 1176.
125. Mann, B. E., Masters, C., Shaw, B. L., and Stainbank, R. E., *Chem. Commun.* (1971), 1103.
126. Tolman, C. A., *J. Am. chem. Soc.,* **92** (1970), 2956.
127. Pauling, P. J., Porter, D. W., and Robertson, G. B., *J. chem. Soc. (A)* (1970), 2728.
128. Bancroft, G. M., Mays, M. J., and Prater, B. E., *J. chem. Soc. (A)* (1970), 956.
129. Fluck, E., and Kuhn, P., *Z. anorg. allg. Chem.,* **350** (1967), 263.
130. Angelici, R. T., and Ingemanson, C. M., *Inorg. Chem.,* **8** (1969), 83.

PART 2
Transition Metal Complexes Containing Phosphine Ligands

K. K. CHOW, W. LEVASON and C. A. McAULIFFE

Department of Chemistry, University of Manchester Institute of Science and Technology

6 INTRODUCTION

Since 1857, when Hofmann reported[1] the first phosphine[†] complexes (those formed between triethylphosphine and platinum and gold), the chemistry of the coordination complexes of transition metal ions and phosphines has been increasingly investigated; there are few fields of chemistry in which research activity increases at such a pace.

The last major review of monodentate phosphine complexes to appear was that by Booth[2] in 1964. Phosphido complexes derived from monodentate phosphines have also been reviewed[3]. Phosphido complexes resulting from the cleavage of bidentate phosphines have been dealt with quite recently[4]. It should also be mentioned that the preparation of organophosphine ligands has been reviewed by Maier[5].

In this part of the book we cover all phosphine and phosphido complexes of the transition metals with two exceptions. Complexes derived from PF_3 have been extensively reviewed in recent years[6], and we do not plan to duplicate this. Although we are including phosphine complexes derived from metal carbonyls, we do not include those phosphine adducts of organometallic compounds in which interest mainly centres on the metal–carbon bond.

Phosphine complexes have received a good deal of attention from chemists interested in variations of coordination number and stereochemistry in coordination compounds, as models for biological oxygen transport and nitrogen fixation, and as catalysts in organic chemistry.

The bonding in metal phosphine complexes has been discussed by Pidcock[7], and a report of the instrumental techniques used to investigate the complexes has also appeared[4].

† Please see appendix for abbreviations used in text.

7 LANTHANIDES AND ACTINIDES

Although a number of tertiary phosphine complexes of uranium have been reported, for example $UCl_4(PEt_3)_2$ and $UCl_4(PPr^n_3)_2$ [8], $UO_2X_2(PPh_3)_2$ (X = Cl, Br, I, NO_3) [9], all these have been subsequently shown to be phosphonium salts or phosphine oxide complexes [10-14]. The green $UCl_5 . PPh_3$ obtained [15] from $UCl_5 . (Cl_2C=CClCOCl)$ and PPh_3 in benzene seems to be the only genuine tertiary phosphine complex in this group of metals.

8 TITANIUM, ZIRCONIUM, HAFNIUM

In 1832 Rose[16] reported the isolation of $TiCl_4 . PH_3$; $TiCl_4(PH_3)_2$ has also been prepared[17]. Chatt and Hayter[18] obtained the dark red $TiCl_4(PR_3)_2$ (R = Et, Ph), and found them to be extremely sensitive to moisture. Similar 1:1 $TiCl_4$–phosphine complexes have been prepared[19,20]. Fowles and Walton[21] also obtained $TiCl_4(PPh_3)_2$, but found that $TiBr_4$ was reduced by PPh_3; these same workers failed to obtain any zirconium complexes.

The series $TiCl_4(PR_3)$ ($R_3 = Me_3$, Me_2H, MeH_2, H_3) was prepared by Beattie and Collis[22] by direct reaction at $-127°C$ or in solution. The PMe_3 complex was shown to be monomeric by cryoscopic studies, but the other complexes were extensively dissociated in solution. On the basis of vibrational spectra all were thought to be five-coordinate trigonal-bipyramidal molecules, with axial phosphine in the PMe_3 complex, and equatorial phosphine in the other three cases. In a somewhat similar study, Calderazzo et al.[23] have measured the ^1H and ^{31}P n.m.r. solution spectra of several $TiCl_4$–PR_3 ($R_3 = Me_3$, $PhMe_2$, Cy_3, Ph_3) systems at variable temperatures. The results suggest the presence of fast equilibria in solution involving 1:1 and 1:2 adducts, and free tertiary phosphine. There is also evidence for the probable existence of binuclear species of the type $Ti_2Cl_8(PR_3)$.

Isslieb and Wenschuh[24] prepared $Ti(PCy_2)_2$ by reaction of $TiCl_3 . THF$ or $TiCl_4 . 2THF$ with $LiPCy_2$. The dark brown product ($\mu_{eff} = 0.60$ B.M.) was pyrophoric, was decomposed by water, and reacted with iodine in benzene to give $TiI_2(PCy_2)_2$ and $TiI_3(PCy_2)_2$. The organotitanium compounds $MeTiCl_3 . PPh_3$ [25,26], $TiMe_4 . PMe_3$ and $TiMe_4(PMe_3)_2$ [27] have been isolated.

The only zirconium complexes reported are the red-brown phosphido-bridged $(\pi\text{-}C_5H_5)_2Zr(PR_2)_2$ (R = Et, Bu^n) formed from $(\pi\text{-}C_5H_5)_2ZrBr_2$ and $LiPR_2$ [28]. The dark violet titanium analogues are known, and the proposed structures are

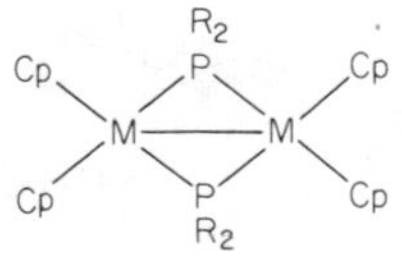

(M = Ti, Zr)

(I)

9 VANADIUM, NIOBIUM, TANTALUM

Werner[29-32] and Hieber et $al.$[33,34] obtained red-brown $[V(CO)_4(PR_3)_2]$ (R = Et, n-Pr, Ph) from the reaction of $V(CO)_6$ with the appropriate phosphine. These complexes had effective magnetic moments of 1.79 B.M. and were assigned $trans$ structures on the basis of i.r. evidence. Orange $[V(CO)_4(PR_3)_2]$ complexes were also obtained with $PPhH_2$ and PPh_2H, and in addition these ligands formed the blue-green diamagnetic phosphido-bridged complexes obtained by the reaction[34]

$$2V(CO)_6 + 2PRR'H \rightarrow [V(CO)_4(PRR')]_2 + H_2 + 4CO$$

The yellow, diamagnetic $[V(CO)_4(PCy_3)_2]_2$(C.N. = 7?) is formed with the PCy_3 ligand[34].

Nitric oxide reacts with $[V(CO)_4(PPh_3)_2]$ in n-hexane to form $[V(CO)_4(NO)(PPh_3)]$, and $[V(CO)_4(PPh_3)_2]$ can be reduced by sodium amalgam in ethanol to the yellow $[V(CO)_5(PPh_3)]^-$ ion, isolated as the Et_4N^+ salt on addition of Et_4NI[29]. Recently, a better method for the preparation of $[V(CO)_5(PPh_3)]^-$ has been reported, and the niobium and tantalum analogues have been isolated in very low yields (about 1 per cent) as the Ph_4As^+ salts[35]. The reaction of $V(CO)_6$ with PPh_3 in ether produces the curious green $[V(Et_2O)_6][V(CO)_5(PPh_3)]_2$ which contains V(II) and V(−I), with μ_{eff} = 4.02 B.M. and θ = −14 K[33].

Phosphine substitutes one carbonyl group in $(\pi\text{-}C_5H_5)V(CO)_4$ on irradiation in benzene solution to give $(\pi\text{-}C_5H_5)V(CO)_3(PH_3)$ [36]. Brown-black $V(PCy_2)_3$ (Cy = cyclohexyl), analogous to the Ti and Cr compounds, is formed from $VCl_3 . 3THF$ and $LiPCy_2$; it reacts with iodine to form $VI(PCy_2)_3$ [24].

The reaction of $V(CO)_6$ with AP and cis-PP produces yellow-green air sensitive solids, $[V(CO)_4L]$; $[V(CO)_4(AP)]$ has μ_{eff} = 1.98 B.M.[37]. They are reduced by Na/Hg to $[V(CO)_5L]^-$ which contain the first examples of monodentate olefin-phosphines in carbonyl complexes.

The phosphine complexes of vanadium in higher oxidation states have received little attention. Isslieb and Bohn[38] obtained red $[VCl_3(PR_3)_2]$ (R = Me, Et, n-Pr) with the trialkyl phosphines, but found that when R = Cy or Ph compounds of varying halide:phosphine ratios were obtained, which appear to be mixtures of the five-coordinate monomers and $[VCl_3(PR_3)]_2$ dimers. The i.r. and electronic spectra of $[VCl_3(PEt_3)_2]$ (μ_{eff} = 2.83 B.M.) are consistent with a trigonal-bipyramidal structure[38,39]. The brown amorphous $[V(NCS)_3(PR_3)(MeOH)_2]$ (Re = Me_3, Et_3, Pr^n_3, Bu^n_3, $PhEt_2$) are produced when $K_3V(NCS)_6$ reacts with PR_3 in methanol[40].

Selbin and Vigee[41] could not obtain a triphenylphosphine complex of $VOCl_2$, and showed that the reported $VOCl_2(PPh_3)_2$ [9] was a phosphine oxide complex. However, $VOCl_2(PPh_3)_2$ has recently been reported to form from $VOCl_2$ and PPh_3 in acetonitrile[42], and e.s.r. evidence has been obtained for the existence of $VOCl_2(PEt_3)_2$ in a solution of the constituents in toluene[43]. Tarama *et al.*[44] have reported $VOCl_3(PPh_3)$ to be a red-brown solid, with $\nu(V{-}O)$ at 1000 cm^{-1}. Triphenylphosphine partially reduces VCl_4, and dark blue crystals of approximate composition $VCl_4(PPh_3)_2$ ($\mu_{eff} = 2.67$ B.M.), which may be $[PPh_3H][VCl_4(PPh_3)]$, have been isolated[45].

The reactions of NbX_4 (X = Cl, Br, I) with PPh_3 were studied by Machin and Sullivan[46], who obtained $NbCl_4(PPh_3)_{1.5}$ as a bright orange-yellow solid. Small amounts of an unidentified product were formed with $NbBr_4$, but NbI_4 did not react. Russian workers have reported the products of heating MCl_5 (M = Nb, Ta) with Bu^n_3P to have the empirical formulae '$MPCl_2$'; $TaCl_5(Bu^n_3P)$ is said to form at room temperature[47]. Green $NbCl_5(PPh_3)$ and brown $NbCl_5(PPh_3)_2$ have been obtained[48]. Some organoniobium complexes, $[CpNb(CO)_3(PPh_3)]$ and $[CpNb(CO)_2(PPh_3)_2]$ have also been isolated[49].

10 CHROMIUM, MOLYBDENUM, TUNGSTEN

10.1 CARBONYL COMPLEXES

The substituted carbonyls of Cr, Mo and W have been extensively investigated (table 8). The reaction of $M(CO)_6$ (M = Cr, Mo, W) with monodentate phosphines, PR_3, produces a range of products $M(CO)_{6-n}(PR_3)_n$ ($n = 1-4$), depending upon the nature of M and R, and upon the experimental conditions. The literature up to mid-1965 has been reviewed by Dobson et al.[91]. There are two main routes to the phosphine-substituted carbonyls of these elements.

(a) Reaction of $M(CO)_6$ with the phosphine, either under reflux in a suitable solvent—diglyme, benzene, hexane—or by heating the reactants in a sealed tube in the absence of a solvent. Ultraviolet irradiation of the reactants has also proved a useful method in a number of cases.

(b) Displacement of a substituent from a partially substituted carbonyl; for example, the displacement of cycloheptatriene, norbornadiene, or mesitylene. This often provides a better route to a specific complex than direct synthesis from the hexacarbonyl. This method is also used to prepare highly substituted complexes such as $M(CO)_2(PMe_3)_4$ [51] that cannot be obtained directly.

Purification of the products is achieved by chromatography, recrystallisation or vacuum sublimation. The characterisation of these complexes and the use of various spectroscopic techniques to investigate the nature of the M—P and the M—C≡O bonds have been the subject of a large number of papers (see, for example, references 61, 91-4).

Only one complex of stoichiometry $[M(CO)_5(PR_3)]$ is possible, but for the more highly substituted complexes two geometric isomers of each stoichiometry can exist, namely cis- and trans-$[M(CO)_4(PR_3)_2]$, fac- and mer-$[M(CO)_3(PR_3)_3]$[†], and cis- and trans-$[M(CO)_2(PR_3)_4]$ (table 8).

In general, these phosphine substituted carbonyls are white or yellow solids which are fairly stable in air, although a number of complexes which are liquid at room temperature have also been made; for example $[M(CO)_5(PBu^n_3)]$ and mer-$[Cr(CO)_3(PMe_3)_3]$.

†

fac-$[M(CO)_3(PR_3)_3]$ mer-$[M(CO)_3(PR_3)_3]$

The kinetics of the substitution reaction have been reviewed by Werner[62] and Angelici[95].

Organometallic phosphines $(Me_3E)_3P$ (E = Si, Ge, Sn) produce monosubstituted carbonyl complexes[96,97]. Phosphine, PH_3, reacts with $M(CO)_6$ in benzene or hexane to produce *cis*-$[M(CO)_4(PH_3)_2]$ (M = Cr, Mo, W)[86] (also obtained from $M(CO)_4$(norbornadiene) and PH_3[88]), and reacts with $[Cr(CO)_3(B_3N_3Me_6)]$ in hexane to form $[Cr(CO)_3(PH_3)_3]$. Recently, Becker and Ebsworth[98] reported that KPH_2 in dimethylether at $0°C$ extracts a proton from the Mo complex to form $[Mo(CO)_5(PH_2)]^-$ and $[(CO)_5MoPH_2Mo(CO)_5]^-$. Another example of this type of coordinated phosphine reaction is the report[99] that $[Mo(CO)_5P(NMe_2)_3]$ reacts with HX (X = Cl, Br, I) with subsequent rupture of the P—N bonds. In an autoclave HCl produces $[Mo(CO)_5PCl_3]$, whilst in *n*-pentane solution $[Mo(CO)_5PCl(NMe_2)_2]$, $[Mo(CO)_5PCl_2(NMe_2)]$ and $[Mo(CO)_5PCl_3]$ are all produced. Hydrogen bromide produces $[Mo(CO)_5PBr_3]$ in 50 per cent yield, and HI forms $[Mo(CO)_5PI_2(NMe_2)]$ and the unstable $[Mo(CO)_5PI_3]$. From the reaction of $M(CO)_6$ (M = Cr, Mo) and either $PH(C_6F_5)_2$ or $(C_6F_5)_2P—P(C_6F_5)_2$, $[M(CO)_5PH(C_6F_5)_2]$ are formed, which can be metallated by *n*-BuLi to give $[M(CO)_5\{P(C_6F_5)_2Li\}]$[80].

Preliminary X-ray data have been reported for the compounds $[M(CO)_4(PEt_3)_2]$ (M = Mo, W)[53,59], and the structure of $[Cr(CO)_5PPh_3]$ has been determined[100]. It is octahedral, and an interesting feature is that the Cr—C distance is considerably shorter (184 pm) for the carbonyl group *trans* to PPh_3 than for the four *cis* carbonyl groups (188 pm). A similar difference was observed for the analogous phosphite complex.

'Mixed' ligand complexes have attracted increasing interest in recent years. They are essentially of two kinds: (i) those containing different phosphines, for example *trans*-$[M(CO)_4(PBu''_3)(PPh_3)]$ (M = Cr, Mo, W)[61,91], which, like the analogous $[M(CO)_4(PR_3)_2]$ complexes, are pale coloured solids; (ii) those containing other ligands such as 2,2'-bipyridyl or 1,10-phenanthroline, which are highly coloured. Type (ii) complexes are obtained from $[M(CO)_4L]$ (L = bipy, phen)[101-3]; for example $[Mo(CO)_4bipy]$ or $[Mo(CO)_3bipy]_2$ react with PPh_3 to form violet $[Mo(CO)_3(bipy)PPh_3]$, whilst in a refluxing mesitylene solution of PPh_3, green $[Mo(CO)_2(bipy)(PPh_3)_2]$ is obtained. The dicarbonyl complexes take up SO_2 reversibly in CH_2Cl_2 solution to form red or purple $[M(CO)_2L(PPh_3)(SO_2)]$[104]. An interesting new precursor to the dicarbonyl derivatives is $[(\pi\text{-allyl})Mo(MeCN)_2(CO)_2X]$ (X = halogen), which reacts with PPh_3 to give (II), from which the MeCN is easily displaced by a number of ligands to give $[Mo(PPh_3)_2L_2(CO)_2]$ (L = py, DMF; L_2 = bipy, phen, en, etc.)[105].

(II)

TABLE 8 SUBSTITUTED CARBONYLS OF CHROMIUM, MOLYBDENUM AND TUNGSTEN

Ligand	Metal	Preparation†	Products‡	Comments	Ref.
PMe_3	Cr,Mo,W	a,b	1,t2,c2,f3,m3,c4		51–7
PEt_3	Cr,Mo,W	a,b	1,t2,c2,f3,m3		52,53,58–60
PBu^n_3	Cr,Mo,W	a	1,t2,c2	Mo–1 compound is liquid at room temperature	55,61–65
PPh_3	Cr,Mo,W	a,b and $[M(CO)_5]^-$ + PPh_3	1,t2,c2,f3		58,63,66–72
PCy_3	Cr,Mo,W	a and $[M(CO)_5C(OMe)R]^+$ PCy_3	1,t2		64,89
$P(CH_2CH_2CN)_2$	Cr	a,b	1		63,67,73
$PPhMe_2$	Cr,Mo,W	a,b	1,c2,t2,f3,m3	dipole moments and n.m.r. studies indicate some isomerism of 2 in solution for Cr only	56
$PPhEt_2$	Cr,Mo,W	a,b	t2,c2,f3		58,74
$PPhBu^n_2$	Cr,Mo,W	a	1,t2,c2,m3,f3	Mo–1 compound is dark liquid	55,61,75
PPh_2Me	Cr,Mo,W	a	1		55,61
PPh_2Et	Cr,Mo	b	1,f3		58,61
PPh_2Pr^n	Mo	a	1		61
PPh_2Pr^i	Cr,Mo,W	a	1		61

Ligand	Metal	Method[†]	Isomer[‡]	Notes	Ref.
PPh_2Bu^n	Cr,Mo,W	a	$1,t2,c2$		55,61
PPh_2Bu^t	Cr,Mo,W	a	1		61
$P(NMe_2)_3$	Cr,Mo,W	a,b	$1,t2,c2$	W-1 compound unknown	54,76
$PMe(NMe_2)_2$	Cr,Mo,W	a,b	$1,t2,c2$	W-1 not isolated pure	77
PCl_3	Cr,Mo,W	a,b	$1,t2,c2,f3,m3,c4$	for Mo-3 compounds / $\nu(C\!-\!O)$ rises along series / $R_3 = Ph_3 < Ph_2Cl < PhCl_2 < Cl_3$	52,53,66
PPh_2Cl	Cr,Mo,W	a,b	$1,c2,f3$		66,78,79
$PPhCl_2$	Mo	b	$f3$		66
$PMeBr_2$	Mo	a,b	$1,c2,f3$		52,53
$P(CH_2Cl)_3$	Mo	b	$c2$		57
$PH(C_6F_5)_2$	Cr,Mo,W	a,b	$1,2,3$		80
$PPh(C_6F_5)_2$	Mo		1		81
$PPh_2(C_6F_5)$	Mo		1		81
$P(p\text{-}C_6H_4F)_3$	Cr,Mo,W		$1,c2,t2$		90
$PHPh_2$	Cr,Mo,W	a	$1,c2,f3$	M-3 is purple	78,79
AP	Cr,Mo,W	b	1	AP rearranged to give cis-PP complex	82,83
PP	Cr,Mo,W	b	1		83
BPE	Cr	a,b	$1,2$	cis and trans PP used	84
DPA	Cr,Mo,W	b	$c2,t2,f3,m3$	also $Cr(CO)_3(BPE)(MeCN)$	85
PH_3	Cr,Mo,W	a,b	$1,c2,f3,4$		50,87–9

† Refers to methods (a) and (b) discussed in text.

‡ $1 = M(CO)_5PR_3$, $c2 = $ *cis*-$M(CO)_4(PR_3)_2$, $t2 = $ *trans*-$M(CO)_4(PR_3)_2$, $m3 = $ *mer*-$M(CO)_3(PR_3)_3$, $f3 = $ *fac*-$M(CO)_3(PR_3)_3$, $c4 = $ *cis*-$M(CO)_2(PR_3)_4$; when no letter attached, for example '2' means $M(CO)_4(PR_3)_2$, isomer not stated.

Ligands with N or S donor atoms, such as $Ph_2PCH_2CH_2NMe_2$ and $Ph_2PCH_2CH_2SMe$ (SPE) have been studied by Dobson *et al.*[106,107]. SPE behaves as a bidentate S–P donor in $[Mo(CO)_4(SPE)]$ and $[Mo(CO)_2(SPE)_2]$, but as a monodentate P donor in $[Mo(CO)_3(bipy)(SPE)]$.

Bennett *et al.*[82,83] found that 2-allylphenyldiphenylphosphine (AP), $CH_2=CHCH_2C_6H_4PPh_2$, reacts with $[M(CO)_4(C_7H_8)]$ (M = Cr, Mo, W) in refluxing benzene to yield yellow crystals, which analyse as $MC_{25}H_{19}O_4P$, but which were shown by i.r. and n.m.r. studies to contain the unsaturated phosphine ligand 2-propenyl-phenyldiphenylphosphine, $CH_3CH=CHC_6H_4PPh_2$ (PP). The same workers[83] prepared both *cis* and *trans* isomers of PP (abbreviated *cis*PP; *trans*PP) and found that both reacted with $[M(CO)_4(C_7H_8)]$ to give $[M(CO)_4(PP)]$ without isomerisation. By spectral comparisons the principal product formed in the isomerisation of AP was identified as $[M(CO)_4(cisPP)]$, which is interesting, since preliminary studies indicated that *cis*PP is the less stable isomer, and its formation from AP has been ascribed to a faster rate of formation than that of *trans*PP. An X-ray analysis by Luth *et al.*[108,109] has confirmed the structure proposed for $[Mo(CO)_4(cisPP)]$ on the basis of spectroscopic studies (III). The Mo–P bond length is 151.7 pm and the C=C bond length in the complex is 140 pm, considerably longer than a normal C=C bond length. Isslieb and Haftendorn[84] prepared 3-butenyldiethylphosphine, $CH_2=CHCH_2CH_2PEt_2$ (BPE),

(III)

and found that it reacted with $Cr(CO)_6$ to give $[Cr(CO)_5(BPE)]$, in which only the phosphorus is coordinated, but with $Cr(CO)_4(MeCN)_2$ and $Cr(CO)_3(MeCN)_3$ it gave $[Cr(CO)_4(BPE)]$ and $[Cr(CO)_3(MeCN)(BPE)]$, respectively, which are thought to contain coordinated C=C on the basis of i.r. evidence. Diphenylphosphinoacetylene (DPA), $Ph_2PC\equiv CPh$, functions only as a P donor ligand; the triple bond does not coordinate[85]. *Trans*-$[Cr(CO)_4(DPA)_2]$ and *fac*-$[M(CO)_3(DPA)_3]$ are formed by displacing cycloheptatriene from the appropriate cycloheptatriene-substituted carbonyls. At room temperature the chromium compound gave only the *fac* isomer, but at 85°C a mixture of *fac* and *mer* isomers, readily separable by chromatography, was obtained. Interestingly, DPA also produces the rearranged *trans*-$[Cr(CO)_4(DPA)_2]$ from $[(norbornadiene)Cr(CO)_4]$ instead of the expected *cis* isomer[85].

From the reaction of Ph_2PCN, $PhP(CN)_2$, and Me_2PCN with $[M(CO)_4(C_7H_8)]$ (M = Cr, Mo) yellow crystalline complexes of the form $[M(CO)_4L]_2$ were obtained. On the basis of i.r. and n.m.r. evidence structure (IV) was proposed[110].

Diphosphines, R_4P_2, react with Group VIA hexacarbonyls to produce either diphosphine-bridged derivatives[4] or phosphido-bridged [(CO)$_4$M(PR$_2$)$_2$M(CO)$_4$] complexes. The latter are formed at higher temperatures (240–60°C) and are normally deep orange or red[111-14]. In addition to the simple phosphido derivatives, a few

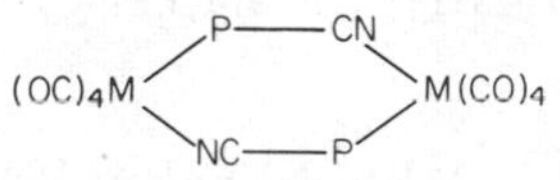

(IV)

heteronuclear derivatives are known; for example [(CO)$_4$Cr(PMe$_2$)$_2$Mo(CO)$_4$] and [(CO)$_4$M(PPh$_2$)$_2$Fe(CO)$_3$] (M = Cr, Mo, W) formed from [M(CO)$_5$PPh$_2$Cl] and [Fe(CO)$_4$PPh$_2$H][115]. The unusual complex (V) is formed from [(π-allyl)PdCl]$_2$

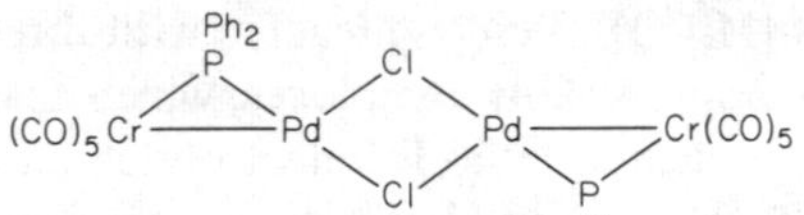

(V)

and [Cr(CO)$_5$(PPh$_2$H)] in THF[116]. The simple phosphido complexes, which have structure (VIA), react readily with PR$_3$ (R = Ph, Et, Cy) on irradiation in benzene to give the disubstituted products (VIB). With a 1:1 PR$_3$:M$_2$(CO)$_8$(PR$_2$)$_2$ ratio

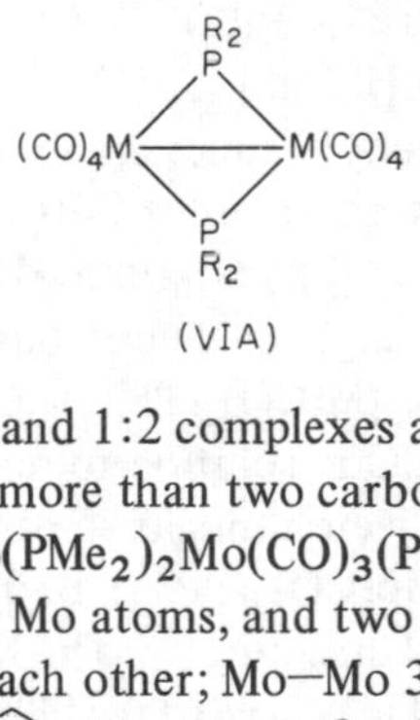

(VIA) 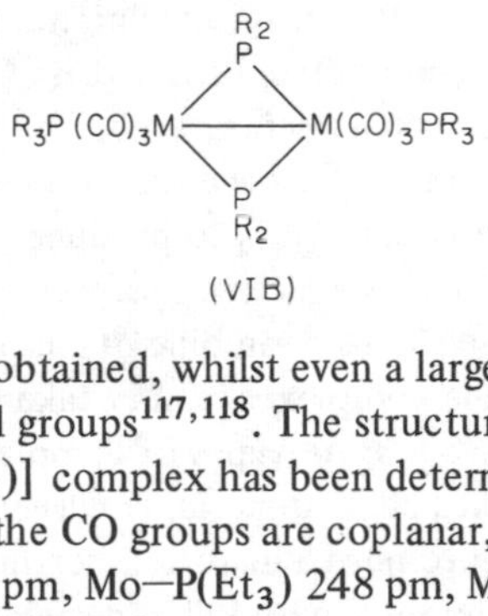   (VIB)

a mixture of 1:1 and 1:2 complexes are obtained, whilst even a large excess of PR$_3$ does not replace more than two carbonyl groups[117,118]. The structure of the [(PEt$_3$)(CO)$_3$Mo(PMe$_2$)$_2$Mo(CO)$_3$(PEt$_3$)] complex has been determined[119]. The four P atoms, the two Mo atoms, and two of the CO groups are coplanar, with the PEt$_3$ ligands *trans* to each other; Mo–Mo 309 pm, Mo–P(Et$_3$) 248 pm, Mo–P(Me$_2$) 245 pm, (Me$_2$)P$\widehat{M}$oP(Me$_2$) = 102°, Mo$\widehat{P}$Mo = 78°.

An unusual complex containing a bridging hydrogen atom was obtained by Hayter[113] from NaCpMo(CO)$_3$ and PClMe$_2$, (VII); the structure was determined by Doedens and Dahl[120]. Hayter[113] also obtained a complex having three phosphido-

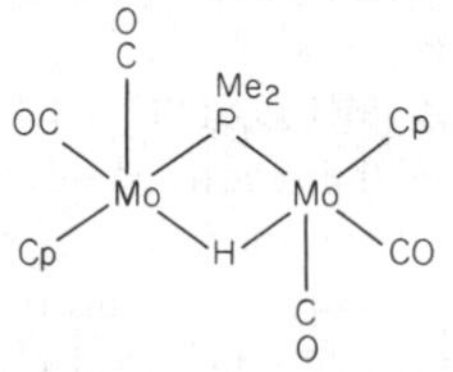

(VII) 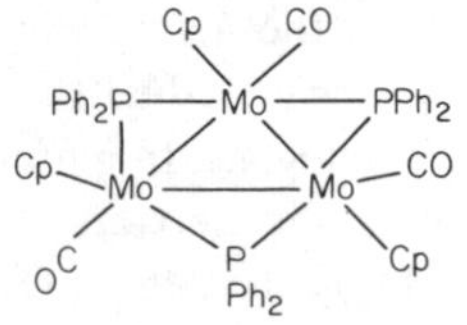   (VIII)

bridges which probably has structure (VIII), although the i.r. spectrum (band at 1852 cm^{-1}) does not distinguish between bridging and terminal CO groups. A rather different kind of phosphido complex, $K_3[M(CO)_3(PPh_2)_3]$ (M = Cr, Mo, W), is formed[121] on reacting $[M(CO)_3(NH_3)_3]$ with $KPPh_2$.

10.2 CARBONYL HALIDES

These compounds, which have recently been reviewed by Colton *et al.*[122] are almost completely confined to molybdenum and tungsten. Lewis and Whyman[123] found that halogen oxidation of triphenylphosphine-substituted carbonyls of these elements produced phosphine oxide complexes, but that under controlled conditions $[M(CO)_3(PPh_3)_3]$ or *trans*-$[M(CO)_4(PPh_3)_2]$ reacted with three equivalents of bromine or iodine to form orange-yellow complexes, which analysed as $M(CO)_3(PPh_3)_2X_3$[124]. These diamagnetic, 1:1 electrolytes exhibited an i.r. band attributable to the PPh_3H^+ ion, and hence they were formulated as $[PPh_3H][M(CO)_3(PPh_3)X_3]$ (X = Br, I). Tsang *et al.*[125] obtained the iodo-complexes as $Bu^n_4N^+$ salts by reaction of PPh_3 with $[Bu^n_4N][M(CO)_4I_3]$, and this method was extended to the chloro- and bromo-analogues by Colton and coworkers[126,127]. The latter investigators found that two distinct isomers of each complex were formed, and they were able to identify one type of isomer with the complexes prepared by Lewis and Whyman. Recently, the anionic $[Cr(CO)_4(PR_3)X]^-$ (R = p-C_6H_4F, p-C_6H_4Cl) have been obtained from $[Cr(CO)_5X]^-$ and PR_3[128].

The relationships between a number of carbonyl halides are shown in figure 1. Colton and coworkers[130-2] have shown that the compounds $[M(CO)_4X_2]$ (X = Cl, Br) react with PPh_3 in CH_2Cl_2 to produce yellow $[M(CO)_3(PPh_3)_2X_2]$ as air-stable diamagnetic complexes which, on refluxing in CH_2Cl_2, or upon bubbling N_2 through the solutions, lose CO to form blue diamagnetic $[M(CO)_2(PPh_3)_2X_2]$. The reaction is reversible; the blue complexes readily take up CO in solution to reform the yellow $[M(CO)_3(PPh_3)_2X_2]$. Both types of complexes were thought to be seven-coordinate, the blue complexes being dimeric. However, $[Mo(CO)_2(PPh_3)_2Br_2]$ is reported to be mononuclear and to have a distorted octahedral structure[138]. The $M(CO)_4I_2$ compounds react differently with PPh_3[127]; when M = W, $[W(CO)_3(PPh_3)_2I_2]$ is formed, which loses CO to form $[W(CO)_2(PPh_3)_2I_2]$ on heating *in vacuo*, whereas the chloro- and bromo-derivatives lose CO on boiling in solution.

$[Mo(CO)_3(PPh_3)_2I_2]$ has not been prepared, although the $AsPh_3$ and $SbPh_3$ analogues are known[127]. Conversely, thiocyanato-complexes, $[M(CO)_2(PPh_3)_2(NCS)_2]$, are known, but not for the $AsPh_3$ and $SbPh_3$ derivatives[133].

Moss and Shaw[134] have studied the reactions of $[M(CO)_4(PR_3)_2]$ (R_3 = Et_3, $PhMe_2$, $PhEt_2$) with halogens. In a number of cases the reactions resemble those of PPh_3 described above, but there are some notable differences. Chlorine oxidises *cis*-$[Mo(CO)_4(PPhMe_2)_2]$ to $[Mo_2(CO)_4(PPhMe_2)_4Cl_2]$ and the bromo- and iodo-analogues are readily obtained on heating $[Mo(CO)_3(PPhMe_2)_2X_2]$ (X = Br, I). The blue $[Mo(CO)_2(PEt_3)_2Br_2]$ absorbs CO in the solid state to form yellow

[Mo(CO)$_3$(PEt$_3$)$_2$Br$_2$], a reaction which will only occur in solution for the PPh$_3$ complex. A new type of complex, [M(CO)$_2$(PPhMe$_2$)$_3$X$_2$], was formed from [M(CO)$_4$X$_2$] and PPhMe$_2$ (X = Cl, Br)[134]; these are seven-coordinate, mononuclear species[139], which crystallise with a molecule of alcohol that cannot be removed on

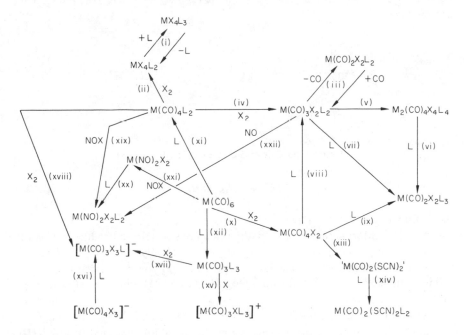

(Some reactions have been ommitted for clarity.)

Figure 1 Some carbonyl halide and nitrosyl derivatives of molybdenum and tungsten[124-7,129,130-7]
(i) M = W; L = PPhMe$_2$. (ii) M = Mo, W; L = PPhMe$_2$, PPhEt$_2$. (iii) M = Mo, W; L = PPh$_3$ or M = Mo; L = PEt$_3$. (iv), (v), (vi) M = Mo; L = PPhMe$_2$; X = Cl, Br, I. (vii) M = Mo; L = PPhMe$_2$. (viii) M = Mo, W; L = PPh$_3$; X = Br, Cl *or* M = W; L = PPhMe$_2$. (ix) M = W; L = PPhMe$_2$; X = I (x) M = Mo, W; X = Cl, Br(−78°C); X = I (u.v.). (xi) M = Mo, W; L = PPhMe$_2$, PPh$_3$. (xii) M = Mo, W; L = PPhMe$_2$. (xiii) M = Mo, W. (xiv) L = PPh$_3$. (xv) M = Mo, W; L = PPhMe$_2$; X = I. (xvi), (xvii) M = Mo, W; L = PPh$_3$; X = Cl, Br, I. (xviii) M = Mo, W; L = PPh$_3$; X = Br. (xix) M = Mo, W; X = Br. (xx) L = PPh$_3$. (xxi) X = Cl. (xxii) L = PPh$_3$

heating *in vacuo*. The action of excess halogen removes all the CO to give complexes such as [W(PPhMe$_2$)$_2$Cl$_4$] (see figure 1), which are discussed below.

10.3 NITROSYLS

Cotton and Johnson[135] prepared the dark green [M(NO)$_2$(PPh$_3$)$_2$Cl$_2$] (M = Mo, W) by reaction between M(NO)$_2$Cl$_2$ and PPh$_3$, and Johnson[136] obtained the bromo-analogues and [Mo(NO)$_2$(PPh$_3$)$_2$I$_2$]. All are simple octahedral monomers[136]. Colton *et al.*[137] reacted nitric oxide with $[M(CO)_3(PPh_3)_2X_2]$ and found that various *cis* isomers (figure 2) were formed, which could be separated by chromatography. The *cis* configuration of all the complexes was indicated by the presence of two $\nu(N-O)$ frequencies in the i.r. spectra. In contrast, Cotton and Johnson's method above is stereospecific, usually producing only isomer (a)[137]. Violet $[Cr(NO)_2(PPh_3)_2I_2]$ has been reported[140].

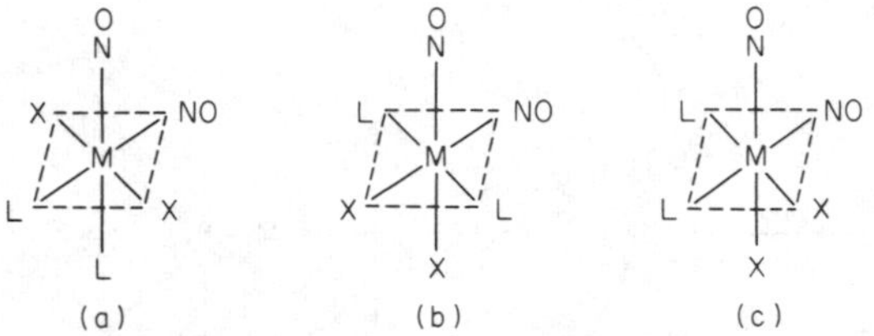

Figure 2 Possible *cis* isomers of $[M(NO)_2 L_2 X_2]$

10.4 HALIDES

10.4.1 Chromium

Chromium(II) chloride reacts with triethylphosphine in THF to produce a deep blue solution, which probably contains $CrCl_2 . (PEt_3)_2$, but only the colourless $CrCl_2 . PEt_3$ has been isolated as a solid[141]. The latter is probably polymeric. Similar rather insoluble complexes are produced from PMe_3, $HPMe_2$ and H_2PMe, and chromium(II) salts[142].

The grey-green $Cr(PCy_2)_3$ produced[24,143] from $LiPCy_2$ and $CrCl_3$. 3 THF has μ_{eff} = 2.93 B.M., indicating the presence of metal–metal interaction. Chromium(III) halides form three types of complex with tertiary phosphines: green $[CrCl_3(PR_3)_2]_2$ (R = Et, *n*-Bu) formed from anhydrous $CrCl_3$ and PR_3 in benzene[141,144]; violet $[CrCl_3(PR_3)]_n$ formed on melting $CrCl_3$ with PPh_3 or PCy_3 [141], or by warming the green complexes with 1:1 $CH_2Cl_2 : CCl_4$ [144]; and the red-brown $[CrBr_3(PR_3)_2]$ (R = Et, *n*-Bu)[141]. No complex could be isolated with PMe_3, and $PPhEt_2$ and PPh_2Et did not react. The green $[CrCl_3(PR_3)_2]_2$ are non-electrolytes, $\mu_{eff} \sim 3.8$ B.M., with high dipole moments (~8D), which would seem to rule out a highly symmetric structure. It has not proved possible to assign structures from spectral results[144]. On heating these green complexes become purple at ~110°C, but slowly regain their original colour on cooling. The violet $[CrCl_3(PR_3)]_n$ are insoluble in non-coordinating solvents, so molecular weight values have not been determined, but the solid reflectance spectra are characteristic of octahedral Cr(III), making halide-bridged polymeric structures probable. Violet $[CrCl_3(HPEt_2)_3]$, μ_{eff} = 3.84 B.M., is synthesised from $HPEt_2$ [145] but $HPPh_2$ and $HPCy_2$ do not react[146]. Primary phosphines H_2PPh [147] and H_2PCy [148] form dark blue complexes, $[CrCl_3(H_2PPh)_3]$ and $[CrCl_3(H_2PCy)_2]_2$. The green $[CrCl_3(PR_3)_2]_2$ react with Ph_4PCl or Ph_4AsCl to yield salts of the $[CrCl_4(PR_3)_2]^-$ ion, which have been assigned a *trans* octahedral structure on i.r. evidence[144].

A large number of phosphines react with $K_3Cr(NCS)_6$ to produce red or violet complexes analogous to Reinecke's salt. These are of three types: $[R_3PH]_3[Cr(NCS)_6]$ (R = Cy, *t*-Bu, *i*-Pr), $[R_3PH][Cr(NCS)_4(PR_3)_2]$ (R_3 = Me_3, Et_3, Et_2H, Cy_2H, Et_2Ph), and the rare $[R_3PH]_2[Cr(NCS)_5(PR_3)]$ (R_3 = Ph_2Et)[145,149-52]. Triphenylphosphine does not react with $K_3Cr(NCS)_6$ [149]. Salts with other large cations can be obtained by exchange reactions, and on passing the salts through a cation exchange resin some of the unstable free acids have been obtained[149]. It is not possible to replace more NCS groups to give $[Cr(NCS)_3(PR_3)_3]$ [150]. The i.r. and visible spectra have been studied by Clark *et al.*[151] and by Thomas[152], who deduced a spectrochemical series for the different phosphine ligands in these complexes. Issleib and Biermann[153] have prepared some molybdenum analogues; the general trends observed for chromium were found; for example whilst PEt_3 and PBu^n_3 gave $[Mo(SCN)_4(PR_3)_2]^-$, PCy_3 produced only $[Cy_3PH]_3[Mo(SCN)_6]$, and PPh_3 did not react.

10.4.2 Molybdenum and tungsten

Reduction of $MoCl_3$. 3THF by Na/Hg or magnesium in the presence of the appropriate phosphine leads to unusually low oxidation state complexes[154]. With $HPPh_2$ in an inert atmosphere $[Mo(HPPh_2)_6]$ is formed, whilst $PPhMe_2$ gives $[Mo(PPhMe_2)_4]$ in argon, but in nitrogen a solution that may contain *cis*-$[Mo(N_2)_2(PPhMe_2)_4]$ is obtained. A complex thought to be *cis*-$[Mo(N_2)_2(PPhMe_2)_4]$ has been isolated by the reduction of $[MoCl_4(PPhMe_2)_2]$ in THF in the presence of two moles of $PPhMe_2$ under nitrogen[155]. Under analogous conditions the PPh_2Me complex gives *trans*-$[Mo(N_2)_2(PPh_2Me)_4]$. The reduction of *trans*-$[WCl_4(PPhMe_2)_2]$ under nitrogen produces *cis*-$[W(N_2)_2(PPhMe_2)_4]$, whilst under CO and H_2, *fac*-$[W(CO)_3(PPhMe)_3]$ and $[WH_6(PPhMe_2)_3]$ result, respectively[156]. Molybdenum(II) chloride reacts with PPh_3 to form $[\{Mo_6Cl_8\}Cl_3(PPh_3)_3]Cl$, a 1:1 electrolyte; in the presence of pyridine or ethanol $[\{Mo_6Cl_8\}Cl_2(PPh_3)_2L_2]Cl_2$ (L = py, EtOH) are obtained[157]. Fowles *et al.*[158] reported $[MoBr_3(PPh_3)_2(MeCN)]$, μ_{eff} = 3.89 B.M., several years ago, and recently it has been found[154] that THF is readily displaced from $[MoCl_3(THF)_3]$ to form $[MoCl_3(PR_3)_x(THF)_{3-x}]$ (x = 3 for $PPhMe_2$; x = 2 for $PPhEt_2$).

A much larger number of Mo(IV) and W(IV) complexes are known. Fowles *et al.*[158,159] originally reported $[MCl_4(PPh_3)_2]$ (M = Mo, W), and Butcher and Chatt[160] characterised a series of maroon complexes, $[MoCl_4(PPh_2R)_2]$ (R = Me, Et, *n*-Pr), with magnetic moments in the range 2.22-2.61 B.M., consistent with d^2, Mo(IV) species, and possessing *trans* configurations on the basis of i.r. evidence. Moss and Shaw[161] found that excess Cl_2 or Br_2 decomposed $[M(CO)_4(PR_3)_2]$ (R_3 = $PhMe_2$, $PhEt_2$) to form Mo(IV) and W(IV) complexes, $[MX_4(PR_3)_2]$; these were shown by infrared spectral measurements, and in the case of $[WCl_4(PPhMe_2)_2]$ by an X-ray study[162], to have *trans* structures. In benzene solution the $PPhMe_2$ complexes were capable of adding a further molecule of phosphine to form red $[MX_4(PPhMe_2)_3]$. $\frac{1}{2}C_6H_6$, which were non-electrolytes and probably seven-coordinate. On heating *in vacuo* they revert to the $[MX_4(PPhMe_2)_2]$ complexes.

The molybdenum compounds $[MoCl_4(PR_3)_2]$ (R_3 = Ph_2Me, Ph_2Et) are reduced by ethanolic $NaBH_4$ in the presence of excess PR_3 to $[MoH_4(PR_3)_2]$, diamagnetic yellow compounds, which exhibit characteristic i.r. and n.m.r. spectra, the latter indicating four hydrogen atoms per molybdenum[163]. In contrast, *trans*-[$WCl_4(PPhMe_2)_2$] is reduced to white [$WH_6(PPhMe_2)_3$] [164,165] which dissolves in dilute hydrochloric acid to reform *trans*-[$WCl_4(PPhMe_2)_2$] and evolves hydrogen.

Molybdenum(IV) oxo complexes have recently attracted a great deal of attention. The reaction of sodium molybdate with alkylaryl phosphines in ethanol in the presence of a small amount of HCl produced[166] a series of six-coordinate diamagnetic compounds, [$MoOCl_2(PR_3)_3$] (R_3 = Ph_2Me, $PhMe_2$, Ph_2Et, $PhEt_2$). The presence of a terminal Mo=O group was indicated by the presence of a strong i.r. band at 940–956 cm^{-1}. Diphosphines displaced the PR_3 ligands to form [$MoOCl(diphos)$]Cl, but with a calculated amount of VPP, *cis*-1,2-bis(diphenylphosphino)ethylene, [$MoOCl_2(PR_3)(VPP)$] is formed. Butcher and Chatt[160] found that if the reaction of $MoCl_4(EtCN)_2$ with PR_3 was carried out in ethanol, [$MoOCl_2(PR_3)_3$] (R_3 = $PhMe_2$, $PhEt_2$, $PhPr^n_2$, Ph_2Et, Ph_2Me) were formed. Metathesis with the appropriate alkali halide or pseudohalide yielded the corresponding Br, I, CNS, OCN, and also the [$MoOICl(PPhMe_2)_3$] complex. The products were blue or green, monomeric diamagnetic non-conductors, which have so far[160] shown no resemblance to the [$ReOCl_3(PR_3)_2$] analogues.

There are two forms of [$MoOCl_2(PPhMe_2)_3$]: a green, unstable isomer, and a blue form. Originally[160], the complexes were assigned meridional structures, with *trans* chlorines in the green complex and *cis* chlorines in the blue, but X-ray analysis has shown the relationship to be more complicated.

The blue isomer was shown to have structure (IX)[167], and it was assumed that the

cis-mer-[$MoOCl_2(PPhMe_2)_3$]   *cis-mer*-[$MoOCl_2(PPhEt_2)_3$]

(IX)   (X)

green isomer probably had *trans* chlorines with the oxygen *trans* to a phosphine group. Unfortunately, crystals of the green isomer suitable for X-ray determination have not been obtained, but Chatt *et al.*[168] have determined the structure of [$MoOCl_2(PPhEt_2)_3$](X), which is green. Surprisingly, this has also a meridional configuration with *cis* chlorines, just as in the blue form of [$MoOCl_2(PPhMe_2)_3$]. The differences in the structure are due to differing amounts of distortion, and it appears that ligand repulsion is the important factor. The problem will not be finally solved until the structure of green [$MoOCl_2(PPhMe_2)_3$] has been determined, but it seems that a new kind of isomerism, called 'distortional' isomerism by the workers involved[168], may be present and that the blue and green form of *cis-mer-*[MoOCl$_2$(PPhMe$_2$)$_3$] are distortional isomers.

Excess triphenylphosphine reacts with MoOCl$_3$(MeCN)$_2$ to produce light-green [MoOCl$_3$(PPh$_3$)$_2$], μ_{eff} = 1.72 B.M. and ν(Mo=O) at 950 cm^{-1} [169]; but WOCl$_4$ seems to be partially reduced by PPh$_3$ to a green solid which, though not properly characterised, has been assigned a magnetic moment of 1.51 B.M.[170].

11 MANGANESE, TECHNETIUM, RHENIUM

11.1 MANGANESE

11.1.1 Manganese carbonyls

There has been some confusion in the literature about the formulation of the products from the reaction of $Mn_2(CO)_{10}$ with tertiary phosphines. Hieber and Freyer[171] reported that PPh_3 and PEt_3 produced paramagnetic $[Mn(CO)_4(PR_3)]$, but that PCy_3 produced diamagnetic $[Mn(CO)_4(PCy_3)]_2$, which appeared to dissociate reversibly on heating in xylene solution to give the paramagnetic monomer. More recent work has failed to substantiate the existence of these monomers. Photochemical reaction of $Mn_2(CO)_{10}$ and various phosphines[172,173] produced only $[Mn(CO)_4(PR_3)]_2$. Under different reaction conditions monosubstituted $[Mn_2(CO)_9(PR_3)]$ complexes can be obtained[174,175]. Wawersik and Basolo[175] showed that the Mn—Mn bond is not broken in the reaction of $Mn_2(CO)_{10}$ with PPh_3, and an X-ray study of the PEt_3 complex[176], $\mu_{eff} = 1.23$ B.M., has shown it to have structure (XI). Mass spectroscopic[177] and i.r.[178] studies also support the formulation of the $[Mn(CO)_4(PR_3)]_2$ complexes as dimers (figure 3).

Tetraphenylbiphosphine reacts with $Mn_2(CO)_{10}$ to produce the phosphido-bridged $[Mn(CO)_4(PPh_2)]_2$, also formed from $[Mn(CO)_5]^-$ and PPh_2Cl [179,180]. In the latter reaction small quantities of a compound identified as $[(CO)_4Mn(PPh_2)(H)Mn(CO)_4]$ were formed[179], the structure (XII) of which has been determined by Doedens et

(XI)   (XII)

al.[181]. Dimethylchlorophosphine reacts with $NaMn(CO)_5$ to give $[Mn_2(CO)_9(PMe_2)_2]$, which loses a carbonyl group on refluxing in toluene solution to give the normal phosphido-bridged complex[180,182]. The proposed structures are (XIII) and (XIV), respectively. Reaction of Me_3SiPPh_2 and $Mn(CO)_5X$ produces Me_3SiX and the phosphido-bridged $[(CO)_4Mn(PPh_2)_2Mn(CO)_4]$, whilst more prolonged reaction

(XIII)   (XIV)

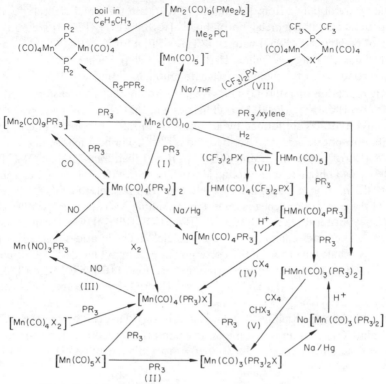

Figure 3 Some reactions of manganese carbonyl compounds with phosphines[171–5,179,180,184,185,189–200,211,214,215]
R = Ph except I = Ph, Et, Cy, p-FC$_6$H$_4$
II = Bu^n, Ph
X = Cl, Br, I except III X = I   IV, V X = Cl
VI X = F, CF$_3$, Me
VII X = F, Cl, Br, I SCF$_3$, SeCF$_3$

results in the formation of trimeric species, for which structure (XV) has been suggested[183].

The [Mn(CO)$_4$(PR$_3$)]$_2$ complexes are reduced by alkali-metal amalgams to M[Mn(CO)$_4$(PR$_3$)] (R = Ph, Et, Cy), from which acids liberate the corresponding [HMn(CO)$_4$(PR$_3$)] [184]. With alkyl halides they form [RMn(CO)$_4$(PR$_3$)], and with Hg(CN)$_2$ they produce Hg[Mn(CO)$_4$(PR$_3$)]$_2$ [184,185]. The zinc and cadmium analogues

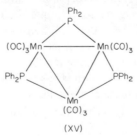

(XV)

of the latter are obtained from $[HMn(CO)_4(PR_3)]$ and the metal dialkyl[186]. A series of compounds containing metal–boron bonds $[R_2BMn(CO)_4(PPh_3)]$ are obtained on reaction of $NaMn(CO)_4(PPh_3)$ with R_2BCl (R = Ph, *n*-Bu, Cl, NR_2, OMe)[187], and Parshall[188] has obtained the yellow $H_3BMn(CO)_4(PPh_3)$. Hieber *et al.*[189] have produced a second type of carbonyl manganate anion, $Na[Mn(CO)_3(PR_3)_2]$ (R = Ph, Cy), by Na/Hg reduction of $[Mn(CO)_3(PR_3)_2X]$, and found that these also reacted with acids (to give $HMn(CO)_3(PR_3)_2$), alkyl halides and $Hg(CN)_2$.

Carbonyl hydrides are formed by acidification of the appropriate anions, preferably with phosphoric acid, as described above[184,189]. Triphenylphosphine reacts with $HMn(CO)_5$ to give *cis*-$[HMn(CO)_4(PPh_3)]$ and then *trans*-$[HMn(CO)_3(PPh_3)_2]$ [190]; the latter is also obtained on refluxing $Mn_2(CO)_{10}$ with PPh_3 for several days in xylene solution (the hydrogen is supplied by the solvent or PPh_3, presumably)[191]. The properties of these hydrides depends upon the phosphine, varying from $[HMn(CO)_4(PEt_3)]$, an extremely air-sensitive liquid, to solid $[HMn(CO)_4(PPh_3)]$, which can be sublimed *in vacuo*[184]. The latter compound is barely acidic[189]. The hydrides reacted with CX_4 or CHX_3 (X = halogen) to form the corresponding carbonyl halides[184]. Carbonyl halides, $Mn(CO)_5X$, are also formed from the reaction of $(CF_3)_2PX$ (X = Cl, Br, I) with $HMn(CO)_5$, but $(CF_3)_2PY$ (Y = CF_3, Me, F) produce a mixture of *cis*- and *trans*-$[HMn(CO)_4\{(CF_3)_2PY\}]$ [192].

o-Styryldiphenylphosphine (SP), o-$CH_2=CHC_6H_4PPh_2$, reacts with $HMn(CO)_5$ to form $[HMn(CO)_4(SP)]$, in which the double bond is uncoordinated; but on heating in cyclohexane under nitrogen this isomerises to a mixture of (XVI) and (XVII), the

(XVI)   (XVII)

major product (80–90 per cent) being (XVI)[193]. A proton is readily extracted from these molecules by $Ph_3C^+BF_4^-$ to produce $[Mn(CO)_4(SP)]^+$, which can also be obtained from $Mn(CO)_5Cl$ (see below).

Substituted carbonyl halides have been investigated by a large number of workers. On heating $[Mn(CO)_5X]$ or $[Mn(CO)_4X]_2$ with tertiary phosphines, $[Mn(CO)_4(PR_3)X]$ and $[Mn(CO)_3(PR_3)_2X]$ (R_3 = Ph_3, Cy_3, *n*-Bu_3, $PhCl_2$, $PhEt_2$) are obtained, respectively[194–8]. The *cis*-$[Mn(CO)_3(PPhEt_2)_2Br]$ is an orange liquid, which on long standing solidifies, apparently forming the *trans* isomer[198]. The reaction of *cis*-$[Mn(CO)_4Br_2]$ with PPh_3 gives *cis*-$[Mn(CO)_4(PPh_3)Br]$, and then *trans*-$[Mn(CO)_3(PPh_3)_2Br]$[199]. These complexes can be obtained by a number of methods, including halogenation of $[Mn(CO)_4(PR_3)]_2$ [173,200]. At 0°C bromine reacts with $[Mn(CO)_4(PPh_3)]_2$ to give *trans*-$[Mn(CO)_4(PPh_3)Br]$, which isomerises in solution at room temperatures to yield the *cis* isomer. The bonding in these compounds has been discussed on the basis of i.r. measurements[201], and recently Vahrenkamp[202] has reported the result of X-ray measurements on *cis*-$[Mn(CO)_4(PPh_3)Cl]$.

Bigorgne *et al.*[203] obtained [Mn(CO)$_4$(PH$_3$)I] and [Mn(CO)$_3$(PH$_3$)$_2$I] by passing PH$_3$ into a hexane solution of Mn(CO)$_5$I, and have discussed the i.r. and Raman spectra of these interesting species. In a series of papers[204,205] Kruck *et al.* report the preparation of cationic complexes [Mn(CO)$_4$(PPh$_3$)$_2$]$^+$ by the reaction of [Mn(CO)$_4$(PPh$_3$)Cl] or [Mn(CO)$_3$(PPh$_3$)$_2$Cl] with carbon monoxide and a Lewis base–ZnCl$_2$, AlCl$_3$, FeCl$_3$–under pressure. [Mn(CO)$_4$(PPh$_3$)$_2$]$^+$ is also formed from [Mn(CO)$_3$(PPh$_3$)$_2$Cl] in benzene in the presence of PPh$_3$ and AlCl$_3$, or by bubbling CO through the solution in the presence of AlCl$_3$[205]. The cations react with alkoxides to form [Mn(CO)$_3$(PR$_3$)$_2$COOR], a reaction reversed by mineral acids. Complexes with carbonylmetallates, for example [Mn(CO)$_4$(PPh$_3$)$_2$][Fe(CO)$_3$NO] and [Mn(CO)$_5$(PPh$_3$)][Co(CO)$_4$], are also known.

Olefin phosphines *cis*-PP, AP, and SP react[206] with Mn(CO)$_5$X to produce neutral, red or yellow [LMn(CO)$_3$X], which have been shown by i.r. and n.m.r. spectroscopy to contain the ligand bonded through the double bond and the phosphorus. The AP and *cis*-PP complexes are not identical showing that in these reactions AP was not isomerised to *cis*-PP, as occurred with the Group VIA carbonyls. In the presence of AlCl$_3$ or FeCl$_3$ these complexes react with carbon monoxide to produce [LMn(CO)$_4$]$^+$ cations, analogous to those obtained by Kruck from monodentate phosphines.

The effect of other ligands upon the bonding mode of thiocyanate in CNS$^-$ complexes is shown in [Mn(CO)$_4$(PPh$_3$)(SCN)] (S-bonded) and [Mn(CO)$_3$(PPh$_3$)$_2$(NCS)] (N-bonded)[207]. Triphenylphosphine reacts with Mn(CO)$_5$(NO$_3$) to yield [Mn(CO)$_4$(PPh$_3$)(NO$_3$)], suggested by Addison and Kilner[208] on the basis of an unusual i.r. spectrum to contain pyramidal N-bonded nitrate. Moelwyn-Hughes *et al.*[209] prepared [Mn(NCO)(CO)$_2$(N$_2$H$_4$)(PPhMe$_2$)$_2$] from which the hydrazine can be replaced by CO or PPhMe$_2$ to give [Mn(NCO)(CO)$_x$(PPhMe$_2$)$_y$] (x = 3, y = 2 or x = 2, y = 3).

In contrast to the reaction with HMn(CO)$_5$[192], (CF$_3$)$_2$PX (X = Cl, Br, I, SCF$_3$, SeCF$_3$) react with Mn$_2$(CO)$_{10}$ to produce phosphido-bridged carbonyl halides which, on the basis of ^{1}H n.m.r., ^{19}F n.m.r. and i.r. spectroscopy, were assigned structure (XVIII)[210,211]. From (CF$_3$)$_2$PP(CF$_3$)$_2$ a related complex (X = P(CF$_3$)$_2$) is produced, whilst NaBH$_4$ reduces the iodo complex to Mn$_2$(CO)$_8${P(CF$_3$)$_2$}H [211]. With LiPR$_2$ (R = Me, Et), Mn$_2$(CO)$_8${P(CF$_3$)$_2$}I produces (XIX), which contains two different phosphido bridges[212]. Tetraphenylbiphosphine forms X(CO)$_3$Mn(PPh$_2$)$_2$Mn(CO)$_3$X with Mn(CO)$_5$X in boiling benzene[213].

(XVIII)   (XIX)

The reaction of [Mn(NO)$_3$(CO)] with PPh$_3$ in benzene produces[214] dark-green, diamagnetic [Mn(NO)$_3$(PPh$_3$)]. Hieber and Tengler[215] obtained a series of nitrosyl carbonyls–Mn(NO)$_3$(PR$_3$), Mn(NO)(CO)$_3$(PR$_3$), Mn(NO)(CO)$_2$(PR$_3$)$_2$ and

$Mn(NO)_2(PR_3)_2X_2$—from the reaction of $[Mn(CO)_4(PR_3)]_2$ or $Mn(CO)_4(PR_3)I$ with NO, and from PR_3 or X_2 (halogens) on $Mn(NO)(CO)_3(PR_3)$. Sodium borohydride reduction of $[Mn(NO)_2(PPh_3)_2Br]$ produced the unusual nitrosyl hydride $[HMn(NO)_2(PPh_3)_2]$ [215]. All these complexes are considerably more stable than the unsubstituted nitrosyl carbonyls. The rate and mechanism of formation of $[Mn(NO)(CO)_2(PR_3)_2]$ and $[Mn(NO)(CO)_3(PR_3)]$ have been studied by Wawersik and Basolo [216,217].

The structures of $[Mn(NO)(CO)_2(PPh_3)_2]$ and $[Mn(NO)(CO)_3(PPh_3)]$ have been shown to be trigonal bipyramidal with axial phosphine ligands [218,219]. The position of the nitrosyl group in the latter complex could not be distinguished by X-ray methods, but i.r. spectroscopy indicates that it occupies an equatorial position.

Finally, mention should be made of $[Ph_3GeMn(CO)_4(PPh_3)]$ [220] and $[Ph_3SnMn(CO)_4(PPh_3)]$ [221]; the latter was one of the first authenticated examples of a heteronuclear metal–metal bond, Sn–Mn = 255 pm, Mn–P = 236 pm [222].

11.1.2 Manganese halides

Manganese(II) halides react with PPh_3 in acetone to give $[PPh_3H]_2[MnX_4]$ (X = Cl, Br, I) [223]. In dry THF, high-spin $[Mn(PPh_3)_2X_2]$ and $[Mn(PPh_3)_4](I_3)_2$ were reported to be formed [224], but this has been disputed by Negoui [225], who reported that '$Mn(PPh_3)_2Br_2$' was identical with an authentic sample of $Mn(PPh_3O)_2Br_2$. The only other non-carbonyl phosphine complexes of manganese are $Mn(PH_2)_2 . 3NH_3$ and $[Mn(PH_2)_4]^{2-}$, formed from $Mn(NH_3)_xX_2$ and KPH_2 [226].

11.2 TECHNETIUM AND RHENIUM

Unsubstantiated early reports on rhenium complexes should be treated with reserve, as more recent work has shown that in a number of cases early workers failed to recognise the presence of oxygen (and possibly nitrogen) in their products.

11.2.1 Carbonyl complexes

Triphenylphosphine reacts with $Re_2(CO)_{10}$ to produce white, diamagnetic $[Re(CO)_4(PPh_3)]_2$ [200,227-32]. Nyman reported $[Re(CO)_3(PPh_3)_2]$ and a diamagnetic $[Re(CO)_4(PPhEt_2)]_2$, which became paramagnetic in solution [229]. Freni *et al.* found that PPh_3 reacts further with $Re_2(CO)_{10}$ in xylene solution to form orange, paramagnetic $[Re(CO)_3(PPh_3)_2]$, and white $[Re(CO)_3(PPh_3)_2]_2$ [228]. Triphenylphosphine substitutes *trans* to the metal–metal bond, but $PPhMe_2$ [233] substitutes carbonyl groups in the *cis* position, whilst PPh_2Me is intermediate in substituting both *cis* and *trans* to the Re—Re bond axis. Moelwyn-Hughes and coworkers [230,233] have shown that a variety of products (namely $[Re_2(CO)_9(PR_3)]$, $[Re(CO)_4(PR_3)]_2$, $[Re(CO)_3(PR_3)_2]$, $[Re(CO)_5(PPhMe_2)]$, and isomers of $[Re_2(CO)_7(PR_3)_3]$), separable by chromatography, are produced from $Re_2(CO)_{10}$ and PPh_2Me or $PPhMe_2$, depending upon the experimental conditions. The structures of these complexes have been discussed in the light of their i.r. and n.m.r. spectra, and of their reactions with HCl and halogens to form carbonyl halides. The most probable structures are shown in figure 4. In addition, very low yields of a cluster complex, thought to be $[Re_4(CO)_{10}(PPh_2Me)_6]$ were obtained[230]. Molecular weight measurements confirmed the Re_4 grouping, but the number of carbonyls is not known for certain.

The reaction of $KPPh_2$ with $Re(CO)_5Cl$ produces $[Re(CO)_4(PPh_2)]_2$ [234], which can also be obtained, along with $[Re(CO)_3(PPh_2)]_3$, from Me_3SiPPh_2 (compare with the manganese compound[183]).

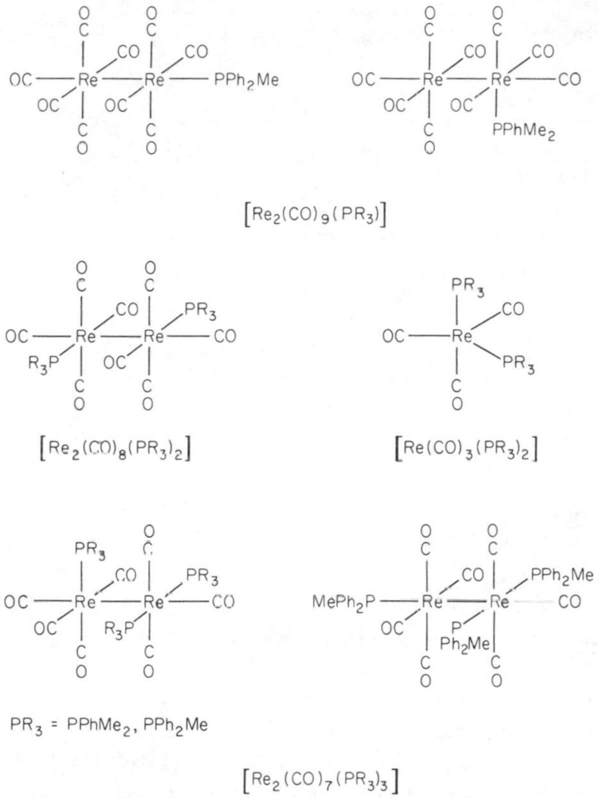

$$[Re_2(CO)_9(PR_3)]$$

$$[Re_2(CO)_8(PR_3)_2] \qquad [Re(CO)_3(PR_3)_2]$$

$PR_3 = PPhMe_2, PPh_2Me$

$$[Re_2(CO)_7(PR_3)_3]$$

Figure 4 Proposed structures of some PPh_2Me and $PPhMe_2$ derivatives of $Re_2(CO)_{10}$[231-3]

Substituted rhenium carbonyl halides, like the manganese analogues, can be obtained by the reaction of hydrogen halides or halogens on the substituted carbonyl, or from the carbonyl halide and PR_3. In addition, they can be prepared by carbonylation of rhenium halide complexes, a reaction unknown in manganese chemistry.

Triphenylphosphine reacts with $[Re(CO)_4X]_2$ or $Re(CO)_5X$ to form $[Re(CO)_4(PPh_3)X]$ [235] and $[Re(CO)_3(PPh_3)_2X]$ [175,177,194,236]. Freni *et al.*[236] found that refluxing $Re(CO)_5I$ with PPh_3 in benzene afforded *cis*-$[Re(CO)_3(PPh_3)_2I]$, which on heating with more PPh_3 in a sealed tube isomerised to *trans*-$[Re(CO)_3(PPh_3)_2I]$. Bromination of $[Re(CO)_4(PPh_3)]_2$ at $0°C$ leads to a mixture of *cis-* and *trans-*[Re(CO)$_4$(PPh$_3$)Br] [200]. An early claim by Hieber and Schuster[237] to have obtained [Re(CO)$_4$(PPh$_3$)$_2$]Cl must be regarded with some scepticism; the product is probably the tricarbonyl. Triphenylphosphine reacts with Tc(CO)$_5$X to afford [Tc(CO)$_3$(PPh$_3$)$_2$X], which is converted by AlCl$_3$ and CO under pressure to [Tc(CO)$_4$(PPh$_3$)$_2$]AlCl$_4$ [238]. With M(CO)$_5$X (M = Tc, Re) tetraphenylbiphosphine forms [X(CO)$_4$M(PPh$_2$)$_2$M(CO)$_4$X] [234]. Kruck *et al.*[205,239] have obtained the

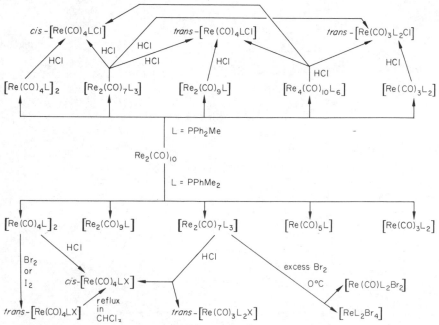

Figure 5 Reactions of PPh$_2$Me and PPhMe$_2$-substituted rhenium carbonyls with HCl and halogens[230,233]

cationic rhenium species [Re(CO)$_4$(PPh$_3$)$_2$]$^+$, and Interrante and Nelson[206] have prepared the rhenium analogues, [LRe(CO)$_3$X] and [LRe(CO)$_4$]$^+$ (L = *cis*-PP, AP, SP), of the manganese olefin phosphine complexes described above.

An interesting reaction is that of [Re(CO)$_6$]ClO$_4$ with PPh$_3$ to produce *trans*-[Re(CO)$_3$(PPh$_3$)$_2$Cl], in which the chloride was shown to have come from reduction of the ClO$_4^-$ by using ^{36}Cl-labelled perchlorate[240]. From the reaction of the rhenium carbonyl complexes of PPhMe$_2$ and PPh$_2$Me with chlorinated solvents, free halogens or hydrogen halides, a number of carbonyl halides were obtained[231-3]. These reactions are shown in figure 5. Another method of obtaining some of these complexes is by carbonylation of a suitable rhenium halide. Freni and Valenti[241] found that CO at high pressure converts Re(PPh$_3$)$_2$X$_2$ to [Re(CO)$_2$(PPh$_3$)$_2$X] and [Re(CO)$_3$(PPh$_3$)$_2$X] (X = Cl, Br) or [Re(CO)(PPh$_3$)$_2$I], whilst Douglas and Shaw[242], using PPhMe$_2$ complexes, reported the series of reactions shown in figure 6; interestingly, the complexes obtained include the seven-coordinate [Re(CO)(PPhMe$_2$)$_3$Cl$_3$].

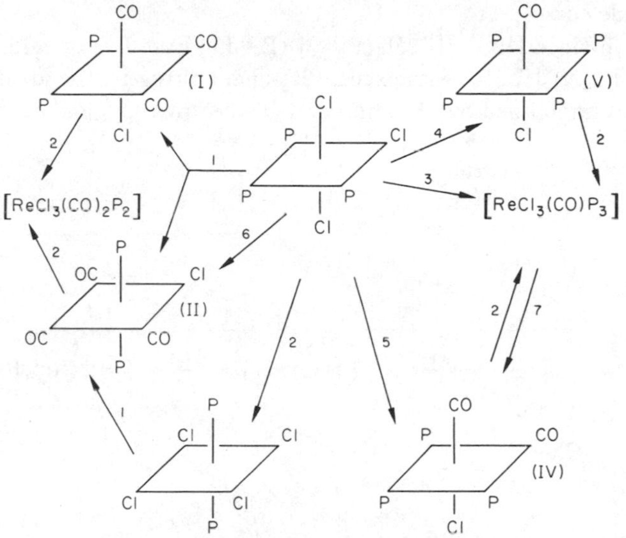

Figure 6 Configurations and interconversions of some rhenium carbonyl–dimethylphenylphosphine complexes
1. CO–2-methoxyethanol. 2. Cl$_2$. 3. CO–ethanol.
4. Diethylaminoethanol. 5. CO–sodium amalgam–THF
6. Formic acid. 7. NaBH$_4$–diglyme
[Reproduced by kind permission of The Chemical Society.]

Some isocyanato complexes of rhenium(I) with PPh$_3$, PPh$_2$Me and PPhMe$_2$ are obtained by reaction of *cis*-[Re(CO)$_4$(PR$_3$)Br] and *trans*-[Re(CO)$_3$(PR$_3$)$_2$X] with anhydrous hydrazine[243].

From HRe(CO)$_5$ Flitcroft *et al.*[244] obtained *cis*-[Re(CO)$_4$H(PR$_3$)] and *trans*-[Re(CO)$_3$H(PR$_3$)$_2$] (compare HMn(CO)$_5$). The reaction between HRe(CO)$_5$ and SP[193] gives only the six-membered Re—C σ-bonded complex, in contrast to HMn(CO)$_5$, but the five-membered ring complex can be obtained as shown below.

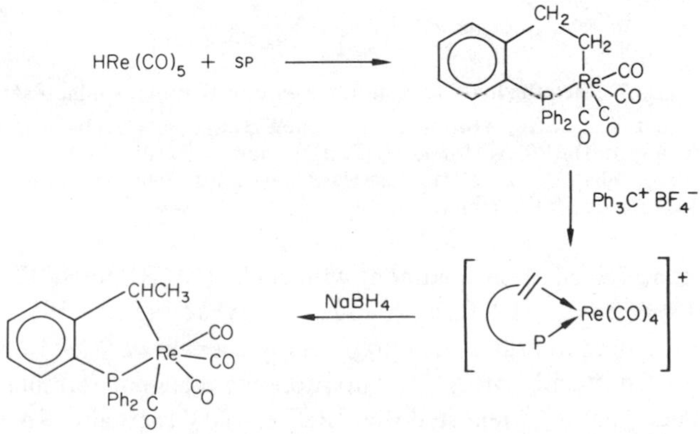

11.2.2 Hydride complexes

Ginsberg[245] obtained $[Et_4N][ReH_8(PR_3)]$ (R = Ph, Et, *n*-Bu) by refluxing isopropanol solutions of $[Et_4N]_2[ReH_9]$ with excess PR_3 under nitrogen. The identity of these complexes was established by ^{1}H n.m.r. and i.r. spectroscopy, and by measurement

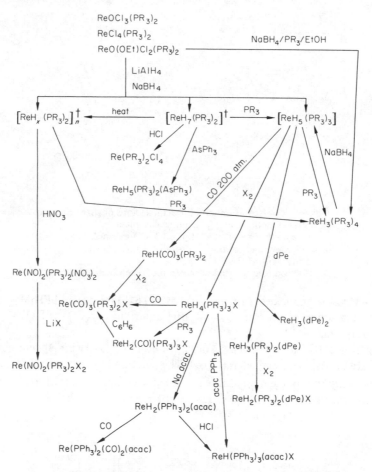

Figure 7 Rhenium hydride complexes with tertiary phosphines[246-55]

† Chatt and Coffey's formulation of these compounds has been used, namely $ReH_7(PR_3)_2$ (for $ReH_5(PPh_3)_2$), and $ReH_x(PPh_3)_2$ (for $ReH_3(PPh_3)_2$). The nature of $ReH_3(PPh_3)_4$ must be in some doubt but it is included for completeness.

of the hydrogen evolved upon treatment with acids; $[Et_4N][ReH_8(PPh_3)]$ has $\nu(Re-H)$ at 1860, 1940, ~1980 cm^{-1}, and $\delta(Re-H)$ at 745 and 695 cm^{-1}.

The lower rhenium hydrides were prepared by Freni *et al.*[246-54], and have been re-examined by Chatt and Coffey[255], who proposed different formulations for them. Freni and Valenti[246] reported that 'Re(PPh$_3$)$_2$I$_2$' (actually, ReO(OEt)(PPh$_3$)$_2$I$_2$)

reacted with $NaBH_4$ in ethanol to produce red $ReH_3(PPh_3)_2 \cdot 2EtOH$, which on re-crystallisation from benzene gave $ReH_3(PPh_3)_2$, and upon reaction with PPh_3 gave $ReH_3(PPh_3)_4$. The hydrogen content was determined by reaction with iodine, since the complexes were not soluble in organic solvents, and hence n.m.r. spectroscopy could not be applied[247]. Malatesta *et al.*[248] found that by varying the conditions $ReH_5(PPh_3)_3$ and $ReH_5(PPh_3)_2$ could be obtained. Chatt and Coffey[255] used $LiAlH_4$ to prepare $ReH_7(PPh_3)_2$, which had identical properties with Malatesta's '$ReH_5(PPh_3)_2$'. The number of protons was established by ^1H n.m.r. and by reaction with HCl, and thus it seems reasonable to conclude that the correct formula for this

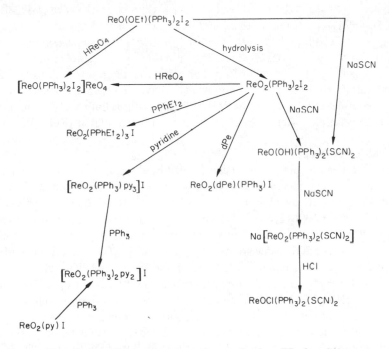

Figure 8 Substitution reactions of some rhenium (V) phosphine complexes[301,302]

complex is $ReH_7(PPh_3)_2$, not $ReH_5(PPh_3)_2$. These workers[255] also obtained a red hydride, $ReH_x(PPh_3)_2$ (Freni's $ReH_3(PPh_3)_2$[246]), but were unable to determine x. The complex $ReH_5(PPh_3)_3$ was confirmed[255]. The presence of hydridic hydrogen in these complexes was established chemically and by i.r. spectroscopy, but when, as in the case of $ReH_x(PPh_3)_2$, the compound is not soluble enough to permit ^1H n.m.r. studies, the determination of the number of hydride ligands is very difficult. Some of the reactions of these hydrides are shown in figure 7.

By reaction of '$ReH_3(PPh_3)_4$' with acids, $[PPh_3H][Re(PPh_3)X_5]$ complexes were obtained, as well as $Re(PPh_3)_2X_4$[249]. With nitric acid, the paramagnetic non-electrolyte $[Re(NO)_2(PPh_3)_2(NO_3)_2]$ was formed[250], from which $[Re(NO)_2(PPh_3)_2X_2]$

(X = halide) could be obtained by metathesis with LiX. On the basis of its i.r. spectrum and dipole moment the nitrato complex was assigned structure (XX).

(XX)

High-pressure carbonylation of 'ReH$_5$(PPh$_3$)$_2$' (that is, [ReH$_7$(PPh$_3$)$_2$]), afforded [ReH(CO)$_3$(PPh$_3$)$_2$] [251], and on treatment with the calculated quantity of halogen, yellow-green ReH$_4$(PR$_3$)$_3$X (X = Br, I, SnCl$_3$) formed [252]. These latter complexes react with CO to give carbonyl halides, and on refluxing in ethanol violet [ReH$_2$(CO)$_2$(PR$_3$)$_3$X] are obtained [252]. The bidentate dPe reacts with ReH$_5$(PPh$_3$)$_3$ to form [ReH$_3$(PPh$_3$)$_2$(dPe)] [253]. Freni *et al.* [254] have also obtained some acetylacetonate complexes by reaction of ReH$_4$(PPh$_3$)$_3$I with sodium acetylacetonate (see figure 8). The complex 'ReH$_4$(PPh$_3$)$_3$' [256] has been reformulated [ReH$_5$(PPh$_3$)$_3$] [248].

11.2.3 Nitrido complexes

Rhenium–nitrogen complexes have been studied by Chatt and coworkers. Rhenium is unique in giving arylimido complexes in addition to the simple nitrido derivatives. The first compound obtained was [ReNCl$_2$(PPh$_3$)$_2$] [257] which had been previously reported [241,258] as 'ReCl$_2$(PPh$_3$)$_2$'. The original preparative method was reaction of a perrhenate, PPh$_3$ and hydrazine dihydrochloride in boiling wet ethanol–benzene [257-9], but better methods are the slow addition of ethanolic Re$_2$O$_7$ to a refluxing mixture of PR$_3$ and N$_2$H$_6$Cl$_2$ in ethanol [259], or by refluxing ReO(OEt)(PPh$_3$)$_2$X$_2$ with N$_2$H$_6$SO$_4$ and PPh$_3$ in ethanol [260]. When the PPh$_3$ is replaced by PPhEt$_2$, PEt$_3$ or PPr$^n_3$, six-coordinate [ReNX$_2$(PR$_3$)$_3$] are obtained [259]. The decomposition of azido complexes also leads to nitrido complexes [261]. The five-coordinate complexes are red, the six-coordinate yellow in colour; the species obtained appears to depend on steric factors, more bulky phosphines producing five-coordination. With PPh$_2$Et and PPh$_2$Pr^n both types have been obtained. All these complexes show i.r. absorptions in the range 1010–1065 cm^{-1}, ν(Re≡N) [261].

Chatt and Heaton [262] have demonstrated that *mer*-[ReNX$_2$(PPhEt$_2$)$_3$] resembles organic nitriles in the coordinative ability of the nitrogen. It reacts with BX$_3$ (X = F, Cl, Br) and PtCl$_2$(PEt$_3$)$_2$ to give brightly coloured adducts [(Et$_2$PhP)$_3$X$_2$ReN → L] (L = BX$_3$, Pt(PEt$_3$)$_2$Cl$_2$).

The structure of [ReNCl$_2$(PPh$_3$)$_2$] is a distorted tetragonal pyramid [263], with Re—N = 160 pm. The six-coordinate [ReNCl$_2$(PPhEt$_2$)$_3$] is a distorted octahedron [264], with Re—N = 178 pm. Although the Re—N (and Re—Cl) distances in these complexes differ markedly, see (XXI) and (XXII), the non-bonded interligand distances are very similar.

The degradation of the benzoylazo-complex [ReCl$_2$(N$_2$COPh)(PPh$_3$)$_2$] in the presence of $PPhMe_2$ or PPh_2Me leads to pale-yellow $[ReCl(N_2)(PR_3)_4]$[265]. The X-ray structural determination by Davis and Ibers[266] shows the N_2 to be *trans* to Cl

(XXI)   (XXII)

(XXIII). From the benzoylazo-complex reaction with CO in refluxing benzene–methanol produced $[Re(CO)_2Cl(N_2)(PPh_3)_2]$[267]; but the complexes $[Re(CO)_3(NH_2)(N_2)(PPhMe_2)]$ and $[Re(CO)_2(NH_2)(N_2)(PPhMe_2)_2]$ reported[268] to be formed by the action of hydrazine on the corresponding carbonyl halides have been shown[269] to be isocyanato complexes of Re(I).

(XXIII)

A wide variety of acceptor molecules react with *trans*-$[ReCl(N_2)(PPh_3)_4]$ to form polynuclear derivatives in which the dinitrogen ligand apparently bonds to the acceptor through the terminal N-atom[270-2]. Among the acceptor molecules are TaF_5, $MoCl_4(PPhEt_2)_2$, $TiCl_3(MeCN)_3$, $ScCl_3$, and $CrCl_3(THF)_3$. The dark violet $[ReCl(PPhMe_2)_4(N_2)CrCl_3(THF)_3]$ has been studied in detail[271]; it is monomeric, and magnetic measurements indicate it contains Cr(III) and Re(I), $\mu_{eff} = 4.07$ B.M. In air it oxidises to the $[ReCl(N_2)(PPhMe_2)_4]^+$ cation, and is decomposed by water.

Green $[Re(NPh)Cl_3(PPhEt_2)_2]$ was obtained[258] by refluxing $[ReOCl_3(PPhEt_2)_2]$ with aniline in benzene solution. Subsequently[259], the reaction was extended to a number of other arylamines, for example p-$H_2NC_6H_4X$ (X = halide, OMe, NH_2). Most of the complexes obtained were green, diamagnetic non-electrolytes.

Alkylimido complexes, $[ReCl_3(NR)(PPh_3)_2]$ (R = Me, Et, n-Pr, Cy), were obtained by the reaction of $ReOCl_3$ with 1,2-disubstituted hydrazines[273]. The structures of $[ReCl_3(NR)(PPhEt_2)_2]$ (R = C_6H_4OMe, C_6H_4COMe) and $[ReCl_3(NMe)(PPh_2Et)_2]$ have been determined by Bright and Ibers[274,275]. Each rhenium is surrounded by a distorted octahedron of ligands, with *trans* phosphines, two *trans* chlorines and a third chlorine *trans* to the NR group (XXIV). In the three complexes X-rayed the Re–N distances are all very similar, 170, 169 and 168 pm, respectively. In the methylimido complex the Re–N–Me is almost linear ($\widehat{ReNMe} = 173°$)[275].

Monoaroyl- and 1,2-diaroylhydrazines react[276] with $[ReOCl_3(PPh_3)_2]$ to give novel complexes of type (XXV), which react with nucleophiles, including tertiary

(XXIV)   (XXV)

phosphines, with opening of the chelate ring to give benzoylazo complexes. These complexes are useful intermediates in the preparation of dinitrogen complexes, as discussed previously.

11.2.4 Halide complexes

The complexes formulated as $ReX_2(PR_3)_2$ [241,258] have been shown to be $[ReNX_2(PR_3)_2]$ or $[ReO(OR')(PR_3)_2X_2]$. The salts, $[Re(RCN)(PPh_3)_2I]$ Y (Y = BPh_4^-, ClO_4^-), were reported to be formed from '$ReI_2(PPh_3)_2$' and RCN[277], but in view of the incorrect formulation of the starting material, this reaction is doubtful.

Rhenium(III) complexes are now well established, but a number of earlier reports have been shown to be erroneous. The yellow-green '$ReCl_3(PPh_3)_2$' [241] is really $[ReOCl_3(PPh_3)_2]$, and the red '$ReCl_3(PPh_3)_2$' [278] is $[ReCl_3(PPh_3O)_2]$ [279,280].

Rhenium trichloride is reported[241,258,278] to give a red complex, $[ReCl_3(PPh_3)]_n$, on reaction with triphenylphosphine, but this was reported to be dimeric on the basis of molecular weight determinations in benzene[280]. Cotton *et al.*[281] obtained a green $[ReCl_3(PPh_3)]_n$ by reaction of $[Re_2Cl_8]^{2-}$ ion with PPh₃, and suggested that n was probably equal to two, although the insolubility of this non-crystalline material prevented confirmation of this. The PEt₃ analogue was isolated as green crystals which were shown by X-ray analysis[282] to be $[Re_2Cl_6(PEt_3)_2]$, in which the Re—Re bond was retained, as was the eclipsed configuration of the parent $Re_2X_8^{2-}$ ion. The PEt₃ groups are *trans* across the molecule; Re—Re = 222 pm, Re—P = 246 pm, Re—Cl = 230 pm (*cis*-P) and 235 pm (*trans*-P).

When β-$ReCl_4$ is reacted with PPh₃ in methanol green $[ReCl_3(PPh_3)]_n$ is obtained[283], and this may be identical to Cotton's product[281]. Subsequently, Walton and co-workers have isolated *trans*-$[ReCl_4(PPh_3)_2]$ and $[ReCl_3(PPh_3)]_2$ from the reaction of β-$ReCl_4$ with PPh₃; and when the reaction is carried out in acetonitrile $[ReCl_3(PPh_3)_2(MeCN)]$ is obtained[284]. This complex contains *trans*-phosphines, *trans*-chlorines and the MeCN *trans* to the third chlorine.

In contrast, the $[Re_2(SCN)_8]^{2-}$ ion reacts[285] with PPh₃ to form $[Re_2(CNS)_8(PPh_3)_2]^{2-}$. This complex has μ_{eff} = 4.1 B.M./Re atom, which is not consistent with a Re—Re bonded structure, and on the basis of conductivity measurements and the i.r. spectrum it is formulated as (XXVI).

By the reaction of PPhEt₂ with $[ReOCl_3(PPhEt_2)_2]$ under various conditions Chatt and Rowe[258] obtained $[ReCl_3(PPhEt_2)]_n$ and $[ReCl_3(PPhEt_2)_3]$. Cotton *et al.*[286] characterised the former as the trimeric $[Re_3Cl_9(PPhEt_2)_3]$, and obtained the bromo analogue from $ReBr_3$ and $PPhEt_2$. The visible spectra were very similar to those of complexes known to contain Re_3 clusters. Triphenylphosphine analogues have also

(XXVI)   (XXVII)

been obtained[286]. The structure (XXVII) of $[Re_3Cl_9(PPhEt_2)_3]$ has been deter-mined[287], with Re—Re = 249 pm, Re—Cl (terminal) = 236 pm and Re—Cl (bridging) = 239 pm.

The complexes $[ReX_3(PR_3)_3]$ (X = Cl, Br; R_3 = PhMe$_2$, Ph$_2$Me, PhPr$''_2$, PhBu$''_2$)[242,258,259,288,289], obtained from $[ReOCl_3(PR_3)_2]$ and PR$_3$, or by reduction of $HReO_4$ with HX in the presence of PR$_3$ have been identified as the meridional isomers. Their i.r.[288] and n.m.r.[288,290] spectra are consistent with a meridional configuration; and the complexes all exhibit T.I.P.[289] with magnetic moments in the range 1.5–2.1 B.M. The fac-$[ReCl_3(PPhMe_2)_3]$ has been characterised, but no preparative details have been published[291].

Aryllithium reagents produce $[ReR'_3(PR_3)_3]$ (R' = aryl) with $[ReCl_3(PR_3)_3]$, but with $[ReOCl_3(PR_3)_2]$ reduction occurs to give poor yields of $[ReR'_3(PR_3)_3]$. Interest-ingly, in this reaction $[ReNCl_2(PR_3)_2]$ retain the nitrogen to give $[ReNR'_2(PR_3)_2]$[292].

Monomeric alkanonitrile complexes of rhenium(III), $[ReX_3(RCN)(PPh_3)_2]$ (X = Cl, Br; R = alkyl), were obtained on heating $trans$-$[ReOX_3(PPh_3)_2]$ with PPh$_3$ and RCN[293]. These orange-yellow complexes readily undergo substitution reactions to yield, for example, $[ReCl_3(py)_2(PPh_3)]$ and are oxidised by air to $[ReOCl_3(PPh_3)(PPh_3O)]$. Halogenated solvents (CCl_4, CH_2Cl_2, CBr_4, etc.) produce $[ReX_4(PPh_3)_2]$, $[ReX_3X'(PPh_3)_2]$ and $[ReX_2X'_2(PPh_3)_2]$(X, X' = Cl, Br; X ≠ X').

Technetium tetrachloride reacts with PPh$_3$ in refluxing ethanol to give emerald-green $[TcCl_4(PPh_3)_2]$, μ_{eff} = 3.92 B.M.[294]. The first Re(IV) complex reported was $[ReI_4(PPh_3)_2]$[278], and the chloro- and bromo-analogues, $[ReX_4(PR_3)_2]$ (R_3 = Ph$_3$, PhEt$_2$), were obtained by pyrolysis of $[PR_3H]_2[ReX_6]$, by oxidising $[ReX_3(PPh_3)_2]$ with halogen in CCl_4[295], or from $[ReCl_4(RCN)_2]$ and PR$_3$[296]. Rhenium penta-chloride reacts with PPh$_3$ in CH_2Cl_2 to form $[ReCl_4(PPh_3)]$[280].

By reaction of $trans$-$[ReOX_3(PPh_3)_2]$ with carboxylic acids, Rouschias and Wilkinson[296] obtained a series of products containing RCO_2^- groups, two of which have recently been characterised by X-ray methods[297,298]. The RCO_2^- moiety behaves as a bridging group in $[Re_2OCl_5(EtCO_2)(PPh_3)_2]$ (XXVIII) and $[Re_2OCl_3(EtCO_2)(PPh_3)_2]$ (XXIX). The short Re—Re distances of ~250 pm clearly indicate the presence of metal–metal bonds.

The reduction of $HReO_4$ with HCl in ethanol in the presence of PR$_3$ has been

(XXVIII)   (XXIX)

shown to give the rhenium(V) complexes, $[ReOCl_3(PR_3)_2]$ [258]. With PPhEt$_2$, green trans-$[ReOCl_3(PPhEt_2)_2]$, a blue cis-$[ReOCl_3(PPhEt_2)_2]$ and a violet form were reported[258]. The violet complex has been shown to be a solid solution of trans-$[ReCl_4(PPhEt_2)_2]$ and trans-$[ReOCl_3(PPhEt_2)_2]$ [295]. The structure of the green $[ReOCl_3(PPhEt_2)_2]$ has been confirmed as trans by an X-ray structure determination; Re—O = 186 pm[299]. The blue cis-$[ReOCl_3(PPhEt_2)_2]$ changes to the green trans isomer in hot solvents[258]. Triphenylphosphine gives three forms of $[ReOCl_3(PPh_3)_2]$ [280], and complexes with trialkylphosphines are also known[295]. Direct preparation of the latter gives poor yields, but the displacement of PPh$_3$ from $[ReOCl_3(PPh_3)_2]$ offers an easy route to the trialkylphosphine complexes[295]. The $[ReOX_3(PR_3)_2]$ (X = Br, I, SCN) analogues have also been obtained[258,295], and a polymeric product, thought to be $\{ReOF_3(PPh_3)\}_n$ has recently been synthesised[276].

Oxoalkoxy complexes, $[ReO(OR')X_2(PR_3)_2]$, have been mentioned briefly above. They are obtained from $[ReOX_3(PR_3)_2]$ on boiling with ethanol[258], and can be obtained directly from HReO$_4$, PR$_3$, EtOH and HX, when insufficient HX is present to form $[ReOX_3(PR_3)_2]$ [280]. The ease of formation of the alkoxy complex depends upon X. When X = I, merely washing $[ReOI_3(PR_3)_2]$ with ethanol is sufficient to convert it to $[ReO(OEt)I_2(PR_3)_2]$, but when X = Cl, the compound must be refluxed with EtOH for about 2 h for complete reaction[280].

These alkoxy species react[300] with β-diketones to give (for example with $[ReOCl_2(OEt)(PPh_3)_2]$ and acac) $[ReOCl_2(PPh_3)(acac)]$, $[ReCl_2(PPh_3)_2(acac)]$, and $[ReCl(PPh_3)(acac)_2]$. Freni et al.[301,302] have reported a number of reactions, which are included in figure 8. These Re(V)-oxo species all show a strong i.r. band at 930–986 cm^{-1}, assignable to $\nu(Re=O)$.

Two isomers of $[ReOCl_3(PPh_3)]^-$ were obtained by Grove and Wilkinson[303], whilst Gehrke et al.[304,305] have obtained the unusual $(DOPT)_2ReCl_6$ (DOPT = MeCOCH$_2$CMe$_2$PPh$_3$) by the reaction of $[ReOCl_3(PPh_3)_2]$ with HCl in acetone, or from ReCl$_5$ and PPh$_3$ in acetone. The reaction of ReCl$_5$ with PPh$_3$ gives a number of other products, depending upon the solvent and the conditions used, including $(DOPT)_2Re_2Cl_8$, $[ReOCl_3(PPh_3)_2]$ and $[ReCl_4(PPh_3)_2]$ [305].

12 IRON, RUTHENIUM, OSMIUM

12.1 IRON

12.1.1 Carbonyls

The direct reaction of $Fe(CO)_5$ with PPh_3, either thermally[306-9,333] or photochemically[310] leads to $[Fe(CO)_4(PPh_3)]$ and $[Fe(CO)_3(PPh_3)_2]$. Manuel and Stone[311] found that the latter complex could also be obtained from $[(cyclooctatetraene)Fe(CO)_3]$ and PPh_3, whilst $[(cyclooctatriene)Fe(CO)_3]$ gave $[Fe(CO)_2(PPh_3)_3]$. The i.r. and Mössbauer spectra are consistent with trigonal-bipyramidal structures[308,312,316,334,335]. The $[Fe(CO)_3(PR_3)_2]$ complexes are only known as the *trans*-isomers, with axial phosphines; the *cis* isomers have not been obtained. Similar mono- and disubstituted products have been obtained with a number of phosphines, including PPh_2Et[313], PEt_3[312,314], PMe_3[314,315], PCy_3, PPr^i_3, PBu_3[314], PCl_3[315], $P(NMe_2)_3$[76] and $P(p-FC_6H_4)_3$[317]. The more bulky phosphines PBu^t_3 and $P(MMe_3)_3$ (M = Ge, Si, Sn) produce only $[Fe(CO)_4(PR_3)]$[318-21]. Phosphine reacts with either $Fe(CO)_5$ or $Fe_2(CO)_9$ to give $[Fe(CO)_4(PH_3)]$[322].

Iron carbonyls, $Fe_2(CO)_9$ and $Fe_3(CO)_{12}$, also react with PR_3 to form mononuclear products, $[Fe(CO)_{5-n}(PR_3)_n]$ (n = 1, 2)[312,317,323]. In addition to these, $Fe_3(CO)_{12}$ reacts with PPh_3 to form $Fe_3(CO)_{11}(PPh_3)$[324], which has been shown to exist in two isomeric forms[325] (XXX). $[Fe_3(CO)_9(PPhMe_2)_3]$ is produced with $PPhMe_2$. It has structure (XXXI) as established by i.r. and Mössbauer spectroscopy,

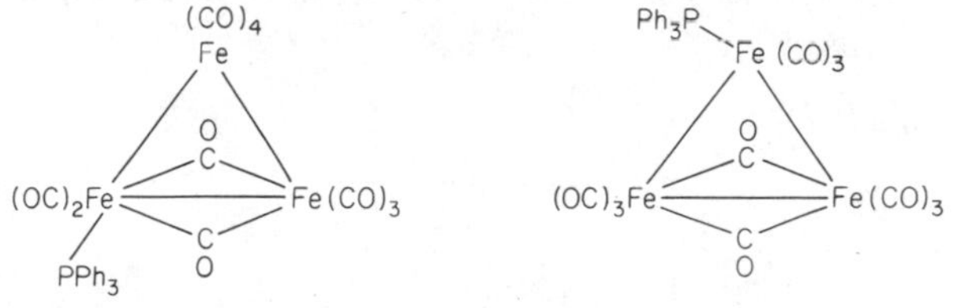

(XXX)

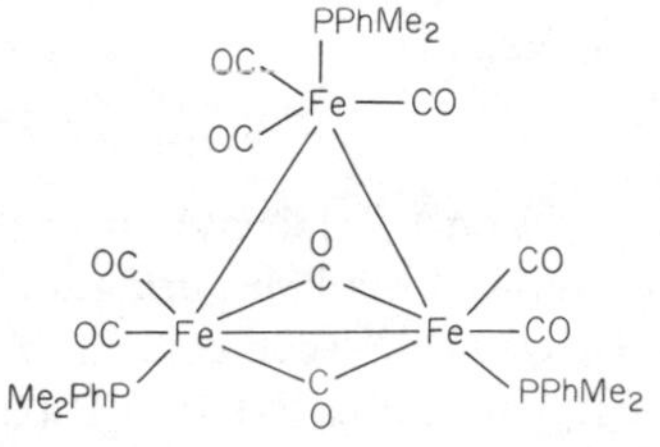

(XXXI)

and by X-ray analysis[326]. This reacts with I_2 at $-70°C$ to form $[Fe(CO)_3(PPhMe_2)I_2]$.

Diphenylphosphine, PPh_2H, reacts with $Fe(CO)_5$ and $Fe_2(CO)_9$ to give

[Fe(CO)$_4$(PPh$_2$H)] and [Fe(CO)$_3$(PPh$_2$H)$_2$], respectively[79]. The former complex has a distorted trigonal-bipyramidal structure with PPh$_2$H in an axial position (Fe—P = 224 pm)[328]. Pentaphenylphosphole reacts with Fe(CO)$_5$ to form [Fe(CO)$_4$L], and with Fe$_3$(CO)$_{12}$ to give [Fe(CO)$_3$L] and [Fe$_2$(CO)$_9$L]; in some cases the C=C bonds may be coordinated[329,330]. The olefin phosphine (SP) forms [Fe(CO)$_3$(SP)] and [Fe(CO)$_2$(SP)$_2$] upon reaction with Fe$_3$(CO)$_{12}$, the former containing bidentate SP, the latter having structure (XXXII)[331]. The Fe—P bond is significantly shorter in the bidentate case than in the P-only donor case. In hexane [Fe(CO)$_3$(SP)] forms a

(XXXII)   (XXXIII)

σ-bonded chelated Fe(II) complex (XXXIII) with HX (X = Cl, Br). These brown or yellow crystals are stable in nitrogen, but decompose rapidly in solution[332]. With DPA, the expected five-coordinate [Fe(CO)$_3$(DPA)$_2$] is formed[81].

Diphosphines readily react with iron carbonyl to produce phosphido-bridged complexes [(CO)$_3$Fe(PR$_2$)$_2$Fe(CO)$_3$] (R = Me, Et, Ph)[114,112]. Both n.m.r.[114] and Mössbauer[336] spectroscopy have been used to detect the phosphido-bridged structure. On treatment of [Fe(CO)$_3$(PMe$_2$)]$_2$ with halogens in CCl$_4$ solution, [Fe(CO)$_3$(PMe$_2$)X]$_2$ are obtained (X = Cl, Br, I)[337]. The iodide has been shown to have structure (XXXIV), with *trans* iodide, and a planar di-μ-phosphido bridge. There

(XXXIV)

is no Fe—Fe bond (Fe—Fe = 359 pm). Triphenyl- or triethylphosphine can substitute [Fe(CO)$_3$(PMe$_2$)]$_2$ without fission of the PMe$_2$ bridges to give [Fe(CO)$_2$(PMe$_2$)(PR$_3$)]$_2$, which probably has *trans*-PR$_3$ ligands[338] (compare Cr, Mo, W). The P(C$_6$F$_5$)$_2$-bridged analogues are afforded by reaction of P(C$_6$F$_5$)$_2$H or P$_2$(C$_6$F$_5$)$_4$ with Fe$_3$(CO)$_{12}$[339], whilst the unusual ligand Ph$_2$PSPh forms a complex with both Ph$_2$P— and PhS— bridges[340]. Heteronuclear phosphido-bridged complexes are formed by the reaction of [Fe(CO)$_4$(PPh$_2$H)] with (π-C$_3$H$_5$)Co(CO)$_3$ and (π-C$_3$H$_5$)Mn(CO)$_4$, whilst with [(π-C$_3$H$_5$)PdCl]$_2$ (XXXV) is formed[341]. It has been X-rayed by Kilbourn and

Mais[342], who found Pd—Fe = 259 pm, Fe—P = 224 pm and P—Pd = 215 pm. The red $[Fe(CO)_3P(CF_3)_2I]_2$ and $[Fe(CO)_3P(CF_3)_2]_2$, formed from $P(CF_3)_2I$ and $Fe(CO)_5$, have $P(CF_3)_2$ bridges[327,343], the latter having an Fe—Fe bond also. This Fe—Fe bond is cleaved by halogens to produce binuclear $[Fe(CO)_3(CF_3)_2PX]_2$, which were shown by ^{19}F n.m.r. spectroscopy to exist in three isomeric forms[344].

$$Ph_2P \diagdown \quad Cl \diagdown \quad Fe(CO)_4$$
$$\begin{array}{ccc} & Pd & Pd \\ (OC)_4Fe & Cl & PPh_2 \end{array}$$

(XXXV)

12.1.2 Carbonyl hydrides

Davison *et al.*[345] found that $[Fe(CO)_{5-n}(PPh_3)_n]$ ($n = 1, 2$) dissolved in 98 per cent H_2SO_4 to give yellow solutions, which were shown by n.m.r. spectroscopy to contain protonated species, although no solids were isolated. Hayter[346] reacted Me_2PCl with $[(\pi-C_5H_5)Fe(CO)_2]_2$ and obtained dark brown $[(\pi-C_5H_5)_2Fe_2H(PMe_2)(CO)_2]$. When $Fe(CO)_5$ or $Fe_3(CO)_{12}$ were reacted with $P(CF_3)_2H$, both *cis*- and *trans*- $[H_2Fe_2(CO)_6\{P(CF_3)_2\}_2]$ were produced, along with the previously known $[Fe(CO)_3P(CF_3)_2]_2$ [347]; but $H_2Fe(CO)_4$ reacted with PPh_3 to give $[Fe(CO)_4(PPh_3)]$ and $[Fe(CO)_3(PPh_3)_2]$ with loss of hydrogen[348]. The loss of hydrogen is in sharp contrast to the reactions of manganese and cobalt carbonyl hydrides, which substitute CO to give $HMn(CO)_4(PPh_3)$ and $HCo(CO)_3(PPh_3)$.

12.1.3 Carbonyl halides

Originally $[Fe(CO)_2(PPh_3)_2X_2]$ and $[Fe(CO)_3(PPh_3)X_2]$ were obtained by careful halogenation of $[Fe(CO)_3(PPh_3)_2]$ and $[Fe(CO)_2(PPh_3)_3]$ [309,349,350]. Carbonyl halides react directly with phosphines to produce analogous products[349,351]. Treatment of *cis*-$Fe(CO)_4I_2$ with PH_3 produces *cis*-$[Fe(CO)_3(PH_3)I_2]$ and $[Fe(CO)_2(PH_3)_2I_2]$ [322]. The third possible route, carbonylation of iron–phosphine complexes was investigated by Booth and Chatt[352], who found that $[FeCl_2(PR_3)_2]$ ($R_3 = PhEt_2$, Ph_2Et, Et_3) reacted readily with carbon monoxide at atmospheric or moderate pressures ($\sim$50 atm = 5 MN m^{-2}), forming $[Fe(CO)_2Cl_2(PR_3)_2]$. Triphenylphosphine complexes did not react, the ease of carbonylation decreasing in the series $PEt_3 > PPhEt_2 > PPh_2Et$ ($> PPh_3$). Dipole moments and i.r. spectra indicate *trans*-octahedral structures. By metathesis the Br, I or NCS complexes can be obtained, but attempts to prepare nitrosyl complexes by reaction with NO gave only oily mixtures. Sodium amalgam, metal hydrides or metal alkyls in the presence of CO reduce the carbonyl halide complexes back to $[Fe(CO)_3(PR_3)_2]$ [313,350]. Hieber *et al.*[353] found that $[Fe(CO)_2(PR_3)_2X_2]$ (R = Ph, Et, Cy) formed cationic $[Fe(CO)_3(PR_3)_2X]^+$ upon treatment with CO in the presence of Lewis acids.

The S- or Se-bridged $[Fe(CO)_3(MR)]_2$ (M = S, Se) substitute CO with PR_3 to form $[Fe_2(CO)_5(MR)_2(PR_3)]$ and $[Fe(CO)_2(MR)(PR_3)]_2$ [354], and these readily add halogens[355]. Photochemical reaction of *trans*-[Fe(CO)$_3$(PEt$_3$)$_2$] with liquid SO$_2$ produces [Fe(CO)$_2$(PEt$_3$)$_2$(SO$_2$)], which undergoes oxidative elimination with HI [356].

Mercuric halides form 1:1 adducts with [Fe(CO)$_3$(PPh$_3$)$_2$] [357], and SnX$_4$ oxidatively adds to [Fe(CO)$_4$(PPh$_3$)] to give [Fe(CO)$_3$(PPh$_3$)X(SnX$_3$)] [358].

12.1.4 Nitrosyls

Tertiary phosphines substitute one or both carbonyls in [Fe(NO)$_2$(CO)$_2$] to give [Fe(NO)$_2$(CO)(PR$_3$)] (R = Ph, Cy, *n*-Bu)[359–61] and [Fe(NO)$_2$(PPh$_3$)$_2$] [362,363]. Hayter and Williams[364] obtained dark red phosphido-bridged [Fe(PR$_2$)(NO)$_2$]$_2$ from [Fe(NO)$_2$(CO)$_2$] and R$_2$P–PR$_2$ (R = Ph, Me), and [Fe(PPh$_2$)(NO)$_2$]$_2$ was obtained directly from KPPh$_2$ and [Fe(NO)$_2$X]$_2$ [365]. The nitrosyl halides [Fe(NO)$_2$X]$_2$ react with tertiary phosphines to form [Fe(NO)$_2$(PR$_3$)X] (R = Ph, Cy; X = Br, I, SCN)[366,367], whilst in a melt [Fe(NO)$_2$Br]$_2$ reacts with PPh$_3$ with disproportionation giving [Fe(NO)$_2$(PPh$_3$)$_2$] and [Fe(NO)(PPh$_3$)$_2$Br$_2$] [366]. Triphenylphosphine reacts with [Fe(NO)(SPh)]$_2$ to give [Fe(NO)$_2$(PPh$_3$)$_2$], but with PCy$_3$ the reaction yields [Fe(NO)$_2$(PCy$_3$)(SPh)] [367]. Sodium cyanide reacts with [Fe(NO)$_2$(PPh$_3$)Br] to form [Fe(NO)$_2$(PPh$_3$)CN] [368].

The [Fe(NO)(CO)$_2$(PPh$_3$)$_2$]X complexes were obtained in moderate yield from *trans*-[Fe(CO)$_3$(PPh$_3$)$_2$] and NOX (X = Cl, Br, NO$_3$), and by metathesis other X-groups were introduced (X = BF$_4$, PF$_6$, I)[369]. Reaction of [Fe(CO)$_3$(PPh$_3$)$_2$] with NOBF$_4$ or NOPF$_6$ in methanol/benzene gives a quantitative yield of these complexes (X = BF$_4$, PF$_6$)[370].

Two carbonyl groups in Hg[Fe(CO)$_3$(NO)]$_2$ can be substituted upon treatment with PPh$_3$, PCy$_3$ or P(NMe$_2$)$_3$ [371,372] to give Hg[Fe(CO)$_2$(NO)(PR$_3$)]$_2$; under more drastic conditions P(NMe$_2$)$_3$ [372] forms [{P(NMe$_2$)$_3$}$_2$Fe(CO)$_2$(NO)][Fe(CO)$_3$(NO)] and [Fe(NO)$_2$P(NMe$_2$)$_3$]$_2$. Complexes of the type [(R$_3$M)Fe(CO)$_2$(NO)(PR$_3'$)] (R = Ph, Cl, Br; R' = Ph, Cy; M = Ge, Sn), Sn[Fe(CO)$_2$(NO)(PR$_3'$)]$_4$ and Cd[Fe(CO)$_2$(NO)(PPh$_3$)]$_2$ are known[371,373]; and [Fe(NO)$_2$(PPh$_3$)Mn(CO)$_5$] is formed from [Fe(NO)$_2$(PPh$_3$)Br] and NaMn(CO)$_5$ [374].

12.1.5 Hydrides

Although iron hydride complexes of diphosphine ligands have been known for some time, those containing monodentate phosphines have only been characterised very recently. Sacco and Aresta[375] reported that NaBH$_4$ reacted with FeCl$_2$. 2H$_2$O and PR$_3$ in ethanol to form 'FeH$_2$(PR$_3$)$_3$', but these are now known[376] to contain Fe(IV), and to have the formulae [FeH$_4$(PR$_3$)$_3$] (R$_3$ = Ph$_2$Et, Ph$_2$Bu). These are yellow crystalline solids, which slowly lose hydrogen on bubbling an inert gas through solutions to form [FeH$_2$(PR$_3$)$_2$]. Some reactions of [FeH$_4$(PPh$_2$Et)$_3$] are shown in figure 9. The yellow, diamagnetic [FeH$_2$(N$_2$)(PPh$_2$Et)$_3$] undergoes a hydrogen migration reaction in sunlight in the solid state[375], a reaction reminiscent of $[Ru(DME)_2]$ and $[Ru(arene)(DME)_2]$ (DME = $Me_2PCH_2CH_2PMe_2$)

$$[FeH_2(N_2)(PPh_2Et)_3] \rightleftharpoons H_2 + [FeH(C_6H_4PPhEt)(N_2)(PPh_2Et)_2]$$

Triethylaluminium abstracts PPh_2Et from $[FeH_2(N_2)(PPh_2Et)_3]$ to give the five-coordinate $[FeH_2(N_2)(PPh_2Et)_2]$. Lithium aluminium hydride in THF reduces $FeCl_2$

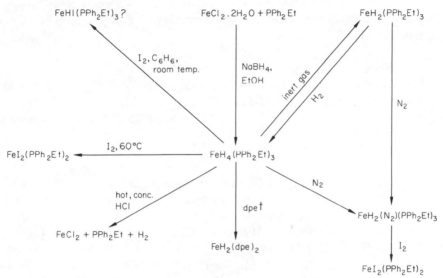

Figure 9 Reactions of $[FeH_4(PPh_2Et)_3]$[376]
† dpe = 1,2-bis(diphenylphosphino)ethane

in the presence of PMe_3 to $[FeH_2(PMe_3)_4]$, which is thought to have a *cis*-octahedral structure[377].

12.1.6 Halides

(1) Ferrous Complexes

Iron(II) phosphine complexes are rather unstable, dissociating readily in solution, especially in hydroxylic solvents. Triethylphosphine does not react with $FeCl_2$ in ethanol[378], but in benzene unstable colourless needles, presumed to be $[Fe(PEt_3)_2Cl_2]$, were obtained[352]. Naldini[379] claimed to have prepared $[Fe(PPh_3)_2X_2]$ and $[Fe(PPh_3)_4]CdI_4$, but Horrocks *et al.*[380] were unable to reproduce these preparations. By refluxing FeX_2 with PR_3 (R_3 = Ph_3, Ph_2Et, $PhEt_2$) in benzene, $[Fe(PR_3)_2X_2]$ were obtained[352,380]. These complexes all have magnetic moments in the range 4.77–5.25 B.M., though for the same complexes different values have been obtained by different workers[145,352,379,380]. The magnetic moments are consistent with a tetrahedral configuration, and this has been confirmed in the case of $[Fe(PPh_3)_2Br_2]$ by visible spectroscopy[380].

The complexes of secondary and primary phosphines show some unusual differences. With PCy_2H is formed the colourless, tetrahedral $[Fe(PCy_2H)_2Cl_2]$, $\mu_{eff} = 5.12$ B.M., but PEt_2H forms red $[Fe(PEt_2H)_2Cl_2]$, $\mu_{eff} = 3.61$ B.M., which is consistent with a square-planar structure[145]. Phenylphosphine forms $[Fe(PPhH_2)_4]X_2$ (X = Cl, Br), which are orange-red square-planar complexes, $\mu_{eff} = 3.48$ B.M.[146]. Cyclohexylphosphine, $PCyH_2$, is anomalous in forming $[Fe(PCyH_2)_3Br_2]$, a non-electrolyte, $\mu_{eff} = 3.48$ B.M., although it does form a square-planar complex $[Fe(PCyH_2)_2Cl_2]$ with ferrous chloride[148]. These complexes would repay further investigation. Lithium dicyclohexylphosphide reacts with $FeBr_2$ to form $Fe(PCy_2)_2$ and $LiFe(PCy_2)_3$[384].

(2) Ferric Complexes

Anhydrous $FeCl_3$ and PPh_3 in ether afford $[Fe(PPh_3)_2Cl_3]_2$, $\mu_{eff} = 5.94$ B.M., which are possibly halogen-bridged dimers[385]. Heating $FeCl_3 \cdot 6H_2O$ with PPh_3, or reaction in alcohols, gives $[PPh_3H][FeCl_4]$[385]. $[FeCl_3(PPh_3)]$ has been prepared by the unusual route

$$Fe_3(CO)_{12} + PPh_3 \xrightarrow[CHCl_3]{reflux} FeCl_3 \cdot PPh_3$$

where the chloroform functions both as solvent and chlorinating agent[386]. This complex has been assigned a tetrahedral configuration[387]. There is a brief report[388] of the formation of $[FeCl_3(PCy_3)]$.

Displacement of NH_3 from $Na_3[Fe(CN)_5NH_3]$ leads to the formation of $Na_3[Fe(CN)_5(PR_3)]$ ($R_3 = Ph_3$, Bu_3, Ph_2H, PhH_2)[389,390]. Bromine oxidation gives the corresponding $Na_2[Fe(CN)_5(PR_3)]$. The PPh_3 complex has $\mu_{eff} = 2.47$ B.M. The compounds have been studied by i.r. and Mössbauer spectroscopy[389,391], the reactions with various metal ions examined and some free acids, $H_3[Fe(CN)_5(PR_3)] \cdot xH_2O$ (R = Ph, Bu, Cy), obtained[392].

12.2   RUTHENIUM AND OSMIUM

12.2.1 Carbonyls

Dark red $Ru_3(CO)_9(PR_3)_3$ are formed from $Ru_3(CO)_{12}$ and PR_3 (R = Ph, Bu^n)[393-6,399]. A kinetic study[397] of the reaction of $Ru_3(CO)_{12}$ with various phosphines produced evidence that the intermediate substitution products $Ru_3(CO)_{11}(PR_3)$ and $Ru_3(CO)_{10}(PR_3)_2$ were not stable, and these have not been prepared. Phosphine reacts with $Ru(CO)_2Cl_2$ to form[398] a compound thought to be $Ru_3(CO)_8(PH_3)_4$, although in the absence of X-ray data a phosphido-bridged structure cannot be eliminated. Tetramethyldiphosphine and $Ru_3(CO)_{12}$ afford[393] the Me_2P-bridged $[Ru(CO)_3(PMe_2)]_2$, and $[Ru(CO)_3P(C_6F_5)_2]_2$ is obtained by an analogous method to the iron complex[339]. The carbido complex, $Ru_6C(CO)_{17}$, undergoes monosubstitution with PR_3 (R = Ph, p-FC_6H_4) in hexane, to form deep red air-stable crystals of $Ru_6C(CO)_{16}(PR_3)$[400]. The $Ru_3(CO)_9(PR_3)_3$ complexes are broken down to mononuclear derivatives by a number of reagents (see figure 10). Several of these reactions are discussed below.

In contrast to the ruthenium compound, $Os_3(CO)_{12}$ reacts with PR_3 (R_3 = $PhEt_2$, Ph_2Me, Et_3) to give a mixture of $Os_3(CO)_{11}(PR_3)$, $Os_3(CO)_{10}(PR_3)_2$ and $Os_3(CO)_9(PR_3)_3$, readily separable by chromatography[404]. These complexes were initially obtained as byproducts of the reaction of $Os_3(CO)_{12}$ with $Au(PPh_3)Cl$ [412],

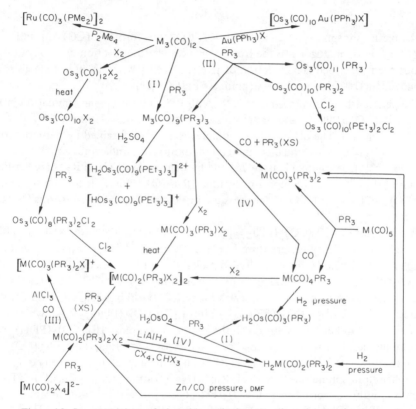

Figure 10 Some reactions of phosphine-substituted carbonyls of ruthenium and osmium[393-7,401-11,414-16,417]
M = Ru, Os
PR_3 = PPh_3, except I R = Ph, *n*-Bu, II, R_3 = Et_3, $PhEt_2$, Ph_2Me,
    III R_3 = Ph_3, Et_3, Cy_3, Ph_2Et,
    IV M = Ru only

which leads to $[Os_3(CO)_{10}Au(PPh_3)X]$ [405]; ruthenium–gold complexes do not seem to be sufficiently stable to be isolated. Chlorine reacts with $Os_3(CO)_{10}(PEt_3)_2$ to form $Os_3(CO)_{10}(PEt_3)_2Cl_2$, and $Os_3(CO)_8(PPh_3)_2Cl_2$ (XXXVI) is obtained from $Os_3(CO)_{10}Cl_2$ and PPh_3 [404]. The trisubstituted $Os_3(CO)_9(PR_3)_3$ [404,413], which contain one phosphine on each osmium atom[404], undergo reactions with halogens or PR_3 leading to mononuclear derivatives. The $Os_3(CO)_{12-n}(PR_3)_n$ complexes dissolve in concentrated sulphuric acid to give solutions containing the dihydrido species $[H_2Os_3(CO)_{12-n}(PR_3)_n]^{2+}$, and by the addition of NH_4PF_6, $[H_2Os_3(CO)_9(PEt_3)_3][PF_6]_2$ and $[HOs_3(CO)_9(PR_3)_3][PF_6]$ can be isolated[401].

$$Ph_3P(CO)_2Os \overset{\displaystyle \overset{(CO)_4}{Os}}{\underset{Cl}{\overset{Cl}{\diagup \diagdown}}} Os(CO)_2PPh_3$$

(XXXVI)

The olefin phosphine, SP, reacts with $Ru_3(CO)_{12}$ to form $[Ru(CO)_3SP]$ and $[Ru(CO)_2(SP)_2]$, analogous to the iron complexes[331]; the reaction with $Os_3(CO)_{12}$ has not been described. The reaction of $[Ru(CO)_3(SP)]$ with HX in hexane gives a σ-bonded Ru(II)–carbon complex (compare $Fe(CO)_3SP$)[332].

The pentacarbonyls, $M(CO)_5$, react[407] with PPh_3 on ultraviolet irradiation in THF to form $[M(CO)_4(PPh_3)]$, and under pressure with two equivalents of PPh_3 to form $[M(CO)_3(PPh_3)_2]$. The disubstituted complexes are also obtained by reduction of $M(CO)_2(PPh_3)_2X_2$ with zinc dust and carbon monoxide, under pressure in DMF solution[416,417] (see figure 12, page 76) and by heating $M_3(CO)_9(PR_3)_3$ (R = *n*-Bu, Ph) with PR_3 [394,402]. The monosubstituted $[Ru(CO)_4(PPh_3)]$ is formed on heating $Ru_3(CO)_9(PPh_3)_3$ with CO under pressure, but $Os_3(CO)_9(PPh_3)_3$ reforms $Os_3(CO)_{12}$ by eliminating PPh_3 [402].

The structure of $[Os(CO)_3(PPh_3)_2]$ is trigonal bipyramidal with axial phosphines, there being two slightly different molecular geometries[418]. These five-coordinate d^8 complexes readily add halogen or hydrogen halide to form six-coordinate $[M(CO)_2(PR_3)_2XY]$ (XY = halogens or hydrogen halide)[356,416,417], and react with mercury halides to form cationic species $[MHgX(CO)_3(PPh_3)_2]HgX_3$ [420]. Sulphur dioxide reacts with $[Ru(CO)_3(PPh_3)_2]$ to form $[Ru(CO)_2(PPh_3)_2SO_2]$, which is oxidised by air in benzene solution to the sulphato complex, $[Ru(CO)_2(PPh_3)_2SO_4]$, also obtainable directly from the carbonyl and H_2SO_4 [421]. Reaction of $Ru(CO)_3(PPh_3)_2$ with oxygen yielded $[Ru(CO)_2(PPh_3)_2CO_3]$, not the expected dioxygen complex[421].

Triphenylphosphine reacts with $H_2Ru(CO)_4$ to form white $[H_2Ru(CO)_2(PPh_3)_2]$, which is rapidly converted to $[Ru(CO)_2(PPh_3)_2Cl_2]$ in $CHCl_3$[408]. Conversely, $[Ru(CO)_2(PR_3)_2Cl_2]$ (R = Ph, Et) are reduced to the dihydrides by $LiAlH_4$ [408], which probably have structure (XXXVII). Only one CO group in $H_2Os(CO)_4$ is

(XXXVII)

substituted by PPh_3, resulting in the colourless $[H_2Os(CO)_3(PPh_3)]$, which probably has a *cis* configuration. The more reactive Bu^n_3P gives $[H_2Os(CO)_2(PBu^n_3)_2]$ directly[422], and both $[H_2M(CO)_2(PPh_3)_2]$ can be obtained by hydrogenation under pressure of [M(CO)$_3$(PPh$_3$)$_2$] [407]. The interesting H$_2$Os$_2$(CO)$_6$(PPh$_3$)$_2$ is formed from H$_2$Os$_2$(CO)$_8$ and PPh$_3$ [423].

12.2.2 Carbonyl halides

The chemistry of the carbonyl halide phosphine compounds of these two elements is complex. In addition to the normal preparative routes from the metal carbonyl

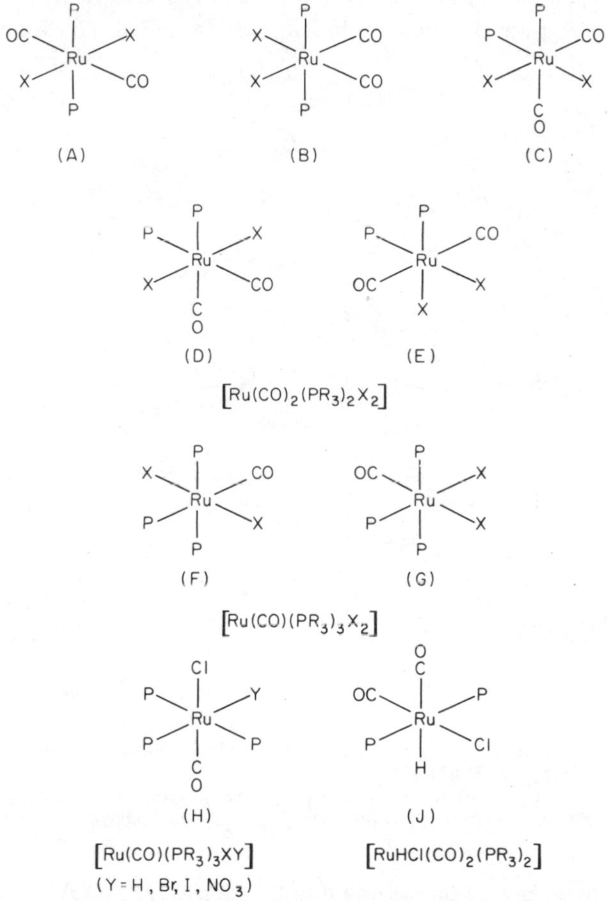

(A)　　　　　　(B)　　　　　　(C)

(D)　　　　　　(E)

$$\left[Ru(CO)_2(PR_3)_2X_2\right]$$

(F)　　　　　　(G)

$$\left[Ru(CO)(PR_3)_3X_2\right]$$

(H)　　　　　　(J)

$$\left[Ru(CO)(PR_3)_3XY\right]\qquad\left[RuHCl(CO)_2(PR_3)_2\right]$$
(Y = H, Br, I, NO$_3$)

Figure 11　Some isomers of [Ru(CO)$_2$(PR$_3$)$_2$X$_2$], [Ru(CO)(PR$_3$)$_3$X$_2$] and [Ru(CO)$_2$(PR$_3$)$_2$XY]

halide or its derivatives and PR$_3$, many ruthenium and osmium phosphine complexes form carbonyl or hydrido carbonyl complexes on boiling in alcoholic solution. The ruthenium complexes have been more extensively studied.

The isolation of [Ru(CO)$_2$(PR$_3$)$_2$I$_2$] (R = Ph, Cy) from PR$_3$ and [Ru(CO)$_2$I$_2$]$_n$ was briefly reported in 1956–7[424] and has been reexamined in detail by Hieber and John[425]. Diamagnetic, non-conducting [Ru(CO)$_2$(PR$_3$)$_2$I$_2$] complexes were obtained by reaction of PR$_3$ (R$_3$ = Ph$_3$, Cy$_3$, Et$_3$, PhEt$_2$, Ph$_2$Et) with [Ru(CO)$_2$I$_2$]$_n$ in

benzene; from which Cl_2 or Br_2 displaced I_2 to form the chloro and bromo analogues[425]. On the basis of ^1H n.m.r. and i.r. spectroscopy, and dipole moment
measurements, the various isomers produced were identified. Four structural
isomers, (A), (B), (C), and (D) are produced from $PPhEt_2$ (figure 11), whereas the
other phosphines produce only (A) and (B). The *trans*-carbonyl complex, (E), was
not obtained[425]. Monocarbonyl complexes $[Ru(CO)(PR_3)_3X_2]$ were isolated for
PEt_3 and $PPhEt_2$; $[Ru(CO)(PPhEt_2)_3I_2]$ was shown to have structure (G)[425,426].
$[Ru(CO)_3(PPh_3)X_2]$ was obtained by Bruce *et al.*[396], and also by Johnson and

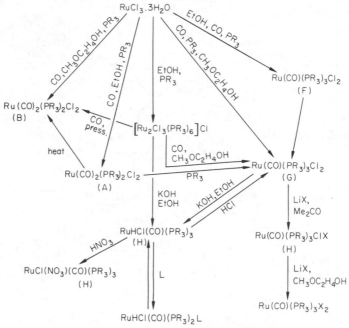

Figure 12   Ruthenium carbonyl and hydridocarbonyl complexes of
$PPhEt_2$ and $PPhMe_2$[416,428-31,434]
(A), (B), etc. refer to configurations in figure 11. L = $PPhMe_2$,
$PPhEt_2$, $PPhPr^n_2$, $PPhBu^n_2$, $PPh(OMe)_2$, PEt_3, $AsPhMe_2$, $AsPhEt_2$

coworkers[403]; these change on heating into the dimeric $[Ru(CO)_2(PPh_3)X_2]_2$, also
obtained by reaction of $[Ru_3(CO)_9(PPh_3)_3]$ with the stoichiometric quantity of
halogen[402,403]. The addition of PPh_3 to $[Ru(CO)_2X_4]^{2-}$ or $[Ru(CO)_3X_3]^-$ produces
$[Ru(CO)_2(PPh_3)_2X_2]$ [409,410], which are mixtures of isomers, as judged by their i.r.
spectra.

A different approach is carbonylation of $RuCl_3 . 3H_2O$ in alcohols under various
conditions, followed by addition of PR_3 ($R_3 = Ph_3$, $PhMe_2$, $PhEt_2$, $PhPr_2$, $PhBu_2$,
Et_3) (figure 12), or by carbonylation of $[Ru_2Cl_3(PR_3)_6]Cl$ [416,427-31]. The product
depends upon the reaction conditions, the solvent and the phosphine concerned.
The *trans*-$[Ru(CO)_2(PR_3)_2Cl_2]$ complexes, (A), are less stable than the *cis* isomers,

(B), into which they readily isomerise upon heating. Under different conditions, $[Ru(CO)(PR_3)_3Cl_2]$ (F, G) are obtained[430]. These last complexes form $[RuHCl(CO)(PR_3)_3]$ on boiling with alcoholic KOH, the reaction being reversed by HCl [430]. Bromo and iodo analogues of $[Ru(CO)(PR_3)_3Cl_2]$ can be obtained by metathesis, the ease of replacement of Cl depending upon the *trans* ligand: Cl *trans* to PR_3 is substituted more rapidly than Cl *trans* to Cl or CO, which allows the isolation of some intermediate $[Ru(CO)(PR_3)_3ClX]$ (X = Br, I, SCN) complexes[430]. The hydrido complexes, $[RuHCl(CO)(PR_3)_3]$, form on boiling $[Ru_2Cl_3(PR_3)_6]Cl$ with alcohol in the presence of KOH[430-2]. Nitric or carboxylic acids substitute the hydride to give, for example, $[Ru(CO)(NO_3)Cl(PR_3)_3]$, without change of configuration. However, on treatment with HX (X = Br, I), $[RuHCl(CO)(PR_3)_3]$, (H), gives mixtures of $[Ru(CO)(PR_3)_3XCl]$ and $[Ru(CO)(PR_3)_3X_2]$, (G); the formation of the latter is surprising, since the second chlorine replaced is *trans* to CO, which usually hinders replacement[431].

James and Markham[433] found that a solution of $[RuHCl_2(PPh_3)_2]$ in dimethylacetamide (DMA) takes up CO to form two complexes, $[RuHCl(CO)_2(PPh_3)_2]$ (J) and $[Ru(CO)_2Cl_2(PPh_3)_2]$ (B). In DMA, $[RuCl_2(PPh_3)_3]$ also takes up CO to form $[Ru(CO)_2(PPh_3)_2Cl_2]$ (E).

The *trans*-labilising effect of the hydride ligands in $[RuHCl(CO)(PR_3)_3]$ ($R_3 = Et_3$, PhEt$_2$, PhPr$^n_2$, PhBu$^n_2$) was demonstrated by Douglas and Shaw[434] who studied the equilibria

$$RuHCl(CO)(PR_3)_3 + L \rightleftharpoons RuHCl(CO)(PR_3)_2L + PR_3$$

L = AsPhEt$_2$, AsPhMe$_2$, PPhBu$^n_2$, PPhPr$^n_2$, PPhEt$_2$, PEt$_3$, P(OEt$_3$)$_2$, PPhMe$_2$, PPh(OMe)$_2$

by observing the hydride ^1H n.m.r. spectra of the complexes in the presence of various L. The relative affinities for Ru increase along the series given. Shaw et al.[435] used $[Ru(CO)_2Cl_2(PBu^t_2H)]$ (two forms, A and B) to determine $^2J(P-M-P)$ from ^1H n.m.r. spectra; PBu$^t_2$H is suitable for this study, since the ^1H n.m.r. spectra are relatively simple due to the small value of $^3J(H-Me)$.

Triphenylphosphine reacts with $[Ru(H_2O)(CO)Cl_2]$ to form $[Ru(H_2O)(CO)(PPh_3)_2Cl_2]$, analogous to the reaction of $Ru(CO)_2Cl_2$ and PPh_3 giving $[Ru(CO)_2(PPh_3)_2Cl_2]$ [436,437].

Hieber et al.[353] showed that $[Ru(CO)_2(PR_3)_2Br_2]$ ($R_3 = Ph_3$, Et_3, Cy_3, Ph_2Et) react with Lewis acids and CO to give $[Ru(CO)_3(PR_3)_2Br]^+$. Stannous chloride is reported to form $[Ru_2Cl_3(SnCl_3)(CO)_2(PPh_3)_3(Me_2CO)_2]$ and $[Ru_2Cl_3(SnCl_3)(CO)(PPh_3)_4]$ [429]. Treatment of $[Ru_3(CO)_9(PPh_3)_3]$ or $[Ru(CO)_4(PPh_3)]$ with carboxylic acids, RCO_2H (R = Et, Me, H), affords $[Ru_2(CO)_4(RCO_2)_2(PPh_3)_2]$, also obtainable from $[Ru_2(CO)_6(RCO_2)_2]$ and PPh_3. Similarly, $[Ru(CO)_3(PPh_3)_2]$ yields $[Ru(CO)_2(PPh_3)_2(RCO_2)_2]$ [438,439].

Hales and Irving obtained $[Os(CO)_2(PR_3)_2X_2]$ [440] by reaction of $[Os(CO)_3X_2]$ with PR_3 ($R_3 = Ph_3$, Ph_2Cl, $PhCl_2$, Cl_3), and the PPh_3 complexes were also obtained as byproducts of the reaction of $Os_3(CO)_{12}$ with $AuCl(PPh_3)$ [405]. Bradford et al.[405]

studied the i.r. and Raman spectra of $[Os(CO)_2(PPh_3)_2X_2]$, and concluded that
they probably had structures analogous to type (B) (figure 11), although type (C)
was not definitely eliminated. Like the ruthenium analogues, $Cs_2[Os(CO)_2X_4]$ are
decomposed by PPh_3 in the presence of formic acid to form $[Os(CO)_2(PPh_3)_2X_2]$,
but $Cs[Os(CO)_3X_3]$ forms both $[Os(CO)_2(PPh_3)_2X_2]$ and $[Os(CO)_3(PPh_3)X_2]$[409,410].
Recently, the dimeric $[Os(CO)_2(PPh_3)Cl_2]_2$ has been obtained, along with *cis-*
$[Os(CO)_4Cl_2]$, by chlorination of $[Os_3(CO)_8(PPh_3)_2Cl_2]$ [404].

Carbonylation of $[Os(CO)_2(PPh_3)_2Br_2]$ in the presence of a Lewis acid gives
$[Os(CO)_3(PPh_3)_2Br]^+$, as for ruthenium, but an unstable $[Os(CO)_4(PR_3)_2]^{2+}$ can also
be formed in the reaction[411,441].

Vaska reported $[Os(PPh_3)_3Br]$ [442], obtained from the reaction of $(NH_4)_2OsBr_6$
with PPh_3 in ethanol, but subsequently showed[443,444] that the product was in fact
$[OsHBr(CO)(PPh_3)_3]$. By performing the reaction in ^{14}C-labelled ethylene glycol it
was confirmed that the CO came from the solvent. Deuterium analogues,
$[OsDX(CO)(PPh_3)_3]$, were also obtained[443], and final confirmation obtained by an
X-ray structure determination of $[OsHBr(CO)(PPh_3)_3]$, (XXXVIII) [445]. Vaska[446]

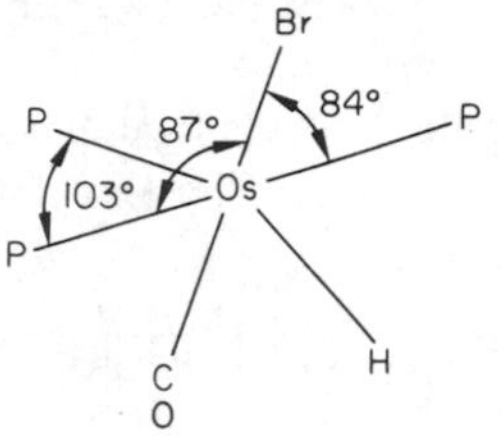

(XXXVIII)

subsequently published an account of a detailed investigation of the preparation of
the complexes $[OsHX(CO)(PPh_3)_3]$ and $[OsHX(CO)(AsPh_3)_3]$ by reaction of
$P(As)Ph_3$ with $(NH_4)_2OsX_6$ in alcohol under varying conditions, and on the basis
of their i.r. spectra concluded that both $[OsHCl(CO)(PPh_3)_3]$ and $[OsH_2(CO)(PPh_3)_3]$
had structure (XXXIX) [447]. By treating $[Os(CO)_3(PPh_3)_2]$ with acids, Laing and

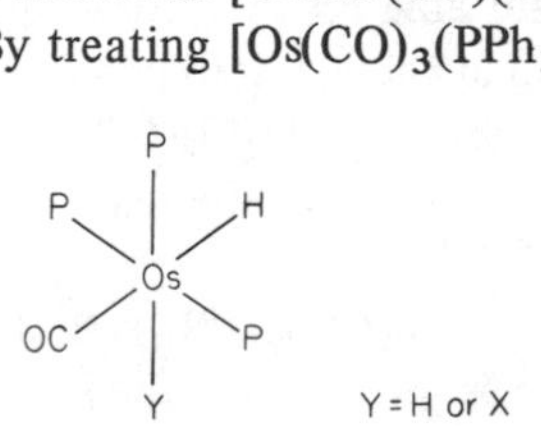

Y = H or X

(XXXIX)

Roper[419] obtained $[OsH(CO)_3(PPh_3)_2]X$ ($X = HCl_2$, Br, ClO_4, BF_4, PF_6). These
complexes were 1:1 electrolytes in nitrobenzene, and i.r. spectra indicate *trans-*
phosphine groups. On heating in ethanol solution a number of these complexes re-
arrange to $[Os(CO)_2(PPh_3)_2X_2]$.

The carbonylation of osmium(III) halo complexes, followed by addition of PR_3
($R_3 = Ph_3$, Ph_2Me, $PhEt_2$) is a convenient route to $[Os(CO)_2(PR_3)_2X_2]$ ($X = Br$).

When X = Cl, carbonylation is difficult in the absence of PPh_3, but upon addition of the ligand $OsCl_3$ it occurs readily in $MeOCH_2CH_2OH$ solution[417]. One equivalent of bromine or iodine reacts with $[Os(CO)_3(PPh_3)_2]$ to form the ionic $[Os(CO)_3(PPh_3)_2X]X$, which slowly lose CO on heating in solution to form $[Os(CO)_2(PPh_3)_2X_2]$ [417]. Reduction of K_2OsCl_6 with PCy_3 in 2-methoxymethanol leads to the formation of $[OsHCl(CO)(PCy_3)_2]$ [448]. It adds pyridine to form $[OsHCl(CO)(PCy_3)_2 . py]$, but attempts to obtain $[OsHCl(CO)(PCy_3)_3]$ have failed, possibly due to the steric hindrance of the cyclohexyl groups.

Chatt *et al.*[449] report that zinc reduction of $[OsCl_3(PR_3)_3]$ $(R_3 = Ph_2Et, Ph_2Pr^n,$ $PhMe_2, PhEt_2, PhPr^n{}_2)$ in the presence of CO gives $[Os(CO)(PR_3)_3Cl_2]$, isomers of structures (F) and (G) (see figure 11), and with $[Os(CO)_2(PR_3)_2Cl_2]$ isomers (A) and (B) were isolated.

Some (organo-Group IVB) carbonyls are known: $[(Me_3L)M(CO)_3(PPh_3)X]$ (L = Si, Ge, Sn; M = Ru, Os) [450].

12.2.3 Nitrosyls

Ruthenium forms more nitrosyl complexes than any other element. Diamagnetic $[Ru(NO)(PR_3)_2X_3]$ $(R_3 = Ph_3, Ph_2Me, PhEt_2, Et_3, Bu^n{}_3)$ have been obtained from $[Ru(NO)(H_2O)_2Cl_3]$ and PR_3 in ethanol, from which the bromo and iodo derivatives were obtained by reaction with LiX [451,452]. A reinvestigation[453] of the reactions has shown that $[Ru(NO)Cl_3]$ reacts with various PR_3 in ethanol or 2-ethoxyethanol to give two structural isomers of $[Ru(NO)(PR_3)_2Cl_3]$. These have configurations (XL) and (XLI). Yellow osmium analogues have recently been

(XL)   (XLI)

obtained[454] from M_2OsX_6 (M = Na, NH_4), NO and PPh_3; these probably have analogous structures to (XL).

Dinitrosyl complexes, $[M(NO)_2(PPh_3)_2]$, are formed from $[M(CO)_2(PPh_3)_2Cl_2]$ and $NaNO_2$ in DMF; $[M(ONO)_2(CO)_2(PPh_3)_2]$ is formed initially, but this isomerises to $[M(NO_2)_2(CO)_2(PPh_3)_2]$, which decomposes to the $[M(NO)_2(PPh_3)_2]$ complex[455]. Strong acids protonate $[Os(NO)_2(PPh_3)_2]$ to give $[OsH(NO)_2(PPh_3)_2]^+$, which reacts with oxygen to form $[Os(OH)(NO)_2(PPh_3)_2]^+$ [455]. The crystal structure of this last complex, isolated as the PF_6^- salt, shows it to contain both linear and bent nitrosyl groups[456]. $[Os(NHOH)(NO)(PPh_3)_2Cl_2]$ is formed[457] on reaction of $[Os(NO)_2(PPh_3)_2]$ with HCl.

The $[Ru(NO)_2(PPh_3)_2Cl]PF_6$ complex (XLII), formed from $NOPF_6$ and $[Ru(NO)(PPh_3)_2Cl]$, also has both linear and bent NO groups[458]. Bright yellow $[Ru(NO)(PPhMe_2)_3Cl_2]X$ (X = Cl, BF_4) are formed as byproducts of the reaction of $[Ru(NO)Cl_3]$ with $PPhMe_2$[453].

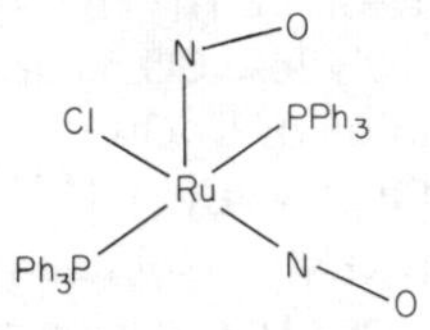

(XLII)

Reduction of $[Ru(NO)(PR_3)_2Cl_3]$ leads to new reactive nitrosyl complexes. Refluxing ethanolic KOH in the presence of excess PR_3 produces $[RuH(NO)(PR_3)_3]$ (R_3 = Ph_3, Ph_2Me, Ph_2Pr^i, Ph_2Cy) as brown crystalline complexes[459]. The $[RuH(NO)(PPh_3)_3]$ complex has a slightly distorted trigonal-bipyramidal structure, with equatorial PPh_3 groups and a linear NO group[460]. Reduction of $[Ru(NO)(PPh_3)_2Cl_3]$ with zinc dust in boiling benzene produced the ruthenium analogue of Vaska's compound, $[Ru(NO)Cl(PPh_3)_2]$, as emerald-green crystals, which readily add O_2 to give $[Ru(O_2)(NO)Cl(PPh_3)_2]$. Similar addition compounds of CO, SO_2, MeI, HX, X_2 and HgX_2 were prepared[461]. The dioxygen adduct can also be prepared by the action of air on $[Ru(CO)(NO)(PPh_3)_2Cl]$[462,464].

Carbonyl nitrosyls, $[Ru(CO)(NO)(PPh_3)_2X]$ (X = OH, Cl, Br, I, N_3, NCO, NCS, HCO_2), have been obtained from $[RuH(CO)(PPh_3)_3Cl]$ and N-nitroso-N-methyl-toluene-*p*-sulphonamide, followed by displacement of Cl[462,464]. The CO is displaced by halogens, affording $[Ru(NO)(PPh_3)_2XY_2]$ (Y = halide). An X-ray analysis shows $[Ru(CO)(NO)(PPh_3)_2I]$ to have structure (XLIII)[463].

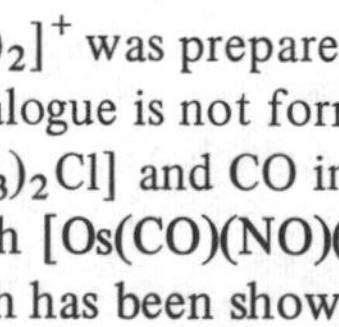

(XLIII)

Cationic $[Ru(CO)_2(NO)(PPh_3)_2]^+$ was prepared from $[Ru_3(CO)_9(PPh_3)_3]$ and NO^+ in methanol; the osmium analogue is not formed in this way, but can be obtained from $[Os(CO)(NO)(PPh_3)_2Cl]$ and CO in the presence of $NaBPh_4$[370].

Mercury(II) chloride reacts with $[Os(CO)(NO)(PPh_3)_2Cl]$ to form $[Os(NO)(HgCl)(PPh_3)_2Cl_2]$, which has been shown to have an octahedral structure with a linear NO group[465]. The unusual $[Os(NO_2)(NO)(PPhMe_2)_3Cl_2]$ and $[Os(NO_2)(NO)(PPhMe_2)_3Cl]^+$ are formed by zinc reduction of $[Os(PPhMe_2)_3Cl_3]$ in the presence of NO[449].

12.2.4 Hydrides

Ruthenium forms a number of rather unstable hydrides, but osmium resembles rhenium and iridium in forming multihydrido complexes. Hydrido carbonyls have already been discussed (12.2.2.). Yamamoto *et al.*[466] found that Et_3Al reduction of $RuCl_3$ in the presence of PPh_3 produce a light yellow diamagnetic complex, tentatively identified as $[H_2Ru(PPh_3)_4]$. This was later confirmed[467], and it was shown that this complex takes up dinitrogen in solution

$$H_2Ru(PPh_3)_4 + N_2 \rightleftharpoons H_2Ru(N_2)(PPh_3)_3 + PPh_3$$

Iodine reacts with $[H_2Ru(PPh_3)_4]$ to give $[Ru(PPh_3)_3I_3]$ or $[RuHI(PPh_3)_3]$, and EtBr forms purple $[RuHBr(PPh_3)_3]$. In solution $[H_2Ru(PPh_3)_4]$ also reacts reversibly with hydrogen, probably to form $[H_4Ru(PPh_3)_3]$. Studies[467] using deuterium have

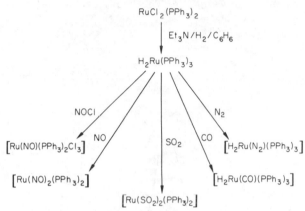

Figure 13 Some reactions of $H_2Ru(PPh_3)_3$

indicated that the ortho position on the phenyl rings is involved in these reactions with intermediates such as (XLIV).

(XLIV)

Knoth[468] independently obtained a complex formulated as $[H_2Ru(N_2)(PPh_3)_3]$ by reaction of $[RuHCl(PPh_3)_3]$ with Et_3Al in ether under nitrogen, but differences in the i.r. spectra may indicate that the structure is different from that obtained by Yamamoto *et al.*[466,467].

The reaction of $[Ru_2Cl_3(PPh_2Me)_6]Cl$ with hydrazine, PPh_2Me and hydrogen under pressure yields $[H_2Ru(PPh_2Me)_4]$[469]. The i.r. and ^1H n.m.r. spectra are consistent with a *cis* octahedral structure. Carbon monoxide at 80°C replaces one phosphine to form $[H_2Ru(CO)(PPh_2Me)_3]$, which seems to contain CO *trans* to H.

The ^{1}H n.m.r. spectra of a series of hydrides, $[H_2Ru(PR_3)_4]$ (R_3 = PhEt$_2$, PhMe$_2$, Ph$_2$Me, Ph$_2$OMe), have been reported; they show the complexes to be *cis* dihydrides except for R_3 = Ph$_2$OMe, which exhibits *cis–trans* isomerism in solution[470]. No preparative details were given.

Eliades *et al.*[471] reported $[H_2Ru(PPh_3)_3]$ as the product from $[RuCl_2(PPh_3)_2]$ and Et$_3$N in benzene under hydrogen, and found it readily reacted with a number of small molecules (figure 13). It was subsequently shown that $[H_2Ru(PPh_3)_3]$ and $[H_2Ru(PPh_3)_4]$ could be obtained by NaBH$_4$ reduction of RuCl$_3$, the product

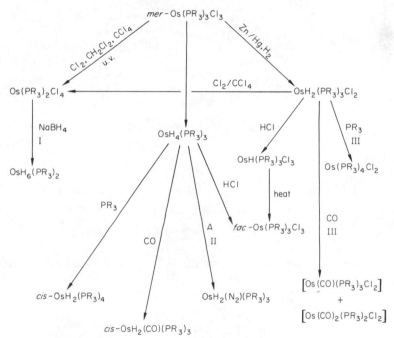

Figure 14 Some reactions of osmium hydrides[475–8]
A   = toluene-*p*-sulphonyl azide
PR$_3$ = PMe$_2$Ph, PEt$_2$Ph, PMePh$_2$, PEtPh$_2$ except I = PMe$_2$Ph; II = PEtPh$_2$;
III = PEt$_2$Ph, PMePh$_2$

obtained depending upon the excess of PPh$_3$ present in solution[472]. An orange form isolated in an atmosphere of nitrogen may contain coordinated dinitrogen.

Benzene–ethanol solutions of $[RuCl_2(PPh_3)_3]$ are effective hydrogenation catalysts for unsaturated substrates. Wilkinson *et al.*[473] showed that the violet-black $[RuHCl(PPh_3)_3]$ is the active catalyst, and isolated it by reaction of NaBH$_4$ with $[RuCl_2(PPh_3)_3]$ in benzene. It can also be synthesised by reaction of RuCl$_3$, HCHO and PPh$_3$ in 2-methoxyethanol[472]. An X-ray structure determination shows it to be a distorted trigonal bipyramid[474], Ru–H = 170 pm (XLV).

The series of complexes $[H_4Os(PR_3)_3]$ (R_3 = Ph$_3$, PhEt$_2$, PhMe$_2$) and $[H_4Os(PPhMe_2)_2L]$ (L = PPhEt$_2$, AsPhMe$_2$) are obtained by reduction of *mer-*

$[OsCl_3(PR_3)_3]$, OsX^{2-}_6/PR_3, and $[OsCl_3(PPhMe_2)_2L]$ with $NaBH_4$ in ethanol, or $LiAlH_4$ in THF[165,472,475,476] (figure 14). These tetrahydrido species are white,

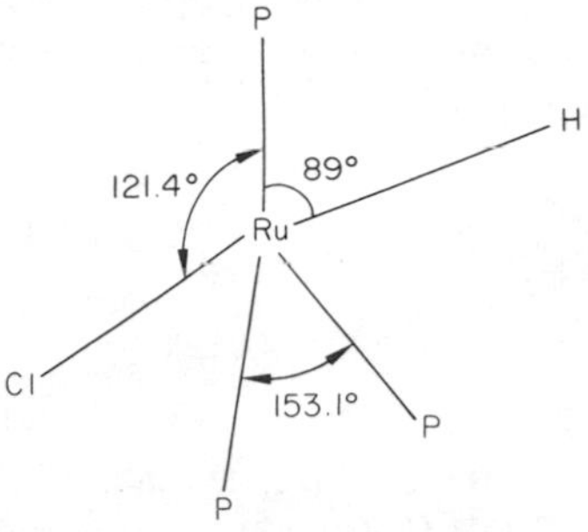

(XLV)

air-stable, crystalline solids. The $[H_4Os(PPhR_2)_3]$ (R = Me, Et) complexes react with hydrogen chloride in methanol to give *fac*-$[OsCl_3(PPhR_2)_3]$. Deuterium–hydrogen exchange in EtOD is slow, but becomes rapid upon the addition of acid; however, attempts to isolate $[H_5Os(PR_3)_3]^+$ species failed[475]. Treatment of $[OsCl_4(PPhMe_2)_2]$ with $NaBH_4$ in EtOH produces a yellow oil, identified as $[H_6Os(PPhMe_2)_2]$[475]. On refluxing $[H_4Os(PR_3)_3]$ (R_3 = $PhMe_2$, $PhEt_2$, Ph_2Me, Ph_2Et) with PR_3, *cis*-$[H_2Os(PR_3)_4]$ were obtained, whilst treatment with CO and toluene-*p*-sulphonyl azide gave *cis*-$[H_2Os(CO)(PR_3)_3]$ and $[H_2Os(N_2)(PPh_2Et)_3]$, respectively[477].

Zinc amalgam reduction of *mer*-$[OsCl_3(PR_3)_3]$ (R_3 = Ph_2Me, Ph_2Et, $PhEt_2$, $PhPr^n_2$) under hydrogen gave $[H_2OsCl_2(PR_3)_3]$, which react with CO, PR_3 and Cl_2 in CCl_4 to give $[Os(CO)(PR_3)_3Cl_2]$ and $[Os(CO)_2(PR_3)_2Cl_2]$, $[OsCl_2(PR_3)_4]$ and $[Os(PR_3)_2Cl_4]$, respectively[478]. Borohydride reduction of $[OsL(PR_3)_3X_2]$ (L = CO, N_2) yield $[OsLH(PR_3)_3X]$, the ^1H n.m.r. spectra of which are consistent with structure (XLVI)[478]. Early reports of $[OsHCl_2(PPhBu^n_2)_3]$[479,480] have since been refuted, the product being a mixture of *mer*-$[OsCl_3(PR_3)_3]$ and $[OsCl_3(N_2)(PR_3)_3]$[481].

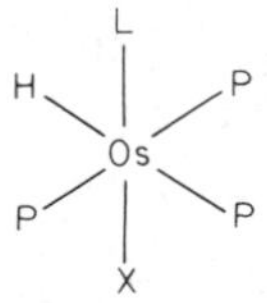

(XLVI)

12.2.5 Dinitrogen complexes

Relatively few dinitrogen–tertiary phosphine complexes of these elements have been reported. Several of these have already been discussed above.

Chatt *et al.*[261] failed to obtain osmium nitrido complexes by the methods used to prepare $[ReNX_2(PR_3)_2]$. The reaction of $OsOCl_3(PPh_3)_2$ with hydrazine salts in boiling ethanol gave $[OsCl_3(NH_3)(PPh_3)_2]$[261], which was shown by Bright and Ibers[482] to have the distorted octahedral structure (XLVII).

The reaction of ammonia or primary amines with $[MX_3(PR_3)_3]$ in ethanol formed *mer*-$[MX_2(am)(PR_3)_3]$ (am = NH_3 or NH_2R) as diamagnetic, orange or yellow solids[483]. The dimethylphenylphosphine complexes are the easiest to obtain and have been well characterised, probably having structure (XLVIII). Secondary and

(XLVII)   (XLVIII)

tertiary alkylamines reduce $[MX_3(PR_3)_3]$ in ethanol, but the products are the alcohol complexes $[MX_2(EtOH)(PR_3)_3]$ [483]. Interestingly, when the alcohol contains small amounts of water, the NHR_2 (R = *n*-alkyl) are dealkylated and $[MX_2(NH_2R)(PR_3)_3]$ result (compare the decarbonylation of alcohols). The reactions with hydrazine leading to $[MCl_2(N_2H_4)(PR_3)_3]$ and $[M_2X_4(N_2H_4)_2(PR_3)_4]$ were also studied[483].

A series of dinitrogen complexes of osmium(II) were obtained by reaction of nitrogen under pressure with the solution produced on reduction of *mer*-$[OsCl_3(PR_3)_3]$ with amalgamated zinc[484]. White, crystalline complexes, $[Os(N_2)(PR_3)_3Cl_2]$ (R_3 = Ph_2Me, Ph_2Et, $PhMe_2$, $PhEt_2$, $PhPr^n_2$, $PhBu^n_2$, Et_3), with $\nu(N\equiv N)$ in the range 2100–2050 cm^{-1}, were obtained, and from i.r. and n.m.r. spectra were assigned meridional structures with N_2 *trans* to halogen. These complexes in solution are very sensitive to oxidation, but in the absence of oxygen they are comparatively inert. The reduction of $[Os(N_2)(PR_3)_3X_2]$ to $[OsHX(N_2)(PR_3)_3]$ has already been described; HCl converts the latter (X = Cl) to $[Os(N_2)(PR_3)_3Cl_2]$ [478]. The kinetics of the dinitrogen substitution in $[Os(N_2)(PR_3)_3X_2]$ by phosphine ligands has been studied by Maples *et al.*[485].

12.2.6 Halides

The complexes originally formulated as $[RuCl(PPh_3)_3]$ and $[OsX(PPh_3)_3]$ [442,486] have since been shown to be hydridocarbonyls[432,443] (section 12.2.2).

Chatt and Hayter[488] prepared Ru(II) and Os(II) complexes, $[M_2Cl_3(PR_3)_6]$ Cl (R_3 = Ph_2Me, Ph_2Et, $PhMe_2$, $PhEt_2$), by reaction of $RuCl_3$ or $(NH_4)_2OsCl_6$ with the phosphine in aqueous ethanol. The orange or yellow crystals are air-stable, diamagnetic and 1:1 electrolytes in nitrobenzene. The ruthenium complexes readily lose the chloride by exchange reactions to give $[Ru_2Cl_3(PR_3)_6]$ X (X = SCN, ClO_4, BPh_4), but PR_3 does not convert them to $[Ru(PR_3)_4X_2]$. Symmetric chloro-bridged structures (XLIX) were proposed[488].

(XLIX)

The amalgamated zinc reduction of *mer*-$[OsCl_3(PR_3)_3]$ under argon gave a green solution which slowly became yellow, and upon adding the appropriate anion $[Os_2Cl_3(PR_3)_6]X$ ($X = PF_6, BF_4, ZnCl_3 \cdot THF$) precipitated[444]. The green intermediate solution reacts with various ligands to form $[OsL(PR_3)_3X_2]$ and $[OsL_2(PR_3)_2X_2]$ ($L = CO, MeCN, PhCN, PhNC, MeNC$)[444].

On heating $[Ru_2Cl_3(PPhEt_2)_6]Cl$ in esters or ketones in the absence of air, unusual bridged derivatives are obtained. In *n*-propyl propionate, dark red crystals of $Ru_2Cl_4(PPhEt_2)_5$ are formed[490,491], whilst in methyl acetate, red $[Ru_2Cl_3(PPhEt_2)_6][RuCl_3(PPhEt_2)_3]$ is the product[490,492]. The environment of the ruthenium in these complexes is distorted octahedral, the Ru—Cl bond lengths varying with the *trans* ligands, with Ru—Cl *trans* to P being larger than Ru—Cl *trans* to Cl. In $[Ru_2Cl_4(PPhEt_2)_5]$ (L)[491] the Ru—Cl (bridging) *trans* to the unique terminal chlorine

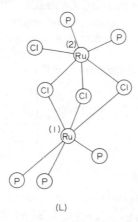

(L)

is 242.6 pm, compared with 247.2–253.5 pm for Ru—Cl (bridging) *trans* to P; Ru—Ru = 336.7 pm. The ionic $[Ru_2Cl_3(PPhEt_2)_6][RuCl_3(PPhEt_2)_3]$ has a distorted meridional-structured anion, the lattice consisting of alternating anions and cations in chains[492]. The cation is similar to the $[Ru_2Cl_4(PPhEt_2)_5]$, with Ru—Cl (bridging) = 249.9–245.8 pm and the Ru—Ru = 344.3 pm. The Ru—P distances in both compounds are ~232 pm for the Ru—P_3 grouping, with two shorter Ru—P distances on Ru(2) in the neutral complex.

A closely related complex is the paramagnetic (0.75 B.M. per Ru atom) $[Ru_2Cl_5(PBu^n_3)_4]$, a mixed Ru(II)–Ru(III) compound[493]. It was obtained by reacting $RuCl_3$ and PBu^n_3 in a 1:2.2 ratio, when $[RuCl_3(PBu^n_3)_2]_2$ crystallised out, and $[Ru_2Cl_5(PBu^n_3)_4]$ was obtained from the mother-liquor[494,495]. The Ru—Ru separation, 311.5 pm, is shorter than in the Ru(II) complexes, and once again the Ru—Cl

$$R_3P \diagdown \quad \diagup Cl \qquad \diagup PR_3$$
$$Cl —\!\!\!— Ru —\!\!\!— Cl —\!\!\!— Ru —\!\!\!— PR_3$$
$$R_3P \diagup \quad \diagdown Cl \diagup \qquad \diagdown Cl$$

(LI)

(bridging) *trans* to P is longer than Ru—Cl (bridging) *trans* to Cl (244.3–250.5 pm as against 237.2–237.6 pm) (LI).

Dark brown [RuCl$_2$(PPh$_3$)$_4$] is formed on shaking hydrated RuCl$_3$ with excess PPh$_3$ in methanol, whilst on refluxing this mixture, [RuCl$_2$(PPh$_3$)$_3$] is obtained[429]. The 'blue solution' of Ru(II), probably [Ru$_5$Cl$_{12}$]$^{2-}$, also forms [RuCl$_2$(PPh$_3$)$_3$] and [Ru$_2$Cl$_3$(PPh$_3$)$_6$] Cl directly from reaction with PPh$_3$[496]. Both complexes are reactive; [RuCl$_2$(PPh$_3$)$_4$] dissolves in benzene, turns green upon exposure to air and rapidly adds CO, norbornadiene, etc.[429]. [RuCl$_2$(PPh$_3$)$_3$] is an active hydrogenation catalyst in C$_6$H$_6$–EtOH solution, due to its conversion to RuHCl(PPh$_3$)$_3$ (q.v.). An X-ray structure determination by La Placa and Ibers[497] confirmed five-coordination with a distorted square-pyramidal structure; *trans* chlorines and *trans* phosphines in the basal plane, and apical phosphine, with one of the ortho-hydrogens blocking the sixth coordinating position. On heating [RuCl$_2$(PPh$_3$)$_3$] in acetone, red RuCl$_2$(PPh$_3$)$_2$(Me$_2$CO) precipitates, whilst reaction in higher ketones gives black [RuCl$_2$(PPh$_3$)$_2$]$_n$[498]. Far i.r. spectra of these products are almost identical, showing both bridging and terminal Cl, and this coupled with reaction with PPh$_3$ to give [RuCl$_2$(PPh$_3$)$_3$] suggests the structure (LII).

(LII)

Reactions of [RuCl$_2$(PPh$_3$)$_3$] with other reagents have been reported: with SO$_2$ to give red [RuCl$_2$(SO$_2$)(PPh$_3$)$_2$][499], with RCN and amines to form [RuCl$_2$(RCN)$_2$(PPh$_3$)$_2$] and [RuCl$_2$(am)$_2$(PPh$_3$)$_2$][498], and with β-diketones and carboxylates to give [Ru(β-diket)$_2$(PPh$_3$)$_2$] and [Ru(OCOR)$_2$(PPh$_3$)$_2$], respectively[498]. With CS$_2$, two products are formed: [RuCl$_2$(CS)(PPh$_3$)$_2$]$_2$ and [RuCl(CS$_2$)(PPh$_3$)]Cl, the former containing a thiocarbonyl, the latter with a π-CS$_2$ group[500]. When R ≠ Ph the [RuCl$_2$(PR$_3$)$_3$] complexes have generally not been isolated[501].

The [OsBr$_2$(PPh$_3$)$_3$] complex was described as a green, diamagnetic product formed on boiling (NH$_4$)$_2$OsBr$_6$ with PPh$_3$ in 2-methoxyethanol[444], and *trans*-[OsCl$_2$(PPhMe$_2$)$_4$] is formed along with the [Os$_2$Cl$_3$(PPhMe$_2$)$_6$]$^+$ by reduction of *mer*-[OsCl$_3$(PPhMe$_2$)$_3$][484]. Reaction of RuCl$_3$ and PHEt$_2$ in ethanol yields [RuCl$_2$(PHEt$_2$)$_4$][502].

Attempts to obtain simple anionic Ru(II) complexes have failed[501], the only one known being contained in [Ru$_2$Cl$_3$(PPhEt$_2$)$_6$][RuCl$_3$(PPhEt$_2$)$_3$]. Thiocarbonyl complexes, M[Ru(CS)(PPh$_3$)$_2$Cl$_3$] (M = Ph$_4$As, Et$_4$N) have been briefly described[503].

Complexes [RuCl$_3$(PPh$_3$)$_2$L] (L = MeOH, EtOH, Me$_2$CO) are formed from RuCl$_3$ and PPh$_3$ in the appropriate solvent[429,444]. Chatt *et al.*[428,504] prepared [RuCl$_3$(PR$_3$)$_3$] from RuCl$_3$, concentrated HCl and R$_3$P in ethanol, using much shorter reaction times than those required to give [Ru$_2$Cl$_3$(PR$_3$)$_6$]Cl (R$_3$ = Ph$_3$, PhMe$_2$, PhEt$_2$, PhBu$^n_2$). Winzer[505] showed that [RuCl$_3$(PPh$_3$)$_3$] could be formed simply by heating RuCl$_3$ with PPh$_3$. Osmium tetroxide in concentrated HX reacts with tertiary phosphines in a stepwise fashion giving $OsX_4(PR_3)_2$, $OsX_3(PR_3)_3$ and $[Os_2X_3(PR_3)_6]X$, successively[504]. The $[OsBr_3(PR_3)_3]$ are also obtained from the chlorides and LiBr, or by reduction of $OsOBr_3(PPh_3)_2$ and excess PR_3 (R_3 = Ph_3, Et_3, Pr^n_3, Bu^n_3, $PhMe_2$, $PhEt_2$, $PhPr^n_2$, $PhBu^n_2$, Ph_2Et)[504]. The osmium(III) complexes are orange or deep red paramagnetic solids (μ_{eff} = 1.9–2.2 B.M.)[504], and their i.r. and ^1H n.m.r. spectra are consistent with meridional structures[288]. The e.s.r. spectra of several Ru(III) and Os(III) complexes have been discussed[506]. When mer-$[OsCl_3(PPhBu^n_2)_3]$ is refluxed with hydrazine hydrate in C_6H_6–EtOH, the fac-$[OsCl_3(PPhBu^n_2)_3]$ can be obtained by precipitating with concentrated HCl[288], and fac-$[OsCl_3(PR_3)_3]$ (R_3 = $PhMe_2$, $PhEt_2$) are obtained as carmine crystals by decomposing $OsH_4(PR_3)_3$ with HCl in methanol[475]. Other M(III) complexes are the halogen-bridged dimers, $[RuCl_3(PR_3)_2]_2$ (R = Pr^n, Bu^n), μ_{eff} = 1.93 B.M. per Ru atom[493], and green $[OsCl_2(PPhMe_2)_3(MeNC)]Cl$, formed in small yield by chlorine oxidation of $[OsCl_2(PPhMe_2)_3(MeNC)]$[484].

Anionic Ru(III) complexes are well established. Excess PR_3 (R_3 = Et_3, $PhEt_2$) and carbonylated $RuCl_3$ solution in ethanol forms $[R_3PH][RuCl_4(PR_3)_2]$[429]. Stephenson[489,501] obtained $M[RuCl_4(PR_3)_2]$ (M = Me_4N, Ph_4As; R_3 = Ph_3, Et_3, $PhMe_2$). The triphenylphosphine complex was formed as red crystals from $[RuCl_3(PPh_3)_2MeOH]$ and MCl in acetone, but metathetical reaction with Br^-, I^- and SCN^- did not yield the expected $[RuX_4(PPh_3)_2]^-$. The same complex was obtained by reaction of $RuCl_2(PPh_3)_{3\ or\ 4}$ with Ph_4As^+, and not the Ru(II) complex which might have been expected; but due to lack of suitable precursors this method cannot be used for other phosphines. However, the $[RuCl_4(PPh_3)_2]^-$ reacts with undiluted PR_3 over several days to give a nearly quantitative yield of $[RuCl_4(PR_3)_2]^-$ (R_3 = Et_3, $PhMe_2$). The complexes are approximately 1:1 electrolytes, with magnetic moments in the range 1.80–2.20 B.M. The anions are easily solvolysed in $MeNO_2$ to $[RuX_3(PR_3)_2(MeNO_2)]$, and it is this solvolysis, rather than isomerism, which is responsible for the colour changes that occur in solution. Spectroscopic studies indicate that the original anion has a *trans* configuration[288,501]. The $[Ph_4As][trans\text{-}OsCl_4(PPhMe_2)_2]$ was isolated from the reaction products of $NaBH_4$ and $[OsCl_4(PPhMe_2)_2]$ (which reaction also produced $[OsH_6(PPhMe_2)_2]$)[475].

Ruthenium(IV) complexes have not been obtained[504], but $[OsX_4(PR_3)_2]$ (X = Cl, Br; R_3 = Pr^n_3, $PhMe_2$, $PhPr^n_2$, $PhBu^n_3$) are obtained by reaction of OsX_4 in concentrated HX with PR_3[504], by oxidation of mer-$[OsCl_3(PR_3)_3]$ with CCl_4[475,504], or, more easily, by chlorine oxidation of mer-$[OsCl_3(PR_3)_3]$ in CCl_4–CH_2Cl_2 under a fluorescent lamp[475]. The Os(IV) complexes have been shown to exhibit T.I.P., with μ_{eff} = 1.4–1.5 B.M., similar to the isoelectronic Re(III) d^4 complexes[289] and the charge-transfer spectra of a number of Os(IV), Ru(III) and Os(III) complexes have been reported and discussed[508]. The $[OsOCl_3(PPh_3)_2]$ and $[OsOBr_3(PPh_3)_2]$ are formed from OsO_4, PPh_3 and HX[404,481] as orange, non-conducting complexes, which were too insoluble for molecular weight determinations.

A number of phosphine complexes of ruthenium are active catalysts[473,509,510], but we shall not discuss this aspect of their chemistry.

13 COBALT, RHODIUM, IRIDIUM

13.1 COBALT

13.1.1 Carbonyls

Dicobalt octacarbonyl was originally reported[306] to react with triphenylphosphine to yield $[Co(CO)_3(PPh_3)]_2$, but other workers[511-13] have characterised the product as $[Co(CO)_3(PPh_3)_2][Co(CO)_4]$. The reaction in polar solvents, at room temperature or below, produces the dark red diamagnetic $[Co(CO)_3(PPh_3)_2][Co(CO)_4]$ [512-14], from which other complexes of the cation can be obtained by exchange, for example $[Co(CO)_3(PPh_3)_2]Y$ $(Y = BPh_4, 2Y^- = HgI_4{}^{2-})$. On the basis of the i.r. spectrum, the cation was suggested to have a trigonal-bipyramidal structure with *trans* phosphines[515] (compare the isoelectronic $Fe(CO)_3(PPh_3)_2$). The reaction of $Co_2(CO)_8$ and PPh_3 in non-polar solvents, or at higher temperatures, produces the neutral dimer $[Co(CO)_3(PPh_3)]_2$ [513,514], which is also formed on heating $[Co(CO)_3(PPh_3)_2][Co(CO)_4]$ above 40°C [514]. The i.r. spectrum suggested a metal–metal bonded structure with *trans* phosphines (LIII).

$$
\begin{array}{c}
\overset{\displaystyle O \;\;\;\; O \;\;\;\; O}{\overset{\displaystyle \|\;\;\;\;\; \|\;\;\;\;\; \|}{\overset{\displaystyle C \;\;\;\; C \;\;\;\; C}{}}} \\
Ph_3P\!-\!\!-\!\!-\!Co\!-\!\!-\!\!-\!Co\!-\!\!-\!PPh_3 \\
\overset{\displaystyle C \;\;\;\; C \;\;\;\; C}{\overset{\displaystyle \|\;\;\;\;\; \|\;\;\;\;\; \|}{\displaystyle O \;\;\;\; O \;\;\;\; O}}
\end{array}
$$

(LIII)

The preparation of both types of complex has been reported with other phosphines, including PEt_3 [227,516], PCy_3 [227,517], $P(p\text{-}C_6H_4X)_3$ $(X = Cl, Br)$ [518], $PBu^n{}_3$ [516,517,519], $PPhMe_2$ [516] and $PPhEt_2$ [516]. Manning[516] showed that the amount of each product depended upon the particular phosphine concerned and upon the reaction time. However, $P(NMe_2)_3$ produces[76] $[Co(CO)_3\{P(NMe_2)_3\}_2][Co(CO)_4]$, and only traces of a product thought to be the neutral dimer. The i.r. spectra of the dimeric complexes have been discussed by a number of workers[516,520,521], and the structure of $[Co(CO)_3(PBu^n{}_3)]_2$ has been determined[522,523]. The molecule consists of an unbridged Co–Co bond (266 pm) with *trans* phosphines, Co–P = 219 pm $\widehat{CoCoP} = 178°$. On the basis of i.r. spectral data it has been suggested[520] that $Co_2(CO)_{8-n}(PEt_3)_n$ can exist in two forms in solution, bridged $n = 1, 2$, and non-bridged $n = 1-3$, and this behaviour has subsequently been observed with other phosphines[516]. The oxidation of $[Co(CO)_3(PPh_3)_2]^+$ to form the salt $[Co(CO)_3(PPh_3)_2][CoCl_4]$ has been reported[512].

Bor and Markó[524], using nujol as the reaction medium, found that $Co_2(CO)_8$ reacted with PPh_3 to give $[Co_2(CO)_7(PPh_3)]$, separable by chromatography, in addition to the higher substitution products; they later extended this work to other phosphines[517]. However, PBu_3^n and PCy_3 do not give the stable monosubstituted derivatives[517], owing to the reaction

$$Co_2(CO)_7(PR_3) \rightleftharpoons Co_2(CO)_8 + Co_2(CO)_6(PR_3)_2$$

Infrared spectral data indicates that the phosphine is *trans* to the Co—Co bond.

On working up the residue from a catalytic system consisting of Grignard reagent and cobalt stearate in the presence of CO and PPh_3, Simon *et al.*[525] obtained, in addition to known cobalt carbonyl phosphines and carbonyl halides, a product thought to be $[Co_2(CO)_2(PPh_3)_6]$. The $[Co(N_2)(PPh_3)_3]$ and $[CoH_2(PPh_3)_3]$ complexes are decomposed by CO to $[Co_2(CO)_2(PPh_3)_6]$, in contrast to $[CoH_3(PPh_3)_3]$, which yields principally $[Co_2(CO)_6(PPh_3)_2]$[526]. There are no other reports of highly substituted products. The reaction with $Co_4(CO)_{12}$ has been little studied, though the monosubstitution product, $Co_4(CO)_{11}(PPh_3)$, was separated by thin-layer chromatography from the reaction products of $Co_4(CO)_{12}$ and PPh_3 in petroleum ether[527]. The only other polynuclear phosphine-substituted carbonyls, seem to be the green $[Co(CO)_2(PR_3)]_n$ ($R_3 = Ph_3$, Ph_2Bu^n, Bu_3^n), formed in the hydroformylation of olefins using $HCo(CO)_3(PR_3)$ as catalyst[528]. These reactions yield paramagnetic, oily products which, on the basis of i.r. spectra and molecular weight determinations, are thought to have the structure (LIV). Polymeric

(LIV)

$[Co(CO)_2(PPh_3)]_n$ had been previously reported to be formed on boiling $[Co(CO)_3(PPh_3)_2][Co(CO)_4]$ with LiX in acetone[529].

Hayter investigated[180] phosphido-bridged cobalt carbonyls, and obtained $[Co(CO)_3(PPh_2)]_2$ and $[Co_3(CO)_7(PMe_2)_2]$ from $Co_2(CO)_8$ and R_4P_2. Diphenylchlorophosphine and $Na[Co(CO)_4]$ also gave $[Co(CO)_3(PPh_2)]_2$, but Me_2PCl gave only red, oily products which could not be purified[180]. The structure of the dicobalt complexes is probably that containing two phosphido bridges and six terminal carbonyls. The structure of the green $Co_3(CO)_7(PMe_2)_2$ is not known, although a Co_3 triangular skeleton with phosphido and carbonyl bridges seems the most likely[180]. Hieber and Duchatsch[530] also reported a polymeric $[Co(CO)_3(PPh_2)]_n$, and suggested n was 4–6.

Red $[Co_2(CO)_5PPh_3As_2]$ is formed on treating $Co_2(CO)_6As_2$, itself formed from $(AsMe)_5$ and $Co_2(CO)_8$, with PPh_3 in refluxing benzene[531]. The essential structure is shown on page 90 (LV), and a remarkable point is the very short As—As bond (227.3 pm). $[Co_6(CO)_{18-n}C_4(PR_3)_n]$ ($n = 1, 2$, R = Ph, Cy; and $n = 2–5$, $R_3 = PhEt_2$)[532] can be obtained from $Co_6(CO)_{18}C_4$, and $CH_3CCo_3(CO)_9$ gives $[CH_3CCo_3(CO)_8(PPh_3)]$ by direct reaction[533], and in the case of the Co_6-PPh_3 complexes by using $Pt(PPh_3)_4$ as a source of PPh_3.

Cobalt carbonyl hydride, $HCo(CO)_4$, reacts directly with PPh_3 to give $[HCo(CO)_3(PPh_3)]$ [534], the reaction being so rapid that the rate could not be determined even at low temperature. It can also be obtained by acidification

(LV)

$(HCl-Et_2O)$ of $Na[Co(CO)_3(PPh_3)]$ [535,536]. It decomposes above $\sim20°C$, which makes it a good deal more stable than $HCo(CO)_4$, which begins to decompose at about $-20°C$; it reacts with heavy metal salts to form $M\{Co(CO)_3(PPh_3)\}_2$ (M = Zn, Cd, Hg, Sn). A second hydride was obtained by acidification of the corresponding sodium salt [537], and recently three other hydrides, $HCo(CO)_4$, which begins to decompose at abov been obtained [519,538] by the pressure hydrogenation of $[Co(CO)_3(PBu^n_3)]_2$ in heptane solution containing PBu^n_3.

The reduction of $[Co(CO)_3(PR_3)]_2$ by Na–Hg in THF proceeds readily and yellow, diamagnetic 1:1 electrolytes, $Na[Co(CO)_3(PR_3)]$, are formed [535]; the disubstituted anions are similarly obtained by reduction of $[Co(CO)_2(PR_3)_2X]$ [537]. Both types of anion react with MeI, $Hg(CN)_2$, etc., to give $MeCo(CO)_{4-n}(PR_3)_n$, $Hg\{Co(CO)_{4-n}(PR_3)_n\}_2$ (n = 1, 2).

As well as by the above method, $Hg\{Co(CO)_3(PR_3)\}_2$ are also obtained by reaction of $Hg\{Co(CO)_4\}_2$ with PR_3 ($R_3 = Bu^n_3$ [539], Ph_3, Et_3, Cy_3 [540], $(NMe_2)_3$ [76], Ph_2H, $PhMe_2$, $PhEt_2$, $(n-octyl)_3$ [541]; they are yellow, air-stable solids. The i.r. spectra of these complexes have been studied [521,541], and the structure of $Hg\{Co(CO)_3(PEt_3)\}_2$ has been determined by X-ray methods [522]. The structure consists of two $Co(CO)_3(PEt_3)$ moieties bridged by the mercury atom, Hg—Co = 249.9(5) pm, identical with that found in $Hg\{Co(CO)_4\}_2$, showing that substitution of a PEt_3 molecule for a *trans* carbonyl group has no measurable effect on the Hg—Co bond. Zinc, cadmium, thallium and tin complexes were obtained analogously by Hieber and Breu [540], and the tin complexes have been examined in more detail by Bonatti *et al.* [542,543]. Stannous chloride and bromide react with $[Co(CO)_3(PR_3)]_2$ or $[Co(CO)_3(PR_3)_2][Co(CO)_4]$ in acetone to yield orange-yellow $X_2Sn\{Co(CO)_3(PR_3)\}_2$, from which the diethyl- or diphenyltin derivatives can be obtained by the Grignard method. The mechanism of the $[Co(CO)_3(PBu^n_3)]_2-SnX_2$ reaction has been studied [544], and some related complexes, $RSnX_2Co(CO)_3(PPh_3)$, described [545]. The cobalt–germanium complex, $[Ph_3GeCo(CO)_3(PPh_3)]$, has also been studied; Co—Ge = 234 pm [546].

13.1.2 Carbonyl halides
The carbonylation of cobalt(II) phosphine complexes leads to carbonyl halide derivatives. In the presence of copper, $[Co(PPh_3)_2I_2]$ is carbonylated to $[Co(CO)_3(PPh_3)_2][CuI_2]$ [512], and under somewhat different conditions

$Co(CO)_2(PPh_3)_2I$ is the product[529]. Booth and Chatt found[352] that $[Co(PEt_3)_2X_2]$ (X = Cl, Br, I) readily absorb CO in solution in organic solvents at ambient temperature and pressure to give dark brown $[Co(CO)(PEt_3)_2X_2]$. These are less stable than the iron analogues, decomposing on exposure to air. Bressan *et al.*[547] found that reaction of $[Co(PR_3)_2X_2]$ in benzene or dichloroethane with CO produces five-coordinate $[Co(PR_3)_2(CO)X_2]$ ($R_3 = Et_3$, Pr^n_3, $PhEt_2$, Ph_2Et, X = Cl, Br, I, NCS; R = Ph, X = Br, I, NCS; R = Cy, X = NCS). More carbon monoxide can be absorbed to form Co(I) complexes, $[Co(CO)_2(PR_3)_2X]$ [547], which can also be synthesised from $Co(CO)_4X$ and PR_3 [529], or by mild halogenation of $Na[Co(CO)_3(PR_3)]$ [548]. Solutions of the five-coordinate $[Co(PR_3)_3(CN)_2]$ also take up CO to form $[Co(CO)(PR_3)_2(CN)_2]$ and $[Co(CO)_2(PR_3)_2(CN)]$ [547].

Tricarbonyl complexes, $[Co(CO)_3(PPh_3)X]$, are obtained by mild halogenation of $Na[Co(CO)_3(PPh_3)]$ with $POCl_3$ at $-70°C$, CF_3I or N-bromosuccinimide[548,549]. Both the $[Co(CO)_3(PPh_3)_2]^+$ and the halides $[Co(CO)_3(PPh_3)X]$ have trigonal-bi-pyramidal structures with equatorial carbonyls[529,550].

The $Co(CO)_2(PPh_3)_2(H_2O)X$ were prepared by halogen oxidation of $[Co(CO)_3(PPh_3)_2][Co(CO)_4]$, followed by addition of water, when the yellow Co(I)–aquo complexes are obtained (X = Cl, Br, I)[551].

13.1.3 Nitrosyls

Tertiary phosphines will substitute carbonyl groups in $Co(NO)(CO)_3$ to give $[Co(NO)(CO)_2(PR_3)]$ and $[Co(NO)(CO)(PR_3)_2]$ ($R_3 = Ph_3$ [363,552,553], Cy_3 [553,554], Et_3 [554], Ph_2Cl [552], $(p\text{-}MeOC_6H_4)_3$ [554], $PhBu_2$ [555], $(CF_3)_3$ [556]). Phosphorus tri-chloride and $PPhCl_2$ give red liquid products, which are inseparable mixtures of the mono- and disubstituted complexes[552]. The bulky $(Me_3E)_3P$ (E = C, Si, Ge, Sn) form only $[(Me_3E)_3PCo(NO)(CO)_2]$ [96,319]. Phosphine also forms the monosubstituted $[Co(NO)(CO)_2(PH_3)]$, an orange-red liquid, by direct reaction[557]. In addition to the normal substitution products, $(CF_3)_3P$ also gives small yields of a trimer, $[(CF_3)_3PCo(NO)(CO)]_3$ [556]. The kinetics of, and solvent effects upon, substitution reactions of $[Co(NO)(CO)_3]$ with PR_3 have been discussed[554,558,559]. The last carbonyl group in $[Co(NO)(CO)(PR_3)_2]$ is not substituted directly, but $[Co(NO)(PR_3)_3]$ (R = Ph, Et) are obtained by sodium amalgam reduction of $[Co(NO)_2(PR_3)X]$ in the presence of PR_3 [560].

Disproportionation occurs when $[Co(NO)_2Cl]_2$ dissolves in molten PPh_3, yielding $[Co(NO)(PPh_3)_3]$, $[Co(OPPh_3)_2Cl_2]$, N_2 and $OPPh_3$ [560]. By dipole moment measurements and i.r. spectroscopy, $[Co(NO)(PPh_3)_3]$ was shown to be tetra-hedral[560]. A phosphido-bridged $[Co(NO)_2(PPh_2)]_2$ is formed by treatment of $[Co(NO)_2Cl]_2$ with $KPPh_2$ [365]; sulphido-bridged $[Co(NO)(PR_3)SR]_2$ are also known[561], as are the monomeric $[Co(NO)_2(PR_3)(SR)]$ [561].

Nitrosyl halides, $[Co(NO)(PR_3)_2X_2]$, are obtained from $Co(PR_3)_2X_2$ and NO in benzene (R = Et) [352]. The darkly coloured, monomeric, diamagnetic $[Co(NO)_2(PR_3)X]$ complexes are formed from $[Co(NO)_2X]_2$ and PR_3 (R = Ph, Et, Cy) [560,562]. The i.r. spectra of a number of them have been studied[360]. The unusual $[Co(NO)_2(PPh_3)Co(CO)_3(PPh_3)]$ is obtained from $[Co(NO)_2(PPh_3)Br]$ and

$Na[Co(CO)_3(PPh_3)]$ [374]. The $[Co(NO)_2(PPh_3)(CN)]$ complex, which results from treating the bromide with NaCN, disproportionates into $[Co(NO)_2(PPh_3)_2][Co(NO)_2(CN)_2]$ [368].

Bianco *et al.* [563] reinvestigated the reaction of $[Co(NO)_2X]_2$ (X = Cl, Br, I) with PPh_3, and found that an equilibrium mixture of $Co(NO)_2(PPh_3)_2X$ and $Co(NO)_2(PPh_3)_2X$ were formed, the equilibrium being displaced in favour of the latter by adding ClO_4^- or BPh_4^-. In the presence of PPh_3 either complex could be reduced by $NaBH_4$ or zinc to $[Co(NO)(PPh_3)_3]$. Shiny black crystals of $[Co(NO)_2(PPh_3)_2]ClO_4$ have been obtained by reaction of $[Co(NO)_2(p\text{-toluidine})_2]ClO_4$ with PPh_3 and NO in CH_2Cl_2 [564]. On heating $[Co(NO)_2(PPh_3)_2]ClO_4$ with KNO_2 the nitrito complex, $[Co(NO)_2(PPh_3)(NO_2)]$, is formed. It also[565] results by disproportionation of nitric oxide in the reaction

$$Co(NO)(PPh_3)_3 + 7NO \xrightarrow[\text{temp.}]{\text{room}}$$

$$Co(NO)_2(PPh_3)(NO_2) + 2PPh_3O + 2N_2O + \tfrac{1}{2}N_2$$

13.1.4 Hydrides and dinitrogen complexes

Cobalt hydrido complexes have recently attracted a great deal of interest following the observations that they react with N_2 to give stable dinitrogen complexes.

Sacco and Rossi[566] found that $NaBH_4$ in ethanol reduces $[Co(PR_3)_2X_2]$ in the presence of excess PR_3 to yield the yellow $[CoH_3(PR_3)_3]$ (R_3 = Ph_3, $PhEt_2$, Ph_2Et). The complexes are unstable in air, soluble in non-polar solvents, and exhibit two i.r. absorptions at 1720–1745 cm^{-1} and 1933–1958 cm^{-1} consistent with an octahedral structure with two *trans* hydrogens. In the solid state or in solution these complexes readily absorb nitrogen

$$CoH_3(PR_3)_3 + N_2 \rightleftharpoons CoH(N_2)(PR_3)_3 + H_2$$

The orange-red crystalline $[CoH(N_2)(PPh_3)_3]$, $\nu(N\equiv N)$ = 2096 cm^{-1}, reacts as shown in figure 15. The i.r. spectrum of this complex does not show $\nu(Co-H)$, but the presence of the hydride ligand is confirmed by various reactions (figure 15)[566].

Phosphines react with π-cyclooctenyl-π-cycloocta-1,5-diene cobalt, $[Co(C_8H_{13})(C_8H_{12})]$, in pentane or benzene under argon to establish several equilibria involving displacement of the organic ligands and the formation of $CoH(PR_3)_3$, $CoH(PR_3)_4$, $CoH_3(PR_3)_3$, and under N_2 to form $CoH(N_2)(PR_3)_3$ (R_3 = Ph_3, Ph_2Me, Ph_2Et, Ph_2Bu, $PhEt_2$, $PhBu_2$, Bu_3) [567].

Yamamoto *et al.*[568] independently prepared $[CoH(N_2)(PPh_3)_3]$, initially formulated as $Co(N_2)(PPh_3)_3$, by reduction of $Co(acac)_3$ with $Al(OEt)Et_2$ under nitrogen

(LVI)

in toluene containing PPh$_3$. This complex did not show a ν(Co–H) absorption. Misono et al.[569] using AlBu$^i_3$ as reducing agent apparently obtained the same complex, but suggested that it was a mixture of [Co(N$_2$)(PPh$_3$)$_3$] and [CoH(N$_2$)(PPh$_3$)$_3$]. Subsequently, Misono et al.[570] showed that [Co(N$_2$)(PPh$_3$)$_3$] and [CoH(N$_2$)(PPh$_3$)$_3$] were identical, the latter being the correct formulation. The structure of [CoH(N$_2$)(PPh$_3$)$_3$] (LVI) has been determined[571,572] and has been shown to be trigonal bipyramidal containing equatorial phosphines. There are two independent molecules in each unit; N–N = 101, 112.3 pm; Co–N = 178, 183 pm.

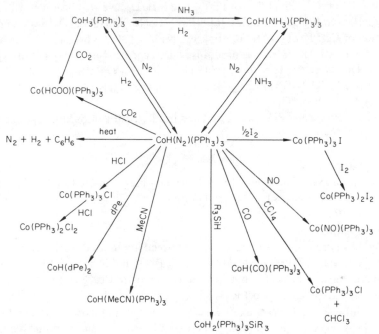

Figure 15 Some reactions of [CoH(N$_2$)(PPh$_3$)$_3$][566,568,573–6]

A number of reactions of [CoH(N$_2$)(PPh$_3$)$_3$] have been studied (figure 15). Several of the products are also obtained from [CoH$_3$(PPh$_3$)$_3$]; for example the green formate, [Co(HCOO)(PPh$_3$)$_3$][573], is formed in the first reported example of CO$_2$ insertion into a transition metal–hydrogen bond. The first reported examples of silyl complexes of the 3d transition metals in a high formal oxidation state are those of [(R$_3$Si)CoH$_2$(PPh$_3$)$_3$] (R = F, OEt), formed from [CoH$_3$(PPh$_3$)$_3$] or [CoH(N$_2$)(PPh$_3$)$_3$] and R$_3$SiH[575]. The preparation and reactions of [CoH$_3$(PR$_3$)$_3$] and [CoH(N$_2$)(PR$_3$)$_3$] have been studied by several other groups[577–83]. Especially interesting is the reduction of N$_2$O, an extremely inert gas at ambient temperature and pressure, to N$_2$ with concomitant formation of PPh$_3$O[568,583]. The [CoH(N$_2$)(PPh$_3$)$_3$] complex is an active catalyst[568,584], and a number of related organic species—MeCo(PPh$_3$)$_3$ and Co(C$_2$H$_4$)(PPh$_3$)$_3$—have also been obtained[568].

Zinc reduction of Co(II) salts in alkaline solution in the presence of PMe_3 leads to the formation of $[CoH(PMe_3)_4]$ [377]. Different dihydrido and dinitrogen complexes have been reported by Speier and Markó[526]. Reduction of $Co(acac)_2$ and PPh_3 with $Al(OEt)Et_2$ under H_2 or N_2 yields $[Co(H_2)(PPh_3)_3]$ and $[Co(N_2)(PPh_3)_3]$, respectively. The reaction of $[Co(N_2)(PPh_3)_3]$ with CCl_4, which yields no $CHCl_3$, with CO, which gives $[Co_2(CO)_2(PPh_3)_6]$, and the fact that very little hydrogen is evolved on thermal decomposition, seems to indicate that it is not the same $[CoH(N_2)(PPh_3)_3]$ reported above (q.v.). Both $[Co(N_2)(PPh_3)_3]$ and $[CoH_2(PPh_3)_3]$ are reported to be more stable than $[CoH(N_2)(PPh_3)_3]$ and $[CoH_3(PPh_3)_3]$. The reactions of $[Co(N_2)(PPh_3)_3]$ with PBu^n_3 and n-BuLi have also been described[585,586].

The dinitrogen-bridged complex $[(Et_2PhP)_3Co(N_2)Co(PPhEt_2)_3]$ was synthesised[587] by reaction of $Co(PPhEt_2)_2Cl_2$, $PPhEt_2$ and Na in a $1:1:2$ ratio under N_2 in THF solution. When a $1:1:3$ ratio was used, black diamagnetic $Na[Co(N_2)(PPhEt_2)_3]$ was produced. The latter reacts with water to give $[CoH(N_2)(PPhEt_2)_3]$, with CO at $-60°C$ to give $Na[Co(CO)(PPhEt_2)_3]$, and with $Co(PPh_3)_3Cl$ to form $[(Et_2PhP)_3Co(N_2)Co(PPh_3)_3]$.

When $HCo(PF_3)_4$ and PH_3 are mixed and irradiated the phosphine complex, $[HCo(PF_3)_3(PH_3)]$, forms[588].

13.1.5 Halides

Cobalt forms an extensive series of phosphine complexes in the Co(II) state, but few Co(I) or Co(III) complexes have been reported.

(1) Cobalt(I)

The formation of $Co(PR_3)_3X$ as an intermediate in the $NaBH_4$ reduction of $Co(PR_3)_2X_2$ (to $CoH_3(PR_3)_3$) was noticed by Rossi and Sacco[589]. Electrolytic reduction or, better, the reaction of powdered zinc with CoX_2 and PR_3 in ethanol was also found to produce $[Co(PR_3)_3X]$ ($R_3 = Ph_3, Ph_2Bu, Ph_2(CH_2Ph)$) as green crystals, soluble in C_6H_6, THF and CH_2Cl_2 [590]. In solution under nitrogen they partially decompose to give $Co(PR_3)_2X_2$, Co and PR_3, but the solutions oxidise rapidly under air, in which the solids are fairly stable. They take up CO to form $Co(CO)_2(PR_3)_2X$, but, unlike the dPe analogues, do not react with hydrogen; CCl_4 converts them to $Co(PR_3)_2X_2$ with formation of C_2Cl_4. The room temperature magnetic moments of $Co(PPh_3)_3X$, 3.11–3.32 B.M., and the electronic spectra both in solution and in the solid state, are consistent with pseudotetrahedral structures[590] (compare the isoelectronic d^8 $Ni(PR_3)_2X_2$).

(2) Cobalt(II)

Jensen[378] reported blue $[Co(PEt_3)_2Cl_2]$ formed from $CoCl_2$ and PEt_3 in ethanol, with a dipole moment of 8.7 D, indicating either a *cis* square-planar or a tetrahedral structure. Subsequently[591] the magnetic moment determination ($\mu_{eff} = 4.39$ B.M.) indicated a tetrahedral configuration. Many other phosphines have been used to prepare $Co(PR_3)_2X_2$ complexes, usually by direct reaction of CoX_2 with PR_3 in alcohol (table 9). Generally they are green, blue or brown, with magnetic moments and electronic spectra indicative of tetrahedral structures. The reported magnetic moments, 4.3–4.8 B.M., are considerably higher than the spin-only value (3.87 B.M.), indicating considerable orbital contribution to the observed value. The i.r. spectra of $Co(PPh_3)_2X_2$ are also consistent with pseudotetrahedral structures[592].

Cotton *et al.*[593] studied the electronic spectra of the tricyclohexyl- and triphenyl-

TABLE 9 PHOSPHINE COMPLEXES OF COBALT(II) HALIDES

PR_3	X	Properties	Ref.
Tetrahedral $Co(PR_3)_2X_2$			
PMe_3	Cl,Br	unstable, also $CoI_2(PMe_3)_{2.5}$	595,602
PEt_3	Cl,Br,I,NCO	μ_{eff} = 4.39–4.74 B.M.; D.M. = 8.7 D(Cl)	352,378, 591,595,602
	NCS	μ_{eff} = 2.2 B.M.; absorbs PEt_3 in solution	598–600,608
PPr^n_3	Cl		378,595
PBu^n_3	Cl	oily product	378
$P(tolyl)_3$	Cl	μ_{eff} = 4.40 B.M.	630
PCy_3	Br,I,NCS	μ_{eff} = 4.40–4.63 B.M.	388,593,598,629
PPh_3	Cl,Br,I,NCS	μ_{eff} = 4.3–4.6 B.M.; D.M. = 7.6–9.6 D	592–7,620, 505,625
PPh_2Et	Cl,Br,NCS	μ_{eff} = 4.4–4.5 B.M.	596,601,604
$PPhEt_2$	Cl,Br	μ_{eff} = 4.5 B.M.	596,602
$PPh(NEt_2)_2$	Cl,NCS	μ_{eff} = 4.34, 4.53 B.M.	628
TMP†	Cl,Br,I,NCS,NO_3	μ_{eff} = 4.41–4.83 B.M.	605
(2-phind)‡	Cl,Br,I	Cl, Br dihydrates	607
Five-coordinate $Co(PR_3)_3X_2$			
PPr_3	NCS	μ_{eff} = 2.05 B.M.; dissociates in solution	601
PPh_2Et	CN	μ_{eff} = 2.0 B.M.; i.r., visible spectra	604
$PPhEt_2$	NCS	μ_{eff} = 1.90 B.M.	601
	CN	μ_{eff} = 2.0 B.M.; i.r., visible spectra	604
(9-phos)§	Br	μ_{eff} = 1.95–2.03; 9-Me, 9-Et compounds	608
(2-phind)‡	Cl,Br,I	Cl is monohydrate	607

† TMP = trimorpholinophosphine
‡ 2-phind = 2-phenylisophosphindoline
§ 9-phos = 9-alkyl-9-phosphafluorenes

phosphine complexes both in solution and in the solid state and the magnetic susceptibilities at varying temperatures, and deduced an order for these ligands in the spectrochemical series. Isslieb and Mitscherling[594] reported green and brown forms of $Co(PPh_3)_2I_2$ both with μ_{eff} = 4.4 B.M. and high dipole moments, 8.08 D and 8.25 D respectively, indicating tetrahedral structures in both cases. The brown form had the expected molecular weight in benzene solution, but the green complex apparently dissociated in acetophenone (molecular weight: found 543.5, calculated 837.3)[594]. Other workers have only reported the brown form[593,595,596].

As can be seen from table 9, the thiocyanate complexes often differ from the halides in their magnetic properties. The $[Co(PR_3)_2(NCS)_2]$ (R = Ph, Cy) are normal, pseudotetrahedral complexes[597,598], but when R = Et unusual behaviour is observed. In solution, the magnetic properties and the electronic and i.r. spectra of $[Co(PEt_3)_2(NCS)_2]$ depend upon the temperature and concentration[599,600], which has been interpreted in terms of a low-spin ⇌ high-spin equilibrium. The high-spin form has been identified as tetrahedral with N-bonded CNS^- groups, and the low-spin form as a dimeric planar species. Solid $[Co(PEt_3)_2(NCS)_2]$ has μ_{eff} = 2.2 B.M., a possible structure being a five-coordinate dimer containing $-SCN-$ bridges[598]. In view of this Boschi *et al.*[601] studied the addition of PR_3 to $Co(PR_3)_2X_2$ to give five-coordinate complexes. In agreement with previous studies[591,602], no evidence for five-coordination was found in the halide systems, or in the thiocyanate complexes with PPh_3 or PCy_3. The $Co(PEt_3)_2(NCS)_2-PEt_3$ system in CH_2Cl_2 contains a five-coordinate complex on the basis of magnetic and spectral evidence, but solid $[Co(PEt_3)_3(NCS)_2]$ could not be isolated[601]. However, when PPr^n_3 and $PPhEt_2$ were used $[Co(PR_3)_3(NCS)_2]$, with magnetic moments consistent with one unpaired electron, were isolated. The importance of steric and electronic factors has been discussed[601,603].

The cyano derivatives are also five-coordinate low-spin complexes, obtained by passing $[Co(PR_3)_2X_2]$ mixed with free phosphine in CH_2Cl_2 solution through an ion exchange resin in the CN^- form[604]. The red complexes are very soluble and only the $PPhEt_2$ and PPh_2Et derivatives have been isolated, although the PEt_3, PPr_3 and PBu_3 complexes were identified in solution by visible spectroscopy. Complexes with more unusual phosphines have also been studied. Trimorpholinophosphine (TMP) forms moisture-sensitive complexes, which probably contain the ligand acting only as a P donor[605], but $P(CH_2CH_2CN)_3$ forms $(CoX_2)_3 \cdot 2P(CH_2CH_2CN)_3$, in which it is bonded to the cobalt only by the nitrogen atoms[606]. Five-coordinate, presumably square-pyramidal complexes, $[CoL_3X_2]$, have been obtained from 2-phenylisophosphindoline[607] and the 9-alkyl-9-phosphafluorenes[608]. On boiling the former complexes in benzene, bis complexes, $[CoL_2X_2]$, separate as blue crystals, but direct preparation of the bis complexes by mixing CoX_2 and 2L is unsatisfactory, mixtures being obtained.

Quagliano *et al.*[609] began a fruitful area of research in preparing complexes of the monoquaternised ditertiary phosphines, $Ph_2PCH_2P^+Ph_2(CH_2Ph)$ and $Ph_2PCH_2CH_2P^+Ph_2(CH_2Ph)$, $[CoLX_3]$ (X = Cl, Br, I). On the basis of magnetic, spectroscopic and X-ray powder diffraction studies these complexes were formulated as high-spin pseudotetrahedral species. Similar complexes have been prepared by Berglund and Meek using ligand (LVII)[610], and Nelson *et al.* have used ligand (LVIII)[611].

(LVII)   (LVIII)

n = 1, 2

Horrocks *et al.*[612–15] studied the ligand exchange rates and isotropic proton shifts of a number of cobalt bisphosphine complexes by n.m.r. spectroscopy, and Pignolet and Horrocks[616,627] investigated the diastereoisomeric complexes $[CoL_2I_2]$ (L = PPh Bu(*p*-anisyl)).

Thermogravimetric measurements on a number of $[Co(PR_3)_2X_2]$ complexes in an oxygen-containing atmosphere have shown that decomposition yields Co_3O_4 and P_2O_5 [617], whilst the slow autoxidation of $[Co(PEt_3)_2Cl_2]$ has been shown to yield $[Co(OPEt_3)_2Cl_2]$ as the sole product[618]. The low-temperature polarised spectrum of $[Co(PPh_3)_2Cl_2]$ has been studied[619].

Anionic complexes $[Co(PPh_3)X_3]^-$ (X = Br, I) were obtained by adding a large cation to a solution of CoX_2–PPh_3 in *n*-butanol or ethyl acetate[593]. They form blue or green crystals, which were shown by magnetic and spectroscopic studies to be pseudotetrahedral[592,593,620].

Displacement of phosphine occurs when $[Co(PR_3)_2X_2]$ and $[Co(PR_3)X_3]^-$ are dissolved in coordinating solvents such as acetonitrile[621]. Even in the presence of a large excess of the phosphine, phosphine-containing species are only partially formed. As the authors point out[621], remarkably little work has been done on the stability of phosphine complexes in solution, and further study should be well worthwhile.

Organometallic complexes $[R_2Co(PR_3)_2]$ have been obtained from $[Co(PR_3)_2Br_2]$ and the appropriate Grignard reagent[596,622]. The mesityl complex has been studied[623,624], and it has a *trans* square-planar structure, with the fifth and sixth coordination positions being blocked by the *o*-methyl groups.

White and Farona have reported the reaction of triphenylphosphine with bis-(dithioacetylacetonato)cobalt(II) to yield $[Co(sacsac)_2(PPh_3)]$, which was assigned a square-pyramidal structure[631].

The work that has been done on primary and secondary phosphine complexes has revealed interesting differences between these and tertiary phosphine complexes.

Isslieb and Doll[145] found that $PHCy_2$ forms 2:1 complexes $[Co(PHCy_2)_2X_2]$, which were non-electrolytes and had magnetic and dipole moments consistent with tetrahedral configurations. Diethylphosphine forms 4:1 complexes $[Co(PHEt_2)_4X_2]$ [145] and 2:1 $[Co(PHEt_2)_2X_2]$ [1349]. The former complexes have magnetic moments corresponding to one unpaired electron and dipole moments (5.6 D) corresponding to an octahedral structure, while on the basis of electronic spectra the latter have been assigned tetrahedral structures[591].

Diphenylphosphine reacts with anhydrous $CoBr_2$ in absolute ethanol to give three different complexes[146]: brown $[Co(PHPh_2)_3Br_2]$, green $[Co(PHPh_2)_3Br]Br$ and yellow $[Co(PHPh_2)_4Br]_2Br_2$. The brown complex, $\mu_{eff} = 2.01$ B.M., is a non-electrolyte and is thus five-coordinate. This was confirmed by an X-ray structure determination by Bertrand and Plymale[632], who also prepared the iodide derivative. The structure of the bromide is intermediate between trigonal bipyramidal and square pyramidal, with two bromines and one phosphorus atom essentially coplanar, the other two phosphines being in axial positions. The green $[Co(PHPh_2)_3Br]Br$ is a 1:1 electrolyte and the cation is tetrahedral[146].

Ethylphosphine reacts directly with $CoCl_2$ in the absence of a solvent to give

$[Co(PH_2Et)_2Cl_2]$, from which half the ethylphosphine can be removed by pumping to give $Co(PH_2Et)Cl_2$ [1349]. The bis ligand complex has $\mu_{eff} = 4.30$ B.M., indicating a tetrahedral structure. With PH_2Cy 3:1 complexes, $Co(PH_2Cy)_3X_2$, result, but the high magnetic moments, ~3.13 B.M., have led to the formulation $[Co(PH_2Cy)_6][CoX_4]$ [148]. Phenylphosphine forms still other different complexes [147]: dark brown $[Co(PH_2Ph)_2X_2]$ (X = Cl, Br), non-electrolytes with $\mu_{eff} = 2.5$ B.M. (Br), which is too low for a tetrahedral configuration and hence a square-planar structure has been proposed. With CoI_2 a yellow $[Co(PH_2Ph)_4I_2]$ is obtained, the magnetic moment (1.83 B.M.) and dipole moment (6.5 D) of which suggest a *cis* octahedral structure [147]. On heating in ethanol the complex loses HI and forms the phosphido complex $[Co(PH_2Ph)_3(PHPh)I]$, $\mu_{eff} = 2.54$ B.M. Table 10 contains a list of some of these complexes.

TABLE 10 PRIMARY AND SECONDARY PHOSPHINE COMPLEXES OF COBALT

Compound	μ_{eff} (B.M.)	Properties	Ref.
$[Co(PHCy_2)_2Br_2]$	4.85	D.M. = 7.89, tetrahedral	145
$[Co(PHEt_2)_4Br_2]$	1.97	D.M. = 5.63, octahedral?	145
$[Co(PHEt_2)_2Cl_2]$		electronic spectra, tetrahedral	1349
$[Co(PHPh_2)_3Br_2]$	2.01 (2.43)	X-ray data, five-coordinate	146,632
$[Co(PHPh_2)_3I_2]$	2.23	five-coordinate	632
$[Co(PHPh_2)_3Br]$ Br	3.37	1:1 electrolyte, tetrahedral	146
$[Co(PH_2Et)_2Cl_2]$	4.30	1:1 electrolyte, tetrahedral	1349
$[Co(PH_2Cy)_6][CoBr_4]$	3.13	octahedral cation	148
$[Co(PH_2Ph)_2Br_2]$	2.5	square planar?	147
$[Co(PH_2Ph)_4I_2]$	1.83	D.M. = 6.5 D; *cis*-octahedral?	147
$[Co(PH_2Ph)_3(PHPh)I]$	2.54		147

These complexes are very varied, but in general there is insufficient evidence available to determine the important factors that produce the particular structures. Most of the complexes were identified by elemental analysis, conductivity and magnetic measurements, aided in some cases by a dipole moment and molecular weight determination. It is unfortunate that with one or two exceptions the electronic spectra have not been reported. The $[Co(PHPh_2)_3X_2]$ complexes have been examined by X-ray methods, but a detailed reexamination of the other complexes, especially spectroscopically would no doubt provide much additional insight into these interesting species.

Lithium dicyclohexylphosphide and potassium diphenylphosphide react with $CoBr_2$ to form the phosphido-bridged $[Co(PCy_2)_2]_2$ ($\mu_{eff} = 4.71$ B.M.) and $[Co(PPh_2)_2]_2$ ($\mu_{eff} = 2.77$ B.M.) [633].

(3) Cobalt(III)

Few complexes in this oxidation state are known; oxidation of Co(II) phosphine complexes is expected to produce phosphine oxide complexes of Co(II)

rather than Co(III) phosphine derivatives. Jensen *et al.*[634] reported that [Co(PMe$_3$)$_2$X$_2$] were oxidised by air to Co(III) complexes, but these could not be isolated pure. However, oxidation of [Co(PR$_3$)$_2$X$_2$] with NOX at $-80°$C produced intensely coloured solids, Co(PR$_3$)$_2$X$_3$[634]; PPh$_3$, PCy$_3$, PBz$_3$ complexes were not oxidised, and with PBu$_3$, unstable violet-black crystals of rather variable composition were obtained. The PEt$_3$ complexes, Co(PEt$_3$)$_2$X$_3$ (X = Cl, Br), are the best characterised, having magnetic moments corresponding to two unpaired electrons, and molecular weight measurements indicate monomeric complexes. The dipole moments are zero, indicating genuine five-coordinate trigonal-bipyramidal structures.

Glacial acetic acid and triphenylphosphine reacts with Na$_5$[Co(SO$_3$)$_2$(CN)$_4$] to form Na[Co(PPh$_3$)$_2$(CN)$_4$] . 3H$_2$O, which has been suggested to have a *trans* configuration[635]. A similar reaction with K$_3$[Co(CN)$_5$Br] has led to the isolation of K$_2$[Co(PPh$_3$)(CN)$_5$] . 3H$_2$O; K[Co(PPh$_3$)$_2$(CN)$_4$] . 2H$_2$O has also been obtained[636].

Cobalt(III) complexes containing phosphines and more esoteric ligands are known; for example the bisdimethylglyoximato complexes, used as a model for the vitamin B$_{12}$ group[637,638], and the tetradentate Schiff-base complexes[639].

13.2 RHODIUM

13.2.1 Carbonyls

The reaction of Rh$_6$(CO)$_{16}$ with PPh$_3$ in chloroform is reported to yield Rh$_6$(CO)$_7$(PPh$_3$)$_9$[640] or Rh(CO)(PPh$_3$)$_2$Cl[641], whilst in benzene in a sealed tube the product is Rh$_6$(CO)$_{10}$(PPh$_3$)$_6$[641]. From Rh$_4$(CO)$_{12}$ and PR$_3$ (R = Ph, *p*-MeC$_6$H$_4$, *p*-FC$_6$H$_4$) are obtained Rh$_4$(CO)$_{12-n}$(PR$_3$)$_n$ (*n* = 1, 2)[641-3] and Rh$_4$(CO)$_{12-n}$(PPh$_3$)$_n$ (*n* = 1–4)[641,643]. In the presence of CO the cluster carbonyls react with PPh$_3$ to give [Rh(CO)$_2$(PPh$_3$)$_2$]$_2$ and [Rh(CO)$_3$(PPh$_3$)]$_2$[641,643,644]. A complex thought to be Rh$_6$(CO)$_8$(PH$_3$)$_8$ is produced from PH$_3$ and [Rh(CO)$_2$Cl]$_2$[398]. The [Rh(CO)$_3$(PR$_3$)]$_2$ complexes are thought to be analogues of [Co(CO)$_3$(PR$_3$)]$_2$[643]. In addition to the method above, [Rh(CO)$_2$(PPh$_3$)$_2$]$_2$ is formed reversibly from [RhH(CO)$_2$(PPh$_3$)$_2$] and CO in solution[645,646]; i.r. spectral evidence suggests structure (LIX).

$$(Ph_3P)_2(CO)Rh \overset{\overset{\displaystyle O}{\overset{\displaystyle \|}{C}}}{\underset{\underset{\displaystyle O}{\underset{\displaystyle \|}{C}}}{- - - - -}} Rh(CO)(PPh_3)_2$$

(LIX)

Inert gases remove CO from the solution of [Rh(CO)$_2$(PPh$_3$)$_2$]$_2$ in CH$_2$Cl$_2$ or EtOH, and red [Rh(CO)(PPh$_3$)$_2$L]$_2$ (L = CH$_2$Cl$_2$, EtOH) are obtained upon concentration[646]. From the reaction of Rh$_4$(CO)$_{12}$ and PBu^n$_3$, [Rh(CO)(PBu^n$_3$)$_3$]$_2$ is obtained[641]. Oxygen reacts with [Rh(CO)$_2$(PPh$_3$)$_2$]$_2$ in benzene to form Rh$_2$(CO)(CO$_2$)(PPh$_3$)$_3$. C$_6$H$_6$[647].

Phosphido-bridged [Rh(CO)$_2$(PPh$_2$)]$_2$ is formed from [Rh(CO)$_2$Cl]$_2$ and KPPh$_2$[648].

13.2.2 Carbonyl derivatives

The chemistry of the phosphine-substituted carbonyl halides of rhodium and iridium is complex and extensive.

Tertiary phosphines react with $[Rh(CO)_2Cl]_2$ to form yellow diamagnetic *trans*-$[Rh(CO)(PR_3)_2Cl]$ ($R = Ph, Cy, p\text{-}MeC_6H_4, p\text{-}MeOC_6H_4, p\text{-}ClC_6H_4$)[424,649-51]. Carbon monoxide is taken up quantitatively in $[Rh(PPh_3)_3Cl]$ solution to form $[Rh(CO)(PPh_3)_2Cl]$, a reaction that is not reversed even by fusion with PPh_3 [652,653].

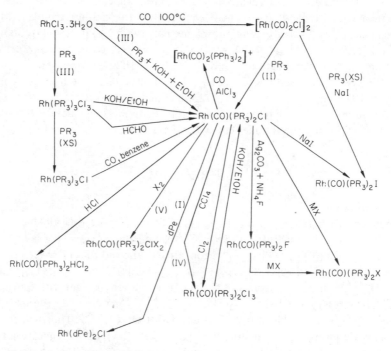

Figure 16 Preparation and some reactions of
$[Rh(CO)(PR_3)_2Cl]$[424,487,649-54,656-8,662-4,669,670] $PR_3 = PPh_3$ except
$I = PPhEt_2$; $II = PPh_3$, PCy_3, $p\text{-}MeC_6H_4$, $p\text{-}ClC_6H_4$, $(C_6F_5)_3P$; $III = PPh_3$, PEt_3,
$PPhMe_2$, $PPhEt_2$, PPh_2Et; $M = Li, Na, K$; $X' = Br, I, CNS$; $X'' = X'$, CNSe, CN,
$NO_2, NO_3, ClO_4, N_3, OH$; $IV = PEt_3$, $V = PBu^n_3$; $X_2 = Br_2, I_2$

The $[Rh(CO)(PPh_3)_2X]$ ($X = Br, I$) derivatives can be obtained from the carbonyl halide and PPh_3 or by reaction of the chloro complex with LiX[654-6]. A different approach is reduction of Rh(III) phosphine complexes. Thus, $[Rh(CO)(PPh_3)_2Cl]$ can be obtained by saturating $RhCl_3 . 3H_2O$ solution in ethanol with CO, followed by addition of PPh_3, or by reaction of $RhCl_3 . 3H_2O$ with excess PPh_3 in alcoholic KOH[487,656]. Refluxing $[Rh(PR_3)_3Cl_3]$ ($R_3 = Ph_2Et, PhEt_2, Et_3$) with alcoholic KOH leads to $[Rh(CO)(PR_3)_2Cl]$, the CO being derived from the alcohol[656] (compare Ru, Os). Unsaturated alcohols react even in the absence of base, whilst in the presence of KOH other organic compounds such as cyclohexanone, dioxan or acetophenone react with $RhCl_3$ and PPh_3 to give $[Rh(CO)(PPh_3)_2Cl]$[656,657]. The most facile synthesis utilises HCHO as reductant[652,658,659]. Halogen oxidation of $[Rh(CO)(PR_3)_2X]$ produces $[Rh(CO)(PR_3)_2XY_2]$ $(Y_2 = halogen)$[651,655,656,660], a reaction reversed by alcoholic KOH, whilst HX $(X = Cl, Br)$ add to give unstable $[Rh(CO)(PR_3)_2HX_2]$ [660,661]. The formation of *trans*-$[Rh(CO)(PPh_3)_2F]$ results from the treatment of $[Rh(CO)(PPh_3)_2Cl]$ with AgF [662], or NH_4F and Ag_2CO_3 [663]. The fluoride is readily replaced by reaction with a large number of univalent anions, for example Cl^-, Br^-, I^-, OH^-, NCS^-, CN^-, NO_2^-, NO_3^- and ClO_4^-[663]. Studies of the pseudo-halide complexes $[Rh(CO)(PPh_3)_2X]$ $(X = NCS, NCSe, NCO)$, have shown that the X ligands are all N-bonded[650,664,665]. Pentafluorophenylphosphines, $PPh_n(C_6F_5)_{3-n}$, give $[Rh(CO)L_2Cl]$ [666]; and PPh_3 can displace $P(C_6F_5)_3$ from $[Rh(CO)\{P(C_6F_5)_3\}_2Cl]$.

Shaw and coworkers[435,667] have prepared numerous similar complexes, $[Rh(CO)(PR_3)_2X]$ $(R_3 = Ph_2Me, Ph_2Et, PhMe_2, PhEt_2, PhPr^n_2, PhBu^n_2, PhBu^t_2, Et_3, Bu^n_3, Me_2Bu^t, Et_2Bu^t, Pr^n_2Bu^t, Pr^nBu^t_2, Bu^n_2Bu^t, Bu^n_2H)$ and studied their n.m.r. spectra. The preparation and reactions of some $[Rh(CO)(PR_3)_2Cl]$ complexes are shown in figure 16.

Three rotamers of *trans*-$[Rh(CO)(PRBu^t_2)_2Cl]$ $(R = Me, Et, Ph)$, and their Ir analogues, have been identified by low-temperature ^{31}P n.m.r. spectroscopy[671]. The *trans* structures of $[Rh(CO)(PR_3)_2X]$ have been confirmed by n.m.r.[667] and i.r.[672] spectroscopy, and by an X-ray study of $[Rh(CO)(PPh_3)_2Cl]$ [673]. Chloride and CO exchange in $[Rh(CO)(PPh_3)_2Cl]$ are immeasurably fast[674,675]; phosphine exchange is also rapid[676,677]

Trans-$[Rh(CO)(PR_3)_2X]$ undergo many oxidative addition reactions: with halogens, X'_2, to form $[Rh(CO)(PR_3)_2XX'_2]$ [651,655,656,660,669]; with alkyl halides, RX', to form $[Rh(CO)R(PR_3)_2XX']$ [669,676,678,679], although in the case of MeX further reaction with CO gives acetyl complexes $[Rh(CO)(MeCO)(PR_3)_2XX']$ [669,676,679]; with BX_3 $(X = Cl, Br)$ to give yellow or brown $Rh(CO)(PPh_3)_2Cl \cdot BX_3$ [680]; and with SO_2 to form yellow-green $[Rh(CO)(SO_2)(PPh_3)_2Cl]$ [681]. The last complex, which loses SO_2 on heating[682] has structure (LX), with Rh–S = 245 pm,

(LX)

Rh–P = 237 pm[683]. Carbon disulphide and triphenylphosphine react with *trans*-$[Rh(CO)(PPh_3)_2Cl]$ in methanol to produce a deep blue solution from which $[Rh(CO)(\pi\text{-}CS_2)(PPh_3)_3]BPh_4$ can be obtained[684].

Sodium amalgam in THF reduces $[Rh(CO)(PPh_3)_2Cl]$ in the presence of CO to form $[Rh(CO)_2(PPh_3)_2]^-$, which reacts with Me_3SnCl to yield $[Me_3SnRh(CO)_2(PPh_3)_2]$ [685] and with benzoic acid to form $[HRh(CO)_2(PPh_3)_2]$ [686]. The anion is also formed from $[Rh(CO)_2(PPh_3)_2]_2$ and Na–Hg and CO in THF[686]. In the presence of CO and $AlCl_3$, $[Rh(CO)(PR_3)_2Cl]$ $(R = Ph, Cy)$, like many other carbonyl halides, form cations originally formulated as $[Rh(CO)_2(PR_3)_2]^+$ [670], although it has been suggested that they are really $[Rh(CO)_3(PR_3)_2]^+$ [687]. The latter can be synthesised from $[RhL(PR_3)_2]^+$ (L = cyclooctadiene, R = Ph) and CO, and on treatment with PPh_3 afford $[Rh(CO)_2(PPh_3)_3]^+$ [687]. The $[Rh(CO)(PR_3)_3]^+$ (R_3 = Ph_2Me, $PhEt_2$) cations are formed on warming $[Rh(CO)(PR_3)_2Cl]$ with PR_3 in methanol, followed by addition of PF_6^- or BPh_4^-. In contrast, when PR_3 is $PPhMe_2$, $[Rh(CO)(PPhMe_2)_4]^+$ is obtained[688]. On passing CO through a mixture of rhodium perchlorate and triphenylphosphine in alcohol, $[Rh(CO)_2(PPh_3)_3]ClO_4$ is formed[689].

Belluco *et al.*[690,691] reported that a new type of carbonyl halide complex, $[Rh(CO)_2(PPh_3)Cl]$, could be isolated from the reaction of $[Rh(CO)_2Cl]_2$ and PPh_3 in a 1:2 ratio (a 1:4 ratio gives the well-known $[Rh(CO)(PPh_3)_2Cl]$), but it has recently been suggested that the product is the dinuclear complex $[Rh(CO)(PPh_3)Cl]_2$ [692,693]; *cis*-$[Rh(CO)_2(PPh_3)Cl]$ may have been obtained, however[693]. The so-called '$Rh(CO)_2(PPh_3)Cl$' takes up SO_2 to form $Rh(CO)_2(PPh_3)(SO_2)Cl$ [694].

Unsaturated tertiary phosphines—$PhP(C{\equiv}CMe)_2$, *cis*-$Ph_2PCH{=}CHPh$ and trisferrocenylphosphine—form $[Rh(CO)(PR_3)_2Cl]$ in which the multiple bonds are not coordinated[695]. The acetylenic DPA forms the dimer $[Rh(CO)(DPA)Cl]_2$ [85]. The hydrogenation and isomerisation of $[Rh(CO)\{Ph_2P(CH_2)_nCH{=}CH_2\}_2Cl]$ (n = 0–3) with H_2 and EtOH or MeOH under reflux has been studied[696]. Hydrogenation is easiest when n = 2, and the proposed mechanism is

A refluxing solution of $RhCl_3$ in alcohol saturated with CO, reacts with SP to yield a compound $[Rh(CO)(SP)Cl_2]$, which has been assigned structure (LXI) on the basis

(LXI)

of spectroscopic studies[697]. This complex reacts with monodentate phosphines to cleave the halogen bridge and form $[Rh(Ph_2PC_6H_4CHCH_3)(CO)(PR_3)Cl_2]$ (R_3 = Ph_3, Ph_2Me, $PhMe_2$). Under different conditions $[Rh(CO)(SP)_2]Cl$ and $[Rh(CO)(PPh_3)(SP)Cl]$ can be isolated[697]. The direct isolation of $[Rh(CO)(SP)_2]Cl$ results from reaction of $[Rh(CO)_2Cl]_2$ with two equivalents of SP. It readily loses CO to form $Rh(SP)_2Cl$ [698].

Tetraphenylbiphosphine reacts with $[Rh(CO)_2Cl]_2$ to form orange phosphido-bridged $[Rh(CO)(PPh_2)Cl]_2$, which is known as the *cis* and *trans* isomers[648].

Vaska and Bath prepared $[HRh(CO)(PPh_3)_3]$ by reduction of $[Rh(CO)(PPh_3)_2Cl]$ with 95 per cent aqueous hydrazine[699]. The complex is diamagnetic, and because of its i.r. and n.m.r. spectra a five-coordinate structure (LXII) has been proposed[447]. This has been confirmed by an X-ray structure determination, Rh—P = 232 pm, Rh—H = 160 pm[700]. The reaction of $[Rh(CO)(PR_3)_2Cl]$ with CO, PR_3 and alkali metal alkoxides (for example, $KOBu^t$) in benzene leads, after hydrolysis, to $[HRh(CO)(PR_3)_3]$ (R_3 = Ph_3, Ph_2Me, $PhEt_2$, Et_3, Bu^n_3)[701]. Under different conditions other products, $[Rh(CO)(PR_3)_2(OH)]$ and $[Rh(CO)_2(PR_3)_2]\cdot2C_6H_6$, are obtained. The reduction of $RhCl_3$ with $NaBH_4$ in the presence of PPh_3 and $HCHO$ in ethanol also leads to $[HRh(CO)(PPh_3)_3]$ [702].

The properties of $[HRh(CO)(PPh_3)_3]$ have been studied in detail by Wilkinson and coworkers[646,703-8]. It is extensively dissociated in organic solvents to form $HRh(CO)(PPh_3)_2$, which is square planar, and this readily takes up CO in solution to form $[HRh(CO)_2(PPh_3)_2]$ and then $[Rh(CO)_2(PPh_3)_2]_2$, the reaction being reversed by bubbling H_2 through the mixture[646]. In solution $[HRh(CO)(PPh_3)_3]$ is an active catalyst for hydrogenation of unsaturated compounds[703], hydroformylation of alkenes[704] and hydrogen exchange reactions. The decay of $[HRh(CO)(PPh_3)_3]$ to an inactive species during the catalytic reactions was observed, and the inactive species has been identified as $[Rh(CO)_2(PPh_3)_2]_2$ [705].

The formation of Rh(III) complexes, $Rh(CO)(PR_3)_2X_3$, (R_3 = Ph_3, $(p-MeC_6H_4)_3$, $PhEt_2$, Et_3, Bu^n_3) by oxidation of $Rh(CO)(PR_3)_2X$ with halogens in CCl_4 has already been mentioned[651,654-6,660,669]. The yellow or red (iodides) products are diamagnetic non-electrolytes. They can also be formed directly from $RhCl_3$ and $2PR_3$ in 2-methoxyethanol on boiling in an atmosphere of CO [656], and $[Rh(CO)(PPhEt_2)_2Br_3]$ forms on boiling $[Rh(CO)(PPhEt_2)_2Br]$ with HBr [656]. Mixed halides $[Rh(CO)(PR_3)_2XY_2]$ (X, Y = halogen; X ≠ Y) are obtained from $[Rh(CO)(PR_3)_2X]$ and Y_2 in CCl_4 [654,660,669]. The structures of $[Rh(CO)(PR_3)_2X_3]$ are known (LXIII)[654,672], but in the case of $[Rh(CO)(PR_3)_2XY_2]$ it is not yet known whether Y ligands are *cis* or *trans* to each other. Belluco *et al.*[691] described $[Rh(CO)_2(PPh_3)X_3]$ as formed by reaction of '$Rh(CO)_2(PPh_3)X$' with halogens, but

(LXII)   (LXIII)

Steele and Stephenson[693] have recently suggested that these complexes be reformulated as $[Rh(CO)(PPh_3)XY_2]_2$.

The CO exchange rate in $[Rh(CO)(PPh_3)_2Cl_3]$ is slow, in contrast to the rate in $[Rh(CO)(PPh_3)_2Cl]$ [709]. The Rh(III) hydridocarbonyls, $[Rh(CO)(PPh_3)_2HX_2]$ (X = Cl, Br) form on treatment of $[Rh(CO)(PPh_3)_2X]$ with HX; they are unstable, readily losing HX in solution[660,710].

13.2.3 Nitrosyls

On reaction with triphenylphosphine $[Rh(NO)_2Cl]_2$ undergoes 'valence disproportionation' to form $[Rh(NO)(PPh_3)_2Cl_2]$ and $Rh(NO)(PPh_3)_3$ [711]. Red $[Rh(NO)(PPh_3)_3]$ is also formed by sodium amalgam reduction of $[Rh(NO)_2Cl]_2$ in the presence of PPh_3 [711], from $RhCl_3 \cdot 3H_2O$ in THF when treated with NO, PPh_3 and zinc[712], or from $RhCl_3 \cdot 3H_2O$ and PPh_3, $NaBH_4$ and N-methyl-N-nitrosotoluene-*p*-sulphonamide[702]. The reaction of $Rh(PPh_3)_3Cl$ with NO in $CHCl_3$ was reported to give $[Rh(NO)_2(PPh_3)_2Cl]$ [713], which decomposed to $[Rh(NO)(PPh_3)_2Cl]$. Halogen oxidation of the latter gave $[Rh(NO)(PPh_3)_2X_3]$ [714]. Both $[Rh(NO)(PPh_3)_2Cl]$ and $[Rh(NO)(PPh_3)_2Cl_3]$ are reported to form on passing NO into solutions of the corresponding carbonyl halides[714]. Other workers obtained a green solid, identified as $[Rh(NO)(NO_2)(PPh_3)_2Cl]$ from the reaction of $[Rh(PPh_3)_3Cl]$ or $[Rh(CO)(PPh_3)_2Cl]$ with NO [715,716].

The yellow or brown $Rh(NO)(PPh_3)_2Cl_2$ are obtained from $[Rh(NO)_2Cl]_2$ [711] or $[Rh(NO)Cl_2]_n$ [717] and PPh_3; metathetical reaction with LiX in the presence of PPh_3 afford $[Rh(NO)(PPh_3)_2X_2]$ (X = Br, I). The $[Rh(NO)(PPh_3)_2Cl_2]$ and $[Ir(NO)(PPh_3)_2Cl_2]$ complexes are isomorphous (see below). Rhodium trichloride can also be used as the starting material for these complexes: NO reacts with $RhCl_3$ in EtOH to yield a paramagnetic species $(Rh(NO)Cl_3?)$ from which addition of PPh_3 precipitates a mixture of $[Rh(NO)(PPh_3)_2Cl_3]$ and $[Rh(NO)(PPh_3)_2Cl_2]$ [718]. A simple preparation of $[Rh(NO)(PR_3)_2X_2]$ (R_3 = Ph_3, Ph_2Me, Et_3, Pr^i_3, Bu^n_3, *p*-MeC_6H_4, *p*-$MeOC_6H_4$, *p*-ClC_6H_4) has been reported[719], and consists of adding an alcoholic solution of RhX_3 to the solution of the phosphine in boiling alcohol, followed by *n*-pentyl nitrite or N-methyl-N-nitrosotoluene-*p*-sulphonamide. It would be interesting to determine the structure of one of these complexes and to compare it with $[Ir(NO)(PPh_3)_2Cl_2]$ (q.v.). The products from the reaction of $[Rh(CO)(PPh_3)_2Cl]$ or $[Rh(PPh_3)_3Cl]$ with NO are also in need of clarification.

A cationic complex, $[Rh(NO)(PPh_3)_3Cl]BF_4$, has been obtained[720] as a red-violet solid from the reaction of $[Rh(PPh_3)_3Cl]$ and $NOBF_4$ in C_6H_6–MeOH in the presence of excess PPh_3. The authors suggest that the Rh–N–O linkage is not linear.

13.2.4 Hydrido complexes

Yellow $[HRh(PPh_3)_4]$ is produced by reduction of RhX_3 (X = Cl, acac) with $AlEt_3$ [466] or $NaBH_4$ [472] in the presence of PPh_3. Hydrazine similarly reduces $[Rh(PPh_3)_3Cl]$ [469]. Hydrogenation of $[RhPh(1,5-cyclooctadiene)(PPh_3)]$ in toluene also affords $[HRh(PPh_3)_4]$ [721,722]. The preparation of $[HRh(PPh_2Me)_4]$ has also been reported[469]. On heating $[HRh(PPh_3)_4]$, only traces of hydrogen are liberated, but acidification produces 80 per cent of the calculated amount of H_2; it reacts with dPe to yield $[HRh(dPe)_2]$ [466]. The i.r. spectrum exhibits $\nu(Rh-H)$ at 2140 cm^{-1}, and the deuterium analogue has $\nu(Rh-D)$ at 1540 cm^{-1} [466,721]. On the basis of their ^1H n.m.r. spectra Dewhirst *et al.*[469] suggested trigonal-bipyramidal and square-pyramidal structures for $[HRh(PPh_3)_4]$ and $[HRh(PPh_2Me)_4]$, respectively, but an X-ray study of $[HRh(PPh_3)_4] \cdot \frac{1}{2}C_6H_6$ [723] has shown that the RhP_4 grouping is tetrahedral, with the hydrogen (position not determined) lying on one of the C_3 axes. The mixed complex $[HRh(PPh_3)_3(AsPh_3)]$, prepared from $[HRh(PPh_3)_3]$ and $AsPh_3$, is also tetrahedral with random occupancy of the tetrahedral positions by P and As, and the H is *trans* to both P and As [724]. Unlike the hydridophosphines of Ru and Co, $[HRh(PPh_3)_4]$ does not take up molecular nitrogen[466].

The orange solid, $[HRh(PPh_3)_3]$, is formed from $[Rh(PPh_3)_3Cl]$ and $AlPr_3^i$ [469,725]. It takes up PPh_3 reversibly[469,472], and n.m.r. evidence suggests a tetrahedral structure

$$HRh(PPh_3)_4 \rightleftharpoons HRh(PPh_3)_3 + PPh_3$$

Both $[HRh(PPh_3)_3]$ and $[HRh(PPh_3)_4]$ react with CO to form $[HRh(CO)(PPh_3)_3]$ [469]. The complex reported to be $HRh(PPh_3)_2$ [726], has now been shown to be $[HRh(PPh_3)_3]$ [727].

Although these hydrido complexes do not react with N_2, a dinitrogen complex, $Rh_2(CO)_2(N_2)(PPh_3)_2Cl_4 \cdot 2NH_4Cl$, has been reported from the reaction of $[Rh(CO)(PPh_3)(acac)]$ and HN_3 in CH_2Cl_2 [728]. Collman *et al.* have also made a brief reference to $[Rh(N_2)(PPh_3)_2Cl]$ [729].

Hydridohalides are much less common than in iridium chemistry. Sacco *et al.*[730,731] found that addition of PPh_2Et to hot alcoholic $RhX_3 \cdot 3H_2O$ under nitrogen led to the formation of $[RhHX_2(PPh_2Et)_3]$ (X = Cl, Br, I) (α-form). Similar reactions occurred in t-BuOH or acetone, and it was shown[731] that the hydrogen came from the water present, not from the solvent, the reaction being

$$RhX_3 \cdot 3H_2O + 4PPh_2Et \rightarrow RhHX_2(PPh_2Et)_3 + OPPh_2Et + HX + 2H_2O$$

Alcoholic solutions of RhX_3 and PPh_2Et are reduced by H_3PO_2 at room temperatures to another form of $RhHX_2(PPh_2Et)_3$ (β-form). On boiling in suspension in ethanol the β-form changes into the α-form. The two forms differ in the energy of the $\nu(Rh-H)$ in the i.r. spectrum, the α-form being at about 2100 cm^{-1}, the β-form being at about 1960–80 cm^{-1}. Two forms of $[RhHCl_2(PPh_3)_3]$ have also been isolated[731]. The formation of $[RhHX_2(PR_3)_3]$ has only been observed with less basic phosphines, such as PPh_3, PPh_2Et, PPh_2Bu [732], but not with trialkylphosphines.

A solution of $Rh(PPh_3)_3Cl$ in CH_2Cl_2 reacts with HCl to give yellow $RhHCl_2(PPh_3)_2 \cdot \frac{1}{2}CH_2Cl_2$, which loses HCl fairly easily to reform $Rh(PPh_3)_3Cl$ [733]. The n.m.r. spectrum indicates (LXIV) is the probable structure. It undergoes hydrogen transfer reactions with ethylene and acetylene to yield $[RhL(PPh_3)_2Cl_2]$ (L = C_2H_5, C_2H_3) [733].

With bulky phosphines, for example $PPr^n_2Bu^t$, $PMeBu^t_2$, $PEtBu^t_2$ and $PPr^nBu^t_2$, five-coordinate $[RhHCl_2(PR_3)_2]$ can be isolated from the reaction with $RhCl_3$ in alcohol[734]. The ^{1}H n.m.r. spectra indicate *trans* structures (LXV), and this was confirmed in the case of $[RhHCl_2(PPr_2Bu^t)_2]$ by an X-ray study[734]. Treatment with NaOMe in MeOH converts them to *trans*-$[Rh(CO)(PR_3)_2Cl]$. The $[RhHCl_2(PPr^n_2Bu^t)_2]$ is a very active hydrogenation catalyst[734].

(LXIV)   (LXV)

Dihydrides $[RhH_2Cl(PR_3)_2]$ are formed by treating $Rh(PR_3)_3Cl$ ($R_3 = Ph_3$, Ph_2Et, Bu^n_3) with H_2 in $CHCl_3$ or C_6H_6 solution[652,731]. These complexes are octahedral in solution due to coordination of solvent. The ^{1}H n.m.r. spectra indicate *cis* hydrides and *cis* phosphines[652]. In dichloromethane $[Rh(PPh_3)_2Cl]_2$ reacts with H_2 to give a different dihydrido complex, thought to have structure (LXVI) [652].

(LXVI)

Solutions of $[Rh(diene)(PR_3)_2]^+$ react with hydrogen to form $[Rh(PR_3)_2H_2L_2]$ (R = Ph; L = solvent), which further react with phosphine to yield $[Rh(PR_3)_4H_2]^+$ ($R_3 = Me_3$, $PhMe_2$) [735], and the ^{1}H n.m.r. spectra of these complexes are consistent with *cis* structures. The $[Rh(PPhMe_2)_4]^+$ complex takes up H_2 to form $[Rh(PPhMe_2)_4H_2]^+$ directly, but $[Rh(PPh_2Me)_4]^+$ does not react, illustrating how sensitive this type of reaction often is to small changes in ligand properties[735].

13.2.5 Some low-valent rhodium complexes

Rhodium(0) complexes, $[Rh(PR_3)_4]$ ($R_3 = Ph_3$, Ph_2Me), were obtained[736] by electrolytic reduction of $Rh(PR_3)_3Cl$ in MeCN in the presence of PR_3. The triphenylphosphine derivative is diamagnetic, but the diphenylmethylphosphine complex has $\mu_{eff} = 1.16$ B.M. The low values are unexpected for a d^9 complex, and are attributed[736] to the existence of a Rh–Rh interaction

$$(R_3P)_4Rh–Rh(PR_3)_4 \rightleftharpoons 2Rh(PR_3)_4$$
$$\text{diamagnetic} \qquad \text{paramagnetic}$$

Keim[725] reported that $Rh(PPh_3)_3Cl$ reacted with Grignard reagents to form $[Rh(PPh_3)_3R]$ (R = Me, Ph). On heating, the methyl complex evolves methane to form $[Rh(C_6H_4PPh_2)(PPh_3)_2]$ (LXVII) [737]. The reactions of (LXVII) with CO, benzoyl chloride and oxalyl chloride have been studied[738]. Diphenylacetylene produces the novel system $[Rh(PPh_3)(Ph_2PC_6H_4C_4Ph_4)]$ [739], which has been shown by an X-ray study[740] to have structure (LXVIII). Other examples of Rh—C σ-bonds are the Rh—SP complex[697] discussed in section 13.2.2, and the Rh(III)–P(o-tolyl)$_3$ complexes discussed later in section 13.2.6.

(LXVII)   (LXVIII)

13.2.6 Rhodium halides and related complexes

(1) Rhodium(I)

The most important Rh(I) phosphine complex is Rh(PPh$_3$)$_3$Cl, formed from RhCl$_3$ and excess PPh$_3$ in alcohol[652,653]. The dark red crystalline [Rh(PPh$_3$)$_3$Cl] dissolves in donor solvents such as DMSO, pyridine and acetonitrile to form Rh(PPh$_3$)$_2$LCl (L = DMSO, py, MeCN)[652,741], and has been reported[652] to be extensively dissociated in solution in benzene or chloroform

$$Rh(PPh_3)_3Cl \rightleftharpoons Rh(PPh_3)_2Cl + PPh_3$$

On heating or long standing dimeric [Rh(PPh$_3$)$_2$Cl]$_2$ is obtained from such solutions[652].

Although [Rh(PPh$_3$)$_3$Cl] is stable indefinitely in air, in solution it reacts with a large range of molecules due to ready loss of PPh$_3$ (see figure 17)[652,653,733,741-4].

In benzene or similar solvents [Rh(PPh$_3$)$_3$Cl] is an efficient hydrogenation catalyst for olefins and acetylenes under ambient conditions. The bromide and the iodide are even better catalysts, but Rh(AsPh$_3$)$_3$Cl is less efficient than [Rh(PPh$_3$)$_3$Cl][652]. The proposed mechanism involves the formation of RhH$_2$(PPh$_3$)$_2$Cl and its subsequent reaction with the unsaturated substrate. Much work has been done to elucidate the catalytic properties of Rh(PPh$_3$)$_3$Cl, but they will not be discussed here further (but see references 510, 652, 733, 744–7, and references therein). The dissociation of [Rh(PPh$_3$)$_3$Cl] in solvents such as CH$_2$Cl$_2$, C$_6$H$_6$, etc., has recently been questioned[745,748,749]. In the absence of air dissociation appears to be less than five per cent, but in the presence of oxygen greater dissociation occurs. It is also an efficient hydroformylation catalyst, probably due to its conversion *in situ* into Rh(CO)(PPh$_3$)$_2$Cl.

The ligand exchange and isomerisation of [Rh(PPh$_3$)$_3$Cl] has been studied by ^{31}P n.m.r. spectroscopy[749]. The Rh(PPh$_2$Me)$_3$Cl complex, prepared from Rh$_2$(C$_2$H$_4$)$_4$Cl$_2$ and PPh$_2$Me in toluene, has a ^{1}H n.m.r. spectrum consistent with a square-planar structure[469], as has [Rh(PPh$_3$)$_3$Cl] itself[750]. The PPh$_2$H complex has been obtained analogously[502].

The unsaturated phosphines, P(CH$_2$CH$_2$CH=CH$_2$)$_3$ [751] and P(o-C$_6$H$_4$CH=CH$_2$)$_3$ [752], form 1:1 complexes [Rh(PR$_3$)Cl]; spectroscopic evidence indicates that the double bonds are coordinated in both cases. An X-ray study[753] of $[RhP(C_6H_4CH=CH_2)_3Br]$ shows the rhodium to be in a trigonal bipyramid of ligands with Rh—Br = 258.7 pm and Rh—P = 217 pm (LXIX).

Dimeric $[Rh(PPh_3)_2Cl]_2$ was obtained by refluxing $Rh(PPh_3)_3Cl$ in benzene[652], and $[Rh\{(C_6F_5)_{3-n}Ph_nP\}_2Cl]_2$ (n = 0, 1, 2) are formed from $RhCl_3$ and $(C_6F_5)_{3-n}Ph_nP$ [666]. The phenylphosphines form the bromo analogues with $RhBr_3$, but the $P(C_6F_5)_3$ complex has not been isolated[666]. The most likely structure for these complexes is the halogen-bridged $(R_3P)_2RhCl_2Rh(PR_3)_2$, although the i.r. spectra[666,67] are anomalous, leading to the proposal that highly unsymmetrical Rh–Cl bridges are present.

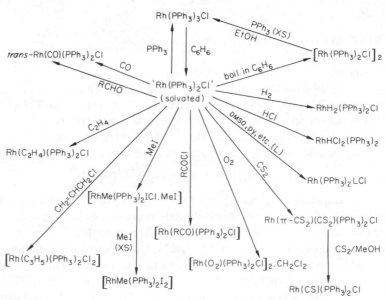

Figure 17 Some reactions of $Rh(PPh_3)_3Cl$ [733,742–5,752,762,767]

Complexes thought to be Rh(I), Rh(III) chlorine-bridged dimers (LXX) were obtained by air oxidation of $RhHCl_2(PPh_2Et)_3$ in benzene, followed by precipitation with hexane[731]. Evidence for this formulation comes from the formation of $Rh_2(PPh_2Et)_4Cl_6$ on chlorine oxidation, and of $Rh(CO)(PPh_2Et)_2Cl$ and $Rh(CO)(PPh_2Et)_2Cl_3$ on treatment with CO. Fusion of $[Rh(CO)_2CN]_n$ with PPh_3

(LXIX) (LXX)

led to the formation of $[Rh(PPh_3)_2CN]_2$ [754], but attempts to prepare
$[Rh(PPh_3)_2N_3]_2$ failed[755].

Recently, Rh(I) cations containing tertiary phosphines have been prepared.
Excess $PPhMe_2$, PPh_2Me or PBu^n_3 will react with $[C_8H_{12}RhCl]_2$ in methanol to
yield $[Rh(PR_3)_4]^+$, isolated as the PF_6^-, ClO_4^- or BPh_4^- salts[735,756]. In the presence
of air $[Rh(PPhMe_2)_4O_2]^+$ was isolated, for which structure (LXXI) has been proposed
(compare $[Rh(dPe)_2O_2]PF_6$) [757]. Both $[Rh(PPh_2Me)_4]PF_6$ and $[Rh(PBu^n_3)_4]BPh_4$
are air stable as solids, and oxygen adducts have not been obtained[758].

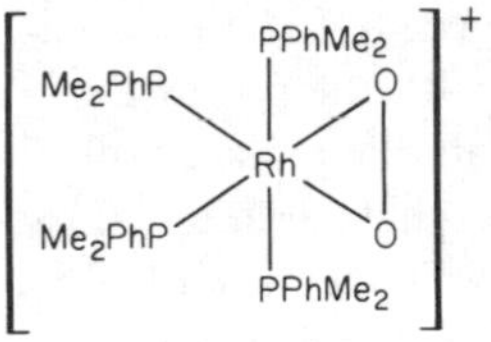

(LXXI)

The diene complex, $[Rh(diene)(PPh_3)_2]BPh_4$ (diene = norbornadiene), may be
hydrogenated in acetone solution to form red-brown needles of $Rh(PPh_3)_2BPh_4$ [759].
The BPh_4^- group is π-bonded to the Rh via one of the phenyl rings (i.r. and n.m.r.
evidence)[760]; a similar structure has been reported for $Rh\{P(OMe)_3\}_2BPh_4$ [761]. These
complexes are very unusual, BPh_4^- being an extremely poor coordinating anion. The
$[Rh(diene)(PR_3)_2]^+$ are useful intermediates for the synthesis of other Rh cationic
species[759]. The cation $[Rh(SP)_2]^+$, formed from $Rh(SP)_2Cl$ and $NaBPh_4$ has been
briefly mentioned[698].

Addition of small molecules to rhodium phosphine derivatives produces Rh(I) and
Rh(III) complexes, which are discussed together here for convenience. Although
Rh(I) phosphine complexes have so far shown no tendency to coordinate molecular
nitrogen, the addition of several other small molecules has been reported. On bubbling
oxygen through a CH_2Cl_2 solution of $Rh(PPh_3)_3Cl$ and allowing to crystallise,
brown crystals are obtained[762]. The structure (LXXII) of these has been determined,
and shown to contain dimeric units with an unusual O_2 bridge. Augustine and Van
Peppen[745] showed that uptake of oxygen by $Rh(PPh_3)_3Cl$ in solution was solvent
dependent. In methanol a complex, whose analysis indicated a formulation

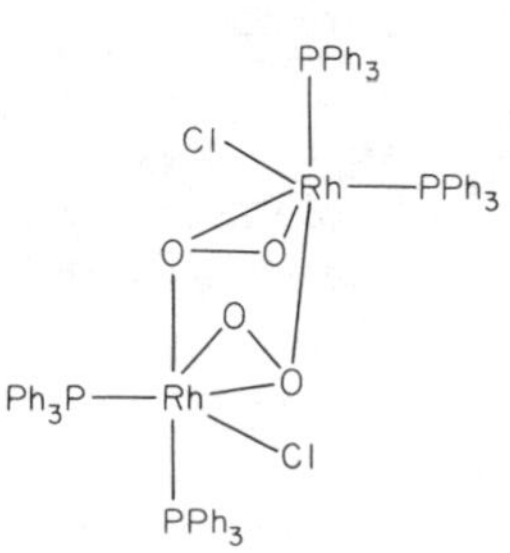

(LXXII) 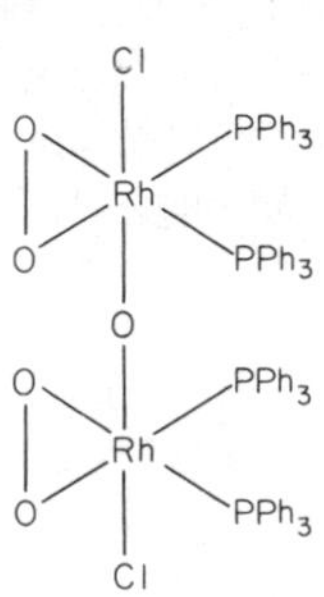   (LXXIII)

$Rh_2(PPh_3)_4Cl_2O_5$, was formed, whilst in ethanol, extensive oxygen uptake and oxidation of the solvent occurred, but small quantities of $[Rh(H_2O)(PPh_3)Cl(O)]_{4-5}$ did precipitate. The latter was suggested to be a mixture of polymers with oxygen bridges. Structure (LXXIII) was proposed for $[Rh_2(PPh_3)_4Cl_2O_5]$ [745]. The $[Rh(PPhMe_2)_4O_2]^+$ cation has already been mentioned, and a complex $[Rh(PPh_3)_3O_2]_n$ is reported [721] as the product of oxygen treatment of $[RhPh(1,5\text{-cyclooctadiene})(PPh_3)]$ in toluene solution (n is probably equal to two).

The addition of P_4 to rhodium complexes has recently been achieved [763]. White phosphorus reacts with $Rh(PR_3)_3Cl$ (R = Ph, $p\text{-MeC}_6H_4$, $m\text{-MeC}_6H_4$) in CH_2Cl_2 or Et_2O at $-78°C$ under nitrogen, to yield yellow $Rh(P_4)(PR_3)_2Cl$. Raman, i.r. and mass spectra are all consistent with the presence of P_4 in these complexes. Carbon monoxide displaces P_4; PEt_3 and dPe replace both the P_4 molecule and the triarylphosphine [763]. Cyanogen reacts with $Rh(PPh_3)_3Cl$ to produce $Rh(PPh_3)_2(C_2N_2)Cl$ but no reaction occurs with $Rh(CO)(PPh_3)_2Cl$ [764]. Baird and Wilkinson found that refluxing $Rh(PPh_3)_3Cl$ with CS_2–MeOH yielded thiocarbonyl complexes, $Rh(CS)(PPh_3)_2Cl$ [742]. It was later shown that in CS_2 solution, $[Rh(PPh_3)_2(CS_2)(\pi\text{-}CS_2)Cl]$, (LXXIV), is the initial product, and that this forms

(LXXIV)

the thiocarbonyl complex only upon addition of MeOH [743]. The structure of $[Rh(CS)(PPh_3)_2Cl]$ is *trans* square planar [765]. Halogens oxidise the thiocarbonyl complexes readily to $Rh(CS)(PPh_3)_2X_3$, but MeI or $HgCl_2$ cause decomposition instead of the expected addition, possibly due to attack on the sulphur [743]. Upon reaction of $Rh(CS)(PPh_3)_2X$ and CS_2 in methanol, followed by addition of $NaBPh_4$, $[Rh(CS)(\pi\text{-}CS_2)(CS_2)(PPh_3)_2]BPh_4$ is formed [684]. In vacuum CS_2 is lost to give $[Rh(CS)(\pi\text{-}CS_2)(PPh_3)_2]BPh_4$. Use of $Rh(PPh_3)_3Cl$ instead of $[Rh(CS)(PPh_3)_2Cl]$ in these reactions leads to the formation of $[Rh(\pi\text{-}CS_2)(CS)_2(PPh_3)_3]BPh_4$ and $[Rh(\pi\text{-}CS_2)(PPh_3)_3]BPh_4$ [684]. Reaction of CS_2 with $Rh(R)(PPh_3)_3$, $RhH(CO)(PPh_3)_3$, $RhH(PPh_2Et)_2Cl_2$ and $Rh(Me)(PPh_3)_2I_2$ have been found to give other rhodium(III) species containing CS or CS_2 ligands [766].

The reaction of Group IVB donor ligands with Rh is receiving increasing attention. Most of the organic complexes fall outside the scope of this review, although several have already been mentioned in passing. Excess MeI reacts with $Rh(PPh_3)_3Cl$ to form $[Rh(Me)(PPh_3)_2I_2]$, one of the few well-established five-coordinate rhodium(III) complexes [767]; it is square pyramidal with an axial methyl group and *trans* groups in the basal plane. Rhodium–silicon complexes, $[RhH(PPh_3)_2(SiR_3)Cl]$ (R_3 = Cl_3, $(OEt)_3$, Cl_2Me, Cl_2Et, Et_3, Me_3, Ph_3 and some Br and I analogues) were obtained from the reaction of $Rh(PPh_3)_3Cl$ with the appropriate silane [768]. Slightly different preparations yielding solvated complexes have also been reported [769], and the structure of $RhHCl(SiCl_3)(PPh_3)_2 \cdot xHSiCl_3$ has been determined[770]. Reports of Rh–Ge and Rh–Sn complexes have also appeared[771].

The tertiary arsine rhodium(II) complexes reported in the literature were later shown to be Rh(III)-hydrido complexes. Recently, genuine Rh(II) complexes with sterically hindered phosphines have been obtained[381]. Rhodium trichloride reacts with $(o\text{-tolyl})_3P$ in ethanol to yield air-stable blue-green $Rh\{(o\text{-tolyl})_3P\}_2Cl_2$ complex. This is paramagnetic, $\mu_{eff} = 2.27$ B.M., and the X-ray powder pattern is very similar to that of the Pd(II) and Pt(II) analogues[381]. A purple modification is obtained by reaction below $0°C$. The blue-green form is thought to be *trans* planar Rh(II), with one *o*-methyl group lying above and below the plane, blocking the other coordination positions and producing pseudo-octahedral coordination. The red Rh $\{(o\text{-tolyl})_2PhP\}_2Cl_2$, formed from $(o\text{-tolyl})_2PhP$, is of uncertain nature, but $(o\text{-tolyl})Ph_2P$ forms $Rh\{(o\text{-tolyl})Ph_2P\}_2Cl$ [381]. Other Rh(II) complexes, $Rh(PR_3)_2Cl_2$ $(R_3 = MeBu^t_2, EtBu^t_2, Pr^nBu^t_2)$, have been prepared by Shaw's group and have been found to have low magnetic moments[734].

Bennett *et al.* [381] found that RhX_3 reacts with $(o\text{-tolyl})_3P$ in refluxing 2-methoxyethanol to form yellow, diamagnetic '$Rh\{(o\text{-tolyl})_3P\}X_2$' (X = Cl, Br), HX being evolved. Under the same conditions $(m\text{-tolyl})_3P$ and $(p\text{-tolyl})_3P$ gave $Rh(PR_3)_3Cl$ or $Rh(CO)(PR_3)_2Cl$. The yellow '$Rh\{(o\text{-tolyl})_3P\}X_2$' react with dPe, diars, py, PPh$_3$, etc., to give such complexes as (LXXV). The pyridine adduct,

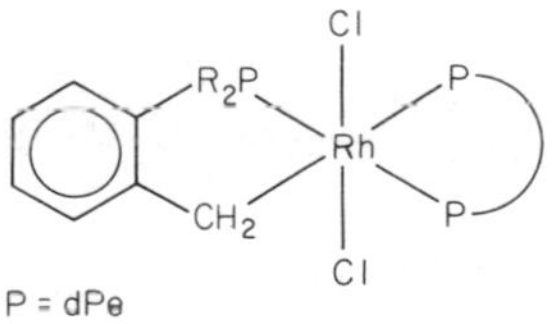

P–P = dPe
R = *o* -tolyl

(LXXV)

'$[Rh\{(o\text{-tolyl})_3P\}(py_2Cl_2)]$' has been examined by X-ray techniques[668]. These complexes contain the deprotonated $(o\text{-tolyl})_3P$ ligand. The σ-bonded complexes, $[Rh(o\text{-CH}_2C_6H_4)P(o\text{-tolyl})_2X_2]$, are thought to be trimeric[381], and they are thermally stable, contrasting with the general lack of organorhodium complexes. It is no doubt significant, however, that organorhodium complexes have been obtained with sterically hindered aryls, for example $[Rh(1\text{-naphthyl})_2(PR_3)_2]$ $(R_3 = Pr^n_3, PhEt_2)$ [382].

Prolonged reaction of $RhCl_3$ with $(o\text{-tolyl})_3P$ in 2-methoxyethanol yields a yellow

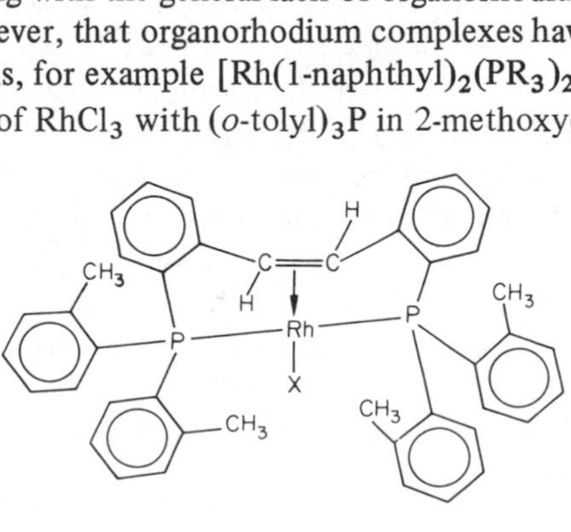

(LXXVI)

'Rh{(o-tolyl)$_3$P}$_2$Cl', from which NaCN displaces a white solid, which has been assigned the formulation (o-tolyl)$_2$PC$_6$H$_4$CH=CHC$_6$H$_4$P(o-tolyl)$_2$, a stilbene derivative, on the basis of analytical and ^{1}H n.m.r. spectral studies. The yellow complex has been assigned the structure (LXXVI). Similar bromo and thiocyanato complexes are also reported[381]. The ligand is formed from two molecules of (o-tolyl)$_3$P coupled through two methyl groups with the loss of four hydrogen atoms.

Rhodium(III)

Chatt *et al.*[772] prepared the orange complexes [Rh(PR$_3$)$_3$Cl$_3$] by the addition of PR$_3$ to hot alcoholic RhCl$_3$ (R$_3$ = Et$_3$, Pr$^n_3$, Bu$^n_3$, Me$_2$Ph, Et$_2$Ph, EtPh$_2$), which, on the basis of dipole moments, were assigned a meridional configuration. Small amounts of another complex, possibly *fac*-[Rh(PEt$_3$)$_3$Cl$_3$], were also isolated[772]. Subsequently, the meridional configuration of [Rh(PMe$_2$Ph)$_3$Cl$_3$] was established by n.m.r. spectroscopy[487]. The *trans* chlorines are more labile than the chlorine *trans* to the phosphine, and on treatment with MX in hot acetone or ethanol, complexes of the type, *mer*-[Rh(PMe$_2$Ph)$_3$ClX$_2$] (LXXVII) are formed. Under reflux,

(LXXVII)

the third chlorine is replaced to give, *mer*-[Rh(PMe$_2$Ph)$_3$X$_3$] (X = NCO, Br, I, N$_3$, NCS).

Evidence was obtained for isomers of [Rh(PMe$_2$Ph)$_3$(CNS)Cl$_2$] with N- and S-bonded thiocyanate. Other complexes of this type are *mer*-[Rh(PR$_3$)$_3$Cl$_3$] (R$_3$ = Me$_3$, Bu$^n_2$Ph, Pr$^n_2$Ph, Bu^{n}Ph$_2$, Pr^{n}Ph$_2$, MePh$_2$)[507,672], and ^{1}H n.m.r. and ^{31}P n.m.r. have been used[487,507,750] to confirm the meridional structure. 2-Phenyliso-phosphindoline behaves as a normal phosphine towards Rh (compare Co), giving [RhL$_3$X$_3$][607]. The bulky phosphine (4-biphenylyl-1-naphthylphenylphosphine) also gives an octahedral complex [RhL$_3$Cl$_3$][383]. The meridional [Rh(PR$_3$)$_3$Cl$_3$] complexes failed to give organorhodium compounds with simple alkyl or aryl Grignards[382,772], but the reaction with hydrazine hydrochloride yielded [Rh$_2$Cl$_6$(N$_2$H$_4$)(PR$_3$)$_4$][772]. Hayter obtained two forms of [Rh(PPh$_2$H)$_3$Cl$_3$], presumably the *mer* and *fac* isomers, but only one form of [Rh(PEt$_2$H)$_3$Cl$_3$][502]. The *fac*-[Rh(PR$_3$)$_3$X$_3$] complexes have been obtained on a number of occasions as byproducts of the preparation of the meridional forms[487,672,772], and their configurations have been established by i.r. and Raman spectra[288,672,898]. The isomerisation of *mer*-[Rh(PEt$_2$Ph)$_3$X$_3$] to the *fac* isomer, on irradiation in solution, has been reported[830]. The mixed phosphine–phosphite complex *mer*-[Rh(Bu$^n_3$P)$_2${P(OMe)$_3$}Cl$_3$] has been studied by X-ray crystallography[1248] confirming the structure deduced from the ^{31}P n.m.r. spectrum.

Binuclear complexes [Rh$_2$X$_6$(PR$_3$)$_4$] (X = Cl; R = Et, Pr^n, Bu^n, *n*-pentyl) are obtained from RhX$_3$ and two equivalents of PR$_3$. Also, [Rh$_2$Cl$_6$(PEt$_3$)$_4$] was obtained from the reaction of $[Rh(PEt_3)_3Cl_3]$ and the calculated amount of $RhCl_3$ in ethanol[772]. Exchange occurs between $[Rh_2Cl_6(PBu^n_3)_4]$ and LiBr in acetone to give the bromine analogue, but iodine complexes have not been obtained pure[772]. Binuclear $[Rh_2Cl_6(PEt)_3]$ was also prepared[772]. The PBu^n_3 complexes have been reinvestigated[750,1228]. Allen and Gabuji[1228] found that Chatt's preparation[772] of $[Rh_2Cl_6(PBu^n_3)_4]$ gave a mixture from which $mer\text{-}[Rh(PBu^n_3)_3Cl_3]$, $[Rh_2Cl_6(PBu^n_3)_4]$ and $[Rh_2Cl_6(PBu^n_3)_3]$ could be separated. On the basis of ^{31}P n.m.r. spectra, the two binuclear complexes were thought to have the structures (LXXVIII, LXXIX) [1228].

(LXXVIII)   (LXXIX)

On irradiation in benzene, $[Rh(CO)(PPr^n_3)_2Cl_3]$ yields $[Rh_2Cl_6(PPr^n_3)_4]$. Brooks and Shaw isolated $[PhMe_2PH][RhCl_4(PPhMe_2)_2]$ in 2–3 per cent yield from the reaction mixture used to prepare *fac*- and *mer*-$[Rh(PMe_2Ph)_3Cl_3]$ [487]. $[Ph_4As][RhCl_4(PPh_3)_2]$ is obtained as orange crystals from $[Rh(PPh_3)_3Cl]$ and excess $[Ph_4AsCl \cdot HCl]$ in acetone[501], or from $[Rh(CO)(PPh_3)_2Cl]$ and excess Ph_4AsCl. Like the ruthenium analogues, treatment of $[RhCl_4(PPh_3)_2]^-$ with undiluted $PPhMe_2$ gave $[RhCl_4(PMe_2Ph)_2]^-$. Exchange with LiX (X = Br, SCN) occurred in solution, but unlike the ruthenium complexes $[RhX_4(PR_3)_2]^-$, they do not undergo solvolysis in nitromethane, MeCN, etc. Spectroscopic studies indicate *trans* structures. Rh(III) complexes containing CS_2, etc. formed by the addition of various small molecules to Rh(I) have already been described.

13.3 IRIDIUM

13.3.1 Carbonyls

Derivatives of $Ir_4(CO)_{12}$ were obtained from reaction of PPh_3 with $[Ir_4(CO)_{11}H]^-$ and $[Ir_8(CO)_{20}]^{2-}$, which afforded $[Ir_4(CO)_{10}(PPh_3)_2]$ and $[Ir_4(CO)_9(PPh_3)_3]$, respectively[773]. X-ray structural studies showed that the Ir_4 cluster was maintained, but that in contrast to the parent $Ir_4(CO)_{12}$, the phosphine-substituted complexes contain bridging carbonyl groups (LXXX)[774].

(LXXX)

Under pressure $Ir_4(CO)_9(PR_3)_3$ (R = Ph, p-MeC$_6$H$_4$) react with CO to give $[Ir(CO)_3(PR_3)]_2$ at 150°C and 100 atm (10 MN m^{-2}), $[Ir_2(CO)_7(PR_3)]$ at 175°C and 450 atm (45.6 MN m^{-2}), and $Ir_4(CO)_{12}$ at higher temperatures[775]. No evidence for intermediate formation of $Ir_2(CO)_8$ was obtained. Trialkylphosphines, PBu$^n_3$ and PEt$_3$, react directly with $Ir_4(CO)_{12}$ on refluxing in toluene to give $[Ir_4(CO)_8(PR_3)_4]$ [775]; on heating with CO under pressure phosphines are lost to produce $[Ir_4(CO)_9(PR_3)_3]$ and $[Ir_4(CO)_{10}(PR_3)_2]$. The $[Ir(CO)_3(PR_3)]_2$ and $[Ir_2(CO)_7(PR_3)]$ complexes presumably have structures analogous to the cobalt complexes[775]. Potassium diphenylphosphide and $[Ir(CO)_3Cl]_n$ react to form $[Ir(CO)_3(PPh_2)]_2$ [648].

13.3.2 Carbonyl halides

The remarkable properties of Vaska's compound, $[Ir(CO)(PPh_3)_2Cl]$, especially its reversible uptake of molecular oxygen, has resulted in much work in this field. We shall deal firstly with the preparation of Vaska's compound and related complexes, and then the oxidative addition reactions will be discussed.

When PPh$_3$ is reacted with $[Ir(CO)_3Cl]_n$ the *trans*-$[Ir(CO)(PPh_3)_2Cl]$ is formed[776]; it may also be synthesised by reduction of IrCl$_3$ or $(NH_4)_2IrCl_6$ in a refluxing alcohol—2-methoxyethanol, glycol, diglycol, etc.—or DMF in the presence of PPh$_3$ [702,777-9] (compare Rh(CO)(PPh$_3$)$_2$Cl). Metathesis yields the Br, I, NCS analogues, but a better method is metathesis from Ir(CO)(PPh$_3$)$_2$F when Ir(CO)(PPh$_3$)$_2$X (X = CN, CNS, CNSe, NO$_2$, CNO, N$_3$, ClO$_4$, HCO$_2$, OPh, OH, etc.) are produced[663]. The preparation of many related complexes with other phosphines has been described. Methods of preparation, in addition to the above, include saturation of IrX$_3$ or M$_2$IrX$_6$ (M = Na, NH$_4$, etc.) solution in a boiling alcohol with CO followed by addition of PR$_3$ [780], reduction of IrHX$_2$(CO)(PR$_3$)$_2$ with bases (dehydrohalogenation)[781], and in a few cases by displacement of other ligands (amines or other phosphines) from iridium carbonyl complexes[782,783]. Thus the following compounds have been obtained: *trans*-$[Ir(CO)(PR_3)_2X]$ (R = Ph, X = Br, I[782,784], NCS, NCSe[665]; R$_3$ = Et$_3$, Ph$_2$Bu, PhMe$_2$, PhEt$_2$, Me$_3$, X = Cl, Br[780,781,783]; R$_3$ = Bu$^t_2$H, X = Cl, Br, I [435]; R = Cy, X = Cl [776]). Upon treatment of Ir(CO)(PPh$_3$)$_2$Cl with PH$_3$ the mixed phosphine complex $[Ir(CO)(PPh_3)(PH_3)Cl]$ is obtained[398].

The usefulness of a particular preparation appears to depend upon the phosphine concerned; for example $[Ir(CO)(PPhEt_2)_2Br]$ is normally obtained from Na$_2$IrBr$_6$, CO and PPhEt$_2$ in 2-methoxyethanol[780], but the employment of PPhMe$_2$ in this same reaction gives inseparable mixtures of $[Ir(CO)(PPhMe_2)_2Br]$ and $[IrHBr_2(CO)(PPhMe_2)_2]$ [781].

The dipole moment of $[Ir(CO)(PPh_3)_2Cl]$ is 3.9 D at 25°C, which suggests a *trans* structure[777], and this has been confirmed by n.m.r. spectroscopy[781,783].

The reactions of *trans*-$[Ir(CO)(PR_3)_2Cl]$, and especially of Vaska's compound itself, have been extensively investigated. Much of the work has been reviewed[784-6] and the following treatment is not intended to be comprehensive. The *trans*-$[Ir(CO)(PR_3)_2X]$ complexes are 'coordinatively unsaturated' square-planar complexes, which readily take up two ligands (Y_2) to form octahedral $[Ir(CO)(PR_3)_2XY_2]$, although this does not necessarily mean that Y_2 is added *trans* across the plane. Some examples are given in figure 18.

The kinetics of the reaction

$$trans\text{-}[Ir(CO)(PPh_3)_2X] + AB \rightarrow [Ir(A)(B)(CO)(PPh_3)_2X]$$
$$(X = Cl, Br, I; AB = H_2, O_2, MeI)$$

have been studied by Chock and Halpern[784(a)], who found the reaction to be second order

$$\text{rate} = k[Ir(CO)(PPh_3)_2X][AB]$$

The stereochemistry of the addition reactions have also been examined. Vaska[784(b)] reported that the products of the reaction of *trans*-$[Ir(CO)(PPh_3)_2Cl]$ with a number of diatomic molecules have structures (LXXXI), that is the product of *cis* addition, and this was confirmed by other workers[672].

In contrast, Deeming and Shaw[809] found that *trans*-$[Ir(CO)(PPhMe_2)_2Cl]$ in benzene solution adds MeI, MeBr, MeCOCl and MeCOBr to give (LXXXII) by *trans*

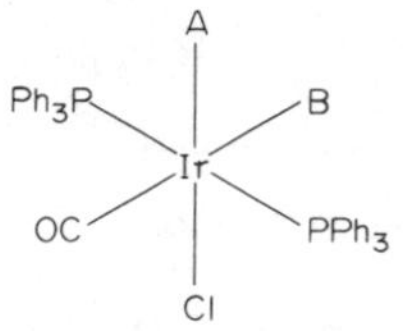

(LXXXI)   (LXXXII)

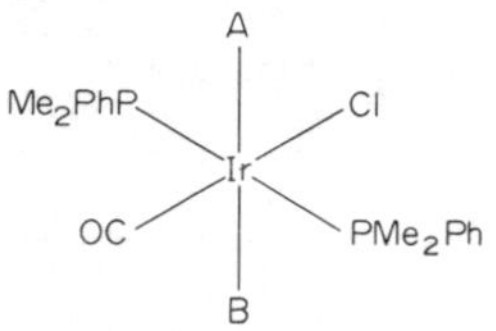

addition. This supported the earlier work by Collman and Sears[783] that addition of HX, X_2, RX and MeCOX to the PPh_2Me complex is kinetically controlled and leads to the *trans* product. The problem is complicated by the fact that isomerisation may occur after the initial addition[783], which illustrates the danger of inferring the mechanism of the addition from the configuration of the final product. Hydrogen bromide in ether adds to $[Ir(CO)(PPh_2Me)_2Cl]$ in benzene to give a mixture of the *cis* and *trans* products[783], which could be the result of a non-stereospecific reaction, or of a stereospecific addition followed by isomerisation. The solvent dependence has also been examined[810,811]; variation of the solvent was found to bring about rearrangement in the addition of allylic halides to Ir complexes[811]. Addition of HX to $[Ir(CO)(PR_3)_2X]$ (PR_3 = aryl phosphine) is stereospecifically *cis* in benzene or chloroform, but gives a *cis–trans* mixture in DMF, MeOH or H_2O; although whether the isomers result from exchange of halide ions before or after addition has not been determined[810]. Normally the PR_3 groups remain *trans* in the product, but in $[Ir(CO)\{C_2(CN)_4\}(PPh_3)_2Br]$ [812] and $[Ir(CO)(C_2F_4)(PPh_3)_2I]$ [813] the PPh_3 groups are mutually *cis*.

Oxygen uptake by Vaska's compound[787] and related complexes has attracted much interest as a model system for biological oxygen uptake. The structure of

Figure 18. Some reactions of *trans*-[Ir(CO)(PPh$_3$)$_2$Cl] [†]

trans-[Ir(CO)(PPh$_3$)$_2$Cl]

	→ Ir(CO)(O$_2$)(PPh$_3$)$_2$Cl [787]
H$_2$	→ Ir(CO)(H$_2$)(PPh$_3$)$_2$Cl [788]
SO$_2$	→ Ir(CO)(SO$_2$)(PPh$_3$)$_2$Cl [681]
(NC)$_2$C=C(CN)$_2$ (TCNE)	→ Ir(CO)(TCNE)(PPh$_3$)$_2$Cl [789]
RC≡CR	→ Ir(CO)(RCCR)(PPh$_3$)$_2$Cl [790]
R$_3$SiH	→ Ir(CO)(H)(R$_3$Si)(PPh$_3$)$_2$Cl [791]
MeI	→ Ir(CO)(Me)(PPh$_3$)$_2$ICl [792]
RCOX[‡] (R = Me, Bz)	→ Ir(CO)(RCO)(PR$_3$)$_2$ClX [793]
RSO$_2$Cl	→ Ir(CO)(RSO$_2$)(PPh$_3$)$_2$Cl$_2$ [794]
R^{f}I	→ Ir(CO)(R^f)(PPh$_3$)$_2$ICl [783]
Na-Hg + CO	→ Na[Ir(CO)$_3$(PPh$_3$)] [685,795]
N$_2$H$_4$-PPh$_3$	→ Ir(CO)H(PPh$_3$)$_2$ [699]
HgCl$_2$	→ Ir(CO)(HgCl)(PPh$_3$)$_2$Cl$_2$ [796]
RCON$_3$	→ Ir(N$_2$)(PPh$_3$)$_2$Cl [778]
BF$_3$	→ Ir(CO)(BF$_3$)(PPh$_3$)$_2$Cl [797]
CH$_2$N$_2$	→ Ir(CO)(CH$_2$Cl)(PPh$_3$)$_2$ [798]
X$_2$	→ Ir(CO)(PPh$_3$)$_2$ClX$_2$ [788]
HX	→ Ir(CO)(PPh$_3$)$_2$(H)ClX [799]
H$_2$SO$_4$	→ Ir(CO)(PPh$_3$)$_2$ClSO$_4$ [421]
S$_4$N$_4$	→ Ir(CO)(PPh$_3$)Cl(S$_4$N$_4$) [800]
R$_3$GeH	→ Ir(CO)(H)$_2$(GeR$_3$)(PPh$_3$)$_2$ [801]
R$_3$SnH	→ Ir(CO)(H)(PPh$_3$)$_2$(R$_3$Sn)Cl [802]
H$_2$S	→ Ir(CO)(H)(HS)(PPh$_3$)$_2$Cl [803]
HCN	→ Ir(CO)(H)(CN)(PPh$_3$)$_2$Cl [803]
CS$_2$	→ Ir(CO)(CS$_2$)(PPh$_3$)$_2$Cl [742]
C$_3$S$_2$	→ Ir(CO)(C$_3$S$_2$)(PPh$_3$)$_2$Cl [804]
Fe(NO$_3$)$_3$ or Cu(NO$_3$)$_2$	→ Ir(CO)(PPh$_3$)$_2$Cl(NO$_3$)$_2$ [805]
HSC$_6$H$_4$X	→ Ir(CO)(H)(SC$_6$H$_4$X)(PPh$_3$)$_2$Cl [806]
SnCl$_2$ + C$_2$H$_2$	→ Ir(CO)(C$_2$H$_2$)(PPh$_3$)$_2$SnCl$_3$ [807]
CO + NaClO$_4$	→ [Ir(CO)$_3$(PPh$_3$)$_2$]ClO$_4$ [808]

[†] References cited are selective and not necessarily the original.
[‡] Does not work for PPh$_3$ complex, but is known for PR$_3$ = Me$_2$PhP, Et$_3$P.

[Ir(CO)(O$_2$)(PPh$_3$)$_2$Cl] showed it to contain oxygen bonded 'sideways on' to the iridium with Ir—O = 207 pm, O—O = 130 ± 3 pm. The corresponding [Ir(CO)(PPh$_3$)$_2$I] takes up oxygen irreversibly and structural studies have shown that in the product, Ir(CO)(O$_2$)(PPh$_3$)$_2$I . CH$_2$Cl$_2$, the O—O distance is 147 ± 2 pm [815]. Interestingly, the 'ν(O—O)' frequency changes very little, 858 cm^{-1}(Cl) and 862 cm^{-1}(I), indicating that it is not a pure O—O stretching vibration. A recent report of the structure of [Ir(CO)(O$_2$)(PPh$_2$Et)$_2$Cl] shows it to have O—O = 146.1 ± 1.4 pm, near to the oxygen distance in the irreversible adduct and yet reversible uptake of oxygen is possible by the PPh$_2$Et complex. Hence, changes in the phosphine and halide ligands can bring about changes in the O—O bond length, and care should be taken in relating these to the reversibility of oxygen uptake[816].

Vaska and Chen[838] found that the rate of oxygen uptake by *trans*-[Ir(CO)(PR$_3$)$_2$Cl], and the stability of the resulting adducts, depends upon the basicity of the phosphine, and upon the size of the R groups. Rate of oxygenation increases in the order PCy$_3$ < P(*p*-ClC$_6$H$_4$)$_3$ < PPh$_3$ < AsPh$_3$ < PPh$_2$Et < P(*p*-MeC$_6$H$_4$)$_3$ < PBu^n$_3$ < PEt$_3$ < P(*p*-MeOC$_6$H$_4$)$_3$. Coordinated oxygen is kinetically more reactive than the free molecule; for example SO$_2$ is readily converted to sulphate, which remains coordinated to the iridium[421,817,818]. Treatment of *trans*-[Ir(CO)(PPhMe$_2$)$_2$Cl] with O$_2$ in benzene yields a solution of [Ir(CO)(O$_2$)(PPhMe$_2$)$_2$Cl], but on standing for two days the solution becomes intense blue-black and contains Me$_2$PhP=O [809].

Addition of HX, X$_2$ or H$_2$ produces the Ir(III) species described below. Green [Ir(CO)(SO$_2$)(PPh$_3$)$_2$Cl] is formed reversibly from Ir(CO)(PPh$_3$)$_2$Cl and SO$_2$ [681]; this diamagnetic complex is structurally similar to the Rh analogue[819], Ir—S = 249 ± 1 pm, Ir—P = 236.9 ± 0.9 pm, and loses SO$_2$ on heating[820]. Metal—metal bonded carbonyl derivatives are obtained by addition of a suitable substrate to [Ir(CO)(PR$_3$)$_2$X], or from [Ir(CO)$_3$(PR$_3$)]$^-$ anions. Nyholm and Vrieze[796] obtained a series of complexes (Ph$_3$P)$_2$(CO)ClYIrHgY (Y = Cl, Br, I, OAc, CN, SCN), and found that the Ir—Hg bond is broken by HX or X$_2$. Silyl iridium complexes have been prepared by hydrosilylation of [Ir(CO)(PPh$_3$)$_2$Cl] [791,821], which leads to [Ir(CO)(R$_3$Si)(H)(PPh$_3$)$_2$Cl] (R$_3$ = Cl$_3$, (OEt)$_3$, EtCl$_2$, PhCl$_2$) (LXXXIII) and Ir(CO)(R$_3$Si)(H)$_2$(PPh$_3$)$_2$. Hooton[822] prepared Ir(HgSiMe$_3$)(SiMe$_3$)$_2$(CO)(PEt$_3$)$_2$ (LXXXIV) from Ir(CO)(PEt$_3$)$_2$Cl and Hg(SiMe$_3$)$_2$. Germyl iridium complexes are obtained analogously[801,822]; for example Me$_3$GeH adds to *trans*-Ir(CO)(PPh$_3$)$_2$Cl to produce [Ir(CO)(Me$_3$Ge)(H)$_2$(PPh$_3$)$_2$], whilst Ph$_3$GeH gives [Ir(CO)(Ph$_3$Ge)(H)(PPh$_3$)$_2$Cl] [801]. Addition of R$_3$SnH (R = Me, Ph) to some *trans*-Ir(CO)(PR$_3$)$_2$X (X = Cl, Br, I) complexes has also been briefly reported[802].

The absorption spectra of a series of *trans*-[Ir(CO)(PR$_3$)$_2$X] (R$_3$ = Ph$_3$, Ph$_2$Bu,

(LXXXIII)

(LXXXIV)

Bz_3, Cy_3, Pr^i_3, Bu^n_3, $(p\text{-}MeC_6H_4)_3$) were studied by Strohmeier and Miller[823]. These workers also examined the formation of solvent complexes[824], and the electron density changes on the Ir in these complexes brought about by changes in X and PR_3 [825].

The catalytic application of Vaska's compound in hydrogenation and isomerisation reactions has also been studied[826-8]. The reduction of *trans*-[Ir(CO)(PPh$_3$)$_2$Cl] by Na–Hg in THF yields Na[Ir(CO)$_3$(PPh$_3$)], which reacts with Hg(CN)$_2$, Ph$_3$PAuCl, Me$_2$SnCl$_2$ and R$_3$SnCl to form the metal–metal bonded $(Ph_3P)(CO)_3Ir\text{-}Hg\text{-}Ir(CO)_3(PPh_3)$, $(Ph_3P)(CO)_3IrAu(PPh_3)$, $Me_2Sn\{Ir(CO)_3(PPh_3)\}_2$ and $R_3SnIr(CO)_3(PPh_3)$, respectively[685].

Wilkinson and coworkers have studied the addition of CS_2, and some related sulphur compounds, to Ir(I) carbonyls[684,742,766,803]. The reaction of *trans*-[Ir(CO)(PPh$_3$)$_2$Cl] with CS_2 and excess PPh$_3$ in MeOH gives a deep violet solution, which precipitates as [Ir(CO)(CS$_2$)(PPh$_3$)$_3$]BPh$_4$ on addition of NaBPh$_4$ [684].

Chlorine oxidation of *trans*-[Ir(CO)(PPh$_3$)$_2$Cl] gives [Ir(CO)(PPh$_3$)$_2$Cl$_3$] [788]. Three configurations (LXXXV; A, B, C) are possible. The [Ir(CO)(PR$_3$)$_2$Cl$_3$] (A)

(A)   (B)   (C)

(LXXXV)

(R_3 = PhEt$_2$, Et$_3$, Bu$^n_3$) complexes were prepared[829] by reaction of PR_3 with chloroiridic acid in boiling 2-methoxyethanol followed by saturation of the boiling solution with CO for several hours. The same reaction, using much shorter reaction times (about twenty minutes) produced isomers of configuration (B). On long standing the (B) isomers were observed to change into the third form, isomer (C)[829], probably due to photochemical reaction[830]. Phenyldimethylphosphine behaves anomalously, producing isomers (B) and (C) by direct reaction with chloroiridic acid, 2-methoxyethanol and CO, but it does not form a complex with structure (A) by this method. (A), however, has been obtained by boiling [Ir(CO)(PPhMe$_2$)$_2$Cl$_3$] (C) with LiCl in 2-methoxyethanol[831]. Direct chlorination of *trans*-[Ir(CO)(PPh$_2$Me)$_2$Cl] produced [Ir(CO)(PPh$_2$Me)$_2$Cl$_3$] (A) [783]. Metathesis using LiX (X = Br, I) converts [Ir(CO)(PPhMe$_2$)$_2$Cl$_3$] (C) into [Ir(CO)(PPhMe$_2$)$_2$X$_3$] (A)[831]. The structures of these complexes were deduced from dipole moment and ^{1}H n.m.r. spectroscopic studies[783,829,831]. The [Ir(CO)(PBu$^t_2$H)$_2$X$_3$] (X = Cl, Br) complexes with configuration (A) have been isolated[435]. Mixed halogen Ir(III) complexes, [Ir(CO)(PR$_3$)$_2$XY$_2$] (X, Y = halide; X ≠ Y) are formed from *trans*-[Ir(CO)(PR$_3$)$_2$X] and Y_2 [672,783]; the Y ligands are probably *trans*[783]. The oxidation of *trans*-[Ir(CO)(PPh$_3$)$_2$X] (X = Cl, Br, I, N$_3$, NCO, NCS) complexes by Ce(IV), Fe(III) or Cu(II) nitrates to [Ir(CO)(PPh$_3$)$_2$X(NO$_3$)$_2$], and by CuX$'_2$ (X$'$ = Cl, Br) to $[Ir(CO)(PPh_3)_2XX'_2]$ have been carried out[805]. The $[Ir(CO)(PPh_3)_2X(NO_3)_2]$ complexes are also formed by reaction of N_2O_4 with $[Ir(CO)(O_2)(PPh_3)_2X]$ [818], and from $[Ir(CO)(PPh_3)_2Cl]$ and concentrated nitric acid[803]. Many of the complexes formed in oxidative addition reactions of *trans*-$[Ir(CO)(PR_3)_2X]$ are Ir(III) complexes. These have already been discussed.

An Ir(II) complex, $[Ir(CO)(PPh_3)_2I_2]$, has been reported[832] to be formed from $K[Ir(CO)_2I_4]$ and PPh_3, but this has not been confirmed.

The monocarbonyl complexes described above are the most important and best known, but recently polycarbonyl species have been obtained. Dicarbonyl complexes, $[Ir(CO)_2(PPh_3)_2X]$, were prepared by Vaska[833] from *trans*-$[Ir(CO)(PPh_3)_2X]$ and CO, and by Angoletta from $K[Ir(CO)_2I_4]$ and PPh_3 or from $[Ir(CO)_2(amine)X]$ and PPh_3 [832].

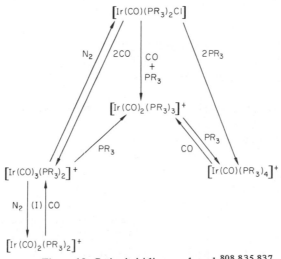

Figure 19 Cationic iridium carbonyls[808,835,837]
$R_3 = PhMe_2$ except I = Cy_3, Pr^i_3

Cationic derivatives, $[Ir(CO)_2(PR_3)_2]^+$ (R = Ph, Cy), were synthesised from $[Ir(CO)(PPh_3)_2Cl]$, CO and $AlCl_3$ [670]. (Note that the Rh analogue has been reformulated as $[Rh(CO)_3(PPh_3)_2]^+$). Malatesta *et al.*[834] prepared $[Ir(CO)_3(PPh_3)_2]ClO_4$ by bubbling CO through a benzene solution of $[Ir(CO)H_2(PPh_3)_2]ClO_4$, and found that it reacted with excess PPh_3 to give $[Ir(CO)_2(PPh_3)_3]ClO_4$. Deeming and Shaw[835] produced a series of five-coordinate cations, $[Ir(CO)_x(PPhMe_2)_{5-x}]^+$ (x = 1 − 3). In suspension in methanol $[Ir(CO)(PPhMe_2)_2Cl]$ takes up CO, and the resulting $[Ir(CO)_3(PPhMe_2)_2]X$ (X = BPh_4, ClO_4) can be precipitated by addition of the appropriate anion. Reaction with $PPhMe_2$ gives the $[Ir(CO)_2(PPhMe_2)_3]^+$ and $[Ir(CO)(PPhMe_2)_4]^+$ cations (figure 19). The structure of $[Ir(CO)_3(PPhMe_2)_2]ClO_4$ has been established by X-ray methods[836] to be (LXXXVI), but the structure of the di- and monocarbonyl cations have not yet been established[835]. The n.m.r. spectra are complex and may indicate the presence of rapidly interconverting species. Other workers[808,837] have independently prepared $[Ir(CO)_3(PR_3)_2]BPh_4$ ($R_3 = Ph_3$, Ph_2Me,

Ph_2Et, Et_3), and found that the complexes do not lose CO, but when $R_3 = Cy_3$ or Pr^i_3, bubbling N_2 through the solution produces an orange colouration, and on heating followed by precipitation with ether, red crystalline $[Ir(CO)_2(PR_3)_2]BPh_4$ were isolated[837]. Dehydrohalogenation of $[Ir(CO)HCl(PPhMe_2)_3]PF_6$ with Et_3N in acetone leads to $[Ir(CO)(PPhMe_2)_4]PF_6$, and the four-coordinate $[Ir(CO)(PR_3)_3]PF_6$ ($R_3 = Ph_2Me$, $PhMe_2$) are obtained from PR_3 and $[Ir(CO)(cyclooctene)_2Cl]_2$[688].

(LXXXVI)

(D)

(E)

(LXXXVII)

Reaction of $[Ir(CO)(PPh_3)_2Cl]$ with $AgClO_4$ in MeCN produces $[Ir(CO)(PPh_3)_2(MeCN)]^+$, which reacts with PR_3 ($R_3 = Ph_3$, Ph_2Me, Ph_2Et) to form $[Ir(CO)(PR_3)_3]^+$; the alkyl substituted phosphine complexes take up O_2 to give $[Ir(CO)(O_2)(PR_3)_3]^+$ [839].

Like the *trans*-$[Ir(CO)(PR_3)_2X]$ derivatives, the five-coordinate cationic complexes undergo oxidative addition reactions[837,840]; for example with H_2, HX, MeX or HgX_2 they give six-coordinate products. Alkoxycarbonyl complexes, $[Ir(CO)(CO_2R)(PR_3)_2X_2]$, $[Ir(CO)_2(CO_2R)(PR_3)]_2$ have been synthesised[834,841,842]. Dry HCl in chloroform decomposes the former type (X = Cl) to yield $[Ir(CO)_2(PR_3)_2Cl_2]^+$ [842]. Impure $[Ir(CO)_2(PPh_3)_2I_2]I$ has also been reported[841]. Shaw and Slade[843] have described the preparation of six-coordinate $[Ir(CO)(PPhMe_2)_3Cl_2]ClO_4$ by reaction of $[Ir(PPhMe_2)_3Cl_2(NO_3)]$ with CO in acetone in the presence of $NaClO_4$, and configuration (D) was assigned to this complex (LXXXVII) on the basis of n.m.r. spectral results. Treatment of $[Ir(PPhMe_2)_3Cl_2(NO_3)]$ with CO in boiling isopropylmethylketone gave another isomer (E) (LXXXVII). Reaction of isomer (D) with NaOMe produced the neutral $[Ir(CO_2Me)(PPhMe_2)_3Cl_2]$, but the corresponding iodo complex yielded a hydrido species, not the alkoxycarbonyl[843].

13.3.3 Hydrido carbonyls

Hydrido complexes in general are more stable and of wider occurrence in iridium chemistry than in rhodium chemistry. Diamagnetic $HIr(CO)(PPh_3)_3$ was prepared by heating *trans*-$[Ir(CO)(PPh_3)_2Cl]$ with hydrazine, preferably with added PPh_3 [699], or by reduction of *trans*-$[Ir(CO)(PPh_3)_2Cl]$ or $K[Ir(CO)I_5]$ and PPh_3 with $NaBH_4$ [844]. Proton n.m.r. spectroscopy indicates that it has an analogous structure to $HRh(CO)(PPh_3)_3$ (q.v.). Whyman[845] has recently found that under pressure a CO–H_2 mixture converts $HIr(CO)(PPh_3)_3$ to the other members of the series, $HIr(CO)_2(PPh_3)_2$ and $HIr(CO)_3(PPh_3)$; $HIr(CO)_3PPh_3$ is also formed by heating $[Ir(CO)_3PPh_3]_2$ or $Ir_4(CO)_9(PPh_3)_3$ with CO–H_2 under pressure. Four-coordinate $HIr(CO)(PPh_3)_2$ was reported to be formed on treatment of $IrH_3(PPh_3)_2$ with CO in benzene suspension, or by treating $[IrH_2(CO)(PPh_3)_2]ClO_4$ with KOH [834]; but other workers[846] have suggested that the product is a mixture of $HIr(CO)(PPh_3)_3$ and an unidentified species. Two forms of $HIr(CO)(PPh_3)_3$ have been reported[834].

The $[HIr(CO)(PPh_3)_3]$ complex reacts with H_2 to form $H_3Ir(CO)(PPh_3)_2$ [847], not $H_3Ir(CO)(PPh_3)_3$ as suggested earlier[848]. The dicarbonyl, $HIr(CO)_2(PPh_3)_2$, is prepared by $NaBH_4$ reduction of *trans*-$[Ir(CO)(PPh_3)_2Cl]$ in the presence of CO [849] or acidification of $Na[Ir(CO)_3(PPh_3)]$ with glacial acetic acid[685], as white crystals, which, unlike $HRh(CO)_2(PPh_3)_2$, do not dimerise with loss of H_2 on refluxing in solution under an inert atmosphere[849]. Reaction with PPh_3 reforms $HIr(CO)(PPh_3)_3$, and H_2 is taken up to form $H_3Ir(CO)(PPh_3)_2$. Two forms of $HIr(CO)_2(PPh_3)_2$ exist in solution in some solvents (i.r. evidence), and the ^1H n.m.r. spectra are strongly solvent and temperature dependent. The $HIr(CO)_2(PR_3)_2$ ($R_3 = (p\text{-}FC_6H_4)_3$, Ph_2Et) complexes were also prepared, but Me_2PhP and Et_2PhP gave intractable oils[849]. Only one form of $HIr(CO)_2(PR_3)_2$ exists in the solid state, but in solution two five-coordinate isomers seem to be present; possible structures of these have been discussed by Yagupsky and Wilkinson[849]. The compounds $HIr(CO)(PPh_3)_3$ and $HIr(CO)_2(PPh_3)_2$ are active catalysts for hydrogenation and hydroformylation reactions[707,847,848]. The reactions of CS_2 with $HIr(CO)(PPh_3)_3$ and $HIr(CO)_2(PPh_3)_2$ have been examined, but the products are very unstable and difficult to characterise[766].

The trihydrido $H_3Ir(CO)(PPh_3)_2$ is formed by $NaBH_4$ reduction of *trans*-$[Ir(CO)(PPh_3)_2Cl]$, whilst an isomeric form results from reduction of $[Ir(CO)H(PPh_3)_2I_2]$ [834]. The latter was assigned a *fac* configuration, and the former a *mer* configuration on the basis of i.r. spectral and dipole moment measurements[834]. $H_3Ir(CO)(PPh_3)_2$ dissolves in $HClO_4$ to give $[H_2Ir(CO)(PPh_3)_2]ClO_4$ [834,844]; the solution takes up CO to give $[H_2Ir(CO)_2(PPh_3)_2]ClO_4$. The $[H_2Ir(CO)_2(PR_3)_2]^+$ cations ($R_3 = Ph_3$, Ph_2Me, Ph_2Et, $PhEt_2$, Et_3, Cy_3, Pr^i_3) were obtained by reaction of H_2 with $[Ir(CO)_3(PR_3)_2]BPh_4$ in acetone–methanol[837], and shown to have structures (F) (LXXXVIII). Reaction with PR_3 has been studied for the PMe_2Ph complex, which gives $[Ir(CO)_2(PMe_2Ph)_3]^+$ [837]. A kinetic study[850] of the reaction of $[H_2Ir(CO)_2(PR_3)_2]BPh_4$ with CO, $P(OMe)_3$, $P(OPh)_3$ and $PPhMe_2$, found that the rate of reaction depends only upon the complex, and is independent of the substituting ligand, which led to the proposal that the mechanism involved reductive elimination of H_2 to form four-coordinate $[Ir(CO)_2(PR_3)_2]^+$, followed by reaction with the ligand. The preparation of $[IrH(CO)_2(PPhMe_2)_3]^{2+}$ has been reported[843], and the structure is thought to involve meridional phosphines and *cis* CO groups.

(F)   (G)   (H)

(LXXXVIII)

Hydridohalides, $[IrH(CO)(PR_3)_2X_2]$, and their deuterium analogues result from addition of HX (or DX) to $[Ir(CO)(PR_3)_2X]$ [777,788], and the hydrido species by boiling $[Ir(CO)(PR_3)_2X_3]$ with alcoholic KOH [829]. Treatment of $[Ir(CO)(PR_3)_2Cl_3]$ ($R_3 = PhMe_2, PhEt_2, Me_3, Et_3$), isomer (B) (LXXXV), with boiling KOH–EtOH yielded $[IrH(CO)(PR_3)_2Cl_2]$ (G), but isomer (A) did not react[781,829]. The same complex (G) is formed by treatment of *trans*-$[Ir(PPhEt_2)_3Cl_3]$ in ethanol with CO at 77°C and 70 atm (7.1 MN m^{-2}), but at 120°C and 78 atm (7.9 MN m^{-2}) isomer (H) is formed. The structures of these species were assigned by proton n.m.r. spectral and dipole moment studies[829]. A dinuclear $[Ir_2H_2(CO)_2(PEt_3)_2Cl_4]$ was obtained from 'chloroiridous' acid, HCl, CO, and PEt$_3$ in boiling 2-methoxyethanol[829]. The stereochemistry of the addition of HX to Vaska-type complexes has already been discussed[783,799,809,810]. Carbon monoxide will convert $[Ir(PEt_3)_3Cl_3]$ to $[IrH(CO)(PEt_3)_2Cl_2]$ in hot acetone or ethanol, but attempts to prepare $[IrH(CO)(PEt_3)_2Cl]^+$ failed[851].

The $[Ir(CO)(PPhMe_2)_4]^+$ cation loses PPhMe$_2$ on treatment with HCl to give (LXXXIX); the same product is formed from $[Ir(CO)_2(PPhMe_2)_3]^+$, but $[Ir(CO)_3(PPhMe_2)_2]^+$ gives the neutral $[IrH(CO)(PPhMe_2)_2Cl_2]$ [840].

(LXXXIX)

Finally, there are dihydrido complexes, $[IrH_2(CO)(PR_3)_2X]$, formed by addition of H$_2$ to Ir(CO)(PR$_3$)$_2$X [447,788,809]. Vaska also showed[852] that HD adds to $[Ir(CO)(PPh_3)_2Cl]$ to yield dihydrido and dideuterio complexes in addition to that containing HD[853]. In the presence of SnCl$_2$, trichlorostannate(II) complexes are formed[853]. The similar addition of hydrogen to $[Ir(CO)(PPhMe_2)_4]^+$ gives $[IrH_2(CO)(PPhMe_2)_3]$, which has meridional phosphines and *cis* hydrogens[840].

13.3.4 Nitrosyls

Violet $[Ir(NO)_2(PPh_3)]$, formally Ir(−II), results from treatment of $[IrH_3(PPh_3)_2]$ ($[IrH_5(PPh_3)_2]$?) with NO [854,855]. It has $\mu_{eff} = 1.4$ B.M. and ν(N–O) at 1645, 1680 cm^{-1}. Neutral orange $[Ir(NO)(PPh_3)_3]$ and $[Ir(NO)(CO)(PPh_3)_2]$ complexes were originally prepared by the same workers. The former, $[Ir(NO)(PPh_3)_3]$, is synthesised from $[Ir(NO)_2(PPh_3)_2]ClO_4$ and PPh$_3$, or by heating $[Ir(NO)(CO)(PPh_3)_2]$ with a large excess of PPh$_3$ in xylene[856]. The $[Ir(NO)(CO)(PPh_3)_2]$ complex results from reaction of $[HIr(CO)(PPh_3)_3]$ and NO [854,855] or N-methyl-N-nitrosotoluene-*p*-sulphonamide [856]. Reed and Roper[856] studied oxidative addition reactions with the d^{10} $[Ir(NO)(PPh_3)_3]$: X$_2$, CX$_4$ or HgX$_2$ produce $[Ir(NO)(PPh_3)_2X_2]$ and $[Ir(NO)(PPh_3)_2X_3]$; MeI in a sealed tube gives $[IrMe(NO)(PPh_3)_2I]$; and HX affords $[IrH(NO)(PPh_3)_2X]$ and $[Ir(NO)(PPh_3)_2X_2]$. Excess HX reacts in a totally different way to form $[Ir(NH_2OH)(PPh_3)_2X_3]$ [856]. The $[IrMe(NO)(PPh_3)_2I]$ compound is the first example of a five-coordinate complex with a methyl group *trans* to another ligand[857]. The isolation of the $[Ir(NO)_2(PPh_3)_3X]$ complexes derives from the treatment of $[Ir(NO)_2(PPh_3)_2]ClO_4$, itself formed from $[IrH_2(PPh_3)_3]ClO_4$ and NO with LiX[854,858]. The best known iridium nitrosyls are $[Ir(NO)(PPh_3)_2X_2]$; formed from $[Ir(NO)_2(PPh_3)_2]ClO_4$ and HX [854,858]; from $[Ir(NO)(PPh_3)_3]$ and X_2 or HX (1 : 2 ratio)[856]; by treating $[Ir(cyclooctadiene)Cl]_2$ with NOCl and refluxing the product in acetone with PPh_3 [717]; or, conveniently, from PPh_3 in refluxing

TABLE 11 IRIDIUM NITROSYLS

Compound	Ir—N (pm)	Ir—P (pm)	IrÑO	N—O (pm)	Ref.
$[Ir(NO)(PPh_3)_3]$	167	231	(linear)	124	861
$[Ir(NO)_2(PPh_3)_2]ClO_4$	177	234	164	121	862
			NÎrN = 154°		
$[Ir(CO)(NO)(PPh_3)_2Cl]BF_4$	197	241	124	116	859,863
$[Ir(CO)(NO)(PPh_3)_2I]BF_4$	189	236	125	117	864
$[Ir(NO)(Me)(PPh_3)_2I]$	191	235	120	123	857
$[Ir(NO)(PPh_3)_2Cl_2]$	194	237	123	103	865
$[Ir(H)(NO)(PPh_3)_3]ClO_4$	168	234 (av)	175	121	866

alcohol treated successively with IrX_3 and N-methyl-N-nitrosotoluene-*p*-sulphon-amide[719]. Treatment of $[Ir(NO)_2 (PPh_3)_2]ClO_4$ or $[Ir(NO)(PPh_3)_3]$ with excess halogen forms $[Ir(NO)(PPh_3)_2X_3]$ [854-6,858]. In the reaction with chlorine the $[Ir(NO)(PPh_3)_2Cl_3]ClO_4$ complex also forms[854].

Violet diamagnetic $[Ir(NO)(CO)(PPh_3)_2X]BF_4$ (X = Cl, I) complexes result from treatment of *trans*-$[Ir(CO)(PPh_3)_2X]$ with $NOBF_4$ [859]. The nitrosyl analogue of Vaska's compound, $[Ir(NO)(PPh_3)_2Cl]^+$, forms on treatment of $[IrH(NO)(PPh_3)_2Cl]$ with $HClO_4$ [860], and being coordinately unsaturated adds CO, PPh_3, etc. to produce five-coordinate complexes. Hydrolysis produces orange $[Ir(OH)(NO)(PPh_3)_2]^+$, and in alcohols $[Ir(NO)(OR)(PPh_3)_2]^+$ form reversibly[860]. Hydridonitrosyls are intermediate in the reaction of $[Ir(NO)(PPh_3)_3]$ with controlled amounts of HX, and with non-complexing acids—$HClO_4$, HBF_4, HPF_6—$[IrH(NO)(PPh_3)_3][anion]$ can be isolated[856]. The structures of several Ir–NO complexes have been determined and some very interesting results have been obtained (table 11).

A novel complex is the oxo-bridged $[\{Ir(NO)(PPh_3)\}_2O]$, formed from *trans*-$[Ir(CO)(PPh_3)_2Cl]$ and $NaNO_2$ in C_6H_6–EtOH–H_2O [867]. X-ray studies have shown it to have structure (XC).

(XC)

Crystals of $[Ir(NO)(PPh_3)_3]$ contain three molecules of the approximately tetra-hedral complex (with a linear NO group) in the unit cell, and exemplify the un-balanced packing of chiral molecules[861]. Much interest in this field has been generated by the discovery that, as well as the common NO^+ complexes, nitric oxide can co-ordinate as NO^-. Thus $[Ir(NO)_2(PPh_3)_2]ClO_4$, for example, can be formulated as $Ir(-I)$ with two NO^+ ligands or as $Ir(I)$ with NO^+ and NO^-, support for both formula-tions coming from some of its reactions[862]. The structure as determined by Mingos and Ibers[862] favours the $Ir(-I)$ d^{10} formulation. An approximately linear $Ir-N-O$ linkage $(175°)$ in the complex $[IrH(NO)(PPh_3)_3]ClO_4$ indicates that the best formula-tion is that of an $Ir(I)$ complex containing an NO^+ group. In contrast, all four $[IrXY(NO)(PPh_3)_2]$ $(X = Y = Cl; X = Me, Y = I)$ and $[Ir(CO)(NO)(PPh_3)_2X]BF_4$ $(X = Cl, I)^{[857,859,863-5]}$ complexes contain bent NO groups $(Ir\widehat{N}O) \simeq 120°)$, and are best considered to be NO^- complexes (compare $[Ru(NO)_2(PPh_3)_2Cl]PF_6$, also an NO^- complex[458]). Some data on these complexes are collected in table 11, and for further information the reader is referred to the extremely interesting discussion on these complexes in the series of papers by Ibers and coworkers[857,863-6] (especially reference 866).

13.3.5 Dinitrogen complexes

Collman and Kang[778] found that *trans*-$[Ir(CO)(PPh_3)_2Cl]$ reacted with azides, $RCON_3$, in $CHCl_3$ at $0°C$ to give golden yellow $[Ir(N_2)(PPh_3)_2Cl]$ [868], and subsequently Collman *et al.*[869] established the mechanism of the reaction. Chatt *et al.*[870] extended this reaction to other iridium phosphine complexes, but found that few of the products could be obtained pure. However, several were prepared, and the $\nu(N\equiv N)$ frequencies measured and compared with the analogous carbonyls; reasons for the instability of the dinitrogen complexes were also discussed[870]. In reactions with such reagents as acetyl chloride the dinitrogen is lost[729,793]. As with the rhodium complexes, reaction of NaN_3 with $[Ir(CO)(PR_3)_2Cl]$ produced the azido complexes $[Ir(CO)(PR_3)_2N_3]$ but these could not be decomposed to nitrido complexes[261] (compare Re).

13.3.6 Hydrido complexes

Iridium forms a large number of hydrido complexes[871] (see figure 20). Carbonyl hydrides have already been discussed (section 13.3.3). Simple hydrido complexes of types $IrH_5(PR_3)_2$, $IrH_3(PR_3)_3$, $IrH_2(PR_3)_3X$, $IrH(PR_3)_3X_2$, $[IrH_2(PR_3)_3]^+$ and $[IrH_2(PR_3)_4]^+$ are known.

The $IrH_5(PR_3)_2$ $(R_3 = PhEt_2, Et_3)$ have been recognised only recently[872], having been previously formulated as trihydrido complexes[873]. The reduction of *mer*-$Ir(PR_3)_3Cl_3$ with $LiAlH_4$ in THF produced $IrH_5(PR_3)_2$ as colourless solids[873]. The pentahydrido formulation is supported by ^1H and ^{31}P n.m.r. spectroscopy, and by the reactions with PR_3 and HCl [872]; taken together with the zero dipole moments, *trans* phosphines are indicated. Reactions with other ligands (see below) produce $IrH_3(PR_3)_2L$, principally as the meridional isomer, although a small quantity of the

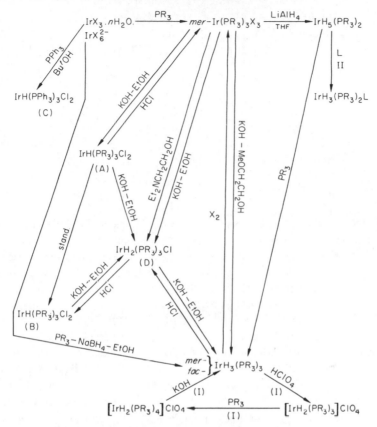

Figure 20 Iridium hydrides[788,834,844,871-80,882-6]
R_3 = Me_2Ph, Ph_3, Et_3, Et_2Ph except I, PR_3 = PPh_3; II, L = $P(OMe)_3$, $SbPh_3$, CO, MeNC, various PR_3. A, B, C, D refer to configurations of isomers (see text)

facial isomer is also formed[872]. Other workers have prepared $IrH_5(PMe_3)_2$ [874]. It has been reported that $NaBH_4$ or $LiAlH_4$ reduction of $IrH(PPh_3)_2X_2$ led to $IrH_3(PPh_3)_2$ [834,844,873], but this may also be the pentahydride. Araneo *et al.* have reported the reaction of '$IrH_3(PPh_3)_2$' with HX, X_2, etc.[875].

Octahedral $IrH_3(PR_3)_3$ (R_3 = Ph_3, Ph_2Me, Ph_2Et, $PhMe_2$, $PhEt_2$) are formed when PR_3 reacts with $IrH_5(PR_3)_2$, or by $LiAlH_4$ or $NaBH_4$ reduction of $IrH(PR_3)_3Cl_2$ [873-80]. Two isomers, *mer-* and *fac-*$IrH_3(PR_3)_3$, are possible, and these have been isolated, although agreement between the properties reported by different workers is not very good. The ^1H n.m.r. spectra of the *fac* isomers are complex, consisting of six peaks[872,873,881], and reports[873,880] that the ^1H n.m.r. spectra of the *mer-*$IrH_3(PR_3)_3$ compounds are symmetrical quartets has been shown to be incorrect[872]. On dissolution in benzene, or on melting, both isomers partially isomerise to the other configuration[873]. When $IrH_5(PR_3)_2$ are treated with the appropriate ligand the $IrH_3(PR_3)_2L$ (L = various PR_3, $P(OMe)_3$, $SbPh_3$, SMe_2, CO, MeNC) result, and the *trans* phosphine configuration predominates in the products[872].

Hydrogen halides convert $IrH_3(PR_3)_3$ to $IrH_2(PR_3)_3X$ and $IrH(PR_3)_3X_2$, and ultimately to $Ir(PR_3)_3X_3$ by halogen addition.

Three isomers of $[IrH(PR_3)_3X_2]$ are possible (XCl: J, K, L). Chatt and coworkers[873]

(J)   (K)   (L)

(XCI)

found that treatment of *mer*-$Ir(PR_3)_3X_3$ with boiling alcoholic KOH, resulted in successive replacement of X by H (the product obtained depends upon the severity of the conditions), with corresponding conversion of the alcohol to the aldehyde or ketone. The $PPhEt_2$ complex gave successively (J), $IrH(PR_3)_3X_2$, and finally a mixture of *mer*- and *fac*-$IrH_3(PR_3)_3$ (figure 18). Hydrogen *trans* to phosphorus is readily replaced: treatment of $IrH(PPhEt_2)_3Cl_2$ (J) with dilute HCl gives vigorous evolution of H_2 and quantitative production of *mer*-$Ir(PPhEt_2)_3Cl_3$ [873]. Conversely, H *trans* to Cl is inert; $IrH_2(PPhEt_2)_3Cl$ (isomer M) gave $IrH(PPhEt_2)_3Cl_2$ (K) on treatment with HCl. The corresponding bromides and iodides result from reaction of $Ir(PR_3)_3Cl_3$ with LiX (X = Br, I) in boiling alcoholic alkali[873]. By analogous routes have been prepared $IrH(PR_3)_3X_2$ (R_3 = Ph_2Me, Ph_2Et, $PhMe_2$, $PhPr^n_2$, $PhBu^n_2$, PPr^n_3, 2-phenylisophosphindoline)[607,873,880,881,883,885]. The preparation of PPh_3 complexes has been reported in a number of papers[788,877-80,882-6], but it is not always clear which isomers were prepared in a particular reaction, especially in some of the earlier work. All three possible isomers are known, the unusual $IrH(PPh_3)_3Cl_2$ (L) resulting from $IrCl_3$, PPh_3 and refluxing aqueous Bu^tOH[788]. The configuration of many of the $IrH(PR_3)_3X_2$ complexes have been assigned by the use of the usual techniques, namely: ^1H n.m.r.[873,880,881] and i.r.[873,885,887] spectroscopy and dipole moment measurements[873,880].

Although three isomers of $IrH_2(PR_3)_3X$ are possible, only isomer (M) (XCII) has been definitely identified, although there have been brief mentions of other isomers (not characterised) in some reports. Most of the routes to these compounds have already been mentioned[853,873,880,881,883-5]; a simple direct method is to reflux *mer*-$Ir(PR_3)_3X_3$, or *mer*-$Ir(PR_3)_3Cl_3$ and LiX' (X' = Br, I), with 2-diethylamino-ethanol in the absence of inorganic base[873]. The formation of $Ir(PR_3)_3X_2Y$ (X ≠ Y = halogen) from the hydridocomplexes is discussed in section 13.3.7.

Powell and Shaw[888] have prepared mixed phosphine complexes, $IrH(PR_3)_2(PPhMe_2)Cl_2$ (XCIII) by making use of the observation that PR_3 *trans* to H in $IrH(PR_3)_3Cl_2$ (J) is labile towards substitution by $PPhMe_2$, but PR_3 *trans* to PR_3 is inert. By following this reaction by ^1H n.m.r. spectroscopy, the rate of substitution of PR_3 was established as $AsEt_3 \gg PEt_3 > PBu^n_3 \gg PPhEt_2 > PPhBu^n_2$;

substitution also proceeds faster in $IrH(PEt_3)_3Br_2$ than in $IrH(PEt_3)_3Cl_2$ [888]. Isomers (K) of $IrH(PR_3)_3Cl_2$ undergo a complex reaction with $PPhMe_2$, more than one phosphine being replaced, whilst $IrH_2(PPhEt_2)_3Cl$ showed no evidence of any reaction [888]. Araneo and Martinengo have reported [883] the $IrH(PPh_3)_2LX_2$ (L = NH_3, py, MeCN, $SbPh_3$) series of complexes.

(M)

(XCII)

(XCIII)

The reaction of Na_3IrCl_6 and PPh_3 is reported [883] to afford $IrH(PPh_3)_2Cl_2$, but this does not seem to have been confirmed. However, genuine five-coordinate $IrH(PR_3)_2X_2$ have been prepared using the bulky $PRBu^t_2$ (R = Me, Et, Pr^n) ligands [889]. These dark purple complexes result on heating $PRBu^t_2$ with $IrCl_6^{2-}$ in isopropanol, and have been assigned square-pyramidal structures. The triplet hydride resonances are the highest yet observed, ~60 τ. It is not possible to add a further molecule of $PRBu^t_2$ to give six-coordinate complexes, but smaller ligands—CO, MeCN, py—do add to form six-coordinate derivatives with the added ligands *trans* to H [889].

A cationic hydride, $[IrH_2(PPh_3)_3]ClO_4$ is obtained on dissolving $IrH_3(PPh_3)_3$ in $HClO_4$ [834,844,877], whilst in the presence of PPh_3 or pyridine $[IrH_2(PPh_3)_2L_2]ClO_4$ result [884]. Treatment of $IrH_3(PPh_3)_3$ with CN^-, NO_2^- or CH_3COO^- afford $IrH_2(PPh_3)_3X$ [884], and with acetylacetone $IrH_2(PPh_3)_2(acac)$ is formed [890]. An unlikely looking $[IrH(PPh_3)_2]^+$ has also been claimed [883]. When excess PBu^n_3 is reacted with $[Ir(cyclooctadiene)Cl]_2$ the *cis*-$[IrH_2(PBu^n_3)_4]^+$ is formed, whilst reaction involving eight moles of $PPhMe_2$ or PPh_2Me affords $[Ir(PR_3)_4]^+$ [891].

Catalysis by iridium hydrides has been reported [892,893], and hydrido species are probably intermediate in the catalysed hydrogenation of alkenes by iridium phosphine complexes.

13.3.7. Halide complexes

Iridium(I) halides, $Ir(PPh_3)_3X$ (X = Cl, Br), have recently been prepared [792,894,895] by treatment of $[Ir(N_2)(PPh_3)_2Cl]$, $[Ir(cyclooctadiene)Cl]_2$ or $[Ir(cyclooctadiene)(PPh_3)Cl]$ with PPh_3 in ligroin under reflux. A series of $[Ir(PR_3)_3Cl]$ (R = C_6D_5, o-DC_6H_4, p-FC_6H_4, p-$MeOC_6H_4$, p-MeC_6H_4) complexes have been prepared [894]. The tendency to dissociate in solution is reported to be less for $[Ir(PPh_3)_3Cl]$ than for $[Rh(PPh_3)_3Cl]$ (q.v.); but like the rhodium analogue the iridium complex readily undergoes oxidative addition reactions and substitution reactions to give, for example, $[Ir(L)(PPh_3)_2Cl]$ (L = CO, PF_3). Hydrogen adds irreversibly to form $[IrH_2(PPh_3)_3Cl]$ (compare the rhodium analogue where H_2 addition to $Rh(PPh_3)_3Cl$ is reversible), and $Ir(PPh_3)_3Cl$ is not a homogeneous hydrogenation catalyst for olefins. Excess chlorine produces both $Ir(PPh_3)_3Cl_3$ and $Ir(PPh_3)_2Cl_4$. On heating

$[Ir(PR_3)_3Cl]$ in benzene, acetone, chloroform or cyclohexane, colourless, monomeric, diamagnetic complexes are formed. These complexes have been identified[894] as of type $[IrH(o\text{-}C_6H_4PR_2)(PPh_3)_2Cl]$ (XCIV), and studies using deuterated phosphines show that the H (D) comes from the ortho position on one of the phenyl rings on the phosphine.

Although $[Ir(PPh_3)_3Cl]$ is not a catalyst, a benzene solution of $[Ir(cyclooctene)_2Cl]_2$ and two moles of PPh_3 is active in the hydrogenation and isomerisation of alkenes, and the active species is $[Ir(PPh_3)_2Cl]$ [896].

A further Ir(I) species is $[Ir(PPh_2Me)_4]^+$, formed from $[Ir(CO)(PPh_3)_2Cl]$ and silver perchlorate in MeCN followed by treatment with PPh_2Me [839]. Rather surprisingly, this complex does not react with O_2 or CO but is oxidised rapidly by H_2, HCl and Cl_2. A brief report of the formation of $[Ir(PPhMe_2)_4O_2]BPh_4$ from $[Ir(PPhMe_2)_4(CO)]PF_6$ and oxygen in acetone has appeared[688].

Both meridional and facial isomers of $[Ir(PR_3)_3X_3]$ are known. On boiling $IrCl_6{}^{2-}$, HCl and PEt_3 in ethanol yellow *mer*-$[Ir(PEt_3)_3Cl_3]$ was formed, accompanied by small quantities of a white substance, probably the *fac* isomer[897]; in addition there is evidence that some $[PHEt_3][Ir(PEt_3)_2Cl_4]$ is obtained. The *mer* isomer can better be prepared in 2-methoxyethanol. In alcohol $IrBr_6{}^{2-}$ produces hydrido complexes with phosphines, but *mer*-$Ir(PPhEt_2)_3Br_3$ is formed in methyl ethyl ketone. Iodo complexes can be prepared from the corresponding chloro analogue by metathesis with NaI in a ketone[885]. By similar routes compounds *mer*-$[Ir(PR_3)_3Cl_3]$ (R_3 = Ph_2Me, $PhMe_2$, $PhEt_2$, $PhPr^n{}_2$, $PhBu^n{}_2$, Et_3, $Pr^n{}_3$, $(p\text{-}MeC_6H_4)Et_2$, 2-phenylisophosphindoline)[504,607,885,897] have been prepared. The facial isomers have generally been prepared in low yield as byproducts of the syntheses of the meridional isomers; early reports that on standing in solution *mer*-$[Ir(PR_3)_3Cl_3]$ isomerised to *fac*-$[Ir(PR_3)_3Cl_3]$ have recently been confirmed[830], and the photochemical nature of the reaction has been demonstrated. When *mer*-$[Ir(PR_3)_3Cl_3]$ (R_3 = $PhMe_2$, $PhEt_2$, $PhPr^n{}_2$, $PhBu^n{}_2$, Et_3) is dissolved in benzene or acetone, and irradiated, the pure *fac* isomer is deposited in high yield[830]. The preparation of $[Ir(PR_3)_3Cl_2X]$ (XCV) (R_3 = $PhMe_2$, $PhEt_2$;

(XCIV)

(XCV)

X = Br, I, CNS) from $[IrH(PR_3)_3Cl_2]$, NaX and nitric acid, or from $[Ir(PPhMe_2)_3Cl_3]$ and NaX, have been reported[885]. The last reaction again demonstrates the lability differences of halide *trans* to PR_3 compared with *trans* halide. Shaw and Slade found that the halide *trans* to phosphorus is readily replaced by nitrate by treatment with $AgNO_3$ [843]. The nitrate ligand is in turn easily displaced by neutral ligands (CO, py, NH_3, β-picoline, γ-picoline) to give cationic complexes, $[Ir(PR_3)_3LX_2]^+$, and by anions to give neutral $[Ir(PR_3)_3YX_2]$ (Y = N_3, NO_2, OMe, Br, I, CN)[843].

Mixed phosphine and phosphine–arsine complexes have been synthesised[880]; *mer*-$[Ir(AsR_3)_3Cl_3]$ (R_3 = PhEt$_2$, Et$_3$) treated with excess PPhMe$_2$ produced

(XCVI)   (XCVII)

$[Ir(AsR_3)(PPhMe_2)_2Cl_3]$, which, on n.m.r. evidence, has *mer* Cl and *trans* PPhMe$_2$ groups (XCVI). Similar treatment of *mer*-$[Ir(PR_3)_3Cl_3]$ (R_3 = PhEt$_2$, Et$_3$) yielded $[Ir(PR_3)(PPhMe_2)_2Cl_3]$, but with a different configuration (XCVII). The physical properties of *mer*-$[Ir(PR_3)_3X_3]$ and *mer*-$[Ir(PR_3)_3X_2Y]$ have been well studied by ^{1}H n.m.r.[843,881], i.r.[288,885] and Raman[898] spectroscopy. Probably because of preparative difficulties the *fac* isomers have been less well studied.

Other iridium(III) complexes that have been reported are $[Ir(PPh_3)_2Cl_3]$, formed from IrH$_3$(PPh$_3$)$_2$ and chlorine, which may have a square-pyramidal structure[899], and the chlorine-bridged dimer $[Ir_2(PEt_3)_4Cl_6]$, formed by boiling chloroiridic acid with 2-methoxyethanol followed by addition of two equivalents of PEt$_3$ per Ir atom[880]. Complexes of PX$_3$ (X = Cl, Br) were reported many years ago[900]; they are claimed to be Ir(PX$_3$)$_3$X$_3$ and Ir(PX$_3$)$_2$X$_3$, but these complexes would repay further study.

Some methyliridium complexes of PPhMe$_2$ have been described[886], but as with rhodium, other simple alkyl or aryl complexes seem to be scarce.

Iridium(IV) complexes have been obtained as violet paramagnetic solids. Vaska[886] mentioned $[Ir(PPh_3)_2Cl_4]$ formed by chlorine oxidation of IrH$_2$(PPh$_3$)$_3$Cl, but few details were included. A better method seems to be chlorine oxidation of $[Ir(PPh_3)_3Cl]$[894] or $[PHR_3][Ir(PR_3)_2Cl_4]$ (R = PhMe$_2$, PhEt$_2$, PPr$^n_3$) in chloroform[504]. The magnetic moments are ~1.9 B.M., and a study of the variation of the magnetic susceptibility with temperature of $[Ir(PPh_3)_2Cl_4]$ showed it to follow the Curie law[894]. The *trans* configuration has been assigned[288], e.s.r. spectra of the PPh$_3$ and PPr$^n_3$ complexes have been measured[506,894], and the charge transfer spectra of $[Ir(PPr^n_3)_2Cl_4]$ discussed[508].

Iridium(III) anions, $[Ir(PR_3)_2Cl_4]^-$ (R_3 = PhMe$_2$, PhEt$_2$, Et$_3$, Pr$^n_3$), have been isolated as pink salts of $[PHR_3]^+$, as byproducts of the preparation of *mer*-$[Ir(PR_3)_3Cl_3]$ or by boiling IrCl$_3$.nH$_2$O with concentrated HCl and PR$_3$ in ethanol[847,885,898], and have been assigned *trans* structures[288,881,898].

Finally, it should be mentioned that complexes of iridium halides, $[Ir(PPh_3)_3X]$, react with CS$_2$ in methanol to give blue $[Ir(\pi\text{-}CS_2)(CS_2)(PPh_3)_3]BPh_4$, and this loses CS$_2$ *in vacuo* to give $[Ir(\pi\text{-}CS_2)(PPh_3)_3]BPh_4$. Pure CS$_2$ and PPh$_3$ react with $[Ir(PPh_3)_3Cl]$ to give a green solution, which when evaporated in the presence of chloroform produces orange $[Ir(CS_2)(PPh_3)_2Cl]$[684]. A much better synthesis of the last compound consists of treating $[Ir(N_2)(PPh_3)_2Cl]$ and CS$_2$ in methanol with PPh$_3$[901].

14 NICKEL, PALLADIUM, PLATINUM

14.1 NICKEL

14.1.1 Carbonyls

Nickel tetracarbonyl reacts with phosphines to form $[Ni(CO)_3(PR_3)]$ and $[Ni(CO)_2(PR_3)_2]$, and very occasionally higher substitution products. Generally,

TABLE 12 NICKEL CARBONYL PHOSPHINES

Phosphine	Known products[†]	Comments	Ref.
PMe_3	1,2,3	rare example of type 3 known	905-7
PEt_3	1,2		905,908
PBu^n_3	1,2		306,909
$P(^noctyl)_3$	2		910
$P(CF_3)_3$	1,2		908,911,912
$P(CH_2CH_2CN)_3$	2	also $Ni_4(CO)_6(PR_3)_4$ (see text)	909
$P(C{\equiv}CPh)_3$	1,2		908
$Ph_2PC{\equiv}CPh$	3	pale green solid	85
PPh_3	1,2		306,905,909,913,914
$P(p\text{-}MeC_6H_4)_3$	1		306,915
PPh_2Et	2		910
$PPhEt_2$	1,2		910
$P(NMe_2)_3$	1,2		51,76,916
$P(NMe_2)_2Cl$	2		916
$P(NMe_2)_2CN$	2		916
PBu^t_3	1⎫	only monosubstitution	96,917
$P(MMe_3)_3$ (M = Si,Ge,Sn,Pb)	1⎬	with these bulky phosphines	
$PPh_2(SnMe_3)$	1,2		918,919
$PPh(SnMe_3)_2$	1		919
PPh_2Bu^n	2		555
$PPhBu^n_2$	2		555
PCl_3	1,2,3	also $Ni(PCl_3)_4$	905,910
PH_3	1	unstable	322,921

† $1 = Ni(CO)_3PR_3$ $2 = Ni(CO)_2(PR_3)_2$ $3 = Ni(CO)(PR_3)_3$

these complexes are prepared by direct reaction of $Ni(CO)_4$ with the appropriate phosphine, although a few other methods have been reported, such as direct reaction of Ni, CO and PPh_3[902], or by carbonylation of a nickel halide phosphine complex[352,903]. Nickel(II) halide complexes of phosphines, $Ni(PR_3)_2X_2$, do not react with CO in solution at room temperature and atmospheric pressure, but at higher temperatures and pressures Ni(0) carbonyl complexes result[352] (compare Fe and Co complexes, which give carbonyl halide complexes). At room temperature

$Ni(PEt_3)_2(NO_3)_2$ reacts with CO, but the product is $Ni(NO)(PEt_3)_2(NO_3)$, not the carbonyl; under more vigorous conditions the dicarbonyl is produced[352]. In a new approach to carbonylation of Ni(0) phosphine complexes, Corain and co-workers[904] report that CO reacts with $Ni(PPh_3)_4$ in dichloromethane or benzene at atmospheric pressure and room temperature to form $[Ni(CO)_2(PPh_3)_2]$ readily, and much more slowly to yield $[Ni(CO)_3(PPh_3)]$.

The formation of $[Ni(CO)_2(PR_3)_2]$ and $[Ni(CO)_3(PR_3)]$ has been achieved with many phosphines (table 12). The direct replacement of more than two carbonyl groups does not occur with most phosphines (PMe_3 is an exception), and the bulky $P(MMe_3)_3$ (M = C, Si, Ge, Sn, Pb) produce only mono-substitution[96,319,917]. With DPA $[Ni(CO)(DPA)_3]$ is obtained[85]. Trifluoro-phosphine[6] and trichlorophosphine, and such phosphines as $PPhCl_2$, $PMeCl_2$, $P(NCO)_3$ and $P(NCS)_3$ are capable of substituting all the carbonyl groups in $Ni(CO)_4$ (see section 14.1.2).

Meriwether et al. found[922] that $P(CH_2CH_2CN)_3$ (L) gave, in addition to $Ni(CO)_2L_2$, a complex identified as $[Ni(CO)L]_n$, which apparently contains bridging carbonyls. An X-ray structure determination[923] showed it to be a cluster compound—the only phosphine carbonyl cluster reported for nickel. The structure of the complex, now formulated as $[Ni_4(CO)_6L_4]$ is shown in (XCVIII).

Ni—Ni = 250.8(4) pm

(XCVIII)

The complexes of types $[Ni(CO)_{4-x}(PR_3)_x]$ (x = 1, 2) are generally colourless, yellow or orange solids or liquids. These compounds have been extensively studied by investigators interested in the nickel–phosphorus bond and in the effect substitution has upon the Ni–CO linkage. A prodigous number of i.r. and Raman studies have been performed (see for example references 905, 908, 920, 922, 924, 925), and the ^{31}P n.m.r. spectra of a large number of these complexes reported[51,910]. The kinetics of the reaction of PR_3 with $Ni(CO)_4$, and exchange between $[Ni(CO)_{4-x}(PR_3)_x]$ and PR_3' have been studied[909,926]. The structure of the substituted carbonyl phosphines have not been determined by X-ray methods, but spectroscopic techniques have unequivocally demonstrated the tetrahedral disposition of the ligands about the nickel atom.

Much of the early interest in the phosphine-substituted carbonyls stemmed from the work of Reppe, who showed that $[Ni(CO)_2(PPh_3)_2]$ is an active catalyst for the polymerisation of olefins and acetylenes[2].

The phosphido-bridged $[Ni_2(PPh_2)_2(CO)_4]$ was prepared by Hayter[114] by heat-ing the diphosphine complex $(OC)_3Ni(P_2Ph_4)Ni(CO)_3$ in mesitylene, when it was found to form black crystals. The corresponding PMe_2-bridged complex could not be obtained by the same route. The diphenylphosphido compound has a small magnetic susceptibility which is temperature independent[927]. The structure (XCIX)

(XCIX)

that has been determined by Jarvis *et al.*[928] has a planar di-μ-phosphido bridge, Ni—P = 219 pm, and a short Ni—Ni bond, 251 pm (compare Ni—Ni = 336 pm in $(\pi\text{-}C_5H_5)Ni(PPh_2)_2Ni(\pi\text{-}C_5H_5)$, which has no metal–metal bond[929]) and $\widehat{NiPNi} \simeq 70°$. Nickel tetracarbonyl reacts with KPRR' (R = Ph; R' = H, Ph) to yield $K[Ni(CO)_3(PRR')]$ as red and orange crystalline solids, respectively[930].

14.1.2 Nickel(0) compounds

Irvine and Wilkinson[931] prepared $Ni(PCl_3)_4$ by direct reaction of PCl_3 with $Ni(CO)_4$; the product was an air-stable yellow solid with an unexpected resistance to reaction with aqueous acids. By analogous methods $Ni(PR_3)_4$ (R_3 = $PhCl_2$, $(NCO)_3$, $(NCS)_3$) were prepared[905,932,933]. Nickel reacts directly with $PMeCl_2$[934] and $PMeBr_2$[935], but apparently not with PCl_3 or $PPhCl_2$[934], to give $Ni(PR_3)_4$; however, $PPhCl_2$ readily displaces $PMeCl_2$ from $Ni(PMeCl_2)_4$[934]. A reversible reaction occurs between $Ni(CO)_4$ and PBr_3 in a sealed tube, but orange-red $Ni(PBr_3)_4$ is easily obtained by treatment of $Ni(PCl_3)_4$ with excess PBr_3[936]. Fluorination of $Ni(PR_3)_4$ (R_3 = Cl_3, $PhCl_2$, $MeCl_2$) with KSO_2F produces the corresponding fluorophosphine complexes (R_3 = F_3, PhF_2, MeF_2)[937]; interestingly, KSO_2F converts the free phosphines to phosphonic fluorides.

Complete replacement of CO from $Ni(CO)_4$ does not occur with organophosphines, but displacement of the organic ligand from some organonickel compounds—$Ni(\pi\text{-}allyl)_2$, $Ni(\pi\text{-}C_5H_5)_2$—or by reaction of $K_2Ni(CN)_4$ in liquid ammonia with the appropriate phosphine, PR_3, produces $Ni(PR_3)_4$ (R_3 = Et_3, Ph_3, Ph_2H)[938–941]. Reaction of $Ni(PMe_3)_2Cl_2$ with KOH in aqueous alcohol produces $Ni(PMe_3)_4$, also formed by reduction of nickel(II) salts in alkaline solution in the presence of PMe_3[942]. The $Ni(PPh_2H)_4$ complex is apparently identical[941] with the compound prepared by Issleib and Wenschuh[146] and formulated by them as $Ni_2(PPh_2)_2(PPh_2H)_2$.

These Ni(0) complexes are generally unstable in air, and dissolve in aqueous acids to give Ni(II) salts[943]. Oxygen reacts with $Ni(PPh_3)_4$ in ethereal solution at −78°C to give $[Ni(PPh_3)_2O_2]$, which decomposes above −35°C[944]. Hidai *et al.*[945] report that $Ni(PPh_3)_4$ undergoes oxidative addition with aryl halides to form $Ni(aryl)(PPh_3)_2X$; in view of the wide range of reactions reported for $Pt(PPh_3)_4$, further work in this field may be expected. The reactions of $Ni(PR_3)_4$ with various phosphines and phosphites (L) to give $Ni(PR_3)_{4-x}L_x$ have recently been studied[946], and an attempt made to establish an order of relative stability for the NiP_4 complexes. All these complexes probably have a tetrahedral structure.

14.1.3 Nitrosyls

Purple $[Ni(NO)_2(PR_3)_2]$ (R = Ph, Pr, Bu) were reported to be formed from $Ni(CO)_2(PR_3)_2$ and NO [947]. The nature of the complexes has presented some problems; the formulation most consistent with the data is $[Ni(NO^+)(NO^-)(PR_3)_2]$ [948]. The reduction of $Ni(PR_3)_2(NO_3)_2$ (R$_3$ = Ph$_3$, Ph$_2$Et, Et$_3$) with CO in benzene produces $Ni(NO)(PR_3)_2(NO_3)$, from which halide (Cl, Br, I) complexes may be obtained by metathetical reactions [352]. Triphenylphosphine reacts directly with $[Ni(NO)Cl_2]_n$ to give $[Ni(NO)(PPh_3)_2Cl]$ [949].

Feltham[950] prepared a series of nitrosyl complexes by (i) reaction of $Ni(PPh_3)_2X_2$ with $NaNO_2$ in THF, which yielded $[Ni(NO)(PPh_3)_2X]$ (X = Cl, Br, I); (ii) treatment of $[Ni(NO)(PPh_3)_2Br]$ with AgX' and PPh$_3$ to give $[Ni(NO)(PPh_3)_2X']$ (X' = OCN, CH$_3$COO, NO$_3$, NO$_2$); (iii) by reaction of pure NO with $Ni(CO)_2(PPh_3)_2$ or $Ni(CO)_3(PPh_3)$ in various solvents to yield $[Ni(NO)(PPh_3)_2X]$ (X = OH, NO$_2$, OMe). Direct reaction of $Ni(NO)(NO_2)$ and PPh$_3$ also yields $[Ni(NO)(PPh_3)_2(NO_2)]$ [952]. The products obtained using nitric oxide as the source of the nitrosyl group are very sensitive to the presence of NO_2 impurities, and Feltham reports[951] that very pure NO is essential in these reactions. He also suggested[951] that the $[Ni(NO)_2(PPh_3)_2]$ of Griffith *et al.*[947] is, in fact, $[Ni(NO)(PR_3)_2(NO_2)]$, but this too has been questioned[948]. Hieber and Bauer[953,954] prepared the $[Ni(NO)(PR_3)_2X]$ (R = Ph, Cy) complexes by direct reaction of $[Ni(NO)X]_4$ with the phosphine. Interestingly, $Ni(PBu^n_3)_2Br_2$ reacts with $NaNO_2$ to give $[Ni(PBu^n_3)_2(NO_2)_2]$, and not the nitrosyl complex[950].

X-ray results indicate[955] that the coordination around $[Ni(NO)(PPh_3)_2(N_3)]$ is approximately tetrahedral; Ni–P = 226, 231 pm, Ni–NO = 169 pm, and $\widehat{NiNO}$ = 153°.

Dimeric $[Ni(NO)(PR_3)X]_2$ complexes were obtained from $[Ni(NO)X]_4$ and PPh$_3$ and PCy$_3$ [953,954]. Halogen bridges are present, and with PCy$_3$ both the *cis* and *trans* isomers (C) have been isolated. In the $[Ni(NO)(PR_3)_2(NCS)]$ complex

(C)

the CNS$^-$ group is N-bonded[367,956]. Brunner[984] studied the reaction of NO with $Ni(PBu^n_3)_2X_2$ (X = Cl, Br, I) in solution in various solvents; evidence for $NO^+[Ni(PBu^n_3)_2X_2]^-$ and $[Ni(NO)(PBu^n_3)_2Br]$ was obtained, and from $Ni(PBu^n_3)_2I_2$ were obtained $[Ni(NO)(PBu^n_3)_2I]$ and $[Ni(NO)(PBu^n_3)I]_2$.

Finally, there is the red-brown phosphido complex, $[Ni(NO)(PPh_2)]_4$, prepared from $[Ni(NO)Br]_4$ and $KPPh_2$ [365].

14.1.4 Hydrido and dinitrogen complexes

Although a hydrido species was shown to be present in the system $NaBH_4$-$Ni(PPr_3)_2Br_2$ by Green *et al.*[958] in 1959, the instability of nickel hydrides has precluded isolation until recently. Reduction of *trans*-$Ni(PCy_3)_2X_2$ by $NaBH_4$ in EtOH–THF yields $[NiH(PCy_3)_2X]$ (X = Cl, Br, CNS), and $[NiH(PPr^i_3)_2Cl]$ is obtained analogously[959,960]. Metathetical replacement leads to the corresponding I and CN complexes. The products are yellow-brown crystals, rapidly oxidised in solution by air, but more stable when solid. The stability of these complexes is attributed to steric factors, the bulky phosphines restricting deformation of the *trans* square-planar structure into tetrahedral, and restricting the approach of other groups[960]. Reaction of *trans*-$[NiH(PR_3)_2Cl]$ (R = Cy, Pr^i) with $NaBH_4$ in 1:1 acetone–EtOH affords *trans*-$NiH(PR_3)_2(BH_4)$[961]. These borohydrido complexes reduce $Ni(PR_3)_2X_2$ and $Pd(PR_3)_2X_2$ (R = Et, Pr^n, Pr^i, Bu^n) to hydrido complexes, which have been studied using n.m.r. spectroscopy[962,963]. Mixed hydrides, $[NiH(PR_3)(PR'_3)X]$ (R = Et, Pr^n, Bu^n; R' = Cy) are obtained from $Ni(PR_3)_2X_2$ and $NiH(PR'_3)_2(BH_4)$[962].

Dinitrogen complexes of nickel are limited to red $[\{Ni(PCy_3)_2\}_2(N_2)]$[964] and orange $[NiH(N_2)(PR_3)_2]$ (R = Et, Bu^n)[965], obtained by reduction of $Ni(acac)_2$ in the presence of the appropriate phosphine, with $AlMe_3$ under N_2, and $AlPr^i_3$ at $-78°C$ respectively. Dinitrogen is readily lost from $[\{Ni(PCy_3)_2\}_2(N_2)]$ to yield the yellow-red $Ni(PCy_3)_2$ on bubbling argon through the solution in toluene. This $Ni(PCy_3)_2$ complex reacts with various acidic compounds–HCl, CH_3COOH, phenol, C_5H_6–to give $[NiH(PCy_3)_2L]$ (L = Cl, CH_3COO, OPh, C_5H_5)[966].

14.1.5 Halides

Nickel(II) forms a large number of phosphine complexes, but few nickel(III) and even fewer Ni(I) complexes are known.

(1) Nickel(I)

Heimbach[967] prepared $Ni(PPh_3)_3X$ by addition of the appropriate halogen to $Ni(PPh_3)_4$, by reaction of $Ni(PPh_3)_4$ with $Ni(PPh_3)_2X_2$ in ether, or by heating $Ni(\pi\text{-allyl})Br$ with PPh_3 in the presence of norbornene in benzene. The yellow or brown products are paramagnetic, $\mu_{eff} = 4.9$ B.M., and decompose in air. Cryoscopic studies in benzene[968] seem to indicate a dissociation

$$Ni(PPh_3)_3X \rightleftharpoons Ni(PPh_3)_2X + PPh_3$$

and $Ni(PPh_3)_2X$ are reported[967] to result from the reaction of $Ni(PPh_3)_2(C_2H_4)$ with $Ni(PPh_3)_2X_2$ in ether. Preliminary X-ray studies indicate a tetrahedral environment of the metal in $Ni(PPh_3)_3X$[968].

(2) Nickel(II)

An extensive series of complexes has been prepared (tables 13, 14, 15) in contrast to the small number of arsine derivatives. The large majority of tertiary phosphine complexes are of the type $[Ni(PR_3)_2X_2]$, but a number of five-coordinate [Ni(PR$_3$)$_3$X$_2$] have been obtained, especially when X = CN. The four-coordinate complexes are of two types: diamagnetic, usually red or yellow planar complexes, and paramagnetic blue or green complexes. A number of phosphines give both types of complex in the solid state, and yet other phosphine complexes exhibit evidence of planar $\rightleftharpoons$ tetrahedral isomerism in solution, although only existing in one form in the solid state. Secondary and primary phosphine complexes show important differences from tertiary phosphines, but, as usual, they have been much less studied, and a great deal of work remains to be done on them. We shall now discuss these topics in more detail.

The preparation of these complexes is normally achieved by direct reaction between the appropriate nickel(II) salt and the phosphine in various alcohols, acetic acid, etc., followed by recrystallisation. In the case of phosphines that give isomeric products, the particular isomer obtained frequently depends upon the solvent and the temperature at which crystallisation occurs. Reaction of the neat liquid phosphines and fusion of the solid phosphines with the nickel(II) salt have also been used.

Trialkylphosphines react with nickel halides to form diamagnetic [Ni(PR$_3$)$_2$X$_2$] (R = Me, Et, *n*-Pr, *i*-Pr, *n*-Bu, *s*-Bu), which are red or brown in colour[595,969,970,972,977]. Jensen[972] showed that several of these complexes had dipole moments of approximately zero, and in conjunction with the observed diamagnetism, this made a *trans* square-planar structure likely. This was confirmed[975,976] by an X-ray study of [Ni(PEt$_3$)$_2$Br$_2$] which showed Ni—Br = 230 pm, Ni—P = 226 pm, $\widehat{BrNiP} \simeq 90°$. Trimethylphosphine differs as a ligand, inasmuch as complexes containing more than two ligands per nickel atom can be obtained[595,970]. Triphenylphosphine does not complex with nickel halides in ethanol[972], but in butanol or acetic acid blue paramagnetic complexes were isolated[980]. An X-ray structural study of [Ni(PPh$_3$)$_2$Cl$_2$] has shown it to be tetrahedral[982], with Ni—Cl = 227 pm, Ni—P = 228 pm, $\widehat{ClNiCl} = 123°$, $\widehat{PNiP} = 117°$; and subsequently [Ni(PPh$_3$)$_2$Br$_2$] was X-rayed[981], showing Ni—Br = 234 pm, Ni—P = 233 pm, $\widehat{BrNiBr} = 126°$, $\widehat{PNiP} = 103°$—the distortion resulted from Br—Br repulsion. By rapid cooling of a concentrated solution of [Ni(PPh$_3$)$_2$Cl$_2$] in dichloromethane, a red diamagnetic solid was prepared[983], which isomerised readily to the blue tetrahedral form. Its instability prevented proper characterisation, but it is probably the hitherto unknown *trans* planar isomer. Browning *et al.*[625] extended the work to (*p*-RC$_6$H$_4$)$_3$P (R = Me, OMe), which also formed tetrahedral complexes.

Most interest has centred upon the mixed alkyl–aryl phosphines, which exist in solution as a mixture of the planar and tetrahedral forms, the amount of each depending upon the solvent and the particular phosphine and X group involved. In the past ten years extensive investigation of these complexes has been undertaken in an attempt to understand the factors involved. Coussmaker *et al.*[971] found that the replacement of Bu^n by Ph in PBu$^n_3$ resulted in a structural change: the PBu$^n_3$ yields planar [Ni(PR$_3$)$_2$X$_2$] complexes, PPhBu$^n_2$ is similar, but PPh$_2$Bu^n gave paramagnetic, blue or brown complexes. Magnetic studies on these paramagnetic complexes in benzene solution were interpreted as showing the presence of an equilibrium between the diamagnetic and paramagnetic isomers[971]. Hayter and Humiec[989,991] studied a series of PPh_2R ($R = Et, Pr^n, Bu^n, Bu^i, Bu^s$, amyl) complexes and found interesting differences between the various halides. The chlorides were planar, the iodides tetrahedral, and both forms of the bromides could be isolated; for example PPh_2Et formed the green $[Ni(PPh_2Et)_2Br_2]$ at room temperature, but at $-78°C$ concentrated solutions of the complex in CS_2 slowly deposit the red isomer. The situation is complicated by the existence of green diamagnetic $[Ni(PPh_2R)_2I_2]$ ($R = Pr^i, Bu^s$) in addition to the tetrahedral form; this system is not completely understood. With PPh_2Bz both isomers were reported with all three halides[996], but X-ray studies[997] on the green 'tetrahedral' $Ni(PPh_2Bz)_2Br_2$ have shown the true state of affairs to be more complex. There are three molecules in the unit cell; one has a *trans* planar arrangement, the other two molecules being related by a centre of symmetry with the nickel tetrahedrally coordinated. The compound is described as an interallogon compound[997]. Of course, there are different dimensions in the two types of molecule: $Ni-Br = 230$ pm (planar), 235 pm (tetrahedral); $Ni-P = 226$ pm (planar), 231 pm (tetrahedral). The higher magnetic moment of the green chloride (3.3 B.M. as against 2.7 B.M. for the allogon bromide) probably indicates that this contains only tetrahedral molecules.

The thermodynamics and kinetics of the planar $\rightleftharpoons$ tetrahedral isomerism in solution for the PPh_2R and $PMe(aryl)(aryl')$ nickel(II) halide complexes have been studied by ^1H n.m.r. and electronic spectroscopy, and by magnetic measurements[990,999,1001]. The electronic and steric factors contributing to the observed isomerism have been discussed by various groups (see references 625, 971, 980, 983, 989, 990, 996, and references therein). The original papers should be consulted for details, but the conclusion may be briefly summarised. The isomerism

$$Ni(PR_3)_2X_2 \rightleftharpoons Ni(PR_3)_2X_2 \qquad (X = Cl, Br, I)$$

paramagnetic	diamagnetic
tetrahedral	square planar

has been observed under various conditions for many nickel(II) halide complexes. Appreciable amounts of both isomers are present under normal conditions only for diarylalkylphosphines. Aryldialkyl- and trialkylphosphines exist as planar diamagnetic complexes both in the solid state and in solution, whilst triarylphosphines are paramagnetic and have tetrahedral structures. This pattern also holds for tricyclohexylphosphine, which produces planar complexes (but not, apparently, with tricyclopropylphosphine)[993]. For any phosphine the tendency towards formation of the tetrahedral isomer increases in the order $Cl < Br < I$, whilst for any halide the tetrahedral form is increasingly stable in the order $R_3P < R_2ArP < RAr_2P < Ar_3P$ (R = alkyl, Ar = aryl). As expected, solvents shift the equilibrium, the formation of the less polar planar isomer being favoured by solvents of low dielectric constant.

Complexes with other nickel(II) salts are limited to pseudohalogens and nitrates (table 13). The red '$Ni(PMe_3)_2(NO_3)_2$'[969] is probably a mixture of

$Ni(PMe_3)_4(NO_3)_2$ and $Ni(NO_3)_2$ [970]. Triethylphosphine forms green paramagnetic $Ni(PEt_3)_2(NO_3)_2$ [972], which is believed to have a tetrahedral structure[975,976]. Nickel nitrate complexes with triarylphosphines seem to be similar[625,980].

Thiocyanates seem to form planar complexes exclusively (table 13). Those formed with PPh_3, $PPhMe_2$, $P(C_6F_5)Me_2$, PPr^n_3 and PPr^i_3 contain N-bonded thiocyanate[629,977,985]. As in the analogous cobalt complexes, the cyanide derivatives are anomalous, and are discussed below in more detail. Orange-red $Ni(PBu^n_3)_2SO_4$ [1002] forms on reaction of excess PBu^n_3 with $NiSO_4.7H_2O$ in absolute ethanol. It is believed to have a *cis* planar structure with a bidentate sulphato ligand.

The four-coordinate bisphosphine nickel(II) complexes have been extensively studied by a number of spectroscopic techniques. ^1H n.m.r. spectra have afforded information on the isomerisation in solution, on isotropic proton shifts and upon ligand exchange rates[613-15]. Pignolet and Horrocks[615] found that for $[M(PPh_3)_2X_2]$ (M = Co, Ni; X = Cl, Br, I) lability increased in the orders $I < Br < Cl$ and $Co < Ni$. Diastereoisomers of $Ni\{PPhMe(p\text{-}MeOC_6H_4)\}_2X_2$ have been identified[616]. Far i.r. spectra have been used to determine the stereochemistry of $[Ni(PR_3)_2X_2]$ compounds[592,992,1003-5]; the metal isotope technique developed by Nakamoto and coworkers[992,1005] was an invaluable aid to assignment of $\nu(Ni-P)$ and $\nu(Ni-X)$. Electronic spectra have been widely used to distinguish between the two types of four-coordinate complexes and to identify five-coordinate complexes. Sacconi has reviewed this aspect thoroughly[1006]. The polarised single crystal spectrum of $[Ni(PPh_3)_2Cl_2]$ has been reported and discussed[1007].

Pseudotetrahedral complexes, $[NiLX_3]$, of quaternised diphosphines $RPh_2\overset{+}{P}(CH_2)_nPPh_2$ (n = 1, 2)[609] and of protonated phosphinopyridines (LXIII)[611] have been prepared. Using the latter ligand the low-spin six-coordinate orange $[NiL_2(CNS)_2](ClO_4)_2$ has been prepared[611].

Jensen and Dahl[970] have prepared red $[Ni(PMe_3)_3X]ClO_4$ (X = Cl, Br) by treatment of $Ni(PMe_3)_3X_2$ solutions in ethanol with $LiClO_4$. $P(CH_2CH_2CN)_2$ has been reported to behave as a bidentate P–N chelate towards Ni(II)[606]. The thermal decomposition of Ni(II) halide complexes of phosphines in an oxygen-containing atmosphere has been studied by thermogravimetric methods[1008].

In recent years there has been increasing interest in the preparation and properties of five-coordinate nickel(II) complexes[1009,1010]. Much of the work has been concerned with multidentate chelates, but certain monodentate phosphines also promote low-spin five coordination (table 14).

Of the trialkylphosphines only PMe_3 produces five coordination with nickel(II) halides[970]. The complexes are blue-black and dissociate reversibly in solution

$$Ni(PMe_3)_3X_2 \rightleftharpoons Ni(PMe_3)_2X_2 + PMe_3$$

On evaporation of the solution the four-coordinate complexes are obtained. Chastain *et al.*[1011] studied $Ni(PMe_3)_3Br_2$ and found it to be essentially diamagnetic with a small paramagnetism, $\mu_{eff} = 0.30$ B.M., attributable to T.I.P. It is fairly air stable, but on long standing it loses PMe_3 to give the square-planar complex. The

TABLE 13 NICKEL(II) PHOSPHINE COMPLEXES. [Ni(PR_3)_2X_2]

Phosphine	X	μ_{eff} (B.M.)	Comments	Ref.
PMe_3	Cl,Br,I	diamagnetic	more PMe_3 taken up to form [NiL_3X]Y and [NiL_3X_2] (X = Cl,Br,I).	595,969,970, 1004
	CN,NO_2			
	NO_3	3.17		
PEt_3	Cl,Br,I	diamagnetic	existence questioned	352,595,
	CNS,CN,CNO	diamagnetic	Br has *trans* planar structure	971-6
PPr^n_3	NO_3	3.1		
	Cl,Br,I,NO_2	diamagnetic		595,972,977
PPr^i_3	NO_3,CNS	diamagnetic		595,625,977
PBu^n_3	Cl,Br,NCS	diamagnetic	N-bonded thiocyanate	595,972,978
PBu^s_3	Cl,Br,I,CN	diamagnetic		595,977
PPh_3	Cl,Br	diamagnetic	both red and green forms of Br reported	306,595,620,
	Cl,Br,I	2.9–3.41	tetrahedral (X-ray of Cl,Br) red, unstable	505
			diamag. Cl reported	
	NO_3	3.0	N-bonded thiocyanate	914,979–83
	NCS	diamagnetic		629
P(p-MeC_6H_4)_3	Cl,Br,I	3.23–3.28		625
	NCS	diamagnetic		
P(p-ClC_6H_4)_3	Cl,Br,I	3.23–3.27		625
P(p-MeOC_6H_4)_3	NO_3	3.24		625
	NCS	diamagnetic		
PPh_2Me	Cl	diamagnetic	planar	989,990
	Br,I	~3.35		
PPh_2Et	Cl,Br	diamagnetic	planar	986,988,989,991
	Br,I	3.2	tetrahedral	352,990,992
	CN		adds ligand to form [NiL_3X_2]	985,987
PPh_2Pr^n	Cl,Br	diamagnetic	planar	989,992
	Br,I	3.1	tetrahedral	990
PPh_2Pr^i	Cl,Br	diamagnetic	planar	989,992
	Br,I	2.8–3.0	also another I isomer	
PCy_2(ol)†	Br		behaves as monodentate P donor	84
PPh_2Bu^n	Cl,Br,NCS	diamagnetic		971,989,990
	Cl,Br,I	3.2–3.35		
PPh_2Bu^i	NO_3	3.4	planar	989,992
	Cl,Br	diamagnetic	tetrahedral	
	Br,I	3.2		

Ligand	X,X'	μ	Comments	Ref.
PPh₂Buˢ	Cl,Br	diamagnetic	tetrahedral; another I isomer	985
PPh₂Buᵗ	Br,I		tetrahedral	989
PPh₂(n-amyl)	Cl	diamagnetic	planar	989
PPh₂Cy	Cl,Br,I	3.1	tetrahedral	983
	Cl,NCS	diamagnetic	wine-coloured Cl isomerises to paramagnetic green on heating	
PPh₂(allyl)	Cl,Br,I	~3.3		996
	Cl,Br,NCS	diamagnetic		
PPh₂Bz	Br,I	3.2–3.3	planar	996–8
	Cl,Br,I,NCS	diamagnetic		
PPhMe₂	Cl,Br,I	2.6–3.2	X-ray shows Br to be allogon	626 985
PPhEt₂	Cl,Br,NCS,NO₂	diamagnetic	NiI₂ forms [Ni(PPhMe₂)₃I₂]	986–8
	Cl,Br	diamagnetic		
PPhBuⁿ₂	CN,CNO,NO₃	diamagnetic	CN adds PPhEt₂ to give [NiL₃(CN)₂]	971
	Cl,Br,I,NCS			
	NO₃	3.2		
PPhCy₂	Cl,Br,I,NCS	diamagnetic		983
PPhBz₂	Cl,Br,I	diamagnetic		996
PPh(NEt₂)₂	NCS	diamagnetic		628
PCy₃	I	3.1	planar; NO₃,ClO₄ not isolated	388,980
		3.3	unstable	983,994,995,1000
P(cyclopropane)₃	Cl	diamagnetic	red soln in C₆H₆, μ_{eff} = 1.3 B.M. (39% td.)	993
PMe₂(CF₃)	Cl,Br,I,NCS		no complexes formed with PMe(CF₃)₂ or P(CF₃)₂ and NiX₂	969
PMe₂(C₆F₅)	NO₃	2.93	planar; will not add to form C.N. = 5, (cf. PMe₂Ph)	985
	Cl,Br,I,NCS	diamagnetic		
PEtCy₂	Cl,Br,NCS	diamagnetic	planar	987
PEt₂Cy	Cl,Br,NCS	diamagnetic	planar	987
(2-phind)‡	Cl,Br,I	3.6–3.7	obtain by boiling solution of NiL₃X₂	988
(TMP)§	NO₃	3.46	also a green NiL₂Br₂, μ_{eff} = 3.0 B.M.	605
	NCS	diamagnetic		
PBz₃	Cl,Br,I,NCS	diamagnetic		595,983,996
	I	3.10		
PR₂(Me₂NC₆H₄)	Cl,Br,I	diamagnetic	R = Me, Et	626
P(CH₂Cl)₃	Br,I		Cl,CNS,NO₃ could not be prepared	625
PMe(p-X'C₆H₄)(p-XC₆H₄)	Cl,Br,I	diamagnetic	various X, X'	999

† ol = but-3-enyl
‡ 2-phind = 2-phenylisophosphindoline
§ TMP = trimorpholinophosphine

TABLE 14 FIVE-COORDINATE. NICKEL(II) COMPLEXES. $[Ni(PR_3)_3X_2]$

Phosphine	X	μ_{eff} (B.M.)	Comments	Ref.
PMe_3	CN,Br,I	0.3 (Br)	small moment due to T.I.P.	970,1011
PEt_3	CN		formed in solution only	987
PBu^n_3	CN			1013
PPh_2Me	CN	diamagnetic		985
$PPhMe_2$	CN,I	diamagnetic	*trans* trigonal-bipyramidal (CN)	985,1014
$PPhEt_2$	CN	diamagnetic	red solid	987
(2-phind)[†]	Cl,Br,I		decompose to $Ni(PR_3)_2X_2$ on boiling in EtOH	988
(9-phos)[‡]	Cl,Br,I	0.65–1.5	origin of unusual magnetic	608,1012
(R = Me,Et)	CNS	0.95,1.2	moments not understood	
	CN	~0.4	$[Ni(9\text{-Me-phos})_2I_2]$ can be prepared	

[†] 2-phind = 2-phenylisophosphindoline.
[‡] 9-phos = 9-alkyl-9-phosphafluorene.

electronic spectrum is compatible with a trigonal-bipyramidal structure[1011]. Alyea and Meek[985] prepared $Ni(PPhMe_2)_3I_2$, but found that $P(C_6F_5)Me_2$ afforded only four-coordinate halide complexes. As with Co(II), 2-phenylisophosphindoline[988] and 9-alkyl-9-phosphafluorenes[608] behave anomalously by readily forming five-coordinate complexes with all three nickel(II) halides. The green $[Ni(2\text{-phind})_3X_2]$ gave a stable solution in chloroform, but on boiling in most solvents the red $[Ni(2\text{-phind})_2X_2]$ precipitated[988]. The 9-alkyl-9-phosphafluorene complexes, NiL_3X_2 (X = Cl, Br, I, CNS), are deeply coloured and slightly dissociated in dichloromethane solution. The peculiar magnetic moments recorded, 0.65–1.5 B.M., are temperature independent and dependent upon the strength of the field, but a satisfactory explanation of this phenomenon has not been forthcoming[608].

In contrast to the general lack of five coordination in the halides, nickel(II) cyanide readily gives $[Ni(PR_3)_3(CN)_2]$. Rigo *et al.*[987] failed in attempts to prepare $[Ni(PR_3)_3X_2]$ (R_3 = Ph_2Et, $PhEt_2$, Et_3, Et_2Cy, Cy_3; X = Cl, Br, I, CNS) by reaction of $[Ni(PR_3)_2X_2]$ with PR_3, although they did observe phosphine exchange. However, the corresponding $[Ni(PR_3)_2(CN)_2]$, which form yellow solutions in dichloromethane, immediately turn red upon addition of PR_3, and visible spectra indicated the presence of $[Ni(PR_3)_3(CN)_2]$ in solution. The $[Ni(PPhEt_2)_3(CN)_2]$ complex was isolated, and a comparison of the reflectance and solution (in the presence of excess PR_3) spectra showed that no marked change in coordination occurs on dissolution. The i.r. spectrum shows that the CN groups are *trans*[987]. The thermodynamic studies of Rigo *et al.*[1015] on the reaction

$$Ni(PR_3)_2(CN)_2 + PR_3 \rightleftharpoons Ni(PR_3)_3(CN)_2$$

established the stability order for the five-coordinate complexes, as determined from the values of the equilibrium constants in ethanol at 20°C; it follows the order $PPr^n_3 \sim PBu^n_3 < PEt_3 < PPhEt_2 \ll PPhEt_2$. Five-coordinate cyano complexes have been isolated with PMe_3[970], PPh_2Me, $PPhMe_2$ and $P(C_6F_5)Me_2$[985]. The structure(CI) of $[Ni(PPhMe_2)_3(CN)_2]$ has been determined by Ibers and Stalick[1014].

(CI)

It is trigonal bipyramidal, with Ni–P = 222, 226 pm, Ni–C = 184, 186 pm.

The 9-alkyl-9-phosphafluorenes (alkyl = Me, Et) complexes with nickel(II) cyanide have identical solution spectra but strikingly different solid-state spectra[608]. X-ray studies[1012] have shown that the 9-Me-9-phosphafluorene complex has a distorted tetragonal-pyramidal structure, but the 9-ethyl complex is trigonal bipyramidal.

Factors which control the formation of five coordination have been discussed by several groups[978,985,987,1016]. The relative importance of steric and electronic factors is disputed. The ability to form five-coordinate species decreases in the order $CN \gg I > Br > Cl$, and $PPhMe_2 > P(C_6F_5)Me_2$; also $PPhMe_2 > PPh_2Me > PPh_2Et$ [985].

Although complexes of type $[Ni(L-L)_2]X_2$ and $[Ni(L-L)_2X]Y$ are well known when L–L is a chelating diphosphine[1017,1018], the monodentate phosphine analogues, NiP_4, are rare. Red $[Ni(PMe_3)_4]X_2$ ($X = NO_3, ClO_4$) have recently been isolated[970]; the perchlorate is a 1:2 electrolyte, but the conductivity value of the nitrate is lower (compare $[Ni(VPP)_2]X_2$, VPP = *cis*-$Ph_2PCHCHPPh_2$)[1019].

The primary and secondary phosphines have been very little studied as complexing agents, and only with PPh_2H is the data satisfactory (table 15). Isslieb

TABLE 15 NICKEL(II) COMPLEXES OF PRIMARY AND SECONDARY PHOSPHINES

Compound	μ_{eff} (B.M.)	Comments	Ref.
$[Ni(PPhH_2)_4Br_2]$		red, unstable	147
$[Ni(PCyH_2)_2Br_2]$	diamagnetic	red, *trans* planar?	148
$[Ni(PEt_2H)_4Cl_2]$			145
$[Ni(PEt_2H)_4Br_2]$	diamagnetic	non-electrolyte; octahedral?	145
$[Ni(PCy_2H)_2Br_2]$	diamagnetic	*trans* planar	145
$[Ni(PPh_2H)_2I_2]$	diamagnetic	brown; *trans* planar	1020
$[Ni(PPh_2H)_3Cl_2 \cdot \frac{1}{2}C_6H_6]$	diamagnetic	dark brown	147,1020
$[Ni(PPh_2H)_3Br_2]$	diamagnetic	dark brown	632,1020
$[Ni(PPh_2H)_3I_2]$	~0.6	dark blue; magnetic moment is temperature independent	632,1020

et al.[145,147,148] have reported preliminary studies on PPh_2H, PEt_2H, PCy_2H and $PCyH_2$, but confirmatory work would appear to be necessary. The $Ni(PEt_2H)_4X_2$ complexes were suggested to contain octahedral Ni, but this is inconsistent with the reported diamagnetism[145]. The yellow product obtained by Isslieb and Wenschuh[146] from $NiBr_2$ and PPh_2H in ethanol, and formulated by them as $Ni(PPh_2H)_2(PPh_2)_2$, has been reformulated as $Ni(PPh_2H)_4$ (see section 14.1.2). Hayter[1020] obtained dark brown $Ni(PPh_2H)_2I_2$, and blue or brown $[Ni(PPh_2H)_3X_2]$ ($X = Cl, Br, I$) from the constituents in dichloromethane. Bertrand and Plymale[632] studied the structure of the five-coordinate products, which are analogous to that of $[Co(PPh_2H)_3Br_2]$ (q.v.). $[Ni(PPh_2H)_3X_2]$ ($X = Cl, Br$) are diamagnetic[1020], but $[Ni(PPh_2H)_3I_2]$ is anomalous. Hayter[1020] reported its magnetic moment as 1.48 B.M. and Bertrand and Plymale[632] found 1.29 B.M. However, a study of the variation of the magnetic susceptibility with temperature by Dick and Nelson[1021] found only a small (~ 0.6 B.M.) residual paramagnetism, approximately independent of temperature, and concluded that $Ni(PPh_2H)_3I_2$ is a normal low-spin five-coordinate nickel(II) complex, the previously reported high values probably being due to paramagnetic impurities.

14.1.6 Other nickel(II) complexes

Nickel(II) phosphido complexes have been reported. The reaction between $Ni(NH_3)_4(NCS)_2$ with KPH_2 in liquid ammonia yields $Ni(PH_2)_2 \cdot xNH_3$, which is transformed by more KPH_2 to $K[Ni(PH_2)_3]$ [1022]. A polymeric $[Ni(PPh_2)_2]_n$ is formed from $NiCl_2$ and Me_3SiPPh_2 [1023]. Potassium diphenylphosphide and $Ni(PEt_3)_2Br_2$ are reported[633] to form $[Ni(PEt_3)_2(PPh_2)]_2$, $\mu_{eff} = 1.13$ B.M., for which structure (CII) has been suggested. Some related complexes are the $[Ni(SPh)_2(PR_3)_2]$ ($R_3 = Ph_3$, $PhMe_2$, Et_3), prepared from $Ni(PR_3)_2Cl_2$ and $NaSPh$ [1025]. Among the products of the reaction of Ph_2BBr with $[Ni(PPh_3)_2(C_2H_4)]$ in ether is $[Ni(PPh_3)_2(BPh_2) \cdot \frac{1}{2}Et_2O]_n$, which is thought to contain BPh_2 bridges[1026].

Simple alkyl and aryl complexes are unstable[986,1024]. Fairly stable $[Ni(Me)(PPr^i_3)_2X]$ have recently been prepared[1027]. They react with MeI to form $[PPr^i_3Me][Ni(PPr^i_3)I_3]$, whilst from PPh_3 in ether the Ni(I) complexes, $[Ni(PPh_3)_3X]$, are obtained. A novel alkyl–nickel(II) complex, (CIII), was obtained from 2-(diphenylphosphino)-benzylpotassium and $Ni(PEt_3)_2Cl_2$ [1028]. Green *et al.* [1029] found that $MeMgBr$ reacts with $Ni(PPh_3)_2Cl_2$ with substitution at the phosphorus, and from which PhH, $PhCH_3$, Ph_2, Ph_2MeP and $PhMe_2P$ are all obtained.

(CII)   (CIII)

14.1.7 Nickel(II) anions

Early workers[306,914] investigating the catalytic properties of $[Ni(PPh_3)_2X_2]$ found that alkyl and aryl halides reacted to form $Ni(PPh_3)_2X_2 \cdot RX$ (or $2RX$). Cotton *et al.*[980,1030] prepared green $[NEt_4][Ni(PPh_3)Br_3]$ and red $[NBu^n_4][Ni(PPh_3)I_3]$ by reaction of NiX_2, PPh_3 and NR_4X in butanol. They recognised that Yamamoto's $Ni(PPh_3)_2Br_2 \cdot Bu^tBr$ complex was a member of the same series, that is $[PPh_3Bu^t][Ni(PPh_3)Br_3]$, but their attempts to prepare $[Ni(PPh_3)Cl_3]^-$ and the PCy_3 analogues were not successful. However, several other complexes have appeared in the patent literature[1031,1032], and Booth[2] has reformulated several of Yamamoto's products[914] as $[RPPh_3][Ni(PPh_3)X_3]$ and $[RPPh_3][Ni(PPh_3)XX'_2]$ ($X \neq X' =$ halogen). The isolation of $[Ni(PPr^i_3)I_3]^-$ has already been mentioned[1027].

The electronic spectra, magnetic and conductance data are all consistent with the ionic formulation and tetrahedral anions[620,980,1030]. The far i.r. spectra of $[Ni(PPh_3)X_3]^-$ ($X = Cl$, Br) have been studied; the chloro complex appears to have a normal tetrahedral structure, but the spectrum of the $[Ni(PPh_3)Br_3]^-$ ion is anomalous, as is the corresponding $[Co(PPh_3)Br_3]^-$ [592]. The crystal structure determination[1033] of $[AsPh_4][Ni(PPh_3)I_3]$ confirmed the pseudotetrahedral structure of the anion, with Ni–I = 255 pm (average), Ni–P = 228 pm.

14.1.8 Nickel(III) complexes

Dark violet $Ni(PEt_3)_2Br_3$ was obtained by bromine oxidation of
$Ni(PEt_3)_2Br_2$ [972,1034]. It has a magnetic moment of ~ 1.7 B.M. and a dipole
moment of ~ 2.5 D. X-ray studies have established the lattice parameters[976,1035],
but its instability has precluded a complete X-ray analysis. Oxidation of
$Ni(PR_3)_2X_2$ with NOX at $-80°C$ yielded[595] $Ni(PR_3)_2X_3$ ($R = Me$, Et, Pr^n, Bu^n;
$X = Cl$, Br), but pure products were difficult to obtain, and all decomposed fairly
rapidly on standing. They are all monomeric, and probably trigonal bipyramidal.

Bromine oxidation of $Ni(PPhMe_2)_2Br_2$ produced the analogous Ni(III)
complex, a monomeric non-electrolyte with $\mu_{eff} = 2.17$ B.M.[1036]. Attempts to
obtain crystals suitable for an X-ray study produced a complex,
$Ni(PPhMe_2)_2Br_3 . 0.5Ni(PPhMe_2)_2Br_2 . C_6H_6$, which contained two molecules of
$[Ni(PPhMe_2)_2Br_3]$ and one of $[Ni(PPhMe_2)_2Br_2]$, accompanied by two molecules
of benzene in the unit cell. The five-coordinate nickel(III) complex has a trigonal-
bipyramidal structure with equatorial bromines[1037], $Ni-P \simeq 227$ pm, $Ni-Br =$
235 pm (average).

14.2 PALLADIUM AND PLATINUM

14.2.1 Carbonyls

In sharp contrast to the other platinum metals, the carbonyl chemistry of these two
elements, especially palladium, is very restricted.

Reaction of $Pd(acac)_2$, PPh_3, $AlEt_3$ and CO at low temperature in toluene
($Pd:PPh_3 \sim 1:5$) leads to $[Pd(CO)(PPh_3)_3]$; a $1:1$ ratio of $Pd:PPh_3$ produces the
red $[Pd(CO)(PPh_3)]_n$ [1038]. In solution $Pd(CO)(PPh_3)_3$ loses PPh_3 to form
$Pd_3(CO)_3(PPh_3)_4$; an equilibrium between these two compounds exists in
solution[1038]. Treatment of $[Pd(PR_3)_2Cl_2]$ ($R = Et$, Ph) with BF_3
affords $[Pd_2(PR_3)_4Cl_2](BF_4)_2$, which reacts with CO to give *trans*-
$[Pd(CO)(PR_3)_2Cl]BF_4$ [1039,1040]. Carbonylation of $[MR(PEt_3)_2X]$ ($M = Pd$,
Pt; $X = Cl$, Br, I; $R = Me$, Et, Ph) produces *trans*-$[M(COR)(PEt_3)_2X]$ [1041], but
cis-$[PtMe_2(PEt_3)_2]$ gave cluster carbonyl[1041].

The platinum cluster carbonyls are of four types: $Pt_3(CO)_3(PR_3)_4$ ($R_3 = Ph_3$,
Ph_2Me, Ph_2Et), $Pt_3(CO)_3(PR_3)_3$ ($R_3 = Ph_2Bz$), and two isomers of formula
$Pt_4(CO)_5(PR_3)_4$. The last complexes were originally formulated, incorrectly, as
$Pt_3(CO)_4(PR_3)_3$ [1042]. The $Pt_3(CO)_3(PR_3)_4$ complexes are obtained by heating
$Pt(CO)_2(PR_3)_2$ in alcoholic solution or, better, by reaction of CO with dilute
ethanolic solutions of Na_2PtCl_4 followed by addition of PR_3 [1042-4]. Phosphines
convert them to $Pt(CO)(PR_3)_3$ or $Pt(PR_3)_4$, or, in the presence of CO with limited
amounts of phosphine to $Pt(CO)_2(PR_3)_2$ [1044]. The structures of $Pt_3(CO)_3(PR_3)_4$
are probably (CIV), whilst the PPh_2Bz complex probably has the more symmetrical
structure (CV)[1044].

In solution the $Pt_3(CO)_3(PR_3)_4$ complexes react with CO with rearrangement
to form $Pt_4(CO)_5(PR_3)_4$, which exist in two isomeric forms. The brown-black

complexes (R_3 = Ph_3, $PhMe_2$) have only bridging carbonyl i.r. absorptions, whilst the red complexes (R_3 = Et_3, Ph_2Bz) show both bridging and terminal carbonyl bands[1044]. Vranka *et al.*[1045] determined the structure of $Pt_4(CO)_5(PPhMe_2)_4$;

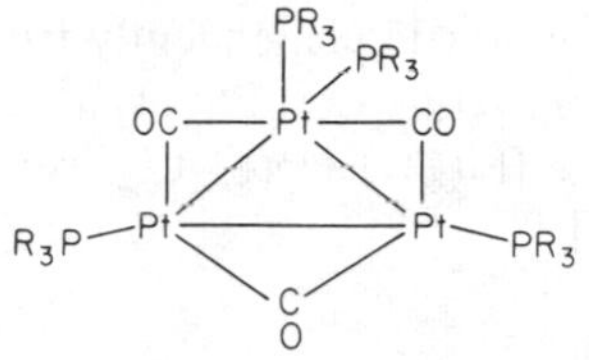

(CIV) 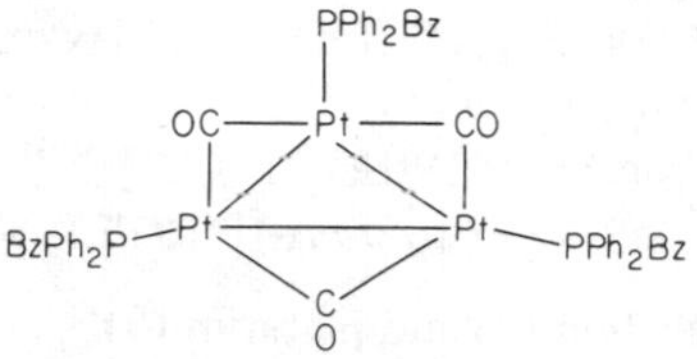   (CV)

the Pt_4 unit is a distorted tetrahedron, Pt–Pt = 275 pm, 279 pm, Pt–P = 227 pm (average) (CVI). The red isomers probably have structure (CVII)[1044]. By photo-chemical decomposition of $[Pt(PPh_3)_2(C_2O_4)]$ in solution under hydrogen the complex, $Pt_4(CO)_3(PPh_3)_5$, is formed[1046].

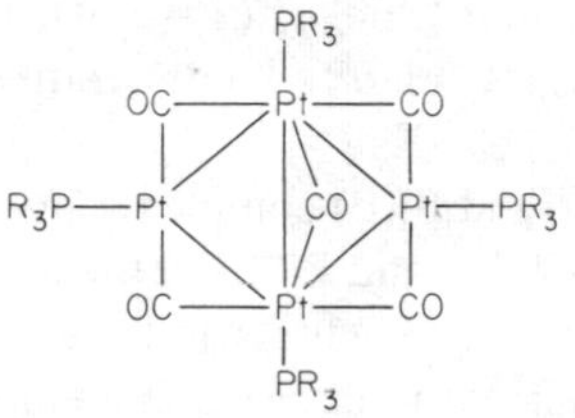

(CVI) 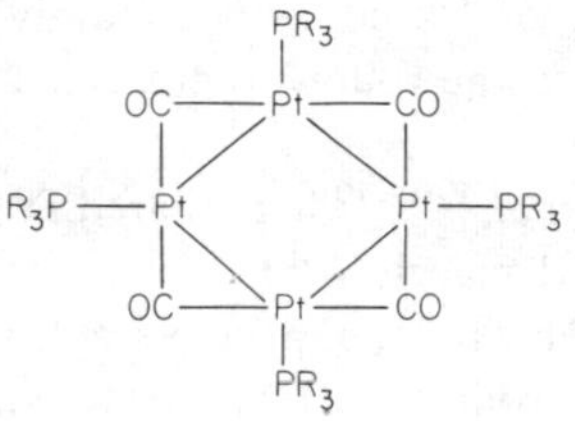   (CVII)

Carbonylation of $Pt(PPh_3)_3$ or $Pt(PPh_3)_4$[1047-9], or reaction of K_2PtCl_4 with CO in the presence of the appropriate quantity of phosphine[1049], leads to the formation of $Pt(CO)(PPh_3)_3$ and $Pt(CO)_2(PPh_3)_2$. Chini and Longoni[1049] prepared analogous complexes from PPh_2Et, PPh_2Pr^i and PPh_2Bz. In solution in the presence of CO complex equilibria exist involving $Pt(CO)(PR_3)_3$, $Pt(CO)_2(PR_3)_2$, $Pt_3(CO)_3(PR_3)_4$, $Pt(PR_3)_3$ and $Pt(PR_3)_2$[1049]. X-ray crystallographic studies by Albano *et al.*[1050-2] have shown that two forms of $Pt(CO)(PPh_3)_3$ exist, both approximately tetrahedral, but differing in the amount of distortion.

Reaction of $[Pt(CO)X_2]_2$ with PCl_3 produced $[Pt(CO)(PCl_3)X_2]$ (X = Br, I, Cl?)[1053]. Chatt *et al.*[1054] treated $[Pt(PR_3)_2Cl_2]$ (R = Et, Pr^n, Bu^n) in benzene with CO at atmospheric pressure, to form colourless *cis*-$[Pt(CO)(PR_3)Cl_2]$; *cis*-$[Pt(CO)(PEt_3)Br_2]$ and *cis*-$[Pt(CO)(PEt_3)I_2]$ were also prepared. Using the same method and by carbonylation of Zeise's salt, K $[Pt(C_2H_4)Cl_3]$, followed by addition of PR_3, Smithies *et al.*[1055] prepared a series of *cis*-$[Pt(CO)(PR_3)X_2]$ (R_3 = Ph_3, $PhMe_2$, $PhCy_2$, Bu_3; X = Cl, I; R_3 = Ph_2Cy, Cy_3, X = Cl; R_3 = $PhBu_2$, X = I). Carbonylation of $Pt(PPh_3)_2F_2$ produced $Pt(CO)_2(PPh_3)_2F_2$[1056]. Cationic carbonyl halides of platinum have recently been obtained. Clark and coworkers[1057] reported a complex '$PtHCl(\pi\text{-}C_2F_4)(PEt_3)_2$' as one of the products of the reaction of *trans*-$[PtH(PEt_3)_2Cl]$ with C_2F_4, but subsequently this was shown[1058] to be $[Pt(CO)(PEt_3)_2Cl]X$ (X = BF_4, SiF_5), the B or Si being furnished by the glass apparatus. The cation has a planar structure, with Pt–Cl = 230 pm, Pt–C = 178 pm, Pt–P = 235 pm[1058]. The *trans*-[Pt(CO)(PR$_3$)$_2$X]BF$_4$ (R = Ph, Et; X = Br, I) species were prepared by cleavage of [Pt(PR$_3$)$_2$X$_2$]$_2$ with CO, or by reaction of *cis*-[Pt(PR$_3$)$_2$X$_2$] with CO and BF$_3$ in benzene at 120° [1039,1059,1060]. The *trans*-[Pt(CO)(PEt$_3$)$_2$Cl]ClO$_4$ complex was obtained independently from *cis*- or *trans*-Pt(PEt$_3$)$_2$Cl$_2$ and CO in acetone[1061]; *trans*-[PtH(CO)(PEt$_3$)$_2$]$^+$ is similarly obtained from *trans*-[PtH(PEt$_3$)$_2$Cl][1061]. The [Pt(CO)(PEt$_3$)$_2$Cl]$^+$ cation is hydrolysed by water to *trans*-[PtH(PEt$_3$)$_2$Cl][1059,1062].

14.2.2 Palladium(0) and platinum(0)

The recent interest in oxidative addition reactions, and in the coordinative reactivity of small molecules, has concentrated mainly upon d^8 complexes (Rh(I) and Ir(I), q.v.). The d^{10} M(PR$_3$)$_x$ (M = Pd, Pt; x = 2, 3, 4) are attracting increasing interest. Work with these complexes is complicated however, by the existence of several species in equilibrium in solution, and a good deal of confusion surrounds the nature of the products of some of the reactions. There is a recent review of this area by Ugo[1063], concerned mainly with Pt(PPh$_3$)$_x$ (x = 3, 4), and as with the Vaska complex, the following treatment is intended to be illustrative rather than comprehensive.

Malatesta *et al.*[1064,1065] prepared Pd(0) and Pt(0) complexes by reduction of M(PR$_3$)$_2$X$_2$ (R = aryl; X = halogen) with N$_2$H$_4$ and PR$_3$ in ethanol, or with ethanolic KOH. The unexpected stability of the products and the fact that these reactions can produce Pt(II) hydrido complexes under certain conditions, resulted in suggestions that the products were really M(II) hydrido complexes[1066]. This was disproved by several groups[1067-9]. Reaction of Pt(PPh$_3$)$_2$Cl$_2$ with 10 per cent alcoholic anhydrous hydrazine produced Pt(PPh$_3$)$_3$, or, in the presence of excess PPh$_3$, Pt(PPh$_3$)$_4$ [1064,1067]. However, (*p*-ClC$_6$H$_4$)$_3$P and (*p*-FC$_6$H$_4$)$_3$P produce only the tris complexes, Pt(PR$_3$)$_3$ [1064,1069]. Alcoholic hydrazine and Pd(PPh$_3$)$_2$I$_2$ produced metallic palladium, but in the presence of excess PPh$_3$, Pd(PPh$_3$)$_3$ was obtained[1065]. Similar complexes have since been synthesised with other phosphines—PPh$_2$Me, PPh$_2$Bz, PPhMe$_2$, P(C$_6$F$_5$)Me$_2$, PMe$_3$, PEt$_3$, PBz$_3$ and P(*p*-XC$_6$H$_4$)$_3$ (X = F, Cl, Me)[1049,1064,1065,1067,1069-73]. The reactive trialkylphosphine complexes have been prepared only recently by displacement of the π-boronyl ligand from Pt(B$_3$H$_7$)(PR$_3$)$_2$ with excess PR$_3$ [1071]. Kruck[1074] prepared Pd(PF$_3$)$_2$(PPh$_3$)$_2$, Pt(PF$_3$)$_2$(PPh$_3$)$_2$ and Pt(PF$_3$)$_3$(PPh$_3$) from M(PF$_3$)$_4$ and PPh$_3$ in ether at −78°C, and mixed phosphine or phosphine–phosphite complexes have also been prepared[1064,1065,1067].

Wheelock *et al.*[1075] using DPAM (diphenylphosphinomethylacetylene), prepared M(DPAM)$_4$ (M = Pd, Pt), whilst from M(PPh$_3$)$_4$ and Ph$_2$PC≡CPPh$_2$ they isolated

(CVIII)

yellow complexes, which have structure (CVIII) in the solid state, but dissociate in solution to mononuclear three-coordinate species.

Most of the studies have been done with the Pt–PPh$_3$ complexes[1063]. The 'parent' Pt(PPh$_3$)$_4$, which presumably has a planar structure is dissociated in solution and Pt(PPh$_3$)$_3$ is obtained on boiling Pt(PPh$_3$)$_4$ in ethanol under nitrogen[1076].

$$Pt(PPh_3)_4 \rightleftharpoons Pt(PPh_3)_3 \rightleftharpoons Pt(PPh_3)_2 \rightleftharpoons [Pt(PPh_3)]_n$$

It can also be obtained by reduction of Pt(PPh$_3$)$_2$X$_2$ [1064,1065], or by boiling [PtH(PPh$_3$)$_3$]HSO$_4$ with alcoholic KOH [1077]. Conducting[1077,1078] the latter reaction in a stream of oxygen, or with added H$_2$O$_2$ produces Pt(PPh$_3$)$_2$ and Ph$_3$PO, but the isolation of pure Pt(PPh$_3$)$_2$ is very difficult. A convenient method for its generation *in situ* is photolysis or thermal decomposition of Pt(PPh$_3$)$_2$CO$_3$ in ethanol[1079], but a monomeric Pt(PPh$_3$)$_2$ cannot be obtained on working up the solution. The Pt(PPh$_3$)$_3$ complex has an approximately trigonal structure[1080], Pt–P = 225–228 pm, the Pt–Pt separation being ~800 pm, ruling out platinum–platinum interactions.

There are also some rather ill-defined bi- and polynuclear complexes. Gillard[1081] isolated a polymeric [Pt(PPh$_3$)]$_n$ from the reaction of Pt(PPh$_3$)$_4$ with dienes in air, and *n*- is probably ~4 (CIX). A red Pt$_3$(PPh$_3$)$_4$ has also been reported[1049]. A red [Pt(PPh$_3$)$_2$]$_n$ forms on melting Pt(PPh$_3$)$_2$ under nitrogen and then rapidly cooling the melt; a trimeric structure (CX) has been proposed[1081]. The same complex

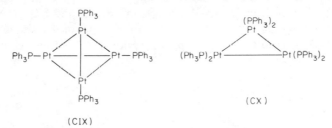

(CIX)

(CX)

is obtained by bubbling nitrogen containing small quantities of oxygen through [Pt(PPh$_3$)$_2$(C$_2$H$_4$)] solution[1077]; very pure nitrogen produces a yellow product. Photolysis of [Pt(PPh$_3$)$_2$CO$_3$] [1079] and [Pt(PPh$_3$)$_2$(C$_2$O$_4$)] [1046] produce Pt$_2$(PPh$_3$)$_4$ compounds, which do not have identical properties[1079]. Clearly more work is needed to elucidate the nature of these complexes.

The reactions of Pt(PPh$_3$)$_4$ in solution are complicated, and in a number of cases there is disagreement on the products obtained. The lower species—Pt(PPh$_3$)$_2$, etc.—are coordinately unsaturated and highly reactive; studies by Ugo *et al.*[1077] indicate that free phosphine reduces the activity of Pt(PPh$_3$)$_2$ and these workers have pointed out the similarity to the poisoning of metal surfaces by phosphines. The Pt(PPh$_3$)$_3$ complex (or Pt(PPh$_3$)$_4$, due to the reaction Pt(PPh$_3$)$_4 \rightleftharpoons$ Pt(PPh$_3$)$_3$ + PPh$_3$) reacts with aqueous or dilute alcoholic acids to form [PtH(PPh$_3$)$_3$]X (X = Cl, NO$_3$, etc.); dry HCl in benzene gives [PtH(PPh$_3$)$_2$Cl][943]. Acids producing only weakly coordinating anions afford only the ionic form, but anions with some intermediate coordinating ability produce both forms[943]. The Pd(0) complexes do not form stable hydrido complexes; the products are Pd(II) salts[943] (compare $Ni(PR_3)_4$). The more reactive $Pt(PEt_3)_3$ reacts with water or ethanol to give $[PtH(PEt_3)_3]^+$ [1072].

Both Pd(0) and Pt(0) take up oxygen irreversibly to give $M(PPh_3)_2(O_2)$ (+ Ph_3PO) [944,1083,1084]. The green palladium complex is rather unstable[944], but $Pt(PPh_3)_2(O_2)$ only decomposes on heating to ~120°C. The $Pt(PPh_3)_2O_2$ complex is diamagnetic, and has $\nu(O-O)$ at ~830 cm^{-1} [944]. X-ray structural studies have been hindered by the instability of the complex, and the accuracy of the reported data is low. The molecule is approximately planar (CXI), with Pt-P

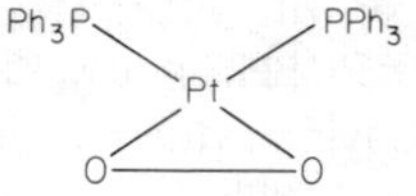

(CXI)

~225-230 pm, and Pt-O ~190-200 pm; but O-O is variously reported as 126 pm ($\widehat{OPtO}$ = 38°) and 145 pm ($\widehat{OPtO}$ = 43°) in $Pt(PPh_3)_2(O_2) \cdot CH_3C_6H_5$ [1085] and $Pt(PPh_3)_2(O_2) \cdot 1.5C_6H_6$ [1086], respectively.

The coordinated oxygen is very reactive. Sulphur dioxide converts the complex to $Pt(PPh_3)_2SO_4$ [818,1087,1088], NO_2 forms $Pt(PPh_3)_2(NO_3)_2$ [818,1087], aldehydes or ketones give (CXII) and NO forms $Pt(PPh_3)_2(NO_2)_2$ [818,1090]. With CO_2 an interesting reaction occurs to give either $Pt(PPh_3)_2CO_3$ or $Pt(PPh_3)_2CO_4$. The reported $PtH_2(PPh_3)_2$, formed from keeping $Pt(PPh_3)_4$ in benzene under nitrogen[1066,1067], is in fact $Pt(PPh_3)_2CO_3$ [1084,1091,1092]. The initial product of the reaction of CO_2 on $Pt(PPh_3)_2O_2$ is reported to be the peroxocarbonate (CXIII), which on

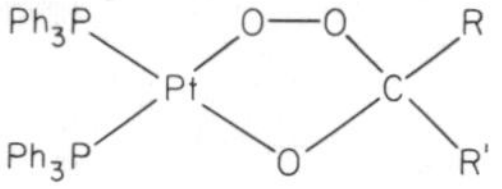

(CXII) 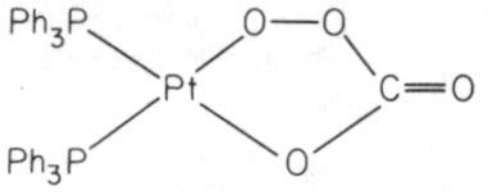   (CXIII)

recrystallisation forms $Pt(PPh_3)_2CO_3$ [1093]. The structure of the latter has been determined[1091]. Carbon disulphide and related molecules have been added to $Pt(PPh_3)_2(O_2)$ to yield $[Pt(PPh_3)_2(O_2)L]$, which may be similar to the products from the reaction with CO_2 [1093]. Iodine in chloroform converts $Pt(PPh_3)_2(O_2)$ to cis-$[Pt(PPh_3)_2I_2]$ [1094]. The mechanism of the $Pt(PPh_3)_3$ catalysed oxidation of PPh_3 has been studied by Halpern and Pickard[1095].

The reaction between $Pt(PPh_3)_n$ species and many other substrates have been reported[1063] (figure 21). Several of these reactions have also been observed with the corresponding palladium complexes; but this area has not been as well

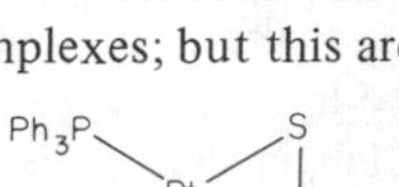

(CXIV)

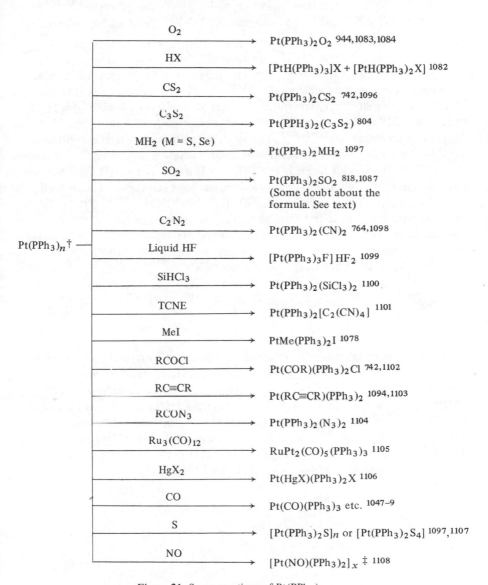

Figure 21 Some reactions of Pt(PPh$_3$)$_n$

† Pt(PPh$_3$)$_n$ (n = 4–1) are related by the equilibria occurring in solution. The active species in solution are almost certainly the lower coordinated species, but which form is actually the reactant is not generally clear. Hence Pt(PPh$_3$)$_n$ is used in this figure.

‡ Apart from this brief report nitrosyls of palladium and platinum seem to be unknown.

studied[1109,1114]. The reaction of $Pt(PPh_3)_3$ with CS_2 in ether produces[742,1096] orange $Pt(PPh_3)_2(CS_2)$, which has structure (CXIV)[1110], Pt–P = 235, 240 pm, $\widehat{SCS} = 136°$, the CS_2 being π-bonded to the platinum. A similar palladium complex has been reported[1111]. The reactions of $M(PPh_3)_4$ with alkenes, alkynes, alkyl and acyl halides, etc, have been much studied, and the structures of several of the products have been elucidated[1063,1112]. The $Pt(PEt_3)_4$ complex is much more reactive than the $Pt(PPh_3)_4$ derivative; for example the former readily adds PhCl to form *trans*-$[Pt(Ph)(PEt_3)_2Cl]$, but the latter reacts only with more active organohalogen compounds[742,1071]. The reactions with some inorganic substrates are shown in figure 21. The true formulation of the SO_2 complex has been disputed. Green $Pt(PPh_3)_2(SO_2)$ is formed from SO_2 and $[Pt(PPh_3)_2(C_2H_4)]$ [1087], but direct reaction of SO_2 with $Pt(PPh_3)_4$ in benzene has been variously reported to yield $Pt(PPh_3)_3(SO_2).1.5C_6H_6$ or $Pt(PPh_3)_2(SO_2).1.5C_6H_6$ [818,1106,1113]. The reactions with H_2S, H_2Se and S are particularly interesting in view of the correlation between the $M(PR_3)_n$ compounds and metal surfaces[1097] and the fact that such compounds are catalyst poisons. Excess sulphur converts $Pt(PPh_3)_4$ to $[Pt(PPh_3)_2S_4]$, which probably has structure (CXV) [1107].

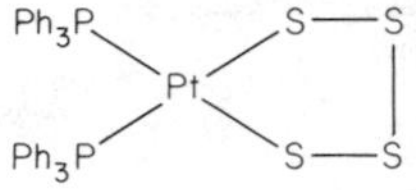

(CXV)

Some metal–metal bonded complexes were obtained by Nyholm and coworkers[1106] from the reaction of $Pt(PPh_3)_{3,4}$ with $Au(PPh_3)X$, HgX_2, Ph_3SnCl and $Cu(PPh_3)_3Cl$. Particularly interesting in this respect are the observations that $Pt(PR_3)_4$ can function as a source of PR_3 in syntheses; for example in the case of $Co_6(CO)_{18}C_4$ reaction gives $Co_6(CO)_{18-n}C_4(PR_3)_n$, without incorporation of $Pt(PR_3)_n$ in the product[552], or in the cases of $Fe_2(CO)_9$, $Ru_3(CO)_{12}$ or $H_2Os(CO)_4$, metal clusters form in which PR_3 and CO exchange between the metals seems to occur[1105]. The products include such compounds as (CXVI)[1105].

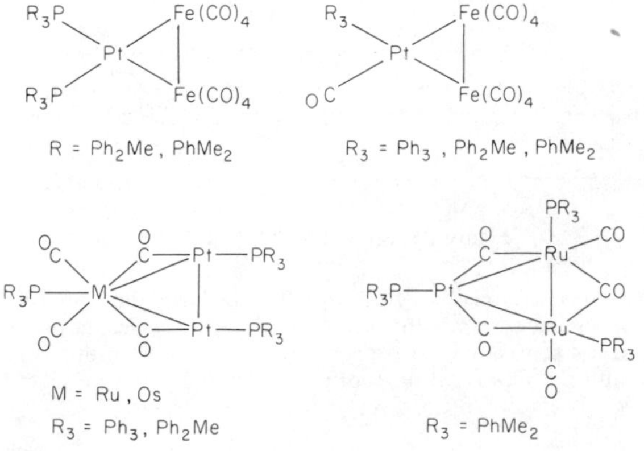

R = Ph_2Me, $PhMe_2$

R_3 = Ph_3, Ph_2Me, $PhMe_2$

M = Ru, Os
R_3 = Ph_3, Ph_2Me

R_3 = $PhMe_2$

(CXVI)

On heating $Pt(PPh_3)_2(COS)$, formed from $Pt(PPh_3)_3$ and COS [742], a bright yellow complex, shown by X-ray analysis[1115] to be $Pt_2S(CO)(PPh_3)_3$ (CXVII), is

(CXVII)

formed; Pt—P = 265 pm, Pt—S = 222, 223 pm, $\widehat{PtSPt}$ = 73°, $\widehat{PtPtS}$ = 53.5° (average). The related complex, $Pt_2S_2(PPh_3)_3$ [1107], formed from *cis*-$[Pt(PPh_3)_2Cl_2]$ and Na_2S in ammoniacal ethanol, probably has a related structure with S replacing CO.

14.2.3 Hydrido complexes

Simple hydrido phosphine complexes of these elements are not known; the reported $H_2Pt(PPh_3)_{3,4}$ [1066,1067] are, in fact, carbonato complexes[1084,1091,1092]. The dihydrido Pt(IV) compound, $PtH_2(PPh_3)_2Cl_2$, was reported[943] to be formed by reaction of $PtH(PPh_3)_2Cl$ with excess HCl in benzene, followed by precipitation with hexane. It was said to be unstable, losing HCl slowly. Recent work has shown it to be merely a different crystalline form of the well-known *trans*-$[PtH(PPh_3)_2Cl]$ [1116] An analogous PEt_3 complex, $PtH_2(PEt_3)_2Cl_2$ [1117,1118], must also be viewed with some doubt.

Hydrido complexes of the divalent elements are well established, those of platinum being much more stable than their palladium analogues. The methods generally used to prepare the platinum complexes (below) are not applicable to palladium, since palladium hydrides decompose in both acidic and basic media[943]. Brooks and Glockling[1119] prepared *trans*-$[PdH(PEt_3)_2X]$ (X = Cl, Br) from *trans*-$[Pd(PEt_3)_2X_2]$ and Me_3GeH, but found that the reaction could not be extended to PPr^n_3 or PPh_3 complexes. A new route discovered by Green *et al.*[963], analogous to that used to prepare nickel hydrides, is the reaction of *trans*-$[Pd(PR_3)_2Cl_2]$ with $NiH(PR'_3)_2BH_4$ (R = Et, Bu^n, R' = Cy, Pr^i). The products are mixtures of *trans* hydrides, since phosphine exchange occurs as well as reduction; the various products were characterised by ^{1}H n.m.r. spectroscopy. The mixtures are not readily separable, neither chromatography nor repeated crystallisation yielding pure products. However, pure colourless *trans*-$[PdH(PR_3)_2X]$ were obtained by a suitable cycle of reactions used in working up the mixture (R = Cy, Pr^i). Oxidative addition of HCl to $Pd(PCy_3)_2$ or $Pd(CO)(PPh_3)_3$ also produced *trans*-$[PdH(PR_3)_2Cl]$ [1120]. Few reactions of these palladium hydrides have been reported; metathesis of *trans*-$[PdH(PEt_3)_2Cl]$ with LiI yielded impure $[PdH(PEt_3)_2I]$; KCN gives $[Pd(PEt_3)_2(CN)_2]$; and HCl or CCl_4 form *trans*-$[Pd(PEt_3)_2Cl_2]$ [1119].

Platinum(II) hydridohalides are formed by reduction of *cis*-$[Pt(PR_3)_2X_2]$ with hydrazine in dilute aqueous or alcoholic solution[1117,1121]. This reaction has been studied in detail and the intermediate complexes characterised[1122]. Alcoholic KOH reduces *cis*-$[Pt(PR_3)_2X_2]$, but not *trans*-$[Pt(PR_3)_2X_2]$, to *trans*-$[PtH(PR_3)_2X]$, aldehyde or ketone being produced simultaneously[1117]. Chatt *et al.*[1117,1121] obtained a series of these complexes (R_3 = Ph_3, Ph_2Et, $PhEt_2$, Et_3, Pr^n_3) as colourless solids. There are several other less satisfactory methods of preparation, including $LiAlH_4$ reduction of either *cis*- or *trans*-$[Pt(PEt_3)_2X_2]$ [1117], hydrogenation of the *cis* dihalide[1117], hydrolysis of the cationic carbonyl[1059] or formic acid reduction of the dihalide[1117]. The action of HX on $Pt(PPh_3)_4$ is a convenient route to the triphenylphosphine complexes[943,1123]. The complexes are much more stable than most transition metal hydrides, but they are converted to $Pt(PR_3)_2X_2$ by halogens, boiling aqueous HX or CX_4 [1117]. The chloride *trans* to the hydride in *trans*-$[PtH(PR_3)_2Cl]$ is labile, and by metathetical exchange *trans*-$[PtH(PR_3)_2X]$ (X = Br, I, NO_2, CNS, CN) are readily obtained[1117]. In contrast to most transition metal hydrides, the platinum complexes are air stable.

The X-ray structural studies on $[PtH(PEt_3)_2Br]$ [1124] and $[PtH(PPh_2Et)_2Cl]$[1125] have confirmed *trans* planar stereochemistry (Pt–P $\sim$ 227 pm); the hydrogen atoms were not located, but the geometry of the molecules indicate that they are indeed *trans* to the halogen. The two phosphorus atoms are mutually *trans*, and the halogen makes up the third position of the square. The Pt–hal bonds are longer than the values predicted by the sum of the radii, which was interpreted as strong evidence for the presence of the hydride[1124,1125]. These complexes have also been studied by ^1H and ^{31}P n.m.r. [1126-8], and i.r. spectroscopy[887,1128,1129]. In *trans*-$[PtH(PEt_3)_2X]$ the ν(Pt–H) frequency decreases along the series of X $NO_3 > Cl > Br > I > NO_2 > CNS > CN$, the order of increasing *trans* effect of X [1129]. There is n.m.r. evidence for N- and S-bonded CNS groups in solutions of *trans*-$[PtH(PEt_3)_2CNS]$ [1127,1128].

Bailar and Itatani[1130] reported the isolation of both *cis* and *trans* isomers of $PtH(PPh_3)_2Cl$, but Clemmit and Glockling[1131] suggested that the *cis* isomer was merely a different crystalline form of *trans*-$[PtH(PPh_3)_2Cl]$, on the basis of i.r. and ^1H n.m.r. spectra. Furlani and coworkers[1132] identified three crystalline modifications of *trans*-$[PtH(PPh_3)_2Cl]$, which were easily interconverted according to the method of purification used. The report of *cis*-$[PtH(PPh_3)_2Cl]$ [1130] is thus most probably incorrect; *cis*-platinum hydrohalides have not been observed with monodentate phosphines, although they are known with bidentate phosphines[1131]. The so-called '$PtH_2(PPh_3)_2Cl_2$' is apparently also the form of *trans*-$PtH(PPh_3)_2Cl$, originally identified as the *cis* isomer[1116].

The insertion of olefins[1117,1133,1134], carbon disulphide[1135], acyl or sulphonyl azides[1136], or the unsaturated phosphine, o-CH_2=$CHC_6H_4PPh_2$, [1137] into Pt–H bonds has recently been reported. The reaction between this olefin–phosphine

(CXVIII)   (CXIX)   (CXX)

(2 moles) and *trans*-PtH(PPh$_3$)$_2$Cl results in a Pt–C σ-bonded complex (CXVIII), whilst in the presence of ClO$_4^-$ and PF$_6^-$ two and three moles of the olefin ligand afford (CXIX) and (CXX), respectively [1137].

The exchange of H for D in *trans*-[PtH(PEt$_3$)$_2$Cl] in D$_2$O–acetone is very slow, but accelerated by addition of DClO$_4$ or DCl [1138].

The only report so far of cationic palladium hydrides is that by Green and Munakata[1139], who prepared [PdH(dPe)PR$_3$]PF$_6$ (R = Cy, *i*-Pr) by reaction of *trans*-[PdH(PR$_3$)$_2$Cl] with dPe and NH$_4$PF$_6$ in C$_6$H$_6$–MeOH.

Cationic hydridocomplexes of platinum, [PtH(PR$_3$)$_2$L]$^+$ (L = neutral ligand) have been prepared by several routes. Dissolution of Pt(PPh$_3$)$_4$ in aqueous acids gives [PtH(PPh$_3$)$_3$]X (X = ClO$_4$, HSO$_4$, Cl, NO$_3$, BF$_4$) [943]. When the anions are weakly coordinating these are the only products, but intermediately coordinating anions (for example Cl$^-$, NO$_3^-$) can produce [PtH(PPh$_3$)$_2$X] under different conditions, whilst with strong donors, for example CN$^-$, the [PtH(PPh$_3$)$_2$X] only can be obtained. Cyanide or thiocyanate ions react with [PtH(PPh$_3$)$_3$]X (X = Cl, NO$_3$, HSO$_4$) to yield *trans*-[PtH(PPh$_3$)$_2$CN] and *trans*-[PtH(PPh$_3$)$_2$CNS], and with BPh$_4^-$ simple anion exchange occurs to give [PtH(PPh$_3$)$_3$]BPh$_4$ [943].

Displacement of chloride in *trans*-[PtH(PR$_3$)$_2$Cl] (R = Ph, Et) occurs in acetone on treatment with a neutral ligand and NaClO$_4$, forming [PtH(PR$_3$)$_2$L]ClO$_4$ (L = PPh$_3$, PEt$_3$, P(OPh)$_3$, P(OMe)$_3$, py, Me$_3$CNC) [1061]. The products are air stable, 1:1 electrolytes and exhibit n.m.r. spectral bands consistent with *trans* structures[1061].

The mixed ligand complex [PtI(dPe)(PEt$_3$)]Cl was obtained from [Pt(GeMe$_3$)(dPe)(PEt$_3$)]Cl and HCl [1140], and subsequently[1141] from *trans*-[PtH(PEt$_3$)$_2$Cl] and dPe in benzene. Use of tertiary phosphines instead of dPe led to [PtH(PEt$_3$)$_2$(PR$_3$)]Cl (R$_3$ = Ph$_3$, Ph$_2$Et, Et$_3$, Bu$_3$) [1141], although as well as displacement of chloride partial displacement of PEt$_3$ was also observed. In addition to tertiary phosphines, AsPh$_3$, SbPh$_3$, NH$_2$Pr^i and substituted pyridines (L) similarly gave [PtH(PEt$_3$)$_2$L]$^+$ cations[1142], isolated as crystalline perchlorates by addition of NaClO$_4$.

The equilibria

$$\textit{trans-}[PtH(PEt_3)_2Cl] + L \rightleftharpoons [PtH(PEt_3)_2L]^+ + Cl^-$$

have been studied [1141,1142], the tendency for the reaction to proceed to the right decreasing PR$_3$ > AsPh$_3$ ≥ SbPh$_3$ ~ N. Only the *cis* isomer of [PtH(PEt$_3$)(AsEt$_3$)$_2$]ClO$_4$ exists, whilst [PtH(PPh$_3$)(AsEt$_3$)$_2$]ClO$_4$ exhibits *cis–trans* isomerism in solution[1143].

(CXXI)

Giustiniani *et al.*[1141] obtained evidence for the formation of
$[(Et_3P)_2HPt(dPe)PtH(PEt_3)_2]Cl_2$ from the reaction of *trans*-[PtH(PEt$_3$)$_2$Cl] with
dPe in dichloroethane. The unusual complex (CXXI) was prepared several
years ago by Chatt and Davidson[1144] from *trans*-[PtH(PEt$_3$)$_2$Cl] and PPh$_2$H.

Hydridocomplexes are also formed by reaction of some silanes, germanes, etc.,
with platinum phosphine complexes (see section 14.2.4).

14.2.4 Complexes containing Group IVB donors

Palladium and platinum form many organocomplexes. The ready preparation of
alkyls and aryls contrasts strongly with the general scarcity of the corresponding
complexes of the preceding elements (Rh, Ir). Most of these complexes fall outside
the scope of this review (see reference 1145), and will only be referred to for
completeness.

Calvin and Coates[1146] prepared organopalladium complexes by the most common
method, that is Pd(PR$_3$)$_2$X$_2$ and RLi or RMgX. The more stable platinum
complexes have been more thoroughly studied. The majority are planar Pt(II)
compounds, but a number of Pt(IV) complexes have been described[1147]. The
spectroscopic properties have been reported[1148-50,1152,1153, and references therein],
and kinetic studies performed on several series of complexes[1151].

A novel way of stabilising M—C σ-bonds (M = Pd, Pt) is exemplified in the
complexes (CXXII), prepared by treating M(PEt$_3$)$_2$Cl$_2$ with 2-(diphenylphosphino)-
benzylpotassium [1028]. Intramolecular Pt—C bond formation occurs on heating
trans-[Pt(PRBu$^t_2$)$_2$X$_2$] (R = Ph, Pr^n, *p*-MeC$_6$H$_4$; X = Cl, Br, I) in 2-methoxy-
ethanol [1195]. The products have been assigned structures (CXXIII, CXXIV and
CXXV) [1195]. The olefin–phosphine complexes have already been described. An
interesting reaction occurs on heating Pd(PPh$_3$)$_4$ with PdCl$_2$ in dimethylsulphoxide,
as Pd(PPh$_3$)$_2$Cl$_2$, *trans*-[PdPh(PPh$_3$)$_2$Cl] and Pd$_3$(PPh$_2$)$_2$(PPh$_3$)$_3$Cl$_2$ have all
been identified as products[1154].

Complexes containing Pd and Pt directly bonded to the other members of
Group IVB have been described in recent years. The trimethylsilyl complex,

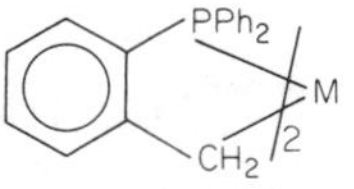

(CXXII)

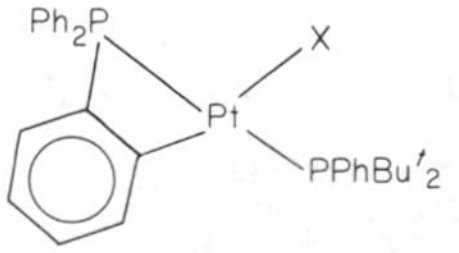

(CXXIII)

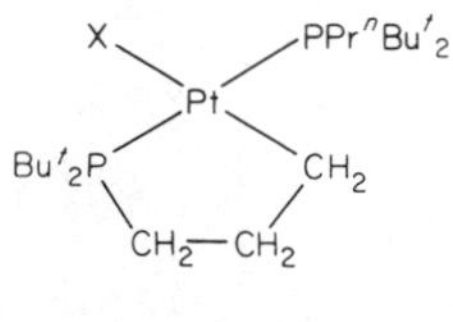

(CXXIV)

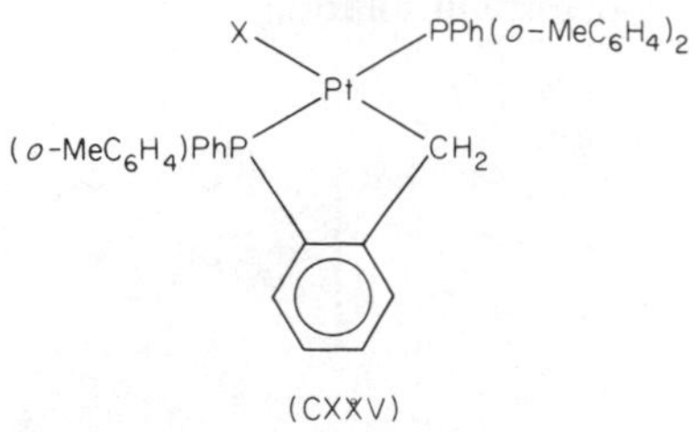

(CXXV)

trans-[Pt(PEt$_3$)$_2$(SiMe$_3$)Cl] was prepared by Glockling and Hooton from Hg(SiMe$_3$)$_2$ and *cis*-[Pt(PEt$_3$)$_2$Cl$_2$] [1140,1155,1156]. Subsequently, Chatt *et al.*[1100,1157] used the reactions of R$'_3$SiLi with *cis*-[Pt(PR$_3$)$_2$Cl$_2$], of R$'_3$SiH with *trans*-[PtH(PR$_3$)$_2$X] or *trans*-[Pt(PR$_3$)$_2$X$_2$], or of HSiCl$_3$, H$_2$SiPh$_2$, etc., with Pt(PPh$_3$)$_4$, to synthesise platinum–silicon bonds. Other silyl complexes have been described by Baird[1158], and by Bentham *et al.*[1159].

Germanium complexes have also attracted a good deal of work[1155,1156,1159-62]; indeed germyl complexes were obtained earlier than the silyl analogues[1163]. The structure of *cis*-[Pt(GePh$_2$OH)(PEt$_3$)$_2$Ph] has been determined[1164].

The best-known platinum–tin compounds are the SnCl$_3$ complexes first described by Wilkinson and coworkers[1165], but few phosphine-containing complexes are known[1166,1167]. Belluco *et al.*[1168,1169] have prepared platinum–lead compounds. A diphenylboron complex, Pt(Ph$_2$B)(PEt$_3$)$_2$Cl, has also been synthesised[1170].

14.2.5 Halide complexes

These two elements form an extensive series of complexes in the divalent state, and platinum also forms quadrivalent compounds. The Pd(II) and Pt(II) derivatives have been extensively studied; much of the pioneering investigations of Chatt and his coworkers in the coordination complexes of phosphines and arsines was done with these elements. They have been used to prepare derivatives of many unusual phosphines for organic characterisation purposes (especially Pd). The Pt(IV) compounds are all of the type *cis*- or *trans*-Pt(PR$_3$)$_2$X$_4$, but for M(II) a number of different types are known. The most important are the *cis*- and *trans*-M(PR$_3$)$_2$X$_2$, and the halogen-bridged M$_2$(PR$_3$)$_2$X$_4$; but other types, namely [M(PR$_3$)X$_3$]$^-$, [M$_2$(PR$_3$)$_4$X$_2$]$^{2+}$, [M(PR$_3$)$_3$X]$^+$ and [M(PR$_3$)$_4$]$^{2+}$ have been prepared, as well as phosphido species. There is also a multitude of complexes containing other ligands—amines, isocyanides, phosphites, etc., most of which have not been mentioned in this review.

(1) cis- and trans-M(PR$_3$)$_2$X$_2$

These complexes (table 16) are generally white, yellow, or orange in colour, air stable and of well-defined melting point. They are diamagnetic planar d^8 complexes, which have been extensively studied spectroscopically. Originally, configurations were assigned on the basis of dipole moment measurements, values of ~1 D being observed for the *trans* complexes and 9–13 D for the *cis* isomers. This has largely been replaced by spectroscopic methods. The early studies on the platinum complexes proved the presence of both isomers in the products obtained, but *cis* palladium complexes are rare, and have been obtained only relatively recently (the *cis*-Pd(SbR$_3$)$_2$X$_2$ are much easier to obtain[1210]). Chatt and Wilkins[1210] attempted to study the *cis–trans* isomerism of [Pd(PR$_3$)$_2$Cl$_2$] (R = Et, Pr^n, Bu^n) by measuring the dielectric constants of their solutions, but found that *cis* isomers, if any, were present in too small amounts to be detected. In the platinum series *cis* and *trans* isomers form readily by reaction of Na$_2$PtX$_4$ with PR$_3$ [1182]. Pure *trans* isomers were obtained from the mixture in benzene solution by adding a trace of PR$_3$, which affords the equilibrium mixture of the isomers, followed by

TABLE 16 PHOSPHINE COMPLEXES OF PALLADIUM(II) AND PLATINUM(II) HALIDES

Phosphine	$Pd(PR_3)_2X_2$	Ref.	$Pt(PR_3)_2X_2$	Ref.
PMe_3	*trans* Cl,Br,I *cis* Cl,Br	1004,1171,1175,1176	*trans* Cl,I *cis* Cl,Br,I	1004,1171,1172 1175,1177
PEt_3	*trans* Cl,Br,I,NO_3,CN,CNS,CNO *cis* CNO	2,1004,1178–81 973,1040	*trans* Cl,Br,I,NO_2,NO_3,CNS,CN,CNO† *cis* Cl,Br,I,NO_2,NO_3,CNS,CNO	973,1004,1177,1178, 1182,1183,1184,1190, 1212
PPr^n_3	*trans* Cl,Br	8,1004,1180	*trans* Cl,Br,I *cis* Cl,Br	1182,1184–7, 1212
PBu^n_3	*trans* Cl,Br,I,NO_2,CNS N_3	1004,1180 1188,1189	*trans* Cl,NO_2 *cis* Cl,NO_2,$\frac{1}{2}$SO_4	1182,1188 1212
$P(n\text{-}C_5H_{11})_3$	*trans* Cl	1180	*trans* Cl	1187
PCy_3	*trans* CNO, N_3	973,1189	*trans* CNO	973
$PPhMe_2$	*trans* Br,I *cis* Cl	1191–4,1197	*trans* I *cis* Cl,I,CNS	1194,1195 1197,1212
$PPhEt_2$	*trans* Cl,Br,I,NO_3	988,1179	*trans* Cl,I *cis* Cl	988,1182,1212
$PPhPr^n_2$			*trans* Cl *cis* Cl	1212
$PPhBu^n_2$	*trans* Cl *cis* Cl	1196,1197	*trans* Cl *cis* Cl	1212
$PPh(o\text{-}MeC_6H_4)_2$			*trans* Cl	1195,1212
$PPh(NEt_2)_2$	Cl‡	628	Cl‡	628
$PPh(CF_3)_2$			*trans* Cl	1198
$PPh(C_6F_5)_2$	*trans* Cl,Br,I	1174	*trans* Cl,Br,I	1174
$PPh(allyl)_2$			*trans* Cl	1148
PPh_2Me	*trans* Cl,N_3	1200,1201	*cis* Cl,N_3	1053,1201
PPh_2Et	*trans* Cl	1197	*cis* Cl	1212

Ligand	Pd complex (isomer, X)	Ref	Pt complex (isomer, X)	Ref
PPh_2Pr^n	*trans* Cl	1197	*cis* Cl	1212
PPh_2Bu^n	*trans* Cl	1197	*trans* Cl	1212
PPh_2Bu^t			*cis* Cl	1195
$PPh_2(CF_3)$	*trans* Cl,Br,I	1202	*trans* Cl; *trans* Cl,Br; *cis* Cl,Br	1195; 1198,1202
$PPh_2(C_6F_5)$	*trans* Cl,Br,I	1174	*trans* Cl,Br,I; *cis* Cl,Br	1174,1198
PPh_3	*trans* Cl,Br,I,CNS,N_3,CNO	57,973,1004,1040, 1197	*trans* Cl,Br,I; *trans* N_3,CNS,CNO; *cis* Cl,Br,I,NO_2	57,505,973,1064, 1182; 1203,1206,1207
$PPr^nBu^t_2$	*cis* NO_2	1181,1203–7		
$PBu^t_2(p\text{-}MeC_6H_4)$			*trans* Cl,Br,I	1195
$PMe(CF_3)_2$			*trans* Cl	1195
$PMe_2(CF_3)$			*trans* Cl; *cis* Cl	1172; 1172
$P(C_6F_5)_3$	*trans* Cl,Br	1174	*trans* Cl,Br,I	1174
$P(C_6H_3F_2)_3$			*trans* Cl,Br	1174
$P(CH_2Cl)_3$	*trans* Cl,I	57		
(2-phind)	*trans* Cl,Br,I,NO_2	988	*trans* I; *cis* Cl,Br	988
(9-phos)†	*trans* Cl,Br,I,CN	608	*trans* Cl,Br,I,CN; *cis* Cl	608
DPA	$Pd(DPA)_2Cl_2$	85	$Pt(DPA)_2Cl_2$	85
PPh_2H	Br,I	1208		
PPhEtH	Cl, Br	1209		
PEt_2H	*cis* Cl; Br,I	1209		
PBu^t_2H	*trans* Cl,Br,I,NO_3; *cis* Cl,Br	435	*trans* Cl,Br,I; *cis* Cl,Br	435

† R = Me, Et, Ph.
‡ Isomer not stated.

addition of $Pt_2(PR_3)_2X_4$ to freeze the equilibrium[1186]. The *trans* isomers are obtained by evaporation of the solvent, followed by recrystallisation from light petroleum. The less soluble *cis* isomers are obtained by evaporation of the solution in light petroleum. In this way, Chatt and Wilkins[1186] isolated *cis*- and *trans*-$[Pt(PPr^n_3)_2Cl_2]$ from the mixture of the two isomers. Subsequently, they studied the thermodynamics of the equilibrium

$$cis\text{-}Pt(PEt_3)_2Cl_2 \rightleftharpoons trans\text{-}Pt(PEt_3)_2Cl_2$$

calculated ΔH, ΔG, ΔS, and found that the total bond energy of the *cis* isomer is approximately 10 kcal (42 kJ) greater than the *trans*[1211]. The equilibrium between the two isomers shifts in favour of the *trans* in the series $Cl < Br < I$ [1187]. It has been suggested that the isomerism is photochemically effected[1190], and photochemically-induced isomerism has been reported[830]. The mechanism of the phosphine-catalysed isomerisation of *cis*-$[Pt(PR_3)_2Cl_2]$ (R = *n*-Pr, *n*-Bu) has recently been reexamined[1199]. The reaction is first order and involves a five-coordinate transition state. Jenkins and Shaw[1194] studied the isomerism of Pd and Pt halide complexes of $PPhMe_2$ by ^1H n.m.r. spectroscopy. In the platinum case the complexes were labile, and equilibrium was reached rapidly upon dissolution. For the palladium chloro-complex in chloroform a $\sim$2:1 *cis* : *trans* ratio was observed, but for the bromo and iodo derivatives only the *trans* isomers were detected. Grim *et al.* observed that whilst $P(alkyl)_3$ (Et, Pr, Bu) complexes of palladium existed only as the *trans* isomers, the complexes of PPh_2R and $PPhR_2$ isomerised spontaneously in solution[1197]. It has recently been observed that PMe_3 complexes are best made from Na_2MX_4 and $[Ag(PMe_3)I]_4$ or $[Ag(PMe_3)NO_3]$ [1175].

The structures of several $M(PR_3)_2X_2$ complexes have been determined by X-ray analysis. Messmer and Amma[1183] studied *trans*-$[Pt(PEt_3)_2X_2]$ (X = Cl, Br) and found the Pt, P and X atoms to be essentially square planar, Pt–P = 229.8 pm (Cl) and 231.5 pm (Br), Pt–Cl = 229.4 pm, Pt–Br = 242.8 pm. The structure of *cis*-$[Pt(PMe_3)_2Cl_2]$ is also planar, but with a small distortion towards tetrahedral geometry, Pt–P = 225.6, 223.9 pm, Pt–Cl = 236.4, 238.8 pm [1213]. Similarly, *cis*-$[Pd(PPhMe_2)_2Cl_2]$ exhibits a small distortion from coplanar square geometry[1193]. In all these compounds the M–M distance is far too great ($\sim$600 pm) for any metal–metal interaction.

(CXXVI)

The case of $Pd(PPhMe_2)_2I_2$ is unusual. On treatment[1194] of *trans*-$[Pd(PPhMe_2)_2Cl_2]$ with NaI in acetone, a mixture of dark red needles and yellow prisms is obtained, both analysing as $Pd(PPhMe_2)_2I_2$, and ^1H n.m.r. spectra indicated that both had *trans* structures. On heating, the yellow form changes to the red form at $120-130°C$, but on cooling the molten complex the yellow form is once more obtained. X-ray studies[1191,1192] show the red form to be orthorhombic with *trans*-iodines at 263 pm, *trans* phosphines at 234 pm and an axial iodine (one of the planar iodines of a second Pd complex molecule) at 328 pm; there are also two phenyl protons close to, and below, the square plane (CXXVI). The yellow form is a *trans* planar isomer with 'octahedral' coordination achieved by interaction of phenyl protons[1192].

The preparation of palladium(II) halide complexes of many unusual phosphorus donor ligands has been accomplished, usually for ligand characterisation purposes; they have been mostly synthesised by Mann and coworkers, for example with the ligands (CXXVII)[1229] and with $P(p-BrC_6H_4)(p-Me_2NC_6H_4)Ph$ and

(CXXVII)

$PPh(p-BrC_6H_4)Et$ [1215]. A large number of complexes containing two different phosphines has been prepared, including $[Pd(PPh_3)(PH_3)Cl_2]$; platinum forms the interesting $Pt_3(PPh_3)_3(PH_3)_3I_2$ [398].

The complexes have been extensively studied by spectroscopic techniques in order to determine configurations, to examine the *trans* effect theory and to study the metal–phosphorus bond. Electronic spectra have been used relatively little, probably due to the difficulty of interpreting the results for square-planar complexes of 4d and 5d elements. An early study using *trans*-$[Pt(piperidine)LCl_2]$ established the order of the ligands, L, producing increasing ligand field splitting to be $Et_2S < AsPr^n_3 < piperidine < PPr^n_3 < P(OMe)_3$ [1216].

The i.r. spectra of *trans*-$[M(PR_3)_2X_2]$ are probably the most thoroughly documented of any metal phosphine complexes. The studies were primarily concerned with establishing the configurations of the complexes, and with the assignments of $\nu(M-X)$ and $\nu(M-P)$ [1172,1177,1184]. The increasing availability of Raman spectrophotometers, allowing Raman and i.r. to be used in conjunction, has further increased the value of vibrational spectral studies. The spectra of the *cis*- and *trans*-$[M(PR_3)_2X_2]$ (M = Pd, Pt; R = Me, Et; X = Cl, Br, I) have been reported and discussed in detail[898,1176,1217-19], the principal bands assigned, the frequency ranges for $\nu(M-P)$ and $\nu(M-X)$ located, and the nature of the M–P bond discussed.

^{31}P and ^1H n.m.r. spectra of a large range of complexes have similarly been obtained[435,1194,1197,1212,1220-3], and Fryer and Smith[1224] have reported the ^{35}Cl nuclear quadrupole resonance spectra of some $[M(PR_3)_2Cl_2]$ complexes.

The coordination of ambidentate ligands to Pd, Pt phosphines has been studied using, for example, NO_2 [1203], CNS [1207,1225] and NCO [1226], and it is interesting to note that $Pd(PPh_3)_2(NCS)_2$ and the most stable form of $Pd(AsPh_3)_2(NCS)_2$ are N-bonded, but $Pd(SbPh_3)_2(SCN)_2$ is S-bonded [1207]. Fluoride complexes have been little studied. McAvoy *et al.* [1056] dissolved $Pt(PPh_3)_4$ in liquid HF and obtained a white product formulated as $Pt(PPh_3)_2F_2$. A re-investigation of the reactions has characterised the product as $[PtF(PPh_3)_3]HF_2$, and a palladium complex has similarly been obtained. Liquid hydrogen fluoride converts $Pt(PPh_3)_2X_2$ (X = Cl, Br) to *cis*-$[Pt(PPh_3)_2FX]$ [1099].

The 2-phenylisophosphindoline (2-phind) and the 9-alkyl-9-phosphafluorenes (9-phos) have provided some interesting compounds. Collier and coworkers [988] prepared the normal $[M(2\text{-phind})_2X_2]$ (M = Pd, Pt) from two equivalents of phosphine, and found that a further equivalent was readily taken up to form $M(2\text{-phind})_3X_2$, which in ethanol reacts with sodium picrate to form salts $[M(2\text{-phind})_3X]$ (picrate). It was suggested that the trisphosphine chloride and bromide of both Pt and Pd can exist in isomeric forms; the structure of one isomer of $Pd(2\text{-phind})_3Br_2$ was found to be five coordinate (CXXVIII) [988]. Similar but

(CXXVIII)

less stable trisphosphine complexes were isolated with PEt_2Ph [988]. Palladium chloride forms scarlet crystalline $Pd(9\text{-phos})_3Cl_2$ [608], which appears not to ionise in solution. However, the platinum bromide complex does show some tendency to ionise into $[Pt(9\text{-phos})_3Br]^+Br^-$, the conductivity in nitrobenzene approaching that of a 1:1 electrolyte at infinite dilution. Interestingly, 9-phenyl-9-phospha-fluorene forms only four-coordinate complexes; four coordination only is also observed in the cyano-complexes of both metals with the 9-alkyl-9-phospha-fluorenes [608].

The isolation of cationic complexes $[PtLX(PEt_3)_2]ClO_4$ (X = Cl, Br; L = pyridine, PEt_3, PPh_3, $P(OMe)_3$, $P(OPh)_3$), was achieved [1061] by the reaction of *cis*-$[Pt(PEt_3)_2X_2]$ with L and $NaClO_4$ in acetone. *Trans* structures were indicated by ^1H n.m.r. spectral data for all the complexes except for L = PPh_3. Other cations of the type $[M(PMe_3)_3X]^+$ were obtained [1217] by treating $M(PMe_3)_2X_2$ with an equivalent of $[AgNO_3(PMe_3)]$ in acetone or dichloromethane; AgX is precipitated and $[M(PMe_3)_3X]NO_3$ is isolated from the solution (X = Cl, Br, I; M = Pd, Pt). The $[PdI(PMe_3)_3]^+$ complex was not obtained due to dis-proportionation to $Pd(PMe_3)_4^+$ and $Pd(PMe_3)_2I_2$. Two equivalents of $[AgNO_3(PMe_3)]$ reacted with $[M(PMe_3)_2X_2]$ and led to complexes of the type $[M(PMe_3)_4]^+$ [1217]. The i.r. spectra of these cations obtained as the NO_3^- and BF_4^- salts have been recorded [1217]. There are scattered reports of some other mononuclear cationic phosphine complexes, $[Pd(PPh_3)_3Cl]^+$ [1227], $[Pd(PEt_3)_3Cl]^+$ [1040], $[Pt(PPh_3)_3Cl]^+$ [1227],

$[Pt(PMePh_2)_3Cl]^+$ [1208,1212], $[Pd(PMePh_2)_3Cl]^+$ [1228], $[Pt(PPh_2Pr^n)_3Cl]^+$ [1212] and $[Pt(PEt_3)_4]^{2+}$ [1182,1230], and also cations containing other neutral ligands such as ammonia, pyridines, etc. The cation $[Pt(PPh_3)_3F]^+$ is formed from $[Pt(PPh_3)_4]$ and liquid hydrogen fluoride [1099].

Secondary and primary phosphines tend to form binuclear bridged complexes (section 14.2.6). However, Hayter[1208] succeeded in isolating $[Pd(PPh_2H)_2X_2]$ and $[Pd(PPh_2H)_3Br]Br$ by using benzene as solvent. These complexes are stable in the cold, but tend to lose HX on heating to give the phosphido-bridged complexes, the reaction being reversed on passing HX into a suspension of $[Pd(PPh_2)(PPh_2H)X]_2$ in benzene

$$2[Pd(PPh_2H)_2X_2] \rightleftharpoons [Pd(PPh_2)(PPh_2H)X]_2 + 2HX$$

The complexes $[Pd(PPh_2H)_2X_2]$ (X = Br, I) are monomeric in benzene, and appear to exhibit *cis–trans* isomerisation in solution, although only one form of the solid is known[1208]. Bronze $[Pd(PPh_2H)_3Br]Br$ dissociates reversibly in solution

$$[Pd(PPh_2H)_3Br]Br \rightleftharpoons [Pd(PPh_2H)_2Br_2] + PPh_2H$$

but in the presence of a large excess of free ligand it has the conductivity of a 1:1 electrolyte. PEt_2H and $PPhEtH$ give $[Pd(PR_3)_2X_2]$, and from PEt_2H and PEt_3 $[Pd(PEt_2H)(PEt_3)X_2]$ is obtained[1209]. Unlike the PPh_2H complexes, those of PEt_2H do not lose HX spontaneously to give phosphido complexes, but upon treatment with base, for example amines, dimeric complexes are readily formed[1209]. $[PdCl_2(PHMe_2)_2]$ decomposes to the phosphido complex on boiling in benzene [502].

In contrast to tertiary phosphine palladium complexes, it appears that the PEt_2H complexes have *cis* structures in the solid state, and that high proportions of the *cis* isomer are present in solution.

Unsaturated tertiary phosphines form some novel complexes with these metals. The ligand $CH_2=CH(CH_2)_3PPh_2$ produces[1232] bridged complexes with palladium (CXXIX), but platinum forms square-planar complexes in which the ligand behaves

(CXXIX)

as a bidentate. Treatment with excess ligand displaces the olefinic bond to give normal $[PtL_2X_2]$. Similar bidentate behaviour was observed by Isslieb and Haftendorn in the complexes of BPE and BPC [84]. Platinum complexes of AP and *cis*-PP, $PtLBr_2$, were obtained by Interrante *et al*.[83]. AP is not isomerised to *cis*-PP on coordination, since treatment of $[Pt(AP)Br_2]$ with KCN liberates the original ligand, and the ^1H n.m.r. spectrum of $[Pd(AP)Cl_2]$ is similar to that of the free ligand, although limited solubility prevented high-resolution spectra being observed[1233]. The AP complexes of both metals, $[M(AP)X_2]$ (X = Cl, Br), contain bidentate chelating AP (CXXX), but on treatment with NaI or NaCNS complexes are obtained in which AP may or may not be chelating. AP does, however, have more tendency to behave as a chelating ligand than the related arsine (CXXXI).

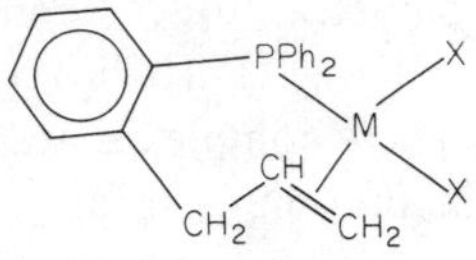

(CXXX) 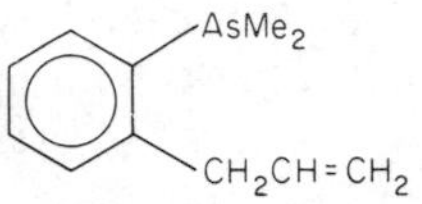   (CXXXI)

$[Pd(AP)I_2]$ may be an infinite polymer in the solid state with single iodine bridges[1233] (compare $[Pd(PMe_2Ph)_2I_2]$ red form).

Bennett *et al.*[1234] prepared $[Pd(SP)Br_2]$, $[Pt(SP)Br_2]$ and $[Pt(MP)Br_2]$, and studied the reaction of these, $[Pt(AP)Br_2]$ and $[Pd(AP)Br_2]$ with bromine. The MP complex and $[Pd(AP)Br_2]$ gave uncharacterised red oils, but the others produced complexes of formula $[PtLBr_4]$ (L = SP and AP) which are thought to be dimeric Pt(IV) complexes (q.v.). The tris-olefin phosphine $P(o-CH_2=CHC_6H_4)_3$ forms[1235] monomeric $[PtLBr_2]$, which, according to i.r. and Raman spectral data, contains the ligand acting as a bidentate, bonded through P and one of the double bonds. ^1H n.m.r. data showed that all three vinyl groups were equivalent indicating rapid exchange between the bonded and non-bonded groups. Bromination of this complex produces a complex of empirical formula $PtLBr_6$, which was formulated as a platinum(II) complex (CXXXII).

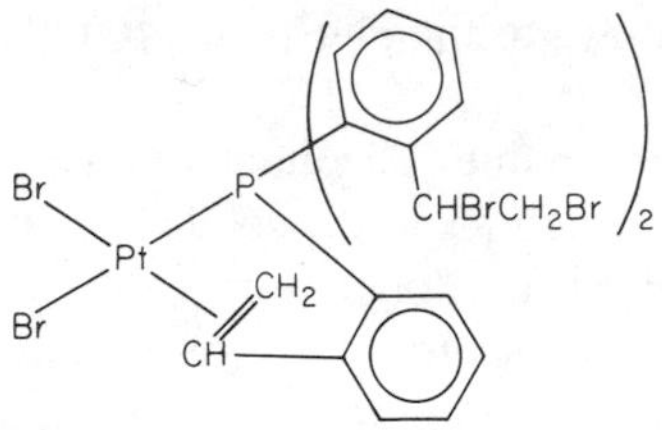

(CXXXII)

The reactions of square-planar platinum(II) complexes, and to a much lesser extent of their palladium analogues, have been extensively studied. The mechanisms and kinetics of their reactions in solution have been elucidated; normally they undergo bimolecular nucleophilic substitution reactions. This aspect of their chemistry has been reviewed[1236]. Chlorophosphine complexes of these two metals undergo coordinated ligand reactions at the phosphorus. Early studies of the PCl_3 complexes[1237-9] reported that they hydrolysed in water and produced orthophosphite complexes with alcohols. Chatt and Heaton[1240] found that the hydrolysis reactions were very complex. They studied the simpler hydrolysis of the monochlorophosphine complexes $[Pt(PR_2Cl)(PR'_3)X_2]$ (R = Ph, Et; R' = Et, Me, Ph), which were prepared by treating the $[Pt(PR'_3)X_2]_2$ complexes with PR_2Cl benzene. The hydrolysis gives successively, *cis*-$[Pt(PR_2OH)(PR'_3)X_2]$,

$[(PR'_3)XPt(R_2PO)_2PtX(PR'_3)]$ and finally two isomeric forms of $[(PR'_3)(OH)Pt(R_2PO)_2Pt(OH)(PR'_3)]$. The phosphinato-bridged complexes are surprisingly stable; the related *p*-toluenesulphinato complexes have also been studied[1241].

14.2.6 Anionic complexes
Substituted anions of Pt(II) with CO, NO, pyridine, etc., are well known, but surprisingly the phosphine analogues have been studied only recently. Chatt[1185] reported $[Pt(bipy)(PPr^n_3)Cl][Pt(PPr^n_3)Cl_3]$ in 1951, but no further reports appeared until 1968, when Goodfellow *et al.*[1257] prepared a series of complexes by treatment of the halogen-bridged $M_2(PR_3)_2Cl_4$ with NPr^n_4Cl in CH_2Cl_2 or $C_2H_4Cl_2$. The complexes were required for solution i.r. spectral studies, and were prepared *in situ*. Subsequently these investigators[1217] described the isolation of the complexes, which vary in colour from yellow to brown, and have molar conductance values in benzene of the order of $\sim 20\ \Omega^{-1}\ cm^{-1}$. All the complexes obtained were of the type $[M(PR_3)X_3]^-$ (M = Pd, Pt; R = Me, Et, X = Cl, Br, I; R = Ph, *n*-Pr, X = Cl). Detailed ^1H n.m.r., i.r. and Raman spectra were reported. Rest[1202] has reported $[Pd(PPh_2CF_3)X_3]^-$ and $[Pt(PPh_2CF_3)I_3]^-$.

14.2.7 Bridged complexes
Binuclear complexes, $M_2(PR_3)_2X_4$, are more important for these elements than for any of the other transition metals. They are obtained simply by treating $M(PR_3)_2X_2$ with K_2MX_4 (M = Pd, Pt; X = halogen) either in alcoholic solution or by melting together[1185,1242,1243]. Other methods include using $C_2H_4Cl_2$ as solvent[1244] or by slurrying in a hydrocarbon solvent[1245]. The structure is (CXXXIII),

(CXXXIII)

and other isomers have not been observed[1242,1243]. The stability of this type of complex falls in the order $PR_3 > AsR_3 > SbR_3$ [1243]. Many of these complexes have been prepared (table 17), and the ^1H n.m.r.[1197,1217], ^{31}P n.m.r.[1212] and i.r.[1217,1246,1250] spectra have been reported. The structure of *trans*-$[Pt_2(PPr^n_3)_2Cl_4]$ has been determined; Pt—P = 223 pm (significant differences in the Pt—Cl lengths depend on whether they are *cis* or *trans* to phosphorus), Pt—Cl = 242.5 pm (*trans*), 231.5 pm (*cis*), and Pt—Cl = 227.9 pm (terminal)[1251].

The halogen bridges can be cleaved by various neutral ligands, for example amines, to give *trans*-$[M(PR_3)LX_2]$ [1252], which have been studied well. Treatment of $[Pt_2(PPr^n_3)_2Cl_4]$ with KCNS in cold acetone leads to one form (α) of the thiocyanate, $[Pt_2(PPr^n_3)_2(CNS)_2Cl_2]$, whilst use of boiling acetone leads to the β form[1253]. It was shown[1253-5] that these had structures (CXXXIV), and more accurate X-ray data has since confirmed this[1256]. The eight-membered rings are approximately planar, and the phosphines exert a *trans* weakening effect, as shown by the lengths of the Pt–N, Pt–S bonds *cis* and *trans* to phosphorus[1256]. The [Pd(PPh$_3$)(N$_3$)$_2$]$_2$ complex contains both terminal and bridging azide[1257]. [Pd(PPh$_3$)(OAc)$_2$]$_2$ has also been characterised[1258].

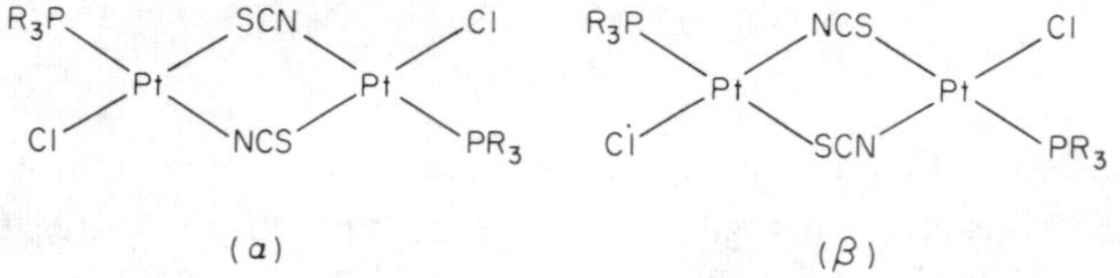

(a)   (β)

(CXXXIV)

TABLE 17 BRIDGED COMPLEXES M$_2$(PR$_3$)$_2$X$_4$

Phosphine	Pd	Pt	Refs.
PMe$_3$	Cl,Br,I	Cl,Br,I	1171,1217,1243,1246
PEt$_3$	Cl,Br,I	Cl,Br,I	1173,1197,1243,1246,1249
PPr$^n_3$	Cl,Br,I	Cl,Br,I	1173,1243,1246,1249
PBu$^n_3$	Cl	Cl	1197,1243,1244,1248
P(n-C$_5$H$_{11}$)$_3$		Cl	1243
P(n-C$_6$H$_{13}$)$_3$		Cl	1243
PPhBu$^n_2$		Cl	1212,1243
PPh$_2$Et	Cl	Cl	1212
PPh$_2$Pr^n		Cl	1212
PPh$_2$CF$_3$	Cl,Br,I	I	1202
PPhMe$_2$	Cl,Br,I	Cl	1194,1197,1212,1244
PPhEt$_2$		Cl	1212
PPhPr$^n_2$		Cl	1212
PPh$_2$Bu		Cl	1212
PPh$_3$	Cl,Br,I	Cl	1197,1246,1249
PPh$_2$Me	Cl		1197
PCl$_3$		Cl	1249
PPh$_2$Cy		Cl	1244

In the preliminary studies by Isslieb's group on secondary and primary phosphines it was found that these ligands tended to lose hydrogen on reaction with palladium salts to give phosphido complexes. From PPh$_2$H and PdCl$_2$ the [Pd(PPh$_2$)(PPh$_2$H)Cl]$_2$ was obtained[146]; it was assigned structure (CXXXV). From PPhH$_2$ and PdCl$_2$ in benzene, a dark brown product, believed to be [(H$_2$PhP)$_2$ClPd(PPh)PdCl(PPhH$_2$)$_2$] was isolated[147], and with PCyH$_2$ the [(H$_2$CyP)$_2$ClPd(PCyH)$_2$PdCl(PCyH$_2$)$_2$] compound was obtained[148]. Hayter has shown structure (CXXXV) to be incorrect[1208], and that the product obtained with

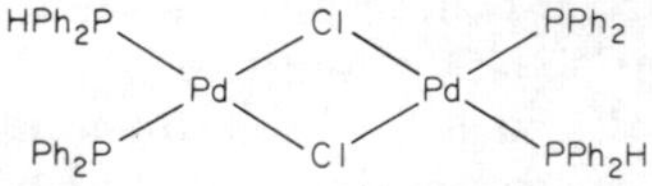

(CXXXV)

PPh_2H has the *trans* phosphido-bridged structure (CXXXVI). He also showed that under different conditions it is possible to isolate $[Pd(PPh_2H)_2Cl_2]$ and $[Pd(PPh_2)_2(PPh_2H)_2]$ as well. The (CXXXVI) complex reacts with dPe to yield (CXXXVII). Hayter[1209,1259] subsequently investigated related complexes of PEt_2H and $PEtH_2$, but, unlike the case of PPh_2H, these ligands only eliminate HX on

(CXXXVI)

(CXXXVII)

reaction of the mononuclear complexes with base. A preliminary X-ray study[1259] indicated that $[Pd_2(PEt_2)_2(PEt_2H)_2Cl_2]$ did indeed have the *trans*-PEt_2-bridged structure originally proposed on the basis of chemical evidence.

Chatt and Davidson[1144] prepared platinum complexes with bridging PR_2 groups (R = alkyl, aryl) by reaction of $[Pt(PR_3)Cl_2]_2$ with $LiPR_2$. These workers found that only *trans* isomers could be isolated, as with the arsenido complexes but in contrast to the corresponding sulphur-bridged derivatives. On boiling in toluene, $[Pt(PPh_2)(PPh_3)Cl]_2$ decomposed to produce $[Pt(PPh_2)Cl]_n$, presumably a polymeric complex. Various related $-SR$-bridged complexes have been prepared[1144,1260,1261] for both metals, and they exhibit interesting differences to the phosphido complexes, in that both *cis* and *trans* forms of $[M(SR)(PR_3)Cl]_2$, as well as complexes containing bridging thiol and halide groups are obtainable.

Dinuclear cations $[Pt_2(PBu^n_3)_4X_2](BX_4)_2$ were obtained from $Pt(PBu^n_3)_2X'_2$ and BX_3[1262]. Interestingly, if $X' \neq X$, the bridges contain X atoms; that is, *cis*-$Pt(PBu^n_3)_2Cl_2$ and BBr_3 give $[Pt_2(PBu^n_3)_4Br_2]^{2+}$[1262]. Clark *et al.*[1039,1040] prepared $[Pd_2(PR_3)_4Cl_2]^{2+}$ (R = Ph, Et) and $[Pt_2(PR_3)_4X_2]^{2+}$ (R = Ph, Et; X = Cl, Br, I) using BF_3 and $M(PR_3)_2X_2$ in chloroform. The PMe_3 complexes of general formula $[Pt_2(PMe_3)_4X_2](BF_4)_2$ are obtained[1217] from reaction of $Pt(PMe_3)_2X_2$ with $AgBF_4$ in acetone. Dixon and Hawke[1227] prepared the other members of the palladium series with PPh_3 and PEt_3, apart from $[Pd_2(PEt_3)_2I_2]^{2+}$, which could not be isolated.

The fluorine analogue, $[Pt_2(PPh_3)_4F_2](ClO_4)_2$, is formed on treating $[Pt(PPh_3)_3F]HF_2$ with $LiClO_4$, and the product obtained by dissolving $Pd(PPh_3)_4$ in liquid HF may be $[Pd_2(PPh_3)_4F_2]F_2$[1099].

14.2.8 Platinum(IV) complexes

Palladium(IV) seems to be too unstable to exist as a component of phosphine complexes, but $[Pt(PR_3)_2X_4]$ complexes are known. Phosphines reduce Pt(IV) salts, but it is possible to obtain $Pt(PR_3)_2X_4$ by halogen oxidation of $Pt(PR_3)_2X_2$[1263]. Chatt[1263] found that *trans*-$[Pt(PPr^n_3)_2Cl_2]$ was oxidised by chlorine in carbon tetrachloride solution to form *trans*-$[Pt(PPr^n_3)_2Cl_4]$, but that similar treatment of *cis*-[Pt(PPr$^n_3$)$_2$Cl$_2$] resulted in a mixture of 70 per cent *cis*- and 30 per cent *trans*-[Pt(PPr$^n_3$)$_2$Cl$_4$]. Chlorine oxidation of the bridged [Pt(PPr$^n_3$)Cl$_2$]$_2$ in CHCl$_3$ gave (CXXXVIII). By halogen oxidation of the corresponding Pt(II) compounds *trans*-

(CXXXVIII)

[Pt(PEt$_3$)$_2$X$_4$] (X = Cl, Br, I), *cis*-[Pt(PEt$_3$)$_2$X$_4$] (X = Cl, Br), *cis*-[Pt(PPhMe$_2$)$_2$Cl$_4$] and *cis*- and *trans*-[Pt(PBu$^n_3$)$_2$Cl$_4$] were obtained[1199, 1220, 1264]. The chloro and bromo complexes are yellow or orange, but the iodo derivatives are black. The configuration of these complexes has been established by ^{31}P n.m.r.[1220], i.r.[288, 1264] and Raman[898] spectroscopy, and the charge transfer spectra of *trans*-[Pt(PEt$_3$)$_2$X$_4$] (X = Cl, Br) have been reported[508]. Alkyl and aryl complexes of Pt(IV) are also known.

Kemmit *et al.*[1099] have prepared Pt(PPh$_3$)$_2$F$_2$Cl$_2$ by decomposition of [Pt(PPh$_3$)$_2$SF$_5$Cl]. Bennett *et al.* have produced Pt(IV) complexes by bromination of [Pt(AP)Br$_2$] and [Pt(SP)Br$_2$][1234].

15 COPPER, SILVER, GOLD

15.1 COPPER

Simple phosphine complexes of copper(II) are unknown; copper(II) salts react with phosphines to produce copper(I) complexes[1265]. The $[Cu_4OX_6(PPh_3)_4]$ (X = Cl, Br) were obtained by displacing MeOH from $[Cu_4OX_6(MeOH)_4]$ by PPh_3 [1266]. The structure consists of an oxygen atom surrounded tetrahedrally by four copper atoms, with bridging chlorines and terminal PPh_3 molecules completing the trigonal-bipyramidal configuration about the copper. The mixed ligand complexes, $[Cu(hfac)_2(PR_3)]$ (hfac = hexafluoroacetylacetonate), in which the oxidation state of the copper has been confirmed as Cu(II) by e.s.r. spectroscopy and magnetic susceptibility measurements, have been obtained recently[1267].

Potassium diphenylphosphide reacts with CuBr to form $CuPPh_2$, which reacts with excess $KPPh_2$ to form $KCu(PPh_2)_2$ [1268]. Ethanol protonates one of the phosphido groups to form $Cu(PPh_2)(PPh_2H)$, which can also be obtained from PPh_2H and $CuPPh_2$. With $LiPCy_2$ and CuCl it is thought that the resulting $[CuPCy_2]_n$ is probably polymeric. This also reacts with excess lithium salt to form $LiCu(PCy_2)_2$ [384]. Abel et al. [1023] found that CuCl reacts with Me_3SiPPh_2 to form $[CuPPh_2]_n$, but that CuBr and CuI gave Ph_4P_2, Cu and Me_3SiX. The polymeric $[CuPPh_2]_n$ is also formed[1023] from CuCl and PPh_2H in the presence of Et_3N. The nature of these complexes is not well understood.

Phosphine reacts with cuprous halides at low temperatures to form $CuX(PH_3)_2$, which readily lose PH_3 to form $CuX(PH_3)$ [1269]. Although $CuH(PEt_3)_2$ and $CuH(PPh_3)_2$ were reported[1270] to be more thermally stable than CuH, some doubt has been cast on these findings by the results of a cryoscopic study[1271] in pyridine of the reaction of CuH with various tertiary phosphines, in which it was found to be impossible to isolate any solid complexes. A hexameric cluster compound, $Cu_6H_6(PPh_3)_6 \cdot DMF$, has been isolated from the reaction of sodium trimethoxyborohydride with $[CuCl(PPh_3)]_4$ in DMF [1272]. An X-ray study shows the structure to be a distorted octahedron of copper atoms with a single PPh_3 molecule bonded to each copper (CXXXIX). The position of the hydride ligands has not been

(CXXXIX)

established. The borohydride complexes are related species. Sodium borohydride will reduce $CuSO_4$ in ethanol containing PPh_3 to form $(PPh_3)_2CuBH_4$; $[CuCl(PPh_3)]_4$ and $[CuCl(PPh_3)_3]$ can also be used[1273-5]. Acids decompose it to Cu(I) phosphine complexes, and, on heating, copper, hydrogen and PPh_3BH_3 are formed[1275]. The structure has been shown to consist of quasi-tetrahedral arrangements around the copper and boron atoms, with two bridging hydride ligands[1276].

By the reaction of $(PPh_3)_2CuBH_4$ with $HClO_4$ or HBF_4, Cariati and Naldini[1277] obtained complexes, which they formulated as $[(PPh_3)_2Cu(BH_4)Cu(PPh_3)_2]X$ (X = BF_4, ClO_4), and on the basis of i.r. measurements they suggested structure (CXL). Excess $HClO_4$ decomposes the complex with formation of

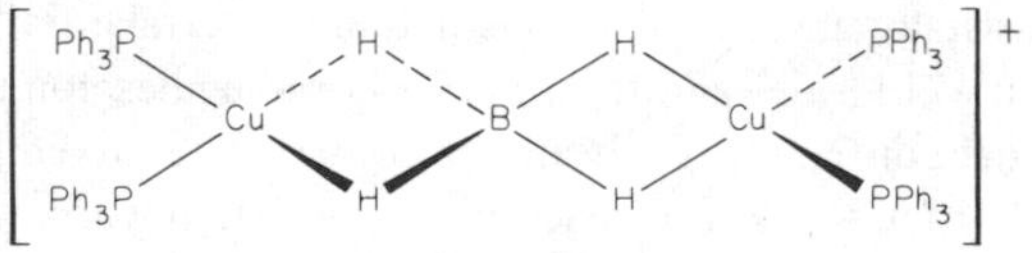

(CXL)

$[Cu(PPh_3)_2(ClO_4)]_2$ [1277]. Subsequently, $Cu(PPh_3)_2B_3H_8$ was prepared using CsB_3H_8 instead of $NaBH_4$ [1275,1278] and $Cu(PPh_3)_2B_5H_8$ [1279], $Cu(PPh_3)_2B_6H_9$ [1279], $Cu(PPh_3)_2B_{11}H_{14}$ [1278] and $Cu(PPh_3)_2B_9H_{14}$ [1278] have also been obtained. The structure of $Cu(PPh_3)_2B_3H_8$ [1280] is shown in (CXLI)[1281].

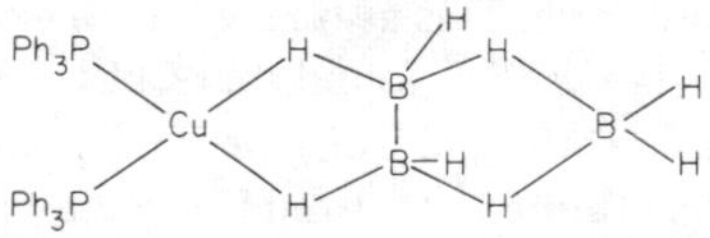

(CXLI)

An early report[1282] of the preparation of $Cu(PPh_3)_2$ by reduction of $[CuCl(PPh_3)_3]$ with hydrazine, and its subsequent decomposition to $[Cu(PPh_3)]_4$ on heating *in vacuo*, has been withdrawn[1283]. Copper(I) halides form phosphine complexes in the ratios $CuX:PR_3$ of 1:1, 1:2, 1:3 and 2:3.

The 1:1 complexes are usually tetrameric $[CuX(PR_3)]_4$, with structures analogous to that of $[CuI(AsEt_3)]$, which has been determined by X-ray crystallography[1284]. From the reaction of PR_3 with CuI in saturated aqueous KI were obtained $[CuI(PR_3)]_4$ (R = Et, *n*-Pr, *n*-Bu, *n*-amyl)[1285]. A 1:1 ratio of $CuX:PPh_3$ in benzene yielded $[CuX(PPh_3)]_4$ [1286,1287], which was also obtained by reducing CuX_2 with PPh_3 in ethanol[1288]. There are reports of other 1:1 complexes with a number of other phosphines, but insolubility prevented determination of the molecular weights[626,1289]. White $[CuX(PCy_3)]_2$ are thought to contain three-coordinate copper (CXLII). A tetramer ⇌ dimer equilibrium is apparently

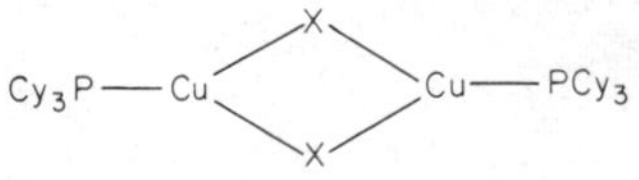

(CXLII)

exhibited by $[(C_5H_{10}N)_2PhPCuBr]$ in solution[1290], whilst $[(C_5H_{10}N)_2CyPCuBr]$ is dimeric[1291]. In contrast, $(Et_2N)_2PhP$ forms a normal tetrameric 1:1 complex[628]. These differences are probably due to steric factors, more bulky phosphines producing less polymeric structures.

A number of 1:2 complexes are known[628,1287], but except for $[CuX(PCy_3)_2]$ [1289], which are monomeric non-electrolytes, there is no evidence for the presence of three-coordinate complexes (compare the corresponding arsine complexes).

The reaction of $Cu(NO_3)_2$ with MeMgX and PPh_3 in ether yields $[CuX(PPh_3)_3]$ [1287], which can also be obtained by reacting CuX or CuX_2 with PPh_3 in ethanol or in a melt[1214,1288]. Copper is in a distorted tetrahedral environment in $[CuCl(PPh_2Me)_3]$ [1292]. Cupric fluoride reacts with PPh_3 in refluxing methanol to give a complex that has stabilised copper(I) fluoride, namely $[CuF(PPh_3)_3]$ [1288].

The 2:3 complexes[1274,1286,1287,1293] are best prepared by reaction of CuX with PPh_3 in a 1:2 ratio, and the unusual structure of $[Cu_2Cl_2(PPh_3)_3]$ (CXLIII), which contains three- and four-coordinate copper, has been determined[1293]. In

(CXLIII)

solution the copper(I)–phosphine systems are complex[1281]. Evidence was obtained for the occurrence of $CuCl(PR_3)_3$ and $(CuCl)_2(PR_3)_3$, and at low temperatures (about $-100°C$) for $(CuCl)_2(PR_3)_4$. When the anion was non-coordinating, for example PF_6^-, the $[Cu(PR_3)_4]^+$ and $[Cu(PR_3)_3]^+$ species were also present.

Copper(II) perchlorate reacts with PPh_3 in ethanol to give $[Cu(PPh_3)_4]ClO_4$ [1294], but with PCy_3, $[Cu(PCy_3)_2]ClO_4$ is formed[1289]; both are 1:1 electrolytes. Cotton and Goodgame obtained[1294] $[Cu(PPh_3)_2(NO_3)]$, which has been shown by Messmer and Palenik[1295] to be a distorted tetrahedron with a bidentate nitrato group. Similar $[Cu(PR_3)_2(NO_3)]$ complexes are also formed by $P(m\text{-tolyl})_3$ and PCy_3, whilst PPh_2Et, PPh_2Me and $PPhMe_2$ form $[Cu(PR_3)_3(NO_3)]$, which contain monodentate nitrato groups[1296].

Pseudohalide complexes have also been studied: $[Cu(PPh_3)_2CN]$ [1297], $[Cu(PPh_3)_2N_3]_2$ [1298,1299] and $[Cu(PPh_3)_2(SCN)]_2$[1298,1300]. In solution, the last two complexes were thought to be monomeric[1298,1301], but crystal structures have shown them to be dimeric with bridging pseudohalides containing copper in a distorted tetrahedral environment ((CXLIV) and (CXLV))[1299,1300].

(CXLIV)   (CXLV)

1:1, 1:2, 1:3 and 1:4 complexes of copper(I) are produced with 2-phenyliso-phosphindoline [1302]. Unlike the analogous 1:1 complexes of this ligand with Au(I), treatment with free halogen results in decomposition, not oxidation.

Phenylphosphine, $PPhH_2$, reacts with CuCl to form white $[CuCl(PPhH_2)]_4$ [147], but $PCyH_2$ forms the dimeric $[CuBr(PCyH_2)]_2$ [148]. The secondary phosphine, PPh_2H, was reported to form $[CuBr(PPh_2H)]_4$ [146], but a reexamination of the reaction of CuX with PPh_2H by Abel *et al.* [1023] has shown the existence of tetrameric 1:1, dimeric 1:2 and 1:3 complexes. No dimeric $[CuCl(PPh_2H)_2]_2$ was obtained, but $[Cu_2Cl_2(PPh_2H)_3]$ was isolated; CuBr and CuI did yield the expected 1:2 complexes. Stannous chloride reacts with $[CuCl(PPh_3)_3]$ to form $[Cu(SnCl_3)(PPh_3)_3]$ [1303].

The diphosphine DPA acts as a monodentate ligand on reaction with CuCl to form the tetrahedral $[L_3CuCl]$ complex [85], whilst the 3-butenylphosphines, $CH_2{=}CHCH_2CH_2PR_2$, form $[CuLCl]_2$ (R = Cy) and $[CuLCl]_4$ (R = Et) [84], in which the double bond is not coordinated. The olefin phosphine (AP) forms tetrahedrally coordinated dimers $[CuX(AP)]_2$, in which AP behaves as a bidentate ligand bonding through P and C=C, in contrast to the arsenic analogue (AA), which behaves as a monodentate As donor [1304]. The related ligand (MP) only complexes with cuprous iodide, forming $[CuI(MP)]_2$.

(AA)

The copper(I) phosphine complexes exhibit a wide range of stoichiometries and structures. The recent interest in the structure of copper complexes (see reference 1265) has resulted in a better understanding of the factors which determine the nature of the products, but more X-ray structure determinations are required to further elucidate these factors. Lippard and Palenik [1305] have produced a valuable discussion of the important factors which determine the stereochemistry of the four-coordinate copper–phosphine complexes.

15.2 SILVER

Silver iodide forms a 1:1 complex with PH_3 analogous to the copper(I) complex [1269]. The only phosphido complex reported is $[AgPPh_2]_n$, formed from PPh_2H and AgCl in liquid ammonia [1023]. The preparations of tetrameric $[AgI(PR_3)]_4$ (R = alkyl) [1306] and monomeric $[AgI(PR_3)_2]$ [626] by reaction of PR_3 with AgI dissolved in concentrated aqueous KI were reported quite some time ago, and $[Ag(NO_3)(PMe_3)]_4$ has recently been reinvestigated [1307]. Complexes between AgI and a number of unusual phosphines (for example, Et_3GePPh_2 [1308], $(C_5H_{10}N)_2PhP$ [1290] and $(C_5H_{10}N)_2CyP$ [1289]) have also been synthesised.

Tertiary arylphosphines have been little studied: Schwekendiek[1309] reported $AgBr(PPh_3)$, but it is not clear if this is polymeric as are the alkylphosphine derivatives. With 2-phenylisophosphindoline (L) $[AgClL]_n$ (n may be four in the solid, but about two in solution) and $[AgXL_2]$ [1302] are formed. With anions of poor coordinating ability—ClO_4^-, BrO_3^-, NO_3^-—$[AgL_3]^+$ and $[Ag(PPh_3)_4]^+$ are obtained[1294,1302]. The complex $[Ag(SCN)(PPr^n_3)]$ is polymeric with $-Ag-SCN-Ag-$ chains and silver–sulphur cross-linking[1310].

Evans *et al.*[1175] reported that $[AgI(PMe_3)]$ could not be prepared from AgI and PMe_3, the product being of variable composition approximating to $AgI(PMe_3)_{0.6}$. However, Dahl and Larsen[1307] obtained pure $[AgI(PMe_3)]$ by washing the crude product with saturated aqueous KI. Cryoscopic measurements in benzene indicate the complex is tetrameric, $[AgI(PMe_3)]_4$, as expected[1307]. $[Ag(PMe_3)NO_3]_4$ also has been prepared, and suggested as an alternative to $[AgI(PMe_3)]_4$ as a source of pure PMe_3[1175].

Muetterties and Alegranti studied[1281] tri-*p*-tolylphosphine complexes in dichloromethane solution at low temperatures by ^{31}P n.m.r. spectroscopy. The species present depended upon the anions present and upon the temperature. When either halide or pseudohalide were present, coordinated $[Ag(PR_3)_3X]$ were found, but with ClO_4^-, BF_4^- or PF_6^-, $[Ag(PR_3)_4]^+$, $[Ag(PR_3)_3]^+$ and $[Ag(PR_3)_2]^+$ were observed under different conditions.

Silver borohydride complexes are known, but have been less studied than the copper analogues; for example $(PPh_3)_2AgBH_4$ [1311], $(PPh_3)_2AgB_3H_8$ and $(PPh_3)_2AgB_9H_{14}$ [1278] have been prepared.

The olefin phosphines AP, BPE, BPC, form complexes $[AgI(AP)]$, $[Ag(NO_3)(AP)]$, $[AgX(AP)_2]$ (X = Cl, Br, I) [1304], $[AgI(BPE)]_4$ and $[AgI(BPC)]_2$ [84], in which the ligands behave as monodentate phosphine donors. With $AgNO_3$ vinyldiphenyl phosphine forms 1:1 and 2:1 complexes, which are thought to contain bidentate ligand[1312].

Other silver complexes include $(PPh_3)_2AgSnCl_3$ [1303], and the non-explosive fulminates $[Ag(NCO)(PR_3)_2]$ (R = Ph, Cy) [1313]. Tertiary alkylphosphines can be stored as their silver salts, from which they can be liberated when required (for example $AgNO_3.PMe_3$, see reference 1175).

15.3 GOLD

15.3.1 Gold(I) complexes

Cahours and Gal[1314] obtained $[AuCl(PEt_3)]$ as long ago as 1870, and monomeric $[AuX(R_3P)]$ complexes have been obtained by a number of workers[1222,1302,1306,1315–17]. These complexes have been studied extensively by n.m.r., i.r. and Raman spectroscopy[1217,1222 and references therein]. The complex $[Au(PPh_3)(CN)]$ contains approximately linear molecules with two-coordinate gold[1318]. A valuable route to gold(I) complexes has recently been described[1319]. It consists of reaction of $[AuCl(PPh_3)]$ with silver acetate to produce $Au(O_2CCH_3)(PPh_3)$, which reacts with HX (X = halide, CN, NCO, N_3, NO_3, COOH, $\frac{1}{2}SO_4$) to form $[AuX(PPh_3)]$. The sulphate, $[PPh_3Au(SO_4)AuPPh_3]$, is apparently covalent.

The nature of the 1:2 complexes, $AuX(PR_3)_2$, has been the cause of some controversy. $[AuI(p\text{-}Me_2NC_6H_4PMe_2)_2]$ ionises in solution[626], and $[Au(2\text{-phenylisophosphindoline})_2]I$ has been shown to be largely ionic in the solid state[1302]. Naldini *et al.*[1315] obtained PPh_3 complexes which they formulated as $[Au(PPh_3)_2]X$ (X = NO_3, ClO_4, BPh_4) and $[Au(PPh_3)_2X]$ (X = Cl, I), but Meyer and Allred[1317] concluded that the latter were better formulated as $[Au(PPh_3)_2]X$. A number of cyano and thiocyanato complexes have been reported[1320].

There is evidence for three-coordinate gold complexes in solution[1281,1321]. In the solid state three- and four-coordinate gold(I) is found in complexes such as $[Au(2\text{-phenylisophosphindoline})_3X]$[1302], $[Au(PPh_3)_4]X$ (X = ClO_4, BPh_4) and $[Au(PPh_3)_3]NO_3$[1315].

The ligand DPA forms a linear 1:1 complex with AuCl[85]. Phosphine does not react with AuCl or AuBr, but gives a 1:1 adduct with AuI[1269]. The $AuCl(PCl_3)$ complex is solvolysed by alcohols but decomposed by water[1322,1323], and is prepared by the reaction of AuCl with excess PCl_3. It has a linear structure with Au–Cl = 233 pm, Au–P = 219 pm and P–Cl = 198, 184, 207 pm[1324].

(1) Metal–metal bonded species

The Ph_3PAu– moiety has been bonded to a number of other elements by reactions such as

$$Ph_3PAuCl + Ph_3SiLi \rightarrow Ph_3PAuSiPh_3 + LiCl \qquad \text{(reference 1158)}$$
$$Ph_3PAuCl + NaMn(CO)_5 \rightarrow Ph_3PAuMn(CO)_5 + NaCl \quad \text{(reference 1326)}$$
$$Ph_3PAuCl + SnCl_2 \rightarrow Ph_3PAuSnCl_3 \qquad \text{(reference 1303)}$$

Among the elements bonded to Ph_3PAu are Mn[1325,1327], Fe[1326], Co[1326], Si[1158], Ge[1158], Sn[1303,1328], Ir[795] and Os[405,1330]. Very few Cu or Ag analogues have been obtained. The complex thought to be $[Cl_2(CO)(PPh_3)_2IrAu(PPh_3)]$[1331] has been shown to be $[Ir(CO)(O_2)Cl(PPh_3)]$[1332].

The structures of $[(Ph_3P)AuMn(CO)_4P(OPh)_3]$ and $[Ph_3PAuMn(CO)_5]$ have been determined. The P–Au distance is surprisingly different: 233 pm against 221 pm, respectively[1333]. The Au–Mn distance in the former is 257 pm, and in $(Ph_3P)AuCo(CO)_4$, Au–Co = 250 pm and P–Au = 223 pm[1334].

(2) Cluster compounds

Cluster compounds of gold have been identified recently. Malatesta *et al.*[1315] reported that $NaBH_4$ reduction of $AuCl(PPh_3)$ did not produce boron-containing compounds (compare Cu, Ag), but yielded a red complex, which analysed as $[Au_5(PPh_3)_4Cl]$. $4H_2O$[1315,1335], from which chloride could be replaced by metathesis. Sodium borohydride reacted with $AuCl(PPh_2Et)$ to form a compound thought to be $[Au_3(PPh_2Et)_2Cl]_4$, from which $[Au_3(PPh_2Et)_2X]_4$ (X = Br, I, SCN) and $[Au_6(PPh_2Et)_4Cl]Y$ (Y = ClO_4, BPh_4, PF_6) could be obtained by exchange[1336].

X-ray crystallographic studies by Mason *et al.*[1337] showed that the complex 'Au$_6$(PPh$_3$)$_4$(SCN)$_2$' is really [Au$_{11}$(PPh$_3$)$_7$(SCN)$_3$]. The corresponding iodide has also been X-rayed[1338]. Mason *et al.*[1337] predict the occurrence of other clusters of the type [Au$_n$L$_{n-1}$]$^{(n-8)+}$, and suggested that a compound analysing as Au$_9$(PPh$_3$)$_8$(PF$_6$)$_3$ belongs to this type, namely Au$_{13}$(PPh$_3$)$_{12}$(PF$_6$)$_5$. Also predicted were Au$_{10}$(PPh$_3$)$_7$X$_2$ and Au$_{12}$(PPh$_3$)$_7$X$_4$[1337].

15.3.2 Gold(III) complexes

In contrast to the monovalent copper and silver complexes, which cannot be oxidised, phosphine complexes of gold(I) readily combine with halogens to give [AuX$_3$(PR$_3$)][1180]. Oxidation with ICl and IBr was investigated in the hope of preparing *cis* and *trans* isomers of [AuX$_2$X'(PR$_3$)], but in every case only one complex was obtained[1180]. The [AuBr$_3$(PMe$_3$)] complex has been shown to be square planar[1339]. The bonding in these complexes has been investigated by i.r., ^{19}F n.m.r. and Mössbauer spectroscopy[1004,1341-3]. The coordination of the ambidentate NCO$^-$, SCN$^-$ and SeCN$^-$ ligands to both Au(I) and Au(III) in phosphine complexes has been shown to be through the N, S and Se atoms respectively[665].

The olefin phosphines AP and SP form linear P-bonded complexes AuLBr, which are oxidised by bromine in benzene to AuLBr$_3$[1329]. The n.m.r. spectra indicate the structure (CXLVI), and this has been confirmed by an X-ray single crystal study. This complex reacts with MeOH or EtOH on heating to perform a ring expansion to (CXLVII)[1340].

(CXLVI)   (CXLVII)

16 ZINC, CADMIUM, MERCURY

The common complexes of these elements are of two types $(R_3P)_2MX_2$ and $[(R_3P)MX_2]_2$, although a number of others have been reported, especially for M = Hg. The tetrahedral $(R_3P)_2MX_2$ and halogen-bridged $[(R_3P)MX_2]_2$, have been studied extensively by i.r. spectroscopy and assignments of $\nu(M-X)$, $\nu(M-P)$, and in some cases $\delta(M-X)$, have been proposed.

Zinc halides form 1:2 complexes with R_3P. These are obtained by direct reaction in water, alcohols or ether,[306,591,626,1344,1348]. The complexes, which have tetrahedral structures[592,1344-8], increase in stability in the series Cl < Br < I. The anions $[Zn(PPh_3)X_3]^-$ have been obtained, and characterised by i.r. spectroscopy[592]. The piperidine–phosphine complexes $Zn(PCyPip_2)_2X_2$ and $(ZnX_2)_2(PPhPip_2)_3$ have been obtained by direct reaction, or by decomposition of the appropriate diphosphine complexes[1291]. Hatfield and Yoke prepared $Zn(PEt_2H)_2X_2$ and $Zn(PEtH_2)X_2$ [1349]. Cadmium forms phosphine complexes in the ratios $CdX_2:R_3P$ 1:2, 1:1 and, rarely, 2:3 [1350]. The 1:2 complexes are tetrahedral and analogous to those of zinc, whilst the 1:1 are halogen-bridged dimers. The complex, $[(Et_3P)CdBr_2]_2$, was shown by crystallographic studies[1350] to have structure (CXLVIII), in which the coordination about the cadmium atom

(CXLVIII)

is tetrahedral. The complexes are non-conductors in nitrobenzene, and molecular weight studies in a variety of solvents indicate the presence of monomer $\rightleftharpoons$ dimer equilibria in these solutions[626].

The 2:3 complexes are of uncertain structure, but according to Evans *et al.*[1350] they are not mixtures of the 1:1 and 1:2 complexes. Complexes of piperidine–phosphine[1291] and of $P(NMe_2)_3$ and $PPh(NMe_2)_2$ [1352] have been described.

The $[Cd(PPh_3)_2X_2]$ (X = Br, I) compounds are converted to phosphonium salts $[RPPh_3]_2[CdX_4]$ on reaction with MeI or EtBr[1351].

Evans *et al.*[1350] described five types of mercury(II) phosphine complex with $HgX_2:R_3P$ ratios of 1:2, 1:1, 2:3, 2:4 and 3:2. The $[(R_3P)_2HgX_2]$ type are the best characterised[1350,1353,1354]; they are simple tetrahedral molecules[592,1344-7]. $Hg(PEt_3)_2Cl_2$ has a conductivity approaching that of a 1:2 electrolyte in aqueous solutions, attributed to the formation of $[Hg(PEt_3)_2(H_2O)_2]Cl_2$. The 1:1 complexes, which are common for mercury, have i.r. spectra consistent with a halogen-bridged structure[1346,1347]; in solution the degree of association depends upon the phosphine and upon the solvent[626]. The other three types were suggested[1350] to have similar halogen-bridged structures, but definite evidence is lacking. In addition a number of mixed cadmium-mercury complexes such as $[(Pr_3P)_2CdHgI_4]$ and $[(Pr_3P)(Pr_3As)CdHgI_4]$ have been obtained[1355], either by boiling solutions of $[(R_3P)_2CdI_2]$ with HgI_2, or by interaction of the appropriate $[(R_3P)_2MI_2]$ complexes.

Aminophosphine complexes of mercury such as $[P(NMe_2)_3HgI_2]_2$[1352] and $\{PhP(NR_2)(NR_2')\}_2HgI_2$[1356] have attracted some interest, whilst Seidal[1290-1] has studied the piperidine-phosphine complexes.

The mixed biphosphine $(C_5H_{10}N)CyP-PCy_2$ reacts with HgI_2 to form[1290] the presumably polymeric phosphido complex $[Cy_2PHgI_2]_n$, which is converted by MeI to $MeCy_2PHgI_2$ and $[Me_2Cy_2P][HgI_3]$.

Attempts to prepare the $[HgI_3PPh_3]^-$ ion by reaction of $[pyH][HgI_3]$ (py = pyridine) with PPh_3 were unsuccessful, $(PPh_3)_2HgI_2$ being obtained instead[1357]. Triphenylphosphine reacts with HgX_2 (X = SCN, NO_3, ClO_4) to produce 1:1 and 1:2 complexes analogous to the halides, but $Hg(CN)_2$ produces only $Hg(CN)_2(PPh_3)_2$[1358,1359]. Infrared spectra indicate that the thiocyanate complex has the structure (CXLIX). In contrast to the parent mercury compound, PPh_3 or PCy_3 complexes of mercury fulminate are not explosive[1313]. The olefin-phosphine, $CH_2=CH(CH_2)_3PPh_2$ (PP), forms $[HgX_2(PP)]_2$ (X = Cl, Br, I), halogen-bridged dimers with the phosphine bonded through the phosphorus only[1232].

(CXLIX)

ABBREVIATIONS

AP

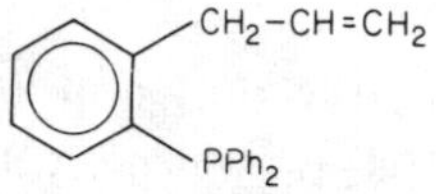

$CH_2-CH=CH_2$

PPh_2

PP

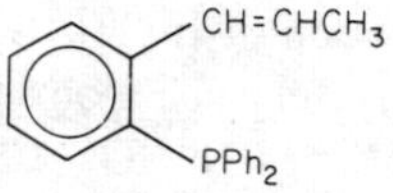

$CH=CHCH_3$

PPh_2

MP

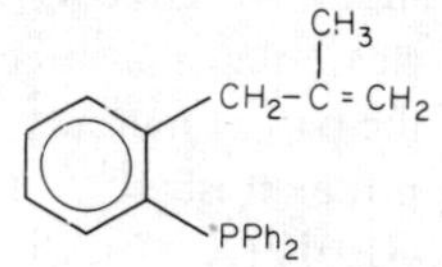

CH_3

$CH_2-C=CH_2$

PPh_2

SP

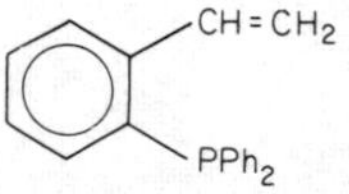

$CH=CH_2$

PPh_2

BPE	$Et_2PCH_2CH_2CH=CH_2$
BPC	$Cy_2PCH_2CH_2CH=CH_2$
DPA	$Ph_2PC\equiv CPh$
DPAM	$Ph_2PC\equiv CMe$
dPe	$Ph_2PCH_2CH_2PPh_2$
pP	$Ph_2P(CH_2)_3CH=CH_2$

2-Phind

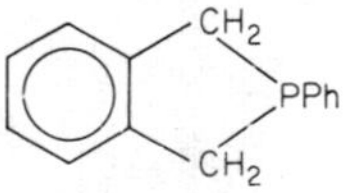

CH_2

PPh

CH_2

9-Phos

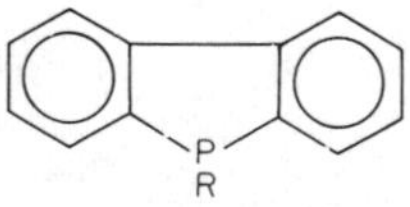

P

R

TMP

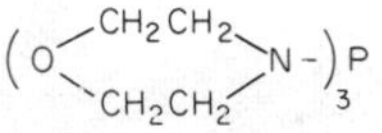

$\left(O \overset{CH_2CH_2}{\underset{CH_2CH_2}{\diagup\diagdown}} N- \right)_3 P$

REFERENCES

1. Hofmann, A. W., *Justus Liebigs Annln Chem.*, **103** (1857), 357.
2. Booth, G., *Adv. inorg. Chem. Radiochem.*, **6** (1964), 1.
3. Hayter, R. G., *Prep. inorg. Reacts.* **2** (1965), 211.
4. Levason, W., and McAuliffe, C. A., *Adv. inorg. Chem. Radiochem.*, **14** (1972), 173.
5. Maier, L., *Prog. inorg. Chem.*, **5** (1963), 27.
6. Nixon, J. F., *Adv. inorg. Chem. Radiochem.*, **13** (1971), 363.
7. Pidcock, A., Part 1 of this book.
8. Allison, J. A. C., and Mann, F. G., *J. chem. Soc.* (1949), 2915.
9. Majumdar, A. K., Mukherjee, A. K., and Bhattachayya, R. G., *J. inorg. nucl. Chem.*, **26** (1964), 386.
10. Hart, F. A., and Newberry, J. E., *J. inorg. nucl. Chem.*, **28** (1966), 1334.
11. Fitzsimmons, B. W., Gans, P., Huyton, B., and Smith, B. C., *J. inorg. nucl. Chem.*, **28** (1966), 915.
12. Majumdar, A. K., and Bhattachayya, R. G., *J. inorg. nucl. Chem.*, **29** (1967), 2359.
13. Day, J. P., and Venanzi, L. M., *J. chem. Soc. (A)* (1966), 1363.
14. Gans, P., and Smith, B. C., *J. chem. Soc.* (1964), 4172.
15. Selbin, J., Ahmad, N., and Pribble, M. J., *J. inorg. nucl. Chem.*, **32** (1970), 3249.
16. Rose, H., *Poggendorff's Annln Phys.*, **24** (1832), 141, 259.
17. Höltje, R., *Z. anorg. allg. Chem.*, **190** (1930), 241.
18. Chatt, J., and Hayter, R. G., *J. chem. Soc.* (1963), 1343.
19. Westland, A. D., and Westland, L., *Can. J. Chem.*, **43** (1965), 426.
20. Beattie, I. R., and Webster, M., *J. chem. Soc.* (1964), 3507.
21. Fowles, G. W. A., and Walton, R. A., *J. chem. Soc.* (1964), 4330.
22. Beattie, I. R., and Collis, R., *J. chem. Soc. (A)* (1969), 2960.
23. Calderazzo, F., Losi, S. A., and Susz, B. P., *Inorg. chim. Acta*, **3** (1969), 329; *Helv. chim. Acta*, **54** (1971), 1156.
24. Issleib, K., and Wenschuh, E., *Chem. Ber.*, **97** (1964), 715.
25. Beerman, C., and Bestian, H., *Angew. Chem.*, **71** (1959), 618.
26. Fowles, G. W. A., Rice, D. A., and Wilkins, J. D., *J. chem. Soc. (A)* (1971), 1920.
27. Müller, J., and Thiele, K. H., *Z. anorg. allg. Chem.*, **362** (1968), 120.
28. Issleib, K., and Häckert, H., *Z. Naturf.*, **21B** (1966), 426.
29. Werner, R. P. M. and Podall, H. E., *French Pat.* 1 322 168; *Chem. Abstr.*, **60** (1964), 3015c.
30. Werner, R. P. M., *U.S. Pat.* 3 254 953; *Chem. Abstr.*, **65** (1966), 8962a.
31. Werner, R. P. M., *U.S. Pat.* 3 294 828; *Chem. Abstr.*, **66** (1967), 57446s.
32. Werner, R. P. M., *Z. Naturf.*, **16B** (1961), 477; **16B** (1961), 478.
33. Hieber, W., Peterhans, J., and Winter, E., *Chem. Ber.*, **94** (1961), 2572.
34. Hieber, W., and Winter, E., *Chem. Ber.*, **97** (1964), 1037.
35. Davison, A., and Ellis, J. E., *J. organometal. Chem.*, **23** (1970), C1.
36. Fischer, E. O., Louis, E., and Schneider, R. J. J., *Angew. Chem.*, **80** (1968), 122.
37. Interrante, L. V., and Nelson, G. V., *J. organometal. Chem.*, **25** (1970), 153.
38. Issleib, K., and Bohn, G., *Z. anorg. allg. Chem.*, **301** (1959), 188.
39. Clark, R. J. H., *The Chemistry of Titanium and Vanadium* (1968), Elsevier, London, p. 116.
40. Hackelwenzel, B., and Thomas, G., *J. less-common Metals*, **23** (1971), 185.
41. Selbin, J., and Vigee, G., *J. inorg. nucl. Chem.*, **30** (1968), 1644.
42. Du Preez, J. G. H., and Gibson, M. L., *Jl. S. Afr. chem. Inst.*, **23** (1970), 184.
43. Henrici-Olivé, G., and Olivé, S., *Angew. Chem.*, **82** (1970), 955.
44. Tarama, K., Yoshida, S., Kanai, H., and Osaka, S., *Bull. chem. Soc. Japan*, **41** (1968), 1271.
45. Bridgland, B. E., Fowles, G. W. A., and Walton, R. A., *J. inorg. nucl. Chem.*, **27** (1965), 383.

46. Machin, D. J., and Sullivan, J. F., *J. less-common Metals,* **19** (1969), 405.
47. Buslaev, Yu. A., Glushkova, M. A., and Ershova, M. M., *Chem. Abstr.,* **70** (1969), 120633a; **74** (1971), 33169t.
48. Desnoyers, J., and Rivest, R., *Can. J. Chem.,* **43** (1965), 1879.
49. Nesmeyanov, A. N., Anisimov, K. N., Kolbova, N. E., and Pasynskii, A. A., *Izv. Akad. Nauk. S.S.S.R. Ser. Khim.* (1966), 2231.
50. Fischer, E. O., and Louis, E., *J. organometal. Chem.,* **18** (1969), P26.
51. Mathieu, R., Lenzi, M., and Poilblanc, R., *Inorg. Chem.,* **9** (1970), 2030.
52. Poilblanc, R., and Bigorgne, M., *C. r. hebd. Séanc. Acad. Sci., Paris,* **252** (1961), 3054.
53. Poilblanc, R., and Bigorgne, M., *Bull. Soc. chim. Fr.* (1962), 1301.
54. Mathieu, R., and Poilblanc, R., *C. r. hebd. Séanc. Acad. Sci., Paris, Ser. C.,* **264** (1967), 1053.
55. Delbeke, F. L., Claceys, E. G., Van der Kelen, G. D., and Eeckhaut, Z., *J. organometal. Chem.,* **25** (1970), 213, 219.
56. Jenkins, J. H., Moss, J. R., and Shaw, B. L., *J. chem. Soc. (A)* (1969), 2796.
57. Jenkins, J. H., and Verkade, J. G., *Inorg. Chem.,* **6** (1967), 2250.
58. Canziani, F., Zingales, F., and Satorelli, N., *Gazz. chim. ital.,* **94** (1964), 841.
59. Huber, M., and Poilblanc, R., *Bull. Soc. chim. Fr.,* (1960), 1019.
60. Poilblanc, R., and Bigorgne, M., *C. r. hebd. Séanc. Acad. Sci., Paris,* **250** (1960), 1064.
61. Grim, S. O., Wheatland, D. A., and McFarlane, W., *J. Am. chem. Soc.,* **89** (1967), 5573.
62. Werner, H., *Angew. Chem.,* **80** (1968), 1017.
63. Magee, T. D., Matthews, C. N., and Wang, T. S., *J. Am. chem. Soc.,* **83** (1961), 3200.
64. Werner, H., and Rascher, H., *Inorg. chim. Acta,* **2** (1968), 181.
65. Green, P. J., and Brown, T. H., *Inorg. Chem.,* **10** (1971), 206.
66. Abel, E. W., Bennett, M. A., and Wilkinson, G., *J. chem. Soc.* (1959), 2323.
67. Matthews, C. N., Magee, T. D., and Wotiz, J. H., *J. Am. Chem. Soc.,* **81** (1959), 2273.
68. Nicholls, B., and Whiting, M. C., *J. chem. Soc.* (1959), 551.
69. Behrens, H., and Klek, W., *Z. anorg. allg. Chem.,* **292** (1957), 151.
70. Hieber, W., Englert, K., and Rieger, K., *Z. anorg. allg. Chem.,* **300** (1959), 295.
71. Hieber, W., and Peterhans, J., *Z. Naturf.,* **14B** (1959), 462.
72. Kraihanzel, C. S., and Cotton, F. A., *J. Am. chem. Soc.,* **84** (1962), 4432.
73. Matthews, C. N., *U.S. Pat.,* 3117 983 (1964); *Chem. Abstr.,* **60** (1964), 6870h.
74. Chatt, J., and Watson, H. R., *J. chem. Soc.* (1961), 4980.
75. White, C., and Mawby, R. J., *Inorg. chim. Acta,* **4** (1970), 261.
76. King, R. B., *Inorg. Chem.,* **2** (1963), 936.
77. Jones, C. E., and Coskran, K. J. *Inorg. Chem.,* **10** (1971), 55.
78. Imperial Chemical Industries, Ltd and Thompson, D. T., *Br. Pat.,* 1156 336 (1969); *Chem. Abstr.,* **71** (1969), 81529v.
79. Smith, J. G., and Thompson, D. T., *J. chem. Soc. (A)* (1967), 1694.
80. Green, M., Taunton-Rigby, A., and Stone, F. G. A., *J. chem. Soc. (A)* (1969), 1875.
81. Hogben, M. G., Grey, R. S., and Graham, W. A. G., *J. Am. chem. Soc.,* **88** (1968), 3457.
82. Bennett, M. A., Interrante, L. V., and Nyholm, R. S., *Z. Naturf.,* **20B** (1965), 633.
83. Interrante, L. V., Bennett, M. A., and Nyholm, R. S., *Inorg. Chem.,* **5** (1966), 2212.
84. Isslieb, K., and Haftendorn, M., *Z. anorg. allg. Chem.,* **351** (1967), 9.
85. King, R. B., and Efraty, A., *Inorg. chim. Acta,* **4** (1970), 319.
86. Fischer, E. O., Louis, E., Bathelt, W., Moser, E., and Müller. J., *Chem. Ber.,* **102** (1969), 2547; *J. organometal. Chem.,* **14** (1968), P9.
87. Fischer, E. O., Louis, E., and Keiter, C. G., *Angew. Chem.,* **81** (1969), 397.
88. Barlow, C. G., and Hollywell, G. C., *J. organometal. Chem.,* **16** (1969), 439.
89. Werner, H., and Prinz, R., *Chem. Ber.,* **99** (1966), 3582.
90. Delbeke, F. T., and Van der Kelen, G. P., *J. organometal. Chem.,* **21** (1970), 155.
91. Dobson, G. R., Stolz, I. W., and Sheline, R. K., *Adv. inorg. Chem. Radiochem.,* **8** (1966), 1.
92. Haynes, L. M., and Stiddard, M. H. B., *Adv. inorg. Chem. Radiochem.,* **12** (1969), 53.
93. Grim, S. O., Wheatland, D. A., and McAllister, P. R., *Inorg. Chem.,* **7** (1968), 161.
94. Mather, G. G., and Pidcock, A., *J. chem. Soc. (A)* (1970), 1226.
95. Angelici, R. J., *Organometal. Chem. Rev. A,* **3** (1968), 173.
96. Schumann, H., and Stelzer, O., *Angew. Chem.,* **80** (1968), 318.

References 179

97. Schumann, H., Stelzer, O., Kuhlmey, J., and Niederreuther, U., *Chem. Ber.*, **104** (1971), 993.
98. Becker, G., and Ebsworth, E. A. V., *Angew. Chem. (int. Ed.)*, **10** (1971), 186.
99. Hoefler, M., and Marre, W., *Angew. Chem. int. Ed.*, **10** (1971), 187.
100. Plastas, M. J., Stewart, J. M., and Grimm, S. O., *J. Am. chem. Soc.*, **91** (1969), 4326.
101. Houk, L. W., and Dobson, G. R., *Inorg. Chem.*, **5** (1966), 2119.
102. Houk, L. W., and Dobson, G. R., *J. chem. Soc. (A)* (1966), 317.
103. Behrens, H., Lehnert, G., and Sauerborn, H., *Z. anorg. allg. Chem.*, **374** (1970), 311.
104. Hull, C. G., and Stiddard, M. H. B., *J. chem. Soc. (A)* (1968), 710.
105. Dieck, H. T., and Friedel, H., *Chem. Commun.* (1969), 411.
106. Ross, E. P., and Dobson, G. R., *J. inorg. nucl. Chem.*, **30** (1968), 2363.
107. Dobson, G. R., Taylor, R. C., and Walsh, T. D., *Chem. Commun.* (1966), 281; *Inorg. Chem.*, **6** (1967), 1929.
108. Luth, H., Truter, M. R., and Robson, A., *Chem. Commun.* (1967), 738.
109. Luth, H., Truter, M. R., and Robson, A., *J. chem. Soc. (A)* (1969), 28.
110. Jones, C. E., and Coskran, K. J., *Inorg. Chem.*, **10** (1971), 1664.
111. Chatt, J., and Thompson, D. T., *J. chem. Soc.* (1964), 2713.
112. Chatt, J., and Thornton, D. A., *J. chem. Soc.* (1964), 1005.
113. Hayter, R. G., *Inorg. Chem.*, **2** (1963), 1031.
114. Hayter, R. G., *Inorg. Chem.*, **3** (1964), 711.
115. Imperial Chemical Industries Ltd, *Neth. Appl.*, 6 611 373 (1967); *Chem. Abstr.*, **67** (1967), 54266t.
116. Thompson, D. T., and Imperial Chemical Industries Ltd, *Br. Pat.*, 1 182 932 (1970); *Chem. Abstr.*, **73** (1970), 4029w.
117. Thompson, D. T., *J. organometal. Chem.*, **4** (1965), 74.
118. Imperial Chemical Industries Ltd, and Thompson, D. T., *Br. Pat.* 1 096 404 (1967); *Chem. Abstr.*, **68** (1968), 95985r.
119. Mais, R. H. B., Owston, P. G., and Taylor, D. T., *J. chem. Soc. (A)* (1967), 1735.
120. Doedens, R. J., and Dahl, L. W., *J. Am. chem. Soc.*, **87** (1965), 2576.
121. Isslieb, K., and Kratz, W., *Z. Naturf.*, **20B** (1965), 1303.
122. Anker, M. W., Colton, R., and Tomkins, I. B., *Rev. pure appl. Chem.*, **18** (1968), 23.
123. Lewis, J., and Whyman, R., *J. chem. Soc.* (1965), 6027.
124. Lewis, J., and Whyman, R., *J. chem. Soc. (A)* (1967), 77.
125. Tsang, W. S., Meek, D. W., and Wojcicki, A., *Inorg. Chem.*, **7** (1968), 1263.
126. Bowden, J. A., and Colton, R., *Aust. J. Chem.*, **22** (1969), 905.
127. Colton, R., and Rix, C. J., *Aust. J. Chem.*, **22** (1969), 305.
128. Allen, A. D., and Barrett, P. F., *Can. J. Chem.*, **46** (1968), 1649.
129. Colton, R., Scollary, G. R., and Tomkins, I. B., *Aust. J. Chem.*, **21** (1968), 15.
130. Colton, R., and Tomkins, I. B., *Aust. J. Chem.*, **19** (1966), 1143.
131. Anker, M. W., Colton, R., and Tomkins, I. B., *Aust. J. Chem.*, **20** (1967), 9.
132. Colton, R., and Tomkins, I. B., *Aust. J. Chem.*, **19** (1966), 1519.
133. Colton, R., and Scollary, G. R., *Aust. J. Chem.*, **21** (1968), 1435.
134. Moss, J. R., and Shaw, B. L., *J. chem. Soc. (A)* (1970), 595.
135. Cotton, F. A., and Johnson, B. F. G., *Inorg. Chem.*, **3** (1964), 1609.
136. Johnson, B. F. G., *J. chem. Soc. (A)* (1967), 475.
137. Anker, M. W., Colton, R., and Tomkins, I. W., *Aust. J. Chem.*, **21** (1968), 1149.
138. Colton, R., unpublished observations quoted in ref. 134.
139. Mawby, A., and Pringle, G. E., unpublished observations quoted in ref. 134.
140. Behrens, H., and Schindler, H., *Z. Naturf.*, **B23** (1968), 1109.
141. Isslieb, K., and Frohlich, H. O., *Z. anorg. allg. Chem.*, **298** (1959), 84.
142. Berman, D. A., *Diss. Abstr.*, **18** (1958), 1240.
143. Isslieb, K., *Proceedings of the 8th International Conference on Coordination Chemistry* (ed. V. Gutmann), Springer, Wien (1964), p. 193.
144. Bennett, M. A., Clark, R. J. H., and Goodwin, A. D. J., *J. chem. Soc. (A)* (1970), 541.
145. Isslieb, K., and Doll, G., *Z. anorg. allg. Chem.*, **305** (1960), 1.
146. Isslieb, K., and Wenschuh, E., *Z. anorg. allg. Chem.*, **305** (1960), 15.
147. Isslieb, K., and Wilde, G., *Z. anorg. allg. Chem.*, **312** (1961), 287.

148. Isslieb, K., and Roloff, H. R., *Z. anorg. allg. Chem.,* **324** (1963), 250.
149. Isslieb, K., and Tzschach, A., *Z. anorg. allg. Chem.,* **297** (1958), 121.
150. Thomas, G., *Z. anorg. allg. Chem.,* **360** (1968), 15.
151. Bennett, M. A., Clark, R. J. H., and Goodwin, A. D. J., *Inorg. Chem.,* **6** (1967), 1625.
152. Thomas, G., *Z. anorg. allg. Chem.,* **362** (1968), 191.
153. Isslieb, K., and Biermann, B., *Z. anorg. allg. Chem.,* **347** (1966), 39.
154. Chatt, J., and Wedd, A. G., *J. organometal. Chem.,* **27** (1971), C15.
155. George, T. A., and Siebold, C. D., *J. organometal. Chem.,* **30** (1971), C13.
156. Bell, B., Chatt, J., and Leigh, G. J., *Chem. Commun.* (1970), 842.
157. Fergusson, J. E., Robinson, B. H., and Wilkins, C. J., *J. chem. Soc. (A)* (1967), 486.
158. Allen, E. A., Feenan, K., and Fowles, G. W. A., *J. chem. Soc.* (1965), 1636.
159. Unpublished work quoted in Allen, E. A., Brisdon, B. J., and Fowles, G. W. A., *J. chem. Soc.* (1964), 4531.
160. Butcher, A. V., and Chatt, J., *J. chem. Soc. (A)* (1970), 2652.
161. Moss, J. R., and Shaw, B. L., *J. chem. Soc. (A)* (1970), 595.
162. Mason, R., and Whimp, P. O., unpublished observations quoted in ref. 161.
163. Pennella, F., *Chem. Commun.* (1971), 158.
164. Moss, J. R., and Shaw, B. L., *Chem. Commun.* (1968), 632.
165. Douglas, P. G., and Shaw, B. L., *J. chem. Soc. (A)* (1970), 334.
166. Atkinson, L. K., Mawby, A. H., and Smith, D. C., *Chem. Commun.* (1970), 1399.
167. Manojlović-Muir, Lj., *Chem. Commun.* (1971), 147; *J. chem. Soc. (A)* (1971), 2797.
168. Chatt, J., Manojlović-Muir, Lj., and Muir, K. W., *Chem. Commun.* (1971), 655.
169. Edwards, D. A., *J. inorg. nucl. Chem.,* **27** (1965), 303.
170. Fowles, G. W. A., and Frost, J. L., *J. chem. Soc. (A)* (1967), 671.
171. Hieber, W., and Freyer, W., *Chem. Ber.,* **92** (1959), 1765.
172. Lewis, J., Nyholm, R. S., Osborne, A. G., Sandhu, S. S., and Stiddard, M. H. B., *Chemy Ind.* (1963), 1398.
173. Osborne, A. G., and Stiddard, M. H. B., *J. chem. Soc.* (1964), 634.
174. Ziegler, M. L., Haas, H., and Sheline, R. K., *Chem. Ber.,* **98** (1965), 2454.
175. Wawersik, H., and Basolo, F., *Chem. Commun.* (1966), 366; *Inorg. chim. Acta,* **3** (1969), 113.
176. Bennett, M. J., and Mason, R., *J. chem. Soc. (A)* (1968), 75.
177. Lewis, J., Manning, A. R., Millar, J. R., and Wilson, J. M., *J. chem. Soc. (A)* (1966), 1663.
178. Parker, D. J., and Stiddard, M. H. B., *J. chem. Soc. (A)* (1966), 695; Lewis, J., Manning, A. R., and Millar, J. R., *J. chem. Soc. (A)* (1966), 845.
179. Green, M. L. H., and Moelwyn-Hughes, J. T., *Z. Naturf.,* **17B** (1962), 783.
180. Hayter, R. G., *J. Am. chem. Soc.,* **86** (1964), 823.
181. Doedens, R. J., Robinson, W. T., and Ibers, J. A., *J. Am. chem. Soc.,* **89** (1967), 4323.
182. Hayter, R. G., *Z. Naturf.,* **18B** (1963), 581.
183. Abel, E. W., and Sabherwal, I. H., *J. organometal. Chem.,* **10** (1967), 491.
184. Hieber, W., Faulhaber, G., and Theubert, F., *Z. Naturf.,* **15B** (1960), 326; *Z. anorg. allg. Chem.,* **314** (1962), 125.
185. Parker, D. J., *J. chem. Soc. (A)* (1969), 246.
186. Carey, N. A. D., and Noltes, J. G., *Chem. Commun.* (1968), 1471.
187. Nöth, H., and Schmid, G., *Z. anorg. allg. Chem.,* **345** (1966), 69.
188. Parshall, G. W., *J. Am. chem. Soc.,* **86** (1964), 361.
189. Hieber, W., Höfler, M., and Muschi, J., *Chem. Ber.,* **98** (1965), 311.
190. Booth, B. L., and Haszeldine, R. N., *J. chem. Soc. (A)* (1966), 157.
191. Ugo, R., and Bonati, F., *J. organometal. Chem.,* **8** (1967), 189.
192. Dobie, R. C., *J. chem. Soc. (A)* (1971), 230.
193. Bennett, M. A., and Watt, R., *Chem. Commun.* (1971), 94.
194. Abel, E. W., and Wilkinson, G., *J. chem. Soc.* (1959), 1501.
195. Angelici, R. J., Basolo, F., and Pöe, A. J., *J. Am. Chem. Soc.,* **85** (1963), 2215.
196. Angelici, R. J., and Basolo, F., *J. Am. chem. Soc.,* **84** (1962), 2495.
197. Beck, W., Nitzschmann, R. E., and Neumair, G., *Angew. Chem.,* **76** (1964), 346.
198. Angelici, R. J., *J. inorg. nucl. Chem.,* **28** (1966), 2687.
199. Angelici, R. J., *Inorg. Chem.,* **3** (1964), 1099.

200. Jolly, P. W., and Stone, F. G. A., *J. chem. Soc.* (1965), 5259.
201. Bennett, M. A., and Clark, R. J. H., *J. chem. Soc.* (1964), 5560.
202. Vahrenkamp, H., *Chem. Ber.,* **104** (1971), 449.
203. Bigorgne, M., Loutellier, A., and Pankowski, M., *J. organometal. Chem.,* **23** (1970), 201.
204. Kruck, T., and Noak, M., *Chem. Ber.,* **96** (1963), 3028; **97** (1964), 1693.
205. Kruck, T., and Höfler, M., *Chem. Ber.,* **96** (1963), 3035; **97** (1964), 2289.
206. Interrante, L., and Nelson, G. V., *Inorg. Chem.,* **7** (1968), 2059.
207. Farona, M. F., and Wojcicki, A., *Inorg. Chem.,* **4** (1965), 1402.
208. Addison, C. C., and Kilner, M., *J. chem. Soc. (A)* (1968), 1539.
209. Moelwyn-Hughes, J. T., Garner, A. W. B., and Howard, A. S., *J. chem. Soc. (A)* (1971), 2370.
210. Grobe, J., and Sheppard, N., *Z. Naturf.,* **23B** (1968), 901.
211. Grobe, J., *Z. anorg. allg. Chem.,* **331** (1964), 63.
212. Grobe, J., Helgerud, J. E., and Stievand, H. S., *Z. anorg. allg. Chem.,* **371** (1969), 123.
213. Hieber, W., and Kummer, R., *Z. Naturf.,* **20B** (1965), 271.
214. Barraclough, C. G., and Lewis, J., *J. chem. Soc.* (1960), 4842.
215. Hieber, W., and Tengler, H., *Z. anorg. allg. Chem.,* **318** (1963), 136.
216. Wawersik, H., and Basolo, F., *Inorg. Chem.,* **6** (1967), 1066.
217. Wawersik, H., and Basolo, F., *J. Am. chem. Soc.,* **89** (1967), 4626.
218. Enermark, J. H., and Ibers, J., *Inorg. Chem.,* **7** (1968), 2339.
219. Enermark, J. H., and Ibers, J., *Inorg. Chem.,* **6** (1967), 1575.
220. Nesmeyanov, A. N., Anisimov, K. N., Kolobova, N. E., and Antonova, A. B., *Izv. Akad. Nauk. S.S.S.R. Ser. Khim.* (1964), 160; *Chem. Abstr.,* **64** (1966), 12720h.
221. Gorsich, R. D., *J. Am. chem. Soc.,* **84** (1962), 2486.
222. Bryan, R. F., *Proc. chem. Soc.* (1964), 232.
223. Naldini, L., and Sacco, A., *Gazz. chim. ital.,* **89** (1959), 2258.
224. Naldini, L., *Gazz. chim. ital.,* **90** (1960), 1337.
225. Negoui, D., *Univ. Buchuresti Ser. Strint. Natur.,* **14** (1965), 145; *Chem. Abstr.,* **67** (1966), 60439t.
226. Uecker, G., and Schmitz-Dumont, O., *Z. anorg. allg. Chem.,* **371** (1969), 318.
227. Hieber, W., and Freyer, W., *Chem. Ber.,* **93** (1960), 462.
228. Freni, M., Giusto, D., and Romiti, P., *J. inorg. nucl. Chem.,* **29** (1967), 761.
229. Nyman, F., *Chem. Ind.* (1965), 604.
230. Moelwyn-Hughes, J. T., Garner, A. W. B., and Gordon, N., *J. organometal. Chem.,* **26** (1971), 373.
231. Haines, L. I. B., and Pöe, A. J., *J. chem. Soc. (A)* (1969), 2826.
232. Cariati, F., Romiti, P., and Valenti, V., *Gazz. chim. ital.,* **98** (1968), 615.
233. Moelwyn-Hughes, J. T., Garner, A. W. B., and Gordon, N., *J. organometal. Chem.,* **21** (1970), 449.
234. Hieber, W., and Opavsky, W., *Chem. Ber.,* **101** (1968), 2966.
235. Zingales, F., Satorelli, U., Canziani, F., and Raveglia, M., *Inorg. Chem.,* **6** (1967), 154.
236. Freni, M., Valenti, V., and Giusto, D., *J. inorg. nucl. Chem.,* **27** (1965), 2635.
237. Hieber, W., and Schuster, L., *Z. anorg. allg. Chem.,* **287** (1956), 214.
238. Hieber, W., Lux, F., and Herget, C., *Z. Naturf.,* **20B** (1965), 1159.
239. Kruck, T., Höfler, M., and Noak, M., *Chem. Ber.,* **99** (1965), 1153.
240. Abel, E. W., and Tyfield, S. P., *Chem. Commun.* (1968), 465; *Can. J. Chem.,* **47** (1969), 4627.
241. Freni, M., and Valenti, V., *Gazz. chim. ital.,* **90** (1960), 1436; *J. inorg. nucl. Chem.,* **16** (1961), 240.
242. Douglas, P. G., and Shaw, B. L., *J. chem. Soc. (A)* (1969), 1491.
243. Moelwyn-Hughes, J. T., Garner, A. W. B., and Howard, A. S., *J. chem. Soc. (A)* (1971), 2361.
244. Flitcroft, N., Leach, J. M., and Hopton, F. J., *J. inorg. nucl. Chem.,* **32** (1970), 137.
245. Ginsberg, A. P., *Chem. Commun.* (1968), 857.
246. Freni, M., and Valenti, V., *Gazz. chim. ital.,* **91** (1961), 1357.
247. Malatesta, L., *Proceedings of the 6th International Conference on Coordination Chemistry.* 'Advances in Chemistry of Coordination Compounds.' (ed. S. Kirschner), Macmillan, New York (1961), p. 475.

248. Malatesta, L., Freni, M., and Valenti, V., *Gazz. chim. ital.,* **94** (1964), 1279.
249. Freni, M., Valenti, V., and Pomponi, R., *Gazz. chim. ital.,* **94** (1964), 521.
250. Freni, M., Giusto, D., and Valenti, V., *Gazz. chim. ital.,* **94** (1964), 797.
251. Freni, M., Giusto, D., and Valenti, V., *J. inorg. nucl. Chem.,* **27** (1965), 757.
252. Freni, M., Giusto, D., Romiti, P., and Zucca, E., *J. inorg. nucl. Chem.,* **31** (1969), 3211.
253. Freni, M., Demichelis, R., and Giusto, D., *J. inorg. nucl. Chem.,* **29** (1967), 1433.
254. Freni, M., Romiti, P., and Giusto, D., *J. inorg. nucl. Chem.,* **32** (1970), 145.
255. Chatt, J., and Coffey, R. S., *Chem. Commun.* (1966), 545; *J. chem. Soc. (A)* (1969), 1963.
256. Malatesta, L., Canziani, F., and Angoletta, M., *Proceedings of the 7th International Conference on Coordination Chemistry* (eds Almquist, W., and Wickel, A. B.), Uppsala (1962), p. 293.
257. Chatt, J., Garforth, J. D., and Rowe, G. A., *Chem. Ind.* (1963), 332.
258. Chatt, J., and Rowe, G. A., *Chem. Ind.* (1962), 92; *J. chem. Soc.* (1962), 4019.
259. Chatt, J., Garforth, J. D., Johnson, N. P., and Rowe, G. A., *J. chem. Soc.* (1964), 1012.
260. Johnson, N. P., *J. chem. Soc. (A)* (1969), 1843.
261. Chatt, J., Falk, C. D., Leigh, G. J., and Paske, R. J., *J. chem. Soc. (A)* (1969), 2288.
262. Chatt, J., and Heaton, B. T., *Chem. Commun.* (1968), 274; *J. chem. Soc. (A)* (1971), 705.
263. Doedens, R. J., and Ibers, J. A., *Inorg. Chem.,* **6** (1967), 204.
264. Corfield, P. W. R., Doedens, R. J., and Ibers, J. A., *Inorg. Chem.,* **6** (1967), 197.
265. Chatt, J., Dilworth, J. R., and Leigh, G. J., *Chem. Commun.* (1969), 687.
266. Davis, B. R., and Ibers, J. A., *Inorg. Chem.,* **10** (1971), 578.
267. Chatt, J., Dilworth, J. R., and Leigh, G. J., *J. organometal. Chem.,* **21** (1970), P49.
268. Moelwyn-Hughes, J. T., and Garner, A. W. B., *Chem. Commun.* (1969), 1309.
269. Moelwyn-Hughes, J. T., Garner, A. W. B., and Howard, A. S., *J. chem. Soc. (A)* (1971), 2361.
270. Chatt, J., Dilworth, J. R., Leigh, G. J., and Richards, R. L., *Chem. Commun.* (1970), 955.
271. Chatt, J., Fay, R. C., and Richards, R. L., *J. chem. Soc. (A)* (1971), 702.
272. Chatt, J., Dilworth, J. R., Richards, R. L., and Sanders, J. R., *Nature, Lond.,* **224** (1969), 1201.
273. Chatt, J., Dilworth, J. R., and Leigh, G. J., *J. chem. Soc. (A)* (1970), 2239.
274. Bright, D., and Ibers, J. A., *Inorg. Chem.,* **7** (1968), 1099.
275. Bright, D., and Ibers, J. A., *Inorg. Chem.,* **8** (1969), 703.
276. Chatt, J., Dilworth, J. R., Leigh, G. J., and Gupta, V. D., *J. chem. Soc. (A)* (1971), 2631.
277. Freni, M., and Valenti, V., *Gazz. chim. ital.,* **90** (1960), 1445.
278. Colton, R., Levitus, R., and Wilkinson, G., *J. chem. Soc.* (1960), 4121.
279. Lock, C. J. L., and Wilkinson, G., *Chem. Ind.* (1962), 40.
280. Johnson, N. P., Lock, C. J. L., and Wilkinson, G., *J. chem. Soc.,* (1964), 1054.
281. Cotton, F. A., Curtis, N. F., and Robinson, W. R., *Inorg. Chem.,* **4** (1965), 1696.
282. Bennett, M. J., Cotton, F. A., Foxman, B. M., and Stokely, P. F., *J. Am. chem. Soc.,* **89** (1967), 2759; Cotton, F. A., and Foxman, B. M., *Inorg. Chem.,* **7** (1968), 2135.
283. Cotton, F. A., Robinson, W. R., and Walton, R. A., *Inorg. Chem.,* **6** (1967), 223.
284. Frew, M. G. B., Tisley, D. G., and Walton, R. A., *Chem. Commun.* (1970), 600.
285. Cotton, F. A., Robinson, W. R., Walton, R. A., and Whyman, R., *Inorg. Chem.,* **6** (1967), 929.
286. Cotton, F. A., Lippard, S. J., and Mague, J. T., *Inorg. Chem.,* **4** (1965), 508.
287. Cotton, F. A., and Mague, J. T., *Inorg. Chem.,* **3** (1964), 1094.
288. Chatt, J., Leigh, G. J., and Mingos, D. M. P., *J. chem. Soc. (A)* (1969), 1674.
289. Gunz, H. P., and Leigh, G. J., *J. chem. Soc. (A)* (1971), 2229.
290. Randall, E. W., and Shaw, D., *J. chem. Soc. (A)* (1969), 2867.
291. Gunz, H. P., unpublished observations quoted in ref. 288.
292. Chatt, J., Garforth, J. D., and Rowe, G. A., *J. chem. Soc. (A)* (1966), 1834.
293. Rouschias, G., and Wilkinson, G., *J. chem. Soc. (A)* (1966), 993.
294. Fergusson, J. E., and Hickford, J. H., *J. inorg. nucl. Chem.* (1966), 2293.
295. Chatt, J., Garforth, J. D., Johnson, N. P., and Rowe, G. A., *J. chem. Soc.* (1964), 601.

296. Rouschias, G., and Wilkinson, G., *Chem. Commun.* (1967), 442.
297. Cotton, F. A., and Foxman, B. M., *Inorg. Chem.,* **7** (1968), 1784.
298. Cotton, F. A., Eiss, R., and Foxman, B. M., *Inorg. Chem.,* **8** (1969), 950.
299. Ehrlich, M. W. W., and Owston, P. G., *J. chem. Soc.,* (1963), 4368.
300. Grove, D. E., Johnson, N. P., Lock, C. J., and Wilkinson, G., *J. chem. Soc.,* (1965), 491.
301. Freni, M., Giusto, D., and Romiti, P., *Gazz. chim. ital.,* **99** (1969), 641.
302. Freni, M., Giusto, D., Romiti, P., and Minghetti, G., *Gazz. chim. ital.,* **99** (1969), 286.
303. Grove, D. E., and Wilkinson, G., *J. chem. Soc. (A)* (1966), 1224.
304. Gehrke, H., Eastland, G., and Leitheiser, M., *J. inorg. nucl. Chem.,* **32** (1970), 867.
305. Gehrke, H., and Eastland, G., *Inorg. Chem.,* **9** (1970), 2722.
306. Reppe, W., and Schwekendiek, W., *Justus Liebigs Annln Chem.,* **560** (1948), 104.
307. Lewis, J., Nyholm, R. S., Osborne, A. G., Sandhu, S. S., and Stiddard, M. H. B., *Chem. Ind.* (1963), 1398.
308. Cotton, F. A., and Parish, R. V., *J. chem. Soc.* (1960), 1441.
309. Clifford, A. F., and Mukherjee, A. K., *Inorg. Chem.,* **2** (1963), 151.
310. Lewis, J., Nyholm, R. S., Sandhu, S. S., and Stiddard, M. H. B., *J. chem. Soc.* (1964), 2825.
311. Manuel, T. A., and Stone, F. G. A., *J. Am. chem. Soc.,* **82** (1960), 366.
312. Reckziegel, A., and Bigorgne, M., *J. organometal. Chem.,* **3** (1965), 341.
313. Manuel, T. A., *Inorg. Chem.,* **2** (1963), 854.
314. Stroheimer, W., and Mueller, F. J., *Chem. Ber.,* **102** (1969), 3613.
315. Tripathi, J. B. P., and Bigorgne, M., *J. organometal. Chem.,* **9** (1967), 307.
316. Singh, S., Singh, P. P., and Rivest, R., *Inorg. Chem.,* **7** (1968), 1236.
317. Delbeke, F. T., Van der Kelen, G. P., and Eckhaut, Z., *J. organometal. Chem.,* **16** (1969), 512.
318. Bigorgne, M., *J. organometal. Chem.,* **24** (1970), 211.
319. Schumann, H., Stelzer, O., and Niederreuther, U., *J. organometal. Chem.,* **16** (1969), P64; *Chem. Ber.,* **103** (1970), 1391.
320. Schumann, H., and Stelzer, O., *J. organometal. Chem.,* **13** (1968), P25.
321. Schumann, H., Stelzer, O., Niederreuther, U., and Roesch, L., *Chem. Ber.,* **103** (1970), 2350.
322. Bigorgne, M., Loutellier, A., and Pankowski, M., *J. organometal. Chem.,* **23** (1970), 201.
323. Braterman, P. S., and Wallace, W. J., *J. organometal. Chem.,* **30** (1971), C17.
324. Angelici, R. J., and Siefert, E. E., *Inorg. Chem.,* **5** (1966), 1457.
325. Dahn, D. J., and Jacobson, R. A., *Chem. Commun.* (1966), 496; *J. Am. chem. Soc.,* **90** (1968), 5106.
326. McDonald, W. S., Moss, J. R., Raper, G., Shaw, B. L., Greatrex, R., and Greenwood, N. N., *Chem. Commun.* (1969), 1295.
327. Emeleus, H. J., and Grobe, J., *Angew. Chem.,* **74** (1962), 466.
328. Kilbourn, B. T., Raeburn, U. A., and Thompson, D. T., *J. chem. Soc. (A)* (1969), 1906.
329. Braye, E. H., and Hubel, W., *Chem. Ind.* (1959), 1250.
330. Braye, E. H., Hubel, W., and Caplier, I., *J. Am. chem. Soc.,* **83** (1961), 4406.
331. Bennett, M. A., Robertson, G. B., Tomkins, I. B., and Whimp, P. O., *Chem. Commun.* (1971), 341.
332. Bennett, M. A., Robertson, G. B., Tomkins, I. B., and Whimp, P. O., *J. organometal. Chem.,* **32** (1971), C19.
333. Clifford, A. F., and Mukherjee, A. K., *Inorg. Synth.,* **8** (1966), 185.
334. Collins, R. L., and Petit, R., *J. chem. Phys.,* **39** (1963), 3433.
335. Burger, K., Korecz, L., and Bor, G., *J. inorg. nucl. Chem.,* **31** (1969), 1527.
336. Gibb, T. C., Greatrex, R., Greenwood, N. N., and Thompson, D. T. *J. chem. Soc. (A)* (1967), 1663.
337. Davies, G. R., Mais, R. B. H., Owston, P. G., and Thompson, D. T., *J. chem. Soc. (A)* (1968), 1251.
338. Thompson, D. T., *J. organometal. Chem.,* **4** (1965), 74.
339. Cooke, M., Green, M., and Kirkpatrick, D., *J. chem. Soc. (A)* (1968), 1507.
340. Job, B. E., McLean, R. A. N., and Thompson, D. T., *Chem. Commun.* (1966), 895.
341. Benson, B. C., Jackson, R., Joshi, K. K., and Thompson, D. T., *Chem. Commun.* (1968), 1506.

342. Kilbourn, B. T., and Mais, R. H. B., *Chem. Commun.* (1968), 1507.
343. Grobe, J., *Z. anorg. allg. Chem.*, **361** (1968), 32.
344. Grobe, J., *Z. anorg. allg. Chem.*, **361** (1968), 47.
345. Davison, A., McFarlane, W., Pratt, L., and Wilkinson, G., *J. chem. Soc.*, (1962), 3653.
346. Hayter, R. G., *J. Am. chem. Soc.*, **85** (1963), 3121.
347. Dobbie, R. C., and Whittaker, D., *Chem. Commun.* (1970), 796.
348. Farmery, K., and Kilner, M., *J. chem. Soc. (A)* (1970), 637.
349. Hieber, W., and Thalhofer, A., *Angew. Chem.*, **68** (1956), 679.
350. Hieber, W., and Muschi, J., *Chem. Ber.*, **98** (1965), 3931.
351. Cohen, I. A., and Basolo, F., *J. inorg. nucl. Chem.*, **28** (1965), 511.
352. Booth, G., and Chatt, J., *J. chem. Soc.* (1962), 2099.
353. Hieber, W., Frey, V., and John, P., *Chem. Ber.*, **100** (1967), 1961.
354. Hieber, W., and Zeidler, A., *Z. anorg. allg. Chem.*, **329** (1964), 92.
355. Hieber, W., and Kaiser, K., *Chem. Ber.*, **102** (1969), 4043.
356. Burt, R., Cooke, M., and Green, M., *J. chem. Soc. (A)* (1969), 2645.
357. Adams, D. M., Cook, D. J., and Kemmit, R. D. W., *Chem. Commun.* (1966), 103; *J. chem. Soc. (A)* (1968), 1067.
358. Kummer, R., and Graham, W. A. G., *Inorg. Chem.*, 7 (1968), 1208.
359. McBride, D. W., Stafford, S. L., and Stone, F. G. A., *Inorg. Chem.*, 1 (1962), 386.
360. Beck, W., and Lottes, K., *Chem. Ber.*, **98** (1965), 2657.
361. Morris, D. E., and Basolo, F., *J. Am. chem. Soc.*, **90** (1968), 2531.
362. Reed, H. W. B., *J. chem. Soc.* (1954), 1931.
363. Malatesta, L., and Araneo, A., *J. chem. Soc.* (1957), 3803.
364. Hayter, R. G., and Williams, L. F., *Inorg. Chem.*, 3 (1964), 717.
365. Hieber, W., and Neumair, G., *Z. anorg. allg. Chem.*, **342** (1966), 93.
366. Hieber, W., and Kramolowsky, R., *Z. anorg. allg. Chem.*, **321** (1963), 94.
367. Hieber, W., Bauer, I., and Neumair, G., *Z. anorg. allg. Chem.*, **335** (1965), 250.
368. Hieber, W., and Fuhrling, H., *Z. anorg. allg. Chem.*, **373** (1970), 48.
369. Crooks, G. R., and Johnson, B. F. G., *J. chem. Soc. (A)* (1968), 1238.
370. Johnson, B. F. G., and Segal, J. A., *J. organometal. Chem.*, **31** (1971), C79.
371. Hieber, W., and Klingshirn, W., *Z. anorg. allg. Chem.*, **323** (1963), 292.
372. King, R. B., *Inorg. Chem.*, 2 (1963), 1275.
373. Casey, M., and Manning, A. R., *J. chem. Soc. (A)* (1971), 256.
374. Hieber, W., and Fuhrling, H., *Z. Naturf. B.*, **25** (1970), 663.
375. Sacco, A., and Aresta, M., *Chem. Commun.* (1968), 1223.
376. Aresta, M., Giannoccaro, P., Rossi, M., and Sacco, A., *Inorg. chim. acta,* 5 (1971), 115.
377. Klein, H. F., *Angew, Chem. (int. Ed.)*, 9 (1970), 904.
378. Jensen, K. A., *Z. anorg. allg. Chem.*, **229** (1936), 282.
379. Naldini, L., *Gazz, chim. ital.*, 90 (1960), 391.
380. Pignolet, L. H., Forster, D., and Horrocks, W. D., *Inorg. Chem.*, 7 (1968), 828.
381. Bennett, M. A., Bramley, R., and Longstaff, P. A., *Chem. Commun.* (1966), 806; Bennett, M. A., Longstaff, P. A., *J. Am. chem Soc.*, **91** (1969), 6266.
382. Chatt, J., and Underhill, A. E., *J. chem. Soc.* (1963), 2088.
383. Harmon, R. E., Parsons, J. L., and Gupta, S. K., *Org. Prep. Proced.*, 2 (1970), 19.
384. Isslieb, K., and Wenschuh, E., *Z. Naturf. B,* 19 (1964), 199.
385. Naldini, L., *Gazz. chim. ital.*, 90 (1960), 1231.
386. Singh, P. P., and Rivest, R., *Can. J. Chem.*, 46 (1968), 1773.
387. Birchall, T., *Can. J. Chem.*, 47 (1969), 1351.
388. Isslieb, K., and Brack, A., *Z. anorg. allg. Chem.*, **277** (1954), 258.
389. Nast, R., and Kruger, K. W., *Z. anorg. allg. Chem.*, **341** (1965), 189.
390. Isslieb, K., and Papp, S., *Z. Chemie Lpz.*, 8 (1968), 188.
391. Fluck, E., and Kuhn, P., *Z. anorg. allg. Chem.*, **350** (1967), 263.
392. Papp, S., and Schonweitz, T., *J. inorg. nucl. Chem.*, **32** (1970), 697.
393. Candlin, J. P., Joshi, K. K., and Thompson, D. T., *Chem. Ind.* (1966), 1960.
394. Piacenti, F., Bianchi, M., Benedetti, E., and Sbrana, G., *J. inorg. nucl. Chem.*, 29 (1967), 1389.
395. Johnson, B. F. G., Johnston, R. D., Josty, P. L., Lewis, J., and Williams, I. G., *Nature, Lond.*, **213** (1967), 901.

396. Bruce, M. I., Gibbs, C. W., and Stone, F. G. A., *Z. Naturf. B*, **23** (1968), 1543.
397. Candlin, J. P., and Shortland, A. C., *J. organometal. Chem.*, **16** (1969), 289.
398. Klanberg, F., and Muetterties, E. L., *J. Am. chem. Soc.*, **90** (1968), 3296.
399. Johnson, R. D., Ph.D. Thesis (1968), Manchester University.
400. Johnson, B. F. G., Lewis, J., and Williams, I. G., *J. chem. Soc. (A)* (1970), 901.
401. Deeming, A. J., Johnson, B. F. G., and Lewis, J., *J. chem. Soc. (A)* (1970), 2697.
402. Piacenti, F., Bianchi, M., Benedetti, E., and Braca, G., *Inorg. Chem.*, **7** (1968), 1815.
403. Johnson, B. F. G., Johnson, R. D., and Lewis, J., *J. chem. Soc. (A)* (1969), 792.
404. Deeming, A. J., Johnson, B. F. G., and Lewis, J., *J. chem. Soc. (A)* (1970), 897.
405. Bradford, C. W., Van Bronswyk, W., Clark, R. J. H., and Nyholm, R. S., *J. chem. Soc. (A)* (1970), 2889.
406. L'Eplattenier, F., and Calderazzo, F., *Inorg. Chem.*, **6** (1967), 2092.
407. L'Eplattenier, F., and Calderazzo, F., *Inorg. Chem.*, **7** (1968), 1290.
408. Cotton, J. D., Bruce, M. I., and Stone, F. G. A., *J. chem. Soc. (A)* (1968), 2162.
409. Cleare, M. J., and Griffith, W. P., *Chem. Ind.* (1967), 1705.
410. Cleare, M. J., and Griffith, W. P., *J. chem. Soc. (A)* (1969), 372.
411. Johnson, R. D., *Adv. inorg. Chem. Radiochem.*, **13** (1970), 471.
412. Bradford, C. W., and Nyholm, R. S., *Chem. Commun.* (1968), 867.
413. Bruce, M. I., Cooke, M., Green, M., and Westlake, D. J., *J. chem. Soc. (A)* (1967), 987.
414. L'Epplatenier, F., *Inorg. Chem.*, **8** (1969), 965.
415. Deeming, A. J., Johnson, B. F. G., and Lewis, J., *J. chem. Soc. (A)* (1970), 2517.
416. Collman, J. P., and Roper, W. R., *J. Am. chem. Soc.*, **87** (1965), 4008.
417. Collman, J. P., and Roper, W. R., *J. Am. chem. Soc.*, **88** (1966), 3504.
418. Stalick, J., and Ibers, J. A., *Inorg. Chem.*, **8** (1969), 419.
419. Laing, K. R., and Roper, W. R., *J. chem. Soc. (A)* (1969), 1889.
420. Collman, J. P., and Roper, W. R., *Chem. Commun.* (1966), 244.
421. Valentine, J., Valentine, D., and Collman, J. P., *Inorg. Chem.*, **10** (1971), 219.
422. L'Epplattenier, F., and Calderazzo, F., *U.S. Pat.* 3505034; *Chem. Abstr.*, **72** (1970), 134788h.
423. Moss, J. R., and Graham, W. A. G., *Chem. Commun.* (1969), 800.
424. Hieber, W., and Heusinger, H., *Angew. Chem.*, **68** (1956), 678; *J. inorg. nucl. Chem.*, **4** (1957), 179.
425. Hieber, W., and John, P., *Chem. Ber.*, **103** (1970), 2161.
426. John, P., *Chem. Ber.*, **103** (1970), 2178.
427. Stone, F. G. A., and Bruce, M. I., *J. chem. Soc. (A)* (1967), 1238.
428. Chatt, J., Shaw, B. L., and Field, A. E., *J. chem. Soc.* (1964), 3466.
429. Stephenson, T. A., and Wilkinson, G., *J. inorg. nucl. Chem.*, **28** (1966), 945.
430. Jenkins, J. M., Lupin, M. S., and Shaw, B. L., *J. chem. Soc. (A)* (1966), 1787.
431. Lupin, M. S., and Shaw, B. L., *J. chem. Soc. (A)* (1968), 741.
432. Chatt, J., and Shaw, B. L., *Chem. Ind.* (1960), 931; (1961), 290.
433. James, B. R., and Markham, L. D., *Inorg. nucl. Chem. Lett.*, **7** (1971), 373.
434. Douglas, P. G., and Shaw, B. L., *J. chem. Soc. (A)* (1970), 1556.
435. Bright, A., Mann, B. E., Masters, C., Shaw, B. L., Slade, R. M., and Stainbank, R. E., *J. chem. Soc. (A)* (1971), 1826.
436. Colton, R., and Farthing, R. H., *Aust. J. Chem.*, **20** (1967), 1283.
437. Colton, R., and Farthing, R. H., *Aust. J. Chem.*, **22** (1969), 2011.
438. Johnson, B. F. G., Johnson, R. D., Lewis, J., and Williams, I. G., *J. chem. Soc. (A)* (1971), 689.
439. Crooks, G. R., Johnson, B. F. G., Lewis, J., Williams, I. G., and Gamlen, G., *J. chem. Soc. (A)* (1969), 2761.
440. Hales, L. A. W., and Irving, R. J., *J. chem. Soc. (A)* (1967), 1932.
441. Hieber, W., and Frey, V., *Z. Naturf.*, **21B** (1966), 704.
442. Vaska, L., *Z. Naturf.*, **15B** (1960), 56.
443. Vaska, L., and Di Luzio, J. W., *J. Am. chem. Soc.*, **83** (1961), 1262.
444. Vaska, L., *Chem. Ind.* (1964), 1402.
445. Oleari, P. L., and Vaska, L., *Proc. chem. Soc.* (1962), 333.
446. Vaska, L., *J. Am. chem. Soc.*, **86** (1964), 1943.
447. Vaska, L., *J. Am. chem. Soc.*, **88** (1966), 4100.

448. Moers, F. G., *Chem. Commun.* (1971), 79.
449. Chatt, J., Melville, D. P., and Richards, R. L., *J. chem. Soc. (A)* (1971), 1169.
450. Ash, M. J., Brookes, A., Knox, S. A. R., and Stone, F. G. A., *J. chem. Soc. (A)* (1971), 458.
451. Fairey, M. B., and Irving, R. J., *J. chem. Soc. (A)* (1966), 475; *Spectrochim. Acta*, 20A (1966), 1757.
452. Chatt, J., and Shaw, B. L., *J. chem. Soc. (A)* (1966), 1811.
453. Townsend, R. E., and Coskran, K. J., *Inorg. Chem.*, 10 (1971), 1661.
454. Araneo, A., Valenti, V., and Cariati, F., *J. inorg. nucl. Chem.*, 32 (1970), 1877.
455. Grundy, K. R., Laing, K. R., and Roper, W. R., *Chem. Commun.* (1970), 1500.
456. Watson, J. M., and Whittle, K. R., *Chem. Commun.* (1971), 518.
457. Grundy, K. R., Reed, C. A., and Roper, W. R., *Chem. Commun.* (1970), 1501.
458. Pierpont, C. G., Van Derveer, D. G., Durland, W., and Eisenberg, R., *J. Am. chem. Soc.* 92 (1970), 4761.
459. Wilson, S. T., and Osborn, J. A., *J. Am. chem. Soc.*, 93 (1971), 3068.
460. Pierpont, C. G., Pucci, A., and Eisenberg, R., *J. Am. chem. Soc.*, 93 (1971), 3050.
461. Stiddard, M. H. B., and Townsend, R. E., *Chem. Commun.* (1969), 1372.
462. Laing, K. R., and Roper, W. R., *Chem. Commun.* (1968), 1556.
463. Hall, D., and Williamson, R. B., unpublished work quoted in ref. 462.
464. Laing, K. R., and Roper, W. R., *J. chem. Soc. (A)* (1970), 2149.
465. Bentley, G. A., Laing, K. R., Roper, W. R., and Waters, J. M., *Chem. Commun.* (1970), 998.
466. Yamamoto, A., Kitazume, S., and Ikeda, S., *J. Am. chem. Soc.*, 90 (1968), 1089.
467. Ito, T., Kitazume, S., Yamamoto, A., and Ikeda, S., *J. Am. chem. Soc.*, 92 (1970), 3011.
468. Knoth, W. H., *J. Am. chem. Soc.*, 90 (1968), 7172.
469. Dewhirst, K. C., Keim, W., and Reilly, C. A., *Inorg. Chem.*, 7 (1968), 546.
470. Meakin, P., Guggenberger, L. J., Jesson, J. P., Gerloch, D. M., Tebbe, P. N., Peet, N. G., and Muetterties, E. L., *J. Am. chem. Soc.*, 92 (1970), 3482.
471. Eliades, T. I., Harris, R. O., and Zia, M. C., *Chem. Commun.* (1970), 1709.
472. Levison, J. J., and Robinson, S. D., *J. chem. Soc. (A)* (1970), 2947.
473. Hallman, P. S., McGarvey, B. R., and Wilkinson, G., *J. chem. Soc. (A)* (1968), 3143.
474. Skapski, A. C., and Troughton, P. G. H., *Chem. Commun.* (1968), 1230.
475. Douglas, P. G., and Shaw, B. L., *Chem. Commun.* (1969), 624; *J. chem. Soc. (A)* (1970), 335.
476. Leigh, G. J., Levison, J. J., and Robinson, S. D., *Chem. Commun.* (1969), 705.
477. Bell, B., Chatt, J., and Leigh, G. J., *Chem. Commun.* (1970), 576.
478. Chatt, J., Melville, D. P., and Richards, R. L., *J. chem. Soc. (A)* (1971), 895.
479. Chatt, J., Leigh, G. J., Mingos, D. M. P., and Paske, R. J., *Chem. Ind.* (1967), 1324.
480. Chatt, J., Leigh, G. J., Mingos, D. M. P., and Paske, R. J., *Chem. Commun.* (1967), 670.
481. Chatt, J., Leigh, G. J., and Mingos, D. M. P., *Chem. Ind.* (1969), 109.
482. Bright, D., and Ibers, J. A., *Inorg. Chem.*, 8 (1969), 1078.
483. Chatt, J., Leigh, G. J., and Paske, R. J., *J. chem. Soc. (A)* (1969), 854.
484. Chatt, J., Leigh, G. J., and Richards, R. L., *Chem. Commun.* (1969), 515; *J. chem. Soc. (A)* (1970), 2243.
485. Maples, P. K., Basolo, F., and Pearson, R. G., *Inorg. Chem.*, 10 (1971), 765.
486. Vaska, L., and Sloane, E. M., *J. Am. chem. Soc.*, 82 (1960), 1263.
487. Brooks, P. R., and Shaw, B. L., *J. chem. Soc. (A)* (1967), 1079.
488. Chatt, J., and Hayter, R. G., *J. chem. Soc.* (1961), 896.
489. Stephenson, T. A., *Inorg. nucl. Chem. Lett.*, 4 (1968), 687.
490. Prince, R. H., and Raspin, K. A., *J. inorg. nucl. Chem.*, 31 (1969), 695.
491. Alcock, N. W., and Raspin, K. A., *J. chem. Soc. (A)* (1968), 2108.
492. Raspin, K. A., *J. chem. Soc. (A)* (1969), 461.
493. Nicholson, J. K., *Angew. Chem. (int. Ed.)*, 6 (1967), 264.
494. Chioccola, G., Daly, J. J., and Nicholson, J. K., *Angew. Chem. (int. Ed.)*, 7 (1968), 131.
495. Chioccola, G., and Daly, J. J., *J. chem. Soc. (A)* (1968), 1981.
496. Gilbert, J. D., Rose, D., and Wilkinson, G., *J. chem. Soc. (A)* (1970), 2765.
497. La Placa, S. J., and Ibers, J. A., *Inorg. Chem.*, 4 (1965), 778.
498. Gilbert, J. D., and Wilkinson, G., *J. chem. Soc. (A)* (1969), 1749.

499. Cenini, S., Fusi, A., and Capparella, G., *J. inorg. nucl. Chem.,* **33** (1971), 3576.
500. Gilbert, J. D., Baird, M. C., and Wilkinson, G., *J. chem. Soc. (A)* (1968), 2198.
501. Stephenson, T. A., *J. chem. Soc. (A)* (1970), 889.
502. Hayter, R. G., *Inorg. Chem.,* **3** (1964), 301.
503. Stephenson, T. A., and Switkes, E., *Inorg. nucl. Chem. Lett.,* **7** (1971), 805.
504. Chatt, J., Leigh, G. J., Mingos, D. M. P., and Paske, R. J., *J. chem. Soc. (A)* (1968), 2636.
505. Winzer, A., *Z. Chem.,* **10** (1970), 438.
506. Hudson, A., and Kennedy, M. J., *J. chem. Soc. (A)* (1969), 1116.
507. Grim, S. O., and Ference, R. A., *Inorg. nucl. Chem. Lett.,* **2** (1966), 205; *Inorg. Chem. Acta.,* **4** (1970), 277.
508. Leigh, G. J., and Mingos, D. M. P., *J. chem. Soc. (A)* (1970), 587.
509. Rose, D., Gilbert, J. D., Richardson, R. P., and Wilkinson, G., *J. chem. Soc. (A)* (1969), 2610.
510. *Discuss. Faraday Soc.,* No. 46 (1968).
511. Sacco, A., and Freni, M., *J. inorg. nucl. Chem.,* **8** (1958), 566.
512. Sacco, A., and Freni, M., *Annali Chim.,* **48** (1958), 218.
513. Hieber, W., and Freyer, W., *Chem. Ber.,* **91** (1958), 1230.
514. McCleverty, J. A., Davison, A., and Wilkinson, G., *J. chem. Soc.* (1965), 3890.
515. Vohler, O., *Chem. Ber.,* **91** (1958), 1235.
516. Manning, A. R., *J. chem. Soc. (A)* (1968), 1135.
517. Szabo, P., Fekete, L., Bor, G., Nagy-Magos, Z., and Markó, L., *J. organometal. Chem.,* **12** (1968), 245.
518. Sacco, A., *Annali Chim.,* **43** (1953), 495; *Chem. Abstr.,* **48** (1954), 5012d.
519. Piacenti, F., Bianchi, M., and Benedetti, E., *Chimica Ind. (Milano),* **49** (1967), 245.
520. Capron-Contigny, G., and Poilblanc, R., *Bull. Soc. chim. Fr.* (1967), 1440.
521. Manning, A. R., and Millar, J. R., *J. chem. Soc. (A)* (1970), 3352.
522. Bryan, R. F., and Manning, A. R., *Chem. Commun.* (1968), 1316.
523. Ibers, J. A., *J. organometal. Chem.,* **14** (1968), 423.
524. Bor, G., and Markó, L., *Chem. Ind.* (1963), 912.
525. Simon, A., Nagy-Magos, Z., Palágye, J., Pályi, G., Bor, G., and Markó, Z., *J. organometal. Chem.,* **11** (1968), 634.
526. Speier, G., and Markó, L., *Inorg. Chim. Acta,* **3** (1969), 126.
527. Cetini, G., Gambino, O., Rossetti, R., and Stanghellini, P. L., *Inorg. Chem.,* **7** (1968), 609.
528. Ferrari, G., and Ugo, R., *Chem. Commun.* (1969), 590.
529. Sacco, A., *Gazz. chim. ital.,* **93** (1961), 542.
530. Hieber, W., and Duchatsch, H., *Z. Naturf.,* **18B** (1963), 1132.
531. Foust, A. S., Foster, M. S., and Dahl, L., *J. Am. chem. Soc.,* **91** (1969), 5633.
532. Dellaca, R. J., Penfold, B. R., Robinson, B. H., Robinson, W. T., and Spenser, J. L., *Inorg. Chem.,* **9** (1970), 2204.
533. Brice, M. D., Penfold, B. R., Robinson, W. T., and Taylor, S. R., *Inorg. Chem.,* **9** (1970), 362.
534. Heck, R. F., *J. Am. chem. Soc.,* **85** (1963), 657.
535. Hieber, W., and Lindler, E., *Z. Naturf.,* **16B** (1961), 137.
536. Hieber, W., and Lindler, E., *Chem. Ber.,* **94** (1961), 1417.
537. Hieber, W., and Duchatsch, H., *Chem. Ber.,* **98** (1965), 2933.
538. Pregaglia, G. F., Andreetta, A., and Ferrari, G. F., *J. organometal. Chem.,* **30** (1971), 387.
539. Bonati, F., Cenini, S., and Ugo, R., *J. chem. Soc. (A)* (1967), 932.
540. Hieber, W., and Breu, R., *Chem. Ber.,* **90** (1957), 1259; **90** (1957), 1270.
541. Manning, A. R., *J. chem. Soc. (A)* (1968), 1018.
542. Bonatti, F., Cenini, S., Morelli, D., and Ugo, R., *Inorg. nucl. Chem. Lett.,* **1** (1965), 107.
543. Bonatti, F., Cenini, S., Morelli, D., and Ugo, R., *J. chem. Soc. (A)* (1966), 1052.
544. Barrett, P. F., and Pöe, A. J., *J. chem. Soc. (A)* (1968), 429.
545. Patmore, D. J., and Graham, W. A. G., *Inorg. Chem.,* **7** (1960), 771.
546. Stalick, J., and Ibers, J. A., *J. organometal. Chem.,* **22** (1970), 213.
547. Bressan, M., Corain, B., Rigo, P., and Turco, A., *Inorg. Chem.,* **9** (1970), 1733.

548. Hieber, W., and Duchatsch, H.. *Chem. Ber.,* **98** (1965), 2530.
549. Hieber, W., and Lindler, E., *Chem. Ber.,* **95** (1962), 273.
550. Pankowski, M., and Bigorgne, M., *C. r. hebd. Séanc. Acad. Sci., Paris, Ser. C,* **264** (1967), 1382.
551. Bowden, J. A., and Colton, R., *Aust. J. Chem.,* **21** (1968), 891.
552. Horrocks, W. De W., and Taylor, R. C., *Inorg. Chem.,* **2** (1963), 723.
553. Hieber, W., and Ellermann, J., *Chem. Ber.,* **96** (1963), 1643.
554. Thorsteinson, E. M., and Basolo, F., *J. Am. chem. Soc.,* **88** (1966), 3929.
555. Van Hecke, G. R., and Horrocks, W. De W., *Inorg. Chem.,* **5** (1966), 1960.
556. Burg, A. B., and Sabherwal, I. H., *Inorg. Chem.,* **9** (1970), 974.
557. Sabherwal, I. H., and Burg, A. B., *Chem. Commun.* (1969), 853.
558. Thorsteinson, E. M., and Basolo, F., *Inorg. Chem.,* **5** (1966), 1691.
559. Reichenbach, G., *J. organometal. Chem.,* **31** (1971), 103.
560. Hieber, W., and Heinicke, K., *Z. anorg. allg. Chem.,* **316** (1962), 305.
561. Hieber, W., and Ellerman, J., *Chem. Ber.,* **96** (1963), 1650.
562. Hieber, W., and Heinicke, K., *Z. Naturf.,* **16B** (1961), 553.
563. Bianco, T., Rossi, M., and Uva, L., *Inorg. Chim. Acta,* **3** (1969), 443.
564. Jackson, T. B., Baker, M. J., Edwards, J. O., and Tutas, D., *Inorg. Chem.,* **5** (1966), 2046.
565. Rossi, M., and Sacco, A., *Chem. Commun.* (1971), 694.
566. Sacco, A., and Rossi, M., *Chem. Commun.* (1967), 316; *Inorg. Chim. Acta,* **2** (1968), 127.
567. Rossi, M., and Sacco, A., *Chem. Commun.* (1969), 471.
568. Yamamoto, A., Kitazume, S., Pu, L. S., and Ikeda, S., *Chem. Commun.* (1967), 79; *J. Am. chem. Soc.,* **93** (1971), 371.
569. Misono, A., Uchida, Y., Saito, T., and Sony, K. M., *Chem. Commun.* (1967), 419; *Bull. chem. Soc. Japan,* **40** (1967), 700.
570. Misono, A., Uchida, Y., Hidai, M., and Araki, M., *Chem. Commun.* (1968), 1044.
571. Enemark, J. H., Davis, B. R., McGinnety, J. A., and Ibers, J. A., *Chem. Commun.* (1968), 96.
572. Davis, B. R., Payne, N. C., and Ibers, J. A., *J. Am. chem. Soc.,* **91** (1969), 1240.
573. Pu, L. S., Yamamoto, A., and Ikeda, S., *J. Am. chem. Soc.,* **90** (1968), 3896.
574. Misono, A., Uchida, Y., Hidai, M., and Kuse, J., *Chem. Commun.* (1969), 208.
575. Archer, N. J., Haszeldine, R. N., and Parish, R. V., *Chem. Commun.* (1971), 524.
576. Yamamoto, A., Pu, L. S., Kitazume, S., and Ikeda, S., *J. Am. chem. Soc.,* **89** (1967), 3071.
577. Lorberth, J., Noth, H., and Rinze, R. V., *J. organometal. Chem.,* **16** (1969), P1.
578. Walker, J. R., and Lee, J. B., *Loughborough Univ. Tech. chem. Dept Summer Final Year Student Project Thesis* (1969); *Chem. Abstr.,* **73** (1971), 31119y.
579. Srivastava, S. C., and Bigorgne, M., *J. organometal. Chem.,* **19** (1969), 241.
580. Misono, A., Uchida, Y., Saito, T., Hidai, M., and Araki, M., *Inorg. Chem.,* **8** (1969), 168.
581. Sacco, A., and Rossi, M., *Inorg. Synth.,* **12** (1970), 18.
582. Misono, A., *Inorg. Synth.,* **12** (1970), 12.
583. Pu, L. S., Yamamoto, A., and Ikeda, S., *Chem. Commun.* (1969), 189.
584. Schunn, R. A., *Inorg. Chem.,* **9** (1970), 2567.
585. Otvos, I., Speier, G., and Markó, L., *Acta chim. hung.,* **66** (1970), 27; *Chem. Abstr.,* **74** (1971), 27611g.
586. Speier, G., and Markó, L., *J. organometal. Chem.,* **21** (1970), P46.
587. Aresta, M., Nobile, C. F., Rossi, M., and Sacco, A., *Chem. Commun.* (1971), 781.
588. Cambell, J. M., and Stone, F. G. A., *Angew. Chem.,* **81** (1969), 120.
589. Rossi, M., and Sacco, A., *Proceedings of the 10th International Conference on Coordination Chemistry* (ed. K. Yamasaki), The Chemical Society of Japan, Tokyo (1967), p. 125.
590. Aresta, M., Rossi, M., and Sacco, A., *Inorg. chim. Acta,* **3** (1969), 227.
591. Hatfield, W., and Yoke, J. T., *Inorg. Chem.,* **1** (1962), 475.
592. Bradbury, J., Forest, K. P., Nuttall, R. H., and Sharp, D. W. A., *Spectrochim. Acta (A),* **23** (1967), 2701.
593. Cotton, F. A., Faust, O. D., Goodgame, D. M. L., and Holm, R. H., *J. Am. chem. Soc.,* **83** (1961), 1780.

594. Isslieb, K., and Mitscherling, B., *Z. anorg. allg. Chem.*, **304** (1960), 73.
595. Jensen, K. A., Nielson, P. H., and Pedersen, C. T., *Acta chem. scand.*, **17** (1963), 1115.
596. Chatt, J., and Shaw, B. L., *J. chem. Soc.* (1961), 285.
597. Cotton, F. A., Goodgame, D. M. L., Goodgame, M., and Sacco, A., *J. Am. chem. Soc.*, **83** (1961), 4157.
598. Nicolini, M., Pecile, C., and Turco, A., *Coord. Chem. Rev.*, **1** (1966), 133.
599. Turco, A., Pecile, C., Nicolini, M., and Martelli, M., *J. Am. chem. Soc.*, **85** (1963), 3510.
600. Nicolini, M., Pecile, C., and Turco, A., *J. Am. chem. Soc.*, **87** (1965), 2379.
601. Boschi, T., Nicolini, M., and Turco, A., *Coord. Chem. Rev.*, **1** (1966), 269.
602. Chatt, J., and Booth, G., unpublished work quoted in ref. 2.
603. Boschi, T., Rigo, P., Pecile, C., and Turco, A., *Gazz. chim. ital.*, **97** (1967), 1391.
604. Rigo, P., Bressan, M., and Turco, A., *Inorg. Chem.*, **7** (1968), 1460.
605. Donaghue, T. T., McMillan, J. A., and Peters, D. A., *J. inorg. nucl. Chem.*, **31** (1969), 3661.
606. Walton, R. A., and Whyman, R., *J. chem. Soc. (A)* (1968), 1394.
607. Collier, J. W., and Mann, F. G., *J. chem. Soc.* (1964), 1815.
608. Allen, D. W., Millar, I. T., and Mann, F. G., *J. chem. Soc. (A)* (1969), 1101.
609. Ercolani, C., Quagliano, J. V., and Vallarino, L. M., *Inorg. chim. Acta*, **3** (1969), 421.
610. Berglund, D., and Meek, D. W., *J. Am. chem. Soc.*, **80** (1968), 518.
611. Dahloff, W. V., Dick, T. R., and Nelson, S. M., *J. chem. Soc. (A)* (1969), 2919.
612. Horrocks, W. De W., and La Mar, G. N., *J. Am. chem. Soc.*, **85** (1968), 3512.
613. La Mar, G. N., *J. phys. Chem.*, **69** (1965), 3212.
614. La Mar, G. N., Horrocks, W. De W., and Allen, L. C., *J. chem. Phys.*, **41** (1964), 2126.
615. Pignolet, L. H., and Horrocks, W. De W., *J. Am. chem. Soc.*, **88** (1966), 5929.
616. Pignolet, L. H., and Horrocks, W. De W., *Chem. Commun.* (1968), 1012.
617. Moedritzer, K., and Millar, R. E., *J. therm. Analysis* **1** (1969), 151.
618. Schmidt, D. D., and Yoke, J. T., *J. Am. chem. Soc.*, **93** (1971), 637.
619. Simo, C., and Holt, S., *Inorg. Chem.*, **7** (1968), 2655.
620. Goodgame, D. M. L., and Goodgame, M., *Inorg. Chem.*, **4** (1965), 139.
621. Sestili, L., Furlani, C., and Festuccia, G., *Inorg. chim. Acta*, **4** (1970), 542.
622. Chatt, J., and Shaw, B. L., *Chem. Ind.* (1969), 675.
623. Owston, P. G., and Rowe, J. M., *J. chem. Soc.* (1963), 3411.
624. Gerloch, M., Lewis, J., Bently, R. B., Mabbs, F. E., and Smail, W. R., *Chem. Commun.* (1969), 119.
625. Browning, M. C., Davies, R. F. B., Morgan, D. J., Sutton, L. E., and Venanzi, L. M., *J. chem. Soc.* (1961), 4816.
626. Cass, R. C., Coates, G. E., and Hayter, R. G., *J. chem. Soc.* (1955), 4007.
627. Horrocks, W. De W., and Pignolet, L. H., *Progress in Coordination Chemistry*, (ed. M. Cais), Elsevier, Amsterdam (1968), p. 203.
628. Lane, A. P., and Payne, D. S., *J. chem. Soc.* (1963), 4004.
629. Pecile, C., *Inorg. Chem.*, **5** (1966), 210.
630. Spacu, P., and Negoui, M., *Ann. Univ. Bucuresti. Ser. Stunt. natur. Chim.*, **18** (1969), 21; *Chem. Abstr.*, **73** (1970), 62141n.
631. White, J. F., and Farona, M. F., *Inorg. Chem.*, **10** (1971), 1080.
632. Bertrand, J. A., and Plymale, D. L., *Inorg. Chem.*, **5** (1966), 879.
633. Isslieb, K., Fröhlich, H. O., and Wenschuh, E., *Chem. Ber.*, **95** (1962), 2742.
634. Jensen, K. A., Nygaard, B., and Pedersen, C. T., *Acta chem. scand.*, **17** (1963), 1126.
635. Maki, N., and Oshima, K., *Bull. chem. Soc. Japan*, **43** (1970), 3970.
636. Watanabe, K., Nishikawa, H., and Shibata, M., *Bull. chem. Soc., Japan*, **42** (1969), 1150.
637. Schrauzer, G. N., and Kohnle, J., *Chem. Ber.*, **97** (1964), 3056.
638. Costa, G., Tauzher, G., and Puxeddu, A., *Inorg. chim. Acta*, **3** (1969), 45.
639. Tauzher, G., Mestroni, G., Puxeddu, A., Costanzo, R., and Costa, G., *J. chem. Soc. (A)* (1971), 2504.
640. Johnson, B. F. G., Lewis, J., and Robinson, P. W., *J. chem. Soc. (A)* (1970), 1100.
641. Booth, B. L., Else, M. J., Fields, R., and Haszeldine, R. N., *J. organometal. Chem.*, **27** (1971), 119.
642. Chini, P., and Martinengo, S., *Chem. Commun.* (1969), 1092.
643. Whyman, R., *Chem. Commun.* (1970), 230.

644. Chini, P., and Martinengo, S., *Inorg. Chim. Acta,* **3** (1969), 315.
645. O'Connor, C., Yagupsky, G., Evans, D., and Wilkinson, G., *Chem. Commun.* (1968), 420.
646. Evans, D., Yagupsky, G., and Wilkinson, G., *J. chem. Soc. (A)* (1968), 2660.
647. Iwashita, Y., and Hayata, A., *J. Am. chem. Soc.,* **91** (1969), 2525.
648. Hieber, W., and Kummer, R., *Chem. Ber.,* **100** (1967), 148.
649. Hieber, W., Heusinger, H., and Vohler, O., *Chem. Ber.,* **90** (1957), 2425.
650. Vallarino, L., *J. chem. Soc.* (1957), 2287.
651. Vallarino, L., *J. inorg. nucl. Chem.,* **8** (1958), 288.
652. Osborn, J. A., Jardine, F. H. J., Young, J. F., and Wilkinson, G., *J. chem. Soc. (A)* (1966), 1711.
653. Bennett, M. A., and Longstaff, P. A., *Chem. Ind.* (1965), 846.
654. Kingston, J. V., and Scollary, G. R., *J. inorg. nucl. Chem.,* **31** (1969), 2557.
655. Reddy, G. K. N., and Leelamani, E. G., *Curr. Sci.,* **34** (1965), 146; *Chem. Abstr.,* **62** (1967), 12733d.
656. Chatt, J., and Shaw, B. L., *J. chem. Soc.* (1966), 1437.
657. Himmele, W., Aquila, W., and Muller, F. J., *Ger. Offen.* 1957300.
658. McCleverty, J. A., and Wilkinson, G., *Inorg. Synth.,* **8** (1966), 214.
659. Evans, D., Osborn, J. A., and Wilkinson, G., *Inorg. Synth.,* **11** (1968), 99.
660. Reddy, G. K. N., and Leelamani, E. G., *Indian J. Chem.,* **4** (1966), 540.
661. Osborn, J. A., Wilkinson, G., and Young, J. F., *Chem. Commun.* (1966), 461.
662. Grinberg, A. A., Singh, M. M., and Varshavski, Yu. S., *Zh. neorg. Khim.,* **13** (1968), 2716.
663. Vaska, L., and Peone, J., *Chem. Commun.* (1971), 418.
664. Jennings, M. A., and Wojcicki, A., *Inorg. Chem.,* **6** (1967), 1854.
665. De Stefano, N. J., and Burmeister, J. L., *Inorg. Chem.,* **10** (1971), 998.
666. Kemmitt, R. D. W., Nichols, D. I., and Peacock, R. D., *Chem. Commun.* (1967), 599; *J. chem. Soc. (A)* (1968), 1898.
667. Mann, B. E., Masters, C., and Shaw, B. L., *J. chem. Soc. (A)* (1971), 1104.
668. Towl, A. D. C., Ph.D. Thesis, University of Sheffield, 1968.
669. Heck, R. F., *J. Am. chem. Soc.,* **86** (1964), 2796.
670. Hieber, W., and Frey, V., *Chem. Ber.,* **99** (1966), 2614.
671. Mann, B. E., Masters, C., Shaw, B. L., and Stainbank, R. E., *Chem. Commun.* (1971), 1103.
672. Bennett, M. A., Clark, R. J. H., and Milner, D. L., *Inorg. Chem.,* **6** (1967), 1647.
673. Watkins, S. F. W., Obi, J., and Dahl, L. F., unpublished work quoted in ref. 26.
674. Wojcicki, A., and Basolo, F., *J. Am. chem. Soc.,* **83** (1961), 523.
675. Gray, H. B., and Wojcicki, A., *Proc. chem. Soc.* (1960), 358.
676. Deeming, A. J., and Shaw, B. L., *J. chem. Soc. (A)* (1969), 597.
677. Fackler, J. P., Jr., *Inorg. Chem.,* **9** (1970), 2625.
678. Baird, M. C., Lawson, D. N., Mague, J. T., Osborn, J. A., and Wilkinson, G., *Chem. Commun.* (1966), 129.
679. Douek, I. C., and Wilkinson, G., *J. chem. Soc. (A)* (1969), 2604.
680. Powell, P., and Nöth, H., *Chem. Commun.* (1966), 637.
681. Vaska, L., and Bath, S. S., *J. Am. chem. Soc.,* **88** (1966), 1333.
682. Mortimer, C. T., and Ashcroft, S. J., *Inorg. Chem.,* **10** (1971), 1326.
683. Muir, K. N., and Ibers, J. A., *Inorg. Chem.,* **8** (1969), 1921.
684. Yagupsky, M. P., and Wilkinson, G., *J. chem. Soc. (A)* (1968), 2813.
685. Collman, J. P., Vastine, F. D., and Roper, W., *J. Am. chem. Soc.,* **88** (1966), 5035.
686. Booth, B. L., Haszeldine, R. N., and Perkins, I., *J. chem. Soc. (A)* (1971), 927.
687. Schrock, R. R., and Osborn, J. A., *J. Am. chem. Soc.,* **93** (1971), 2397.
688. Haines, M., and Singleton, E., *J. organometal. Chem.,* **30** (1971), C81.
689. Reddy, G. K. N., and Susheelamina, C. M., *Chem. Ber.,* **103** (1970), 54.
690. Uguagliati, P., Deganello, G., Busetto, L., and Belluco, U., *Inorg. Chem.,* **8** (1969), 1625.
691. Deganello, G., Uguagliati, P., Crociani, B., and Belluco, U., *J. chem. Soc. (A)* (1969), 2726.
692. Poilblanc, R., and Gallay, J., *J. organometal. Chem.,* **27** (1971), C53.
693. Steele, D. F., and Stephenson, T. A., *Inorg. nucl. Chem. Lett.,* **7** (1971), 877.

694. Palazzi, A., Graziani, L., Busetto, G., Carturan, G., and Belluco, U., *J. organometal. Chem.,* **25** (1970), 249.
695. Mague, J. T., and Mitchener, J. P., *Inorg. Chem.,* **8** (1969), 119.
696. Hartwell, G. E., and Clark, P. W., *Chem. Commun.* (1970), 1115.
697. Bennett, M. A., Gruber, S. J., Hann, E. J., and Nyholm, R. S., *J. organometal. Chem.,* **29** (1971), C12.
698. Bennett, M. A., and Hann, E. J., *J. organometal. Chem.,* **29** (1971), C15.
699. Vaska, L., and Bath, S. S., *J. Am. chem. Soc.,* **85** (1963), 3500.
700. La Placa, S. J., and Ibers, J. A., *Acta. crystallogr.* **18** (1965), 511; *J. Am. chem. Soc.,* **85** (1963), 2501.
701. Gregorio, G., Pregaglia, G., and Ugo, R., *Inorg. Chim. Acta,* **3** (1969), 89.
702. Levison, J. J., and Robinson, S. D., *J. chem. Soc. (A)* (1970), 2947.
703. O'Connor, C., and Wilkinson, G., *J. chem. Soc. (A)* (1966), 2665.
704. Evans, D., Osborn, J. A., and Wilkinson, G., *J. chem. Soc. (A)* (1968), 3131.
705. Yagupsky, M., Brown, C. K., Yagupsky, G., and Wilkinson, G., *J. chem. Soc. (A)* (1970), 937.
706. Yagupsky, M., and Wilkinson, G., *J. chem. Soc. (A)* (1970), 941.
707. Yagupsky, G., Brown, C. K., and Wilkinson, G., *J. chem. Soc. (A)* (1970), 1392.
708. Wilkinson, G., *Bull. Soc. chim. Fr.* (1968), 5055.
709. Brault, A. T., Thorsteinson, E. M., and Basolo, F., *Inorg. Chem.,* **3** (1964), 770.
710. Osborn, J. A., Powell, A. R., and Wilkinson, G., *Chem. Commun.* (1966), 461.
711. Hieber, W., and Heinicke, K., *Z. anorg. allg. Chem.,* **316** (1962), 321; *Z. Naturf.* **16B** (1961), 554.
712. Collman, J. P., Hoffman, N. W., and Morris, D. E., *J. Am. chem. Soc.,* **91** (1969), 5659.
713. Kukushkin, Y. M., and Singh, M. M., *Zh. neorg. Khim.,* **14** (1969), 3167.
714. Kukushkin, Y. M., and Singh, M. M., *Zh. neorg. Khim.,* **15** (1970), 2741.
715. Hughes, W. B., *Chem. Commun.* (1969), 1126.
716. Kiji, J., Yoshikawa, S., and Furukawa, J., *Bull. chem. Soc. Japan,* **43** (1970), 3614.
717. Crooks, G. R., and Johnson, B. F. G., *J. chem. Soc. (A)* (1970), 1662.
718. Baird, M. C., *Inorg. Chim. Acta.,* **5** (1971), 46.
719. Robinson, S. D., and Uttley, M. F., *J. chem. Soc. (A)* (1971), 1254.
720. Busetto, L., Palazzi, A., Ros, R., and Graziani, M., *Gazz. chim. ital.,* **100** (1970), 849.
721. Takesada, M., Yamazaki, H., and Hagihura, N., *Bull. chem. Soc. Japan,* **41** (1968), 270.
722. Lavecchia, M., Rossi, M., and Sacco, A., *Inorg. Chim. Acta,* **4** (1970), 29.
723. Baker, R. W., and Pauling, P. J., *Chem. Commun.* (1969), 1495.
724. Baker, R. W., Ilmaier, B., Pauling, P. J., and Nyholm, R. S., *Chem. Commun.* (1970), 1077.
725. Keim, W., *J. organometal. Chem.,* **8** (1967), P25.
726. Ilmair, B., and Nyholm, R. S., *Naturwissenschaften,* **56** (1969), 415.
727. Ilmair, B., and Nyholm, R. S., *Naturwissenschaften,* **56** (1969), 636.
728. Ukhin, L. Y., and Shvetsov, Y. A., *Izv. Akad. Nauk. SSSR Ser. Khim.* (1969), 2342; *Chem. Abstr.,* **72** (1970), 50447.
729. Collman, J. P., Kubota, M., Vastine, F. D., Sun, Y. J., and Kang, J. W., *J. Am. chem. Soc.,* **90** (1968), 5430.
730. Sacco, A., and Ugo, R., *Chimica Ind., Milano,* **45** (1963), 1096.
731. Sacco, A., Ugo, R., and Moles, A., *J. chem. Soc. (A)* (1966), 1670; *Coord. Chem. Rev.,* **1** (1966), 234.
732. Dewhirst, K. C., *Inorg. Chem.,* **5** (1966), 319.
733. Baird, M. C., Mague, J. T., Osborn, J. A., and Wilkinson, G., *J. chem. Soc. (A)* (1967), 1347.
734. Masters, C., McDonald, W. S., Raper, G., and Shaw, B. L., *Chem. Commun.* (1971), 210.
735. Schrock, R. R., and Osborn, J. A., *J. Am. chem. Soc.,* **93** (1971), 2397.
736. Olson, D. C., and Keim, W., *Inorg. Chem.,* **8** (1969), 2028.
737. Keim, W., *J. organometal. Chem.,* **14** (1968), 179.
738. Keim, W., *J. organometal. Chem.,* **19** (1969), 161.
739. Keim, W., *J. organometal. Chem.,* **19** (1969), 191.
740. Ricci, J. S., and Ibers, J. A., *J. organometal. Chem.,* **27** (1971), 261.

741. Kukuskin, Yu. N., Rubtrova, N. D., and Singh, M. M., *Zh. neorg. Khim.*, **15** (1970), 1879.
742. Baird, M. C., and Wilkinson, G., *Chem. Commun.* (1966), 267; *J. chem. Soc. (A)* (1967), 865.
743. Baird, M. C., Hartwell, G., and Wilkinson, G., *J. chem. Soc. (A)* (1967), 2037.
744. Lawson, D. N., Osborn, J. A., and Wilkinson, G., *J. chem. Soc. (A)* (1966), 1733.
745. Augustine, R. L., and Van Peppen, J. F., *Chem. Commun.* (1970), 495, 497.
746. Blum, J., Roseman, H., and Bergmann, E. D., *Tetrahedron Lett.* (1967), 3665.
747. Heathcock, C. H., and Poulter, S. R., *Tetrahedron Lett.* (1969), 2755.
748. Shriver, D. F., Lehman, D. D., and Wharf, I., *Chem. Commun.* (1970), 1948.
749. Eaton, D. R., and Stuart, S. R., *J. Am. chem. Soc.*, **90** (1968), 4170.
750. Brown, T. H., and Green, P. J., *J. Am. chem. Soc.*, **92** (1970), 2359.
751. Clark, P. W., and Hartwell, G. E., *Inorg. Chem.*, **9** (1970), 1948.
752. Hall, D. I., and Nyholm, R. S., *Chem. Commun.* (1970), 488.
753. Nave, C., and Truter, M. R., *Chem. Commun.* (1971), 1253.
754. Lawson, D. N., Mays, M. J., and Wilkinson, G., *J. Chem. Soc. (A)* (1966), 52.
755. Busetto, L., Palazzi, A., and Ros, R., *Inorg. Chem.*, **9** (1970), 2792.
756. Haines, L. M., *Inorg. Chem.*, **9** (1970), 1517.
757. McGinnety, J. A., Payne, N. C., and Ibers, J. A., *J. Am. chem. Soc.*, **91** (1969), 6301.
758. Haines, L. M., *Inorg. Chem.*, **10** (1970), 1685.
759. Shapley, J. R., Schrock, R. R., and Osborn, J. A., *J. Am. chem. Soc.* (1969), 2816.
760. Schrock, R. R., and Osborn, J. A., *Inorg. Chem.*, **9** (1970), 2339.
761. Notte, M. J., Gafner, G., and Haines, L. M., *Chem. Commun.* (1969), 1406.
762. Bennett, M. J., and Donaldson, P. B., *J. Am. chem. Soc.*, **93** (1971), 3307.
763. Ginsberg, A. P., and Lindsell, W. E., *J. Am. chem. Soc.*, **93** (1971), 2082.
764. Bressan, M., Favero, G., and Turco, A., *Inorg. nucl. Chem. Lett.*, **7** (1971), 203.
765. De Boer, J. L., Rogers, D., Skapski, A. C., and Troughton, P. G. H., *Chem. Commun.* (1966), 756.
766. Commereuc, D., Douek, I. C., Wilkinson, G., *J. chem. Soc.* (1970), 1771.
767. Troughton, P. G. H., and Skapski, A. C., *Chem. Commun.* (1968), 575.
768. Haszeldine, R. N., Parish, R. V., and Parry, D. J., *J. organometal. Chem.*, **9** (1967), 13; *J. chem. Soc. (A)* (1969), 683.
769. de Charentenay, F., Osborn, J. A., and Wilkinson, G., *J. chem. Soc. (A)* (1968), 787.
770. Muir, K. W., and Ibers, J. A., *Inorg. Chem.*, **9** (1970), 440.
771. Glockling, F., and Hill, G. C., *J. organometal. Chem.*, **22** (1970), C48; *J. chem. Soc. (A)* (1971), 2137.
772. Chatt, J., Johnson, N. P., and Shaw, B. L., *J. chem. Soc.* (1964), 2508.
773. Malatesta, L., and Caglio, G., *Chem. Commun.* (1967), 420.
774. Albano, V., Bellon, P. L., and Scatturin, V., *Chem. Commun.* (1967), 730.
775. Whyman, R., *J. organometal. Chem.*, **24** (1970), C35.
776. Hieber, W., and Frey, V., *Chem. Ber.*, **99** (1966), 2607.
777. Vaska, L., and Di Luzio, J. W., *J. Am. chem. Soc.*, **83** (1961), 2784.
778. Collman, J. P., and Kang, J. W., *J. Am. chem. Soc.*, **88** (1966), 3459.
779. Kubota, M., Vrieze, K., Collman, J. P., and Sears, C. T., *Inorg. Synth.*, **11** (1968), 101.
780. Chatt, J., Johnson, N. P., and Shaw, B. L., *J. chem. Soc. (A)* (1967), 604.
781. Deeming, A. J., and Shaw, B. L., *J. chem. Soc. (A)* (1968), 1887.
782. Angoletta, M., *Gazz. chim. ital.*, **89** (1959), 2359.
783. Collman, J. P., and Sears, Jr., C. T., *Inorg. Chem.*, **7** (1968), 27.
784(a). Chock, P. B., and Halpern, J., *J. Am. chem. Soc.*, **88** (1966), 3511.
784(b). Vaska, L., *Acc. chem. Res.*, **1** (1968), 335.
785. Collman, J. P., *Acc. chem. Res.*, **1** (1968), 136.
786. Collman, J. P., and Roper, W. R., *Adv. organometal. Chem.*, **7** (1968), 53.
787. Vaska, L., *Science, N.Y.*, **140** (1963), 809.
788. Vaska, L., and Di Luzio, J. W., *J. Am. chem. Soc.*, **84** (1962), 679.
789. Baddley, W. H., *J. Am. chem. Soc.*, **88** (1966), 4545.
790. Osborn, J. A., Wilkinson, G., and Young, J. F., *Chem. Commun.* (1965), 17.
791. Chalk, A. J., and Harrod, J. F., *J. Am. chem. Soc.*, **87** (1965), 16.
792. Heck, R. F., *J. org. Chem.*, **28** (1963), 604.

793. Kubota, M., and Blake, D. M., *J. Am. chem. Soc.,* **93** (1971), 1368.
794. Collman, J. P., and Roper, W. R., *J. Am. chem. Soc.,* **88** (1966), 180.
795. Collman, J. P., and Roper, W. R., *J. Am. chem. Soc.,* **90** (1968), 2282.
796. Nyholm, R. S., and Vrieze, K., *J. chem. Soc.,* (1965), 5337.
797. Scott, R. N., Shriver, D. F., and Lehman, D. D., *Inorg. Chim. Acta.,* **4** (1970), 73.
798. Mango, F. D., and Dvaretzky, I., *J. Am. chem. Soc.,* **88** (1966), 1654.
799. Vaska, L., *J. Am. chem. Soc.,* **88** (1966), 5325.
800. McCormick, B. J., and Anderson, B. M., *J. inorg. nucl. Chem.,* **32** (1970), 3416.
801. Glockling, F., and Wilbey, M. O., *Chem. Commun.* (1969), 286; *J. chem. Soc. (A)* (1970), 1675.
802. Lappert, M. F., and Travers, N. F., *Chem. Commun.* (1968), 1569.
803. Singer, H., and Wilkinson, G., *J. chem. Soc. (A)* (1968), 2516.
804. Ginsberg, A. P., and Silverthorn, W. E., *Chem. Commun.* (1969), 823.
805. Cash, D. N., and Harris, R. D., *Can. J. Chem.,* **49** (1971), 867.
806. Senoff, C. V., *Can. J. Chem.,* **48** (1970), 2446.
807. Camia, M., Lachi, M. P., Benzoni, L., Zanzotleva, C., and Venture, M. T., *Inorg. Chem.,* **9** (1970), 251.
808. Church, M. J., and Mays, M. J., *Chem. Commun.* (1968), 435.
809. Deeming, A. J., and Shaw, B. L., *J. chem. Soc. (A)* (1969), 1128.
810. Blake, D. M., and Kubota, M., *Inorg. Chem.,* **9** (1970), 989.
811. Deeming, A. J., and Shaw, B. L., *Chem. Commun.* (1968), 751; *J. chem. Soc. (A)* (1969), 1562.
812. McGinnety, J. A., and Ibers, J. A., *Chem. Commun.* (1968), 235.
813. Ibers, J. A., McGinnety, J. A., and Kime, N., *Proceedings of the 10th International Conference on Coordination Chemistry* (ed. K. Yamasaki), The Chemical Society of Japan, Tokyo (1967), p. 93.
814. La Placa, S. J., and Ibers, J. A., *Science, N.Y.,* **145** (1964), 920; *J. Am. chem. Soc.,* **87** (1965), 2581.
815. McGinnety, J. A., Doedens, R. J., and Ibers, J. A., *Science, N.Y.,* **155** (1967), 709; *Inorg. Chem.,* **6** (1967), 2243.
816. Weininger, M. S., Taylor, I. F., and Amma, E. L., *Chem. Commun.* (1971), 1172.
817. Horn, R. W., Weissberger, E., and Collman, J. P., *Inorg. Chem.,* **9** (1970), 2367.
818. Levison, J. J., and Robinson, S. D., *J. chem. Soc. (A)* (1971), 762.
819. La Placa, S. J., and Ibers, J. A., *Inorg. Chem.,* **5** (1966), 405.
820. Ashcroft, S. J., and Mortimer, C. T., *J. organometal. Chem.,* **24** (1970), 783.
821. Chalk, A. J., *Chem. Commun.* (1969), 1207.
822. Hooton, K. A., *J. chem. Soc. (A)* (1971), 1251.
823. Strohmeier, W., and Miller, F. J., *Z. Naturf.,* **B24** (1969), 770.
824. Strohmeier, W., and Onoda, T., *Z. Naturf.,* **B24** (1969), 1185.
825. Strohmeier, W., and Onoda, T., *Z. Naturf.,* **B23** (1968), 1377; **B23** (1968), 1527.
826. Ebhardt, G. G., and Vaska, L., *J. Catal.,* 8 (1967), 183.
827. Vaska, L., and Rhodes, R. E., *J. Am. chem. Soc.,* **87** (1965), 4970.
828. James, B. R., and Memon, N. A., *Can. J. Chem.,* **46** (1968), 217.
829. Chatt, J., Johnson, N. P., and Shaw, B. L., *J. chem. Soc.* (1964), 1625.
830. Brooks, P. R., and Shaw, B. L., *Chem. Commun.* (1968), 919.
831. Shaw, B. L., and Smithies, A. C., *J. chem. Soc. (A)* (1968), 2784.
832. Angoletta, M., *Gazz. chim. ital.,* **90** (1960), 1021.
833. Vaska, L., *Science, N.Y.,* **152** (1966), 769.
834. Malatesta, L., Caglio, G., and Angoletta, M., *J. chem. Soc.* (1965), 6974.
835. Deeming, A. J., and Shaw, B. L., *J. chem. Soc. (A)* (1970), 2705.
836. McDonald, W. S., and Roper, G., unpublished work quoted in ref. 77.
837. Church, M. J., Mays, M. J., Simpson, R. N. F., and Stefanini, F. P., *J. chem. Soc. (A)* (1970), 2909.
838. Vaska, L., and Chen, L. S., *Chem. Commun.* (1971), 1080.
839. Clark, G. R., Reed, C. A., Roper, W. R., Skelton, B. W., and Waters, T. N., *Chem. Commun.* (1971), 758.
840. Deeming, A. J., and Shaw, B. L., *J. chem. Soc. (A)* (1970), 3356.
841. Malatesta, L., Angoletta, M., and Caglio, G., *J. chem. Soc. (A)* (1970), 1836.

842. Deeming, A. J., and Shaw, B. L., *J. chem. Soc. (A)* (1969), 443.
843. Shaw, B. L., and Slade, R. M., *J. chem. Soc. (A)* (1971), 1184.
844. Malatesta, L., Angoletta, M., and Caglio, G., *Proceedings of the 8th International Conference on Coordination Chemistry* (ed. V. Gutman), Springer, Wien (1964), p. 210.
845. Whyman, R., *J. organometal. Chem.,* 29 (1971), C36.
846. Harrod, J. F., Gilson, D. F. G., and Charles, R., *Can. J. Chem.,* 47 (1969), 1431.
847. Burnett, M. G., and Morrison, R. J., *J. chem. Soc. (A)* (1971), 2325.
848. Vaska, L., *Inorg. nucl. Chem. Lett.,* 1 (1965), 89.
849. Yagupsky, G., and Wilkinson, G., *J. chem. Soc. (A)* (1969), 725.
850. Mays, M. J., Simpson, R. N. F., and Stefanini, F. P., *J. chem. Soc. (A)* (1970), 3000.
851. Clark, H. C., and Mittal, R. K., *Can. J. Chem.,* 48 (1970), 119.
852. Vaska, L., *Chem. Commun.* (1966), 614.
853. Taylor, R. C., Young, J. F., and Wilkinson, G., *Inorg. Chem.,* 5 (1966), 20.
854. Malatesta, L., Angoletta, M., and Caglio, G., *Angew. Chem.,* 75 (1963), 1103.
855. Angoletta, M., *Gazz. chim. ital.,* 93 (1963), 1591.
856. Reed, C. A., and Roper, W. R., *Chem. Commun.* (1969), 155; *J. chem. Soc. (A)* (1970), 3054.
857. Mingos, D. M. P., Robinson, W. T., and Ibers, J. A., *Inorg. Chem.,* 10 (1971), 1043.
858. Angoletta, M., and Caglio, G., *Gazz. chim. ital.,* 93 (1963), 1584.
859. Hodgson, D. J., Payne, N. C., McGinnety, J. A., Pearson, R. G., and Ibers, J. A., *J. Am. chem. Soc.,* 90 (1968), 4486.
860. Reed, G. A., and Roper, W. R., *Chem. Commun.* (1969), 1459.
861. Albano, V. G., Bellon, P., and Sansoni, M., *J. chem. Soc. (A)* (1971), 2420.
862. Mingos, D. M. P., and Ibers, J. A., *Inorg. Chem.,* 9 (1970), 1105.
863. Hodgson, D. J., and Ibers, J. A., *Inorg. Chem.,* 7 (1968), 2345.
864. Hodgson, D. J., and Ibers, J. A., *Inorg. Chem.,* 8 (1969), 1282.
865. Mingos, D. M. P., and Ibers, J. A., *Inorg. Chem.,* 10 (1971), 1035.
866. Mingos, D. M. P., and Ibers, J. A., *Inorg. Chem.,* 10 (1971), 1479.
867. Carty, P., Walker, A., Mathew, M., and Palenik, G. J., *Chem. Commun.* (1969), 1376.
868. Collman, J. P., and Norris, N. W., *Inorg. Synth.,* 12 (1970), 8.
869. Collman, J. P., Kubota, M., Sun, J. Y., and Vastine, F., *J. Am. chem. Soc.,* 89 (1967), 169.
870. Chatt, J., Melville, D. P., and Richards, R. L., *J. chem. Soc. (A)* (1969), 2841.
871. Venanzi, L. M., 'Platinum Group Metals and Compounds' *Adv. Chem. Ser.,* 98 (1971), 66.
872. Mann, B. E., Masters, C., and Shaw, B. L., *Chem. Commun.* (1968), 703; (1970), 846; *J. inorg. nucl. Chem.,* 33 (1971), 2195.
873. Chatt, J., Coffey, R. S., and Shaw, B. L., *J. chem. Soc.* (1965), 7391.
874. Barefield, E. K., Parshall, G. W., and Tebe, F. N., *J. Am. chem. Soc.,* 92 (1970), 5234.
875. Araneo, A., Martinengo, S., and Pasquale, P., *Rc. Ist. lomb. Sci. Lett.,* A99 (1965), 797.
876. Angoletta, M., and Araneo, A., *Rc. Ist. lomb. Sci. Lett.,* A97 (1963), 817.
877. Angoletta, M., *Gazz. chim. ital.,* 92 (1962), 811.
878. Hayter, R. G., *J. Am. chem. Soc.,* 83 (1961), 1259.
879. Malatesta, L., Angoletta, M., Araneo, A., and Canziani, F., *Angew. Chem.,* 73 (1961), 273.
880. Angoletta, M., and Caglio, G., *Gazz. chim. ital.,* 99 (1969), 46.
881. Jenkins, J. M., and Shaw, B. L., *J. chem. Soc. (A)* (1966), 1407.
882. Vaska, L., and Di Luzio, J. W., *J. Am. chem. Soc.,* 84 (1962), 4989.
883. Araneo, A., and Martinengo, S., *Gazz. chim. ital.,* 95 (1965), 61.
884. Angoletta, M., and Araneo, A., *Gazz. chim. ital.,* 93 (1963), 1343.
885. Jenkins, J. M., and Shaw, B. L., *J. chem. Soc.,* (1965), 6789.
886. Vaska, L., *J. Am. chem. Soc.,* 83 (1961), 756.
887. Adams, D. M., *Proc. chem. Soc.* (1961), 431.
888. Powell, J., and Shaw, B. L., *J. chem. Soc. (A)* (1968), 617.
889. Masters, C., Shaw, B. L., and Stainbank, R. E., *Chem. Commun.* (1971), 209.
890. Haines, L. M., and Singleton, E., *J. organometal. Chem.,* 25 (1970), C83.
891. Araneo, A., *Gazz. chim. ital.,* 95 (1965), 1431.
892. Coffey, R. S., *Tetrahedron Lett.* (1965), 3809.
893. Nicholson, J. K., and Shaw, B. L., *Tetrahedron Lett.* (1965), 3533.

894. Bennett, M. A., and Milner, D. L., *Chem. Commun.* (1967), 581; *J. Am. chem. Soc.,* **91** (1969), 6983.
895. Cardin, D. J., Lappert, M. F., and Travers, N. F., *Proceedings of the 11th International Conference on Coordination Chemistry* (ed. M. Cais), Elsevier, Amsterdam (1968), p. 821.
896. Van Gaal, H., Cuppers, H. G. A. V., and Van der Ent, A., *Chem. Commun.* (1970), 1694.
897. Chatt, J., Field, A. E., and Shaw, B. L., *J. chem. Soc.* (1963), 3371.
898. Chatt, J., Leigh, G. J., and Mingos, D. M. P., *J. chem. Soc.* (1969), 2972.
899. Angoletta, M., *Gazz. chim. ital.,* **93** (1963), 1343.
900. Geisenheimer, G., *C. r. hebd. Séanc. Acad. Sci, Paris,* **110** (1889), 1004, 1336; **111** (1890), 40; *Anals Chim. Phys.,* **6** (1891), 23, 231.
901. Kubota, M., and Carey, C. R., *J. organometal. Chem.,* **24** (1970), 491.
902. Yamamoto, K., *Bull. chem. Soc. Japan,* **27** (1954), 516.
903. Yamamoto, K., and Oku, M., *Bull. chem. Soc. Japan,* **27** (1954), 382.
904. Corain, B., Bressan, M., and Favero, G., *Inorg. nucl. Chem. Lett.,* **7** (1971), 197.
905. Bigorgne, M., and Zelwer, A., *Bull. Soc. chim. Fr.* (1960), 1986.
906. Mathieu, R., Lenzi, M., and Poilblanc, R., *C. r. hebd. Séanc. Acad. Sci., Paris, Ser. C.,* **266** (1968), 806.
907. Bigorgne, M., *C. r. hebd. Séanc. Acad. Sci., Paris,* **250** (1960), 3484.
908. Bigorgne, M., *J. inorg. nucl. Chem.,* **26** (1964), 107.
909. Meriwether, L. S., and Fiene, M., *J. Am. chem. Soc.,* **81** (1959), 4200.
910. Meriwether, L. S., and Leto, J. R., *J. Am. chem. Soc.,* **83** (1961), 3192.
911. Eneléus, H. J., and Smith, J. D., *J. chem. Soc.* (1958), 527.
912. Burg, A. B., and Mahler, W., *J. Am. chem. Soc.,* **80** (1958), 2334.
913. Rose, J. D., and Statham, F. S., *J. chem. Soc.* (1950), 69.
914. Yamamoto, K., *Bull. chem. Soc. Japan,* **27** (1954), 501.
915. Mitsui Chem. Ind. Co., Yamamoto, K., and Kunizaki, S., *Japan. Pat.* 5087 (1954); *Chem. Abstr.,* **50** (1956), 6508.
916. Nöth, H., and Vetter, H. J., *Chem. Ber.,* **96** (1963), 1479.
917. Schumann, H., and Stelzer, O., *Angew. Chem. (int. Ed.),* **6** (1967), 701.
918. Abel, E. W., Crow, J. P., and Illingworth, S. M., *J. chem. Soc.* (1969), 1631.
919. Schumann, H., Stelzer, O., Niederreuther, U., and Rüsch, L., *Chem. Ber.,* **103** (1970), 1383.
920. Tolman, C. A., *J. Am. chem. Soc.,* **92** (1970), 2953.
921. Sabherwal, I. H., and Burg, A. B., *Inorg. nucl. Chem. Lett.,* **5** (1969), 259.
922. Meriwether, L. S., Colthup, E. C., Fiene, M. L., and Cotton, F. A., *J. inorg. nucl. Chem.,* **11** (1959), 181.
923. Bennett, M. J., Cotton, F. A., and Winquist, B. H. C., *J. Am. chem. Soc.,* **89** (1967), 5366.
924. Bigorgne, M., and Bouquet, G., *C. r. hebd. Séanc. Acad. Sci., Paris, Ser. C.,* **264** (1967), 1485.
925. Manuel, T. A., *Adv. organometal. Chem.,* **3** (1965), 181.
926. Basolo, F., and Pearson, R. G., in *Mechanisms of Inorganic Reactions,* 2nd ed. (1967), Wiley, New York, chapter 7.
927. Ginsberg, A. P., and Koubek, E., *Inorg. Chem.,* **4** (1965), 1517.
928. Jarvis, J. A. J., Mais, R. H. B., Owston, P. G., and Thompson, D. T., *J. chem. Soc. (A)* (1970), 1867.
929. Coleman, J. F., and Dahl, L. F., *J. Am. chem. Soc.,* **89** (1967), 542.
930. Isslieb, K., and Rettkowski, W., *Z. Naturf.,* **B21** (1966), 999.
931. Irvine, J. W., and Wilkinson, G., *Science, N.Y.,* **113** (1951), 742.
932. Wilkinson, G., *Z. Naturf.,* **B9** (1954), 446.
933. Malatesta, L., and Sacco, A., *Annali Chim.,* **44** (1954), 134.
934. Quin, L. D., *J. Am. chem. Soc.,* **79** (1957), 3681.
935. Maier, L., *Angew. Chem.,* **71** (1959), 574.
936. Wilkinson, G., *J. Am. chem. Soc.,* **73** (1951), 5501.
937. Seel, F., Ballreich, K., and Schmutzler, R., *Chem. Ber.,* **94** (1961), 1173.
938. Wilke, G., and Bogdanovič, B., *Angew. Chem.,* **73** (1964), 756.
939. Behrens, H., and Meyer, K., *Z. Naturf.,* **B21** (1966), 489.

940. Behrens, H., and Müller, A., *Z. anorg. allg. Chem.*, **341** (1965), 124.
941. Horner, L., and Kunz, H., *Chem. Ber.*, **104** (1971), 717.
942. Klein, H. F., and Schmidbauer, H., *Angew. Chem. (int. Ed.)*, **9** (1970), 903.
943. Cariati, F., Ugo, R., and Bonati, F., *Inorg. Chem.*, **5** (1966), 1128.
944. Wilke, G., Schott, H., and Heimbach, P., *Angew. Chem.*, **79** (1967), 62.
945. Hidai, M., Kashiwagi, T., Ikeuchi, T., and Uchida, Y., *J. organometal. Chem.*, **30** (1971), 279.
946. Tolman, C. A., *J. Am. chem. Soc.*, **92** (1970), 2956.
947. Griffith, W. P., Lewis, J., and Wilkinson, G., *J. chem. Soc.* (1961), 2259.
948. Johnson, B. F. G., and McCleverty, J. A., *Prog. inorg. Chem.*, **7** (1966), 277.
949. Addison, C. C., and Johnson, B. F. G., *Proc. chem. Soc.* (1962), 305.
950. Feltham, R. D., *Inorg. Chem.*, **3** (1964), 116; *J. inorg. nucl. Chem.*, **14** (1960), 307.
951. Feltham, R. D., *Inorg. Chem.*, **3** (1964), 119.
952. Feltham, R. D., and Carriel, J. T., *Inorg. Chem.*, **3** (1964), 121.
953. Hieber, W., and Bauer, I., *Z. Naturf.*, **16B** (1961), 556.
954. Hieber, W., and Bauer, I., *Z. anorg. allg. Chem.*, **321** (1963), 107.
955. Enemark, J. H., *Inorg. Chem.*, **10** (1971), 1952.
956. Beck, W., and Lottes, K., *Z. anorg. allg. Chem.*, **335** (1965), 258.
957. Brunner, H., *Angew. Chem.*, **79** (1967), 536.
958. Green, M. L. H., Street, C. N., and Wilkinson, G., *Z. Naturf.*, **B14** (1959), 738.
959. Green, M. L. H., and Saito, T., *Chem. Commun.* (1969), 208.
960. Green, M. L. H., Saito, T., and Tanfield, P. J., *J. chem. Soc. (A)* (1971), 152.
961. Green, M. L. H., Munakata, H., and Saito, T., *Chem. Commun.* (1969), 1287.
962. Green, M. L. H., and Munakata, H., *Chem. Commun.* (1970), 881.
963. Green, M. L. H., Munakata, H., and Saito, T., *J. chem. Soc. (A)* (1971), 469.
964. Jolly, P. W., and Jonas, K., *Angew. Chem.*, **80** (1968), 705.
965. Srivastava, S. C., and Bigorgne, M., *J. organometal. Chem.*, **18** (1969), P30.
966. Jonas, K., and Wilke, G., *Angew. Chem.*, **81** (1969), 534.
967. Heimbach, P., *Angew. Chem. (int. Ed.)*, **3** (1964), 648.
968. Porri, I., Gallazze, M. C., and Vitulli, G., *Chem. Commun.* (1967), 228.
969. Beg, M. A., and Clark, H. C., *Can. J. Chem.*, **39** (1961), 595.
970. Jensen, K. A., and Dahl, O., *Acta. chem. scand.*, **22** (1968), 1044.
971. Coussmaker, C. R. C., Hely-Hutchinson, M., Mellor, J. R., Sutton, L. E., and Venanzi, L. M., *J. chem. Soc.* (1961), 2705.
972. Jensen, K. A., *Z. anorg. allg. Chem.*, **229** (1936), 265.
973. Beck, W., and Schuierer, E., *Chem. Ber.*, **98** (1965), 298.
974. Asmussen, R. M., Jensen, K. A., and Soling, H., *Acta. chem. scand.*, **9** (1955), 1391.
975. Giacometti, G., Scatturin, V., and Turco, A., *Gazz. chim. ital.*, **88** (1958), 434.
976. Scatturin, V., and Turco, A., *Proceedings of the 1st International Conference on Coordination Chemistry (Rome, 1957)* Pergamon, London; *J. inorg. nucl. Chem.*, **8** (1958), 447.
977. Giacometti, G., and Turco, A., *J. inorg. nucl. Chem.*, **15** (1960), 242.
978. Giacometti, G., Scatturin, V., and Turco, A., *Nature, Lond.*, **183** (1959), 601.
979. Rigo, P., Corain, B., and Turco, A., *Inorg. Chem.*, **7** (1968), 1623.
980. Cotton, F. A., Faut, O. D., and Goodgame, D. M. L., *J. Am. chem. Soc.*, **83** (1961), 344.
981. Jarvis, J. A. J., Mais, R. H. B., and Owston, P. G., *J. chem. Soc. (A)* (1968), 1473.
982. Garton, G., Henn, D. E., Powell, H. M., and Venanzi, L. M., *J. chem. Soc.* (1963), 3625.
983. Stone, P. J., and Dori, Z., *Inorg. Chim. Acta*, **5** (1971), 434.
984. Brunner, H., *Chem. Ber.*, **101** (1968), 143.
985. Alyea, E. C., and Meek, D. W., *J. Am. chem. Soc.*, **91** (1969), 5761.
986. Chatt, J., and Shaw, B. L., *J. chem. Soc.* (1960), 1718.
987. Rigo, P., Pecile, C., and Turco, A., *Inorg. Chem.*, **6** (1967), 1636.
988. Collier, J. W., Mann, F. G., Watson, D. G., and Watson, H. R., *J. chem. Soc.* (1964), 1803.
989. Hayter, R. G., and Humiec, F. S., *Inorg. Chem.*, **4** (1965), 1703.
990. La Mar, G. N., and Sherman, E. O., *Chem. Commun.* (1969), 809; *J. Am. chem. Soc.*, **92** (1970), 2691.
991. Hayter, R. G., and Humiec, F. S., *J. Am. chem. Soc.*, **84** (1962), 2004.

992. Wang, J. T., Udovich, C., Nakamoto, K., Quattrochi, A., and Ferraro, J. R., *Inorg. Chem.,* **9** (1970), 2675.
993. Shupack, S. I., *J. inorg. nucl. Chem.,* **28** (1966), 2418.
994. Turco, A., Scatturin, V., and Giacometti, G., *Gazz. chim. ital.,* **89** (1959), 2005.
995. Turco, A., Scatturin, V., and Giacometti, G., *Nature, Lond.,* **183** (1959), 601.
996. Browning, M. C., Mellor, J. R., Morgan, D. J., Pratt, S. A. J., Sutton, L. E., and Venanzi, L. M., *J. chem. Soc.* (1962), 693.
997. Kilbourn, B. T., Powell, H. M., and Derbyshire, J. A. C., *Proc. chem. Soc.* (1963), 207.
998. Kilbourn, B. T., and Powell, H. M., *J. chem. Soc. (A)* (1970), 1688.
999. Pignolet, L. H., and Horrocks, W. De W., *J. Am. chem. Soc.,* **91** (1970), 1855.
1000. Bellon, P. L., Albano, V., Bianco, V. D., Pompa, F., and Scatturin, V., *Ric. Sci. Ser. 2, Pt. 2 Rend. Ser. A,* **3** (1963), 1213.
1001. Pignolet, P. L., and Horrocks, W. De W., *J. Am. chem. Soc.,* **91** (1969), 3976.
1002. Blindheim, U., *Inorg. Chim. Acta,* **4** (1970), 507.
1003. Boorman, P. M., and Carty, A. J., *Inorg. nucl. Chem. Lett.,* **4** (1968), 101.
1004. Coates, G. E., and Parkin, C., *J. chem. Soc.* (1963), 421.
1005. Shobatake, K., and Nakamoto, K., *J. Am. chem. Soc.,* **92** (1970), 3332.
1006. Sacconi, L., *Transit. Metal Chem.,* **4** (1968), 199.
1007. Fereday, R. J., Hathaway, B. J., and Dudley, R. J., *J. chem. Soc. (A)* (1970), 571.
1008. Moedritzer, K., and Miller, R. E., *Thermochem. Acta* (1970), 87.
1009. Muetterties, E. L., and Schunn, R. A., *Q. Rev. chem. Soc.,* **20** (1966), 245.
1010. Furlani, C., *Coord. Chem. Rev.,* **3** (1968), 141.
1011. Chastain, B. B., Meek, D. W., Billig, E., Hix, J. E., and Gray, H. B., *Inorg. Chem.,* 7 (1968), 2412.
1012. Allen, D. W., Mann, F. G., Millar, I. T., Powell, H. M., and Watkin, D., *Chem. Commun.* (1969), 1004.
1013. Schrauzer, G. N., and Glockner, P., *Chem. Ber.,* **97** (1964), 2451.
1014. Ibers, J. A., and Stalick, J., *Inorg. Chem.,* **8** (1969), 1090.
1015. Rigo, P., Guastalla, G., and Turco, A., *Inorg. Chem.,* **8** (1969), 375.
1016. Chastain, B. B., Rick, E. A., Pruett, R. L., and Gray, H. B., *J. Am. chem. Soc.,* 90 (1968), 3994.
1017. Booth, G., and Chatt, J., *J. chem. Soc.* (1965), 3238.
1018. McAuliffe, C. A., and Meek, D. W., *Inorg. Chem.,* **8** (1969), 904.
1019. Ramaswamy, H. N., Jonassen, H. B., and Aguiar, A. M., *Inorg. Chim. Acta,* **1** (1967), 141.
1020. Hayter, R. G., *Inorg. Chem.,* **2** (1963), 932.
1021. Dick, T. R., and Nelson, S. M., *Inorg. Chem.,* **8** (1969), 1208.
1022. Schmitz-Dumont, O., Uecker, G., and Schaal, W., *Z. anorg. allg. Chem.,* **370** (1969), 67.
1023. Abel, E. W., McLean, R. A. N., and Sabherwal, I. H., *J. chem. Soc. (A)* (1968), 2371.
1024. Phillips, J. R., Rosevear, D. T., and Stone, F. G. A., *J. organometal. Chem.,* **2** (1964), 455.
1025. Hayter, R. G., and Humiec, F. S., *J. inorg. nucl. Chem.,* **26** (1964), 807.
1026. Cundy, C. S., and Nöth, H., *J. organometal. Chem.,* **30** (1971), 135.
1027. Green, M. L. H., and Smith, M. J., *J. chem. Soc. (A)* (1971), 639.
1028. Longoni, G., Chini, P., Canziani, F., and Fantucci, P., *Chem. Commun.* (1971), 470.
1029. Green, M. L. H., Smith, M. J., Felkin, H., and Swierczeuski, G., *Chem. Commun.* (1971), 158.
1030. Cotton, F. A., and Goodgame, D. M. L., *J. Am. chem. Soc.,* **82** (1960), 2967.
1031. Badische Anilin und Soda Fabrik, Reppe, W., and Schwekendiek, W., *Germ. Pat.* 876 094 (1953); *Chem. Abstr.,* **52** (1958), 10175.
1032. Mitsui Company, *Japan. Pat.* 6625 (1953); *Chem. Abstr.,* **49** (1955), 9690.
1033. Taylor, R. P., Templeton, D. H., Zalkin, A., and Horrocks, W. De W., *Inorg. Chem.,* 7 (1968), 2629.
1034. Jensen, K. A., and Nygard, B., *Acta chem. scand.,* **3** (1949), 474.
1035. Giacometti, G., Scatturin, V., and Turco, A., *Ricerca scient.,* **27** (1957), 2449; *Chem. Abstr.,* **52** (1958), 3586e.
1036. Meek, D. W., Alyea, E. C., Stalick, J. K., and Ibers, J. A., *J. Am. chem. Soc.,* **91** (1969), 4920.

1037. Stalick, J. K., and Ibers, J. A., *Inorg. Chem.,* **9** (1970), 453.
1038. Misono, A., Uchida, Y., Hidai, M., and Kudo, K., *J. organometal. Chem.,* **20** (1969), P7.
1039. Clark, H. C., Dixon, K. R., and Jacobs, W. J., *Chem. Commun.* (1968), 93.
1040. Clark, H. C., and Dixon, K. R., *J. Am. chem. Soc.,* **91** (1969), 596.
1041. Booth, G., and Chatt, J., *J. chem. Soc. (A)* (1966), 634.
1042. Booth, G., Chatt, J., and Chini, P., *Chem. Commun.* (1965), 639.
1043. Booth, G., and Chatt, J., *J. chem. Soc. (A)* (1969), 2131.
1044. Chatt, J., and Chini, P., *J. chem. Soc. (A)* (1970), 1538.
1045. Vranka, R. G., Dahl, L. F., Chini, P., and Chatt, J., *J. Am. chem. Soc.,* **91** (1969), 1574.
1046. Blake, D. M., and Nyman, L. J., *J. Am. chem. Soc.,* **92** (1970), 5359.
1047. Malatesta, L., and Cariello, C., *J. chem. Soc.* (1958), 2333.
1048. Cariati, F., and Ugo, R., *Chimica. Ind., Milano,* **98** (1966), 1288.
1049. Chini, P., and Longoni, G., *J. chem. Soc. (A)* (1970), 1542.
1050. Albano, V. G., Bellon, P. L., and Scatturin, V., 'Metal Carbonyl Derivatives', *Proc. 1st Int. Symp. New Aspects Chem.* (1968); *Chem. Abstr.,* **71** (1969), 129681.
1051. Albano, V. G., Bellon, P. L., and Sansoni, M., *Chem. Commun.* (1969), 899.
1052. Albano, V. G., Ricci, G. H. B., and Bellon, P. L., *Inorg. Chem.,* **8** (1969), 2109.
1053. Irving, R. J., and Magnusson, E. A., *J. chem. Soc.* (1957), 2018.
1054. Chatt, J., Johnson, N. P., and Shaw, B. L., *J. chem. Soc.* (1964), 1662.
1055. Smithies, A. C., Rycheck, M., and Orchin, M., *J. organometal. Chem.,* **12** (1968), 199.
1056. McAvoy, J., Moss, K. C., and Sharp, D. W. A., *J. chem. Soc.* (1965), 1376.
1057. Clark, H. C., and Tsang, W. S., *J. Am. chem. Soc.,* **89** (1967), 529.
1058. Clark, H. C., Corfield, P. W. R., Dixon, K. R., and Ibers, J. A., *J. Am. chem. Soc.,* **89** (1967), 3360.
1059. Clark, H. C., Dixon, K. R., and Jacobs, W. J., *J. Am. chem. Soc.,* **91** (1969), 1346.
1060. Clark, H. C., Dixon, K. R., and Jacobs, W. J., *J. Am. chem. Soc.,* **90** (1968), 2259.
1061. Church, M. J., and Mays, M. J., *J. chem. Soc. (A)* (1968), 3074.
1062. Clark, H. C., and Jacobs, W. J., *Inorg. Chem.,* **9** (1970), 1229.
1063. Ugo, R., *Coord. Chem. Rev.,* **3** (1968), 319.
1064. Malatesta, L., and Cariello, C., *J. chem. Soc.* (1958), 2323; *J. inorg. nucl. Chem.,* **8** (1958), 561.
1065. Malatesta, L., and Angoletta, M., *J. chem. Soc.* (1957), 1186.
1066. Chopoorian, J. A., Lewis, J., and Nyholm, R. S., *Nature, Lond.,* **190** (1961), 528.
1067. Malatesta, L., and Ugo, R., *J. chem. Soc.* (1963), 2080.
1068. Chatt, J., and Rowe, G. A., *Nature, Lond.,* **191** (1961), 1191.
1069. Allen, A. D., and Cook, C. D., *Proc. chem. Soc.* (1962), 218.
1070. Clark, H. C., and Itoh, K., *Inorg. Chem.,* **10** (1971), 1707.
1071. Gerlach, D. M., Kane, A. R., Parshall, G. W., Jesson, J. P., and Muetterties, E. L., *J. Am. chem. Soc.,* **93** (1971), 3543.
1072. Fischer, E. O., and Werner, H., *Chem. Ber.,* **95** (1962), 703.
1073. Mukhedkar, A. J., Green, M., and Stone, F. G. A., *J. chem. Soc. (A)* (1969), 3023.
1074. Kruck, T., and Baur, K., *Z. anorg. allg. Chem.,* **364** (1969), 192.
1075. Wheelock, K. S., Nelson, J. H., and Jonassen, H. B., *Inorg. Chim. Acta,* **4** (1970), 399.
1076. Ugo, R., Cariati, F., and La Monica, G., *Inorg. Synth.,* **11** (1968), 105.
1077. Ugo, R., La Monica, G., Cariati, F., Cenini, S., and Conti, F., *Inorg. Chim. Acta,* **4** (1970), 390.
1078. Ugo, R., Cariati, F., and La Monica, G., *Chem. Commun.* (1966), 868.
1079. Blake, D. M., and Mersecchi, R., *Chem. Commun.* (1971), 1045.
1080. Albano, V., Bellon, P. L., and Scatturin, V., *Chem. Commun.* (1966), 507.
1081. Gillard, R. D., *Chem. Commun.* (1966), 869.
1082. Cariati, F., Ugo, R., and Bonati, F., *Chem. Ind.,* (1964), 1714.
1083. Tokahaski, S., Sonagashira, K., and Hagihara, N., *J. chem. Soc. Japan,* **87** (1966), 610.
1084. Nyman, C. J., Wymore, C. E., and Wilkinson, G., *J. chem. Soc. (A)* (1968), 561.
1085. Cook, C. D., Cheng, P. T., and Nyburg, S. C., *J. Am. chem. Soc.,* **91** (1969), 2123.
1086. Kashinwaga, T., Yasuoka, N., Kasai, N., Kakudo, M., Takahaski, S., and Hagihara, N., *Chem. Commun.* (1969), 743.
1087. Cook, C. D., and Jauhal, G. S., *J. Am. chem. Soc.,* **89** (1967), 3066.
1088. Horn, R. W., Weissberger, E., and Collman, J. P., *Inorg. Chem.,* **9** (1970), 2367.

1089. Ugo, R., Conti, F., Cenini, S., Mason, R., and Robertson, G. B., *Chem. Commun.* (1968), 1498.
1090. Collman, J. P., Kubota, M., and Hosking, J. W., *J. Am. chem. Soc.*, **89** (1967), 4809.
1091. Cariati, F., Mason, R., Robertson, G. B., and Ugo, R., *Chem. Commun.* (1967), 408.
1092. Nyman, C. J., Wymore, C. E., and Wilkinson, G., *Chem. Commun.* (1967), 407.
1093. Hayward, P. J., Blake, D. M., Wilkinson, G., and Nyman, C. J., *Chem. Commun.* (1969), 987; *J. Am. chem. Soc.,* **92** (1970), 5873.
1094. Cook, C. D., and Jauhal, G. S., *Inorg. nucl. Chem. Lett.,* **3** (1967), 31.
1095. Halpern, J., and Pickard, A. L., *Inorg. Chem.,* **9** (1970), 2798.
1096. Baird, M. C., and Wilkinson, G., *Chem. Commun.* (1966), 514.
1097. Morelli, D., Segre, A., Ugo, R., La Monica, G., Cenini, S., Conti, F., and Bonati, F., *Chem. Commun.* (1967), 524.
1098. Argento, B. J., Fitton, P., McKeon, J. E., and Rick, E. A., *Chem. Commun.* (1969), 1427.
1099. Kemmitt, R. D. W., Peacock, R. D., and Stocks, J., *J. chem. Soc. (A)* (1971), 846.
1100. Chatt, J., Eaborn, C., and Kapoor, P. N., *J. chem. Soc. (A)* (1970), 881.
1101. Baddley, W. H., and Venanzi, L. M., *Inorg. Chem.,* **5** (1966), 33.
1102. Cook, C. D., and Jauhal, G. S., *Can. J. Chem.,* **45** (1967), 301.
1103. Allen, A. D., and Cook, C. D., *Can. J. Chem.,* **42** (1964), 1063.
1104. Beck, W., Bauder, M., La Monica, G., Cenini, S., and Ugo, R., *J. chem. Soc. (A)* (1971), 113.
1105. Bruce, M. I., Shaw, G., and Stone, F. G. A., *Chem. Commun.* (1971), 1288.
1106. Layton, A. J., Nyholm, R. S., Pneumaticakis, G. A., and Tobe, M. L., *Chem. Ind.* (1967), 465.
1107. Chatt, J., and Mingos, D. M. P., *J. chem. Soc. (A)* (1970), 1243.
1108. Pneumaticakis, G. A., *Chem. Commun.* (1968), 275.
1109. Fitton, P., and McKeon, J. E., *Chem. Commun.* (1968), 4.
1110. Mason, R., and Rae, A. I. M., *J. chem. Soc. (A)* (1970), 1767.
1111. Kashiwagi, T., Yasuoka, N., Ueki, T., Kasai, N., Kakudo, M., Takahashi, S., and Hagihara, N., *Bull. chem. Soc. Japan*, **41** (1968), 296.
1112. Cenini, S., Ugo, R., and La Monica, G., *J. chem. Soc. (A)* (1971), 409.
1113. Baird, M. C., Hartwell, G., Mason, R., Rae, A. I. E., and Wilkinson, G., *Chem. Commun.* (1967), 92.
1114. Fitton, P., Johnson, M. P., and McKeon, J. E., *Chem. Commun.* (1968), 6.
1115. Skapski, A. C., and Troughton, P. G. H., *Chem. Commun.* (1969), 170; *J. chem. Soc. (A)* (1969), 2772.
1116. Dumler, J. T., and Roundhill, D. M., *J. organometal. Chem.,* **30** (1971), C35.
1117. Chatt, J., and Shaw, B. L., *J. chem. Soc.* (1962), 5075.
1118. Chatt, J., Duncanson, L. A., and Shaw, B. L., *Chem. Ind.* (1958), 859.
1119. Brooks, E. H., and Glockling, F., *J. chem. Soc. (A)* (1967), 1030.
1120. Kudo, K., Hidai, M., *Chem. Commun.* (1970), 1701.
1121. Chatt, J., Duncanson, L. A., and Shaw, B. L., *Proc. chem. Soc.* (1957), 343.
1122. Dobinson, G. C., Mason, R., Robertson, G. B., Ugo, R., Conti, F., Morelli, D., Cenini, S., and Bonati, F., *Chem. Commun.* (1967), 759.
1123. Roundhill, D. M., Tripathy, P. B., and Renoe, B. W., *Inorg. Chem.,* **10** (1971), 727.
1124. Owston, P. G., Partridge, J. M., and Rowe, J. M., *Acta, crystallogr.,* **13** (1960), 246.
1125. Eisenberg, R., and Ibers, J. A., *Inorg. Chem.,* **4** (1965), 773.
1126. Powell, J., and Shaw, B. L., *J. chem. Soc.* (1965), 3879.
1127. Dean, R. R., and Green, J. C., *J. chem. Soc. (A)* (1968), 3047.
1128. Socrates, G., *J. inorg. nucl. Chem.,* **31** (1969), 1667.
1129. Chatt, J., Duncanson, L. A., Shaw, B. L., and Venanzi, L. M., *Discuss. Faraday Soc.,* **26** (1958), 131.
1130. Bailar, J., Jr., and Itatani, M., *Inorg. Chem.,* **4** (1965), 1618.
1131. Clemmit, A. F., and Glockling, F., *J. chem. Soc. (A)* (1969), 2163.
1132. Collamati, I., Furlani, A., and Attioli, G., *J. chem. Soc. (A)* (1970), 1694.
1133. Clark, H. C., and Kurosawa, H., *Chem. Commun.* (1971), 957.
1134. Chatt, J., Coffey, R. S., Gough, A., and Thompson, D. T., *J. chem. Soc. (A)* (1968), 190.

1135. Palazzi, A., Busetto, L., and Graziani, M., *J. organometal. Chem.*, **30** (1971), 273.
1136. Beck, W., Bauder, M., Fehlhammer, W. P., Poellman, P., and Schuechal, H., *Inorg. nucl. Chem. Lett.*, **4** (1968), 143.
1137. Brooks, P. R., and Nyholm, R. S., *Chem. Commun.* (1970), 169.
1138. Halpern, J., and Falk, C. D., *J. Am. chem. Soc.*, **87** (1965), 3523.
1139. Green, M. L. H., and Munakata, H., *Chem. Commun.* (1971), 549.
1140. Glockling, F., and Hooton, K. A., *J. chem. Soc. (A)* (1968), 826.
1141. Giustiniani, M., Dolcetti, G., and Belluco, U., *J. chem. Soc. (A)* (1969), 2047.
1142. Toniolo, T., Giustiniani, M., and Belluco, U., *J. chem. Soc. (A)* (1969), 2666.
1143. Church, J., and Mays, M. J., *J. chem. Soc. (A)* (1970), 1938.
1144. Chatt, J., and Davidson, J. M., *J. chem. Soc.* (1964), 2433.
1145. Cross, R. J., *Organometal. chem. Rev.*, **2** (1967), 97.
1146. Calvin, G., and Coates, G. E., *J. chem. Soc.* (1960), 2008.
1147. Chatt, J., and Shaw, B. L., *J. chem. Soc.* (1959), 705.
1148. Ruddick, J. D., and Shaw, B. L., *J. chem. Soc. (A)* (1969), 2801.
1149. Allen, F. H., and Pidcock, A., *J. chem. Soc. (A)* (1968), 2700.
1150. Adams, D. M., Chatt, J., and Shaw, B. L., *J. chem. Soc.* (1960), 2047.
1151. Basolo, F., Chatt, J., Gray, H. B., Pearson, R. G., and Shaw, B. L., *J. chem. Soc.* (1961), 2207.
1152. Parshall, G. W., *J. Am. chem. Soc.*, **88** (1966), 704.
1153. Parshall, G. W., *J. Am. chem. Soc.*, **86** (1964), 5367.
1154. Coulson, D. R., *Chem. Commun.* (1968), 1530.
1155. Glockling, F., and Hooton, K. A., *Chem. Commun.* (1966), 218.
1156. Glockling, F., and Hooton, K. A., *J. chem. Soc. (A)* (1967), 1066.
1157. Chatt, J., Eaborn, C., Ibekwe, S. D., and Kapoor, P. N., *J. chem. Soc. (A)* (1970), 1343.
1158. Baird, M. C., *J. inorg. nucl. Chem.*, **29** (1967), 367.
1159. Bentham, J. E., Cradock, S., and Ebsworth, E. A. V., *Chem. Commun.* (1969), 528; *J. chem. Soc. (A)* (1971), 587.
1160. Brooks, E. H., and Glockling, F., *J. chem. Soc. (A)* (1966), 1241.
1161. Bentham, J. E., and Ebsworth, E. A. V., *J. chem. Soc. (A)* (1971), 2091.
1162. Wittle, J. K., and Urry, G., *Inorg. Chem.*, **7** (1968), 560.
1163. Cross, R. J., and Glockling, F., *J. chem. Soc.* (1965), 5422.
1164. Gee, R. J. D., and Powell, H. M., *J. chem. Soc. (A)* (1971), 1956.
1165. Young, J. F., Gillard, R. D., and Wilkinson, G., *J. chem. Soc.* (1964), 5176.
1166. Cramer, R. D., Jenner, E. I., Lindsey, R. V., and Stolberg, U. G., *J. Am. chem. Soc.*, **85** (1963), 1691.
1167. Akhtar, M., and Clark, H. C., *J. organometal. Chem.*, **22** (1970), 233.
1168. Degonello, G., Carturan, G., and Belluco, U., *J. chem. Soc. (A)* (1968), 2873.
1169. Carturan, G., Degonello, G., Boshi, T., and Belluco, U., *J. chem. Soc. (A)* (1969), 1142.
1170. Schmid, G., and Nöth, H., *Z. Naturf.*, **20B** (1965), 1008.
1171. Goodfellow, R. J., Evans, J. G., Goggin, P. L., and Duddell, D. A., *J. chem. Soc. (A)* (1968), 1604.
1172. Beg, M. A. A., and Clark, H. C., *Can. J. Chem.*, **38** (1960), 119.
1173. Mann, F. G., and Wells, A. F., *J. chem. Soc.* (1938), 702.
1174. Kemmit, R. D. W., Nichols, D. I., and Peacock, R. D., *Chem. Commun.* (1968), 599; *J. chem. Soc. (A)* (1968), 2149.
1175. Evans, J. G., Goggin, P. L., Goodfellow, R. G., and Smith, J. G., *J. chem. Soc. (A)* (1968), 464.
1176. Park, P. J. P., and Hendra, P. J., *Spectrochim. Acta,* **A25** (1969), 909.
1177. Goggin, P. L., and Goodfellow, R. J., *J. chem. Soc. (A)* (1966), 1462.
1178. Cahours, A., and Gal, H., *C. r. hebd. Séanc. Acad. Sci., Paris,* **70** (1870), 1380.
1179. Hitchcock, C. H. S., and Mann, F. G., *J. chem. Soc.* (1958), 2081.
1180. Mann, F. G., and Purdie, D., *J. chem. Soc.* (1940), 1235.
1181. Norbury, A. H., and Sinha, A. I. P., *Inorg. nucl. Chem. Lett.*, **3** (1967), 355.
1182. Jensen, K. A., *Z. anorg. allg. Chem.*, **229**(1936), 225.
1183. Messmer, G. G., and Amma, E. L., *Inorg. Chem.*, **5** (1966), 1775.
1184. Adams, D. M., Chatt, J., Gerratt, J., and Westland, A. D., *J. chem. Soc. (A)* (1969), 734.
1185. Chatt, J., *J. chem. Soc.*, (1951), 652.

1186. Chatt, J., and Wilkins, R. G., *J. chem. Soc.* (1951), 2532.
1187. Chatt, J., and Wilkins, R. G., *J. chem. Soc.* (1956), 525.
1188. Chatt, J., Duncanson, L. A., Gatehouse, B. M., Lewis, J., Nyholm, R. S., Tobe, M. L., Todd, P. F., and Venanzi, L. M., *J. chem. Soc.* (1959), 4073.
1189. Beck, W., Fehlhammer, W. P., Poellman, P., and Schaechl, H., *Chem. Ber.*, **102** (1969), 1976.
1190. Haake, P., and Hylton, T. A., *J. Am. chem. Soc.*, **84** (1962), 3774.
1191. Bailey, N. A., Jenkins, J. M., Mason, R., and Shaw, B. L., *Chem. Commun.* (1965), 237.
1192. Bailey, N. A., and Mason, R., *J. chem. Soc. (A)* (1968), **2594**.
1193. Martin, L. L., and Jacobson, R. A., *Inorg. Chem.*, **10** (1971), 1795.
1194. Jenkins, J. M., and Shaw, B. L., *J. chem. Soc. (A)* (1966), 770.
1195. Cheney, A. J., Mann, B. E., Shaw, B. L., and Slade, R. M., *Chem. Commun.* (1970), 1176.
1196. Chatt, J., and Mann, F. G., *J. chem. Soc.* (1939), 1622.
1197. Grim, S. O., and Keiter, R. L., *Inorg. Chim. Acta*, **4** (1970), 56.
1198. Beg, M. A. A., and Clark, H. C., *Can. J. Chem.*, **40** (1962), 283.
1199. Haake, P., and Pfeiffer, R. M., *J. Am. chem. Soc.*, **92** (1970), 4996.
1200. Rausch, M. D., and Keiter, R. L., *Inorg. Chim. Acta.*, **4** (1970), 56.
1201. Bowman, K., and Dori, Z., *Inorg. Chem.* (1970), 395.
1202. Rest, A. J., *J. chem. Soc. (A)* (1968), 2212.
1203. Burmeister, J. L., and Timmer, R. C., *J. inorg. nucl. Chem.*, **28** (1966), 1973.
1204. Beck, W., Feldl, K., and Schiuerer, E., *Angew. Chem.*, **77** (1965), 458.
1205. Beck, W., Fehlhammer, W. P., Poellmann, P., and Tobius, R. S., *Inorg. Chim. Acta,* **2** (1968), 467.
1206. Beck, W., Fehlhammer, W. P., Poellmann, P., Schiuerer, E., and Feldl, K., *Chem. Ber.*, **100** (1967), 2335.
1207. Burmeister, J. L., and Basolo, F., *Inorg. Chem.*, **3** (1964), 1587.
1208. Hayter, R. G., *J. Am. chem. Soc.*, **84** (1962), 3046.
1209. Hayter, R. G., and Humiec, F. S., *Inorg. Chem.*, **2** (1963), 306.
1210. Chatt, J., and Wilkins, R. G., *J. chem. Soc.* (1953), 70.
1211. Chatt, J., and Wilkins, R. G., *J. chem. Soc.* (1952), 273.
1212. Grim, S. O., Keiter, R. L., and McFarlane, W., *Inorg. Chem.*, **6** (1967), 1133.
1213. Messmer, G. G., Amma, E. L., and Ibers, J. A., *Inorg. Chem.*, **6** (1967), 725.
1214. Tayim, M. A., Bouldoukian, A., and Awad, F., *J. inorg. nucl. Chem.*, **32** (1970), 3799.
1215. Cule-Davis, W., and Mann, F. G., *J. chem. Soc.* (1944), 276.
1216. Chatt, J., Gamlen, G. A., and Orgel, L. E., *J. chem. Soc.* (1959), 1047.
1217. Duddell, D. A., Goggin, P. L., Goodfellow, R. J., Norton, M. G., and Smith, J. G., *J. chem. Soc. (A)* (1970), 545.
1218. Duddell, D. A., Goggin, P. L., Goodfellow, R. J., and Norton, M. G., *Chem. Commun.* (1968), 879.
1219. Park, P. J. P., and Hendra, P. J., *Spectrochim. Acta*, **A25** (1969), 227.
1220. Pidcock, A., Richards, R. E., and Venanzi, L. M., *J. chem. Soc. (A)* (1966), 1707.
1221. Green, P. J., and Brown, T. H., *Inorg. Chem.*, **10** (1971), 206.
1222. Duddell, D. A., Goggin, P. L., Goodfellow, R. J., Evans, J. G., Rest, A. J., and Smith, J. G., *J. chem. Soc. (A)* (1969), 2134.
1223. Allen, F. M., and Sze, S. N., *J. chem. Soc. (A)* (1971), 2055.
1224. Fryer, C. W., and Smith, J. A. S., *J. chem. Soc. (A)* (1970), 1029.
1225. Turco, A., and Pecile, C., *Nature, Lond.*, **191** (1961), 66.
1226. Norbory, A. H., and Sinha, A. I. P., *J. chem. Soc. (A)* (1968), 1598.
1227. Dixon, K. R., and Hawke, D. J., *Can. J. Chem.*, **49** (1971), 3252.
1228. Allen, F. H., and Gabuji, K. M., *Inorg. nucl. Chem. Lett.*, **7** (1971), 888.
1229. Beeby, M. H., and Mann, F. G., *J. chem. Soc.* (1951), 411.
1230. Parshall, G. W., *Inorg. Synth.*, **12** (1970), 26.
1231. Chatt, J., Duncanson, L. A., and Venanzi, L. M., *J. chem. Soc.* (1956), 2712; (1955), 4461.
1232. Bennett, M. A., Kouwenhoven, H. W., Lewis, J., and Nyholm, R. S., *J. chem. Soc.* (1964), 4570.
1233. Bennett, M. A., Keen, W. R., and Nyholm, R. S., *Inorg. Chem.*, **7** (1968), 556.

1234. Bennett, M. A., Keen, W. R., and Nyholm, R. S., *J. organometal. Chem.,* **26** (1971), 293.
1235. Hall, D. I., and Nyholm, R. S., *J. chem. Soc. (A)* (1971), 1491.
1236. Basolo, F., and Pearson, R. G., *Mechanisms of Inorganic Reactions,* 2nd edn, Wiley, New York (1967), chapter 5.
1237. Schutzenberger, P., *C. r. hebd. Séanc. Acad. Sci., Paris,* **70** (1870), 1414.
1238. Fink, E., *C. r. hebd. Séanc. Acad. Sci., Paris,* **115** (1892), 176; **123** (1896), 603.
1239. Schutzenberger, P., *Bull. Soc. chim. Fr.,* **17** (1871), 482; **18** (1872), 148.
1240. Chatt, J., and Heaton, B. T., *J. chem. Soc. (A)* (1968), 2745.
1241. Chatt, J., and Mingos, D. M. P., *J. chem. Soc. (A)* (1969), 1770.
1242. Mann, F. G., and Purdie, D., *J. chem. Soc.* (1936), 873.
1243. Chatt, J., and Venanzi, L. M., *J. chem. Soc.* (1955), 2787.
1244. Smithies, A. C., Schmidt, J. P., and Orchin, M., *Inorg. Synth.,* **12** (1970), 240.
1245. Goodfellow, R. J., and Venanzi, L. M., *J. chem. Soc.* (1965), 7533.
1246. Goodfellow, R. J., Goggin, P. L., and Venanzi, L. M., *J. chem. Soc. (A)* (1967), 1897.
1247. Duncanson, L. A., and Venanzi, L. M., *J. chem. Soc.* (1960), 3841.
1248. Allen, F. H., Chang, G., Cheung, K. K., Lai, T. F., and Lee, L. M., *Chem. Commun.* (1970), 1297.
1249. Adams, D. M., and Chandler, P. J., *J. chem. Soc. (A)* (1969), 588.
1250. Taylor, M. J., Odell, A. L., and Raethel, H. A., *Spectrochim. Acta,* **A24** (1968), 1855.
1251. Black, M., Mais, R. H. B., and Owston, P. G., *Acta. Crystallogr.,* **B25** (1969), 1760.
1252. Chatt, J., and Venanzi, L. M., *J. chem. Soc.* (1955), 3858; (1957), 2445.
1253. Chatt, J., and Hart, F. A., *J. chem. Soc.* (1961), 1416.
1254. Chatt, J., Duncanson, L. A., Hart, F. A., and Owston, P. G., *Nature, Lond.,* **181** (1958), 43.
1255. Owston, P. G., and Rowe, J. M., *Acta. Crystallogr.,* **13** (1960), 253.
1256. Owston, P. G., Gregory, U. A., Jarvis, J. A. J., and Kilbourn, B. T., *J. chem. Soc. (A)* (1970), 2770.
1257. Goodfellow, R. J., Goggin, P. L., and Duddell, D. A., *J. chem. Soc. (A)* (1968), 504.
1258. Stephenson, T. A., and Wilkinson, G., *J. inorg. nucl. Chem.,* **29** (1967), 2122.
1259. Hayter, R. G., *Nature, Lond.,* **193** (1962), 872.
1260. Chatt, J., and Hart, F. A., *J. chem. Soc.* (1960), 2807; (1963), 2363.
1261. Nyholm, R. S., Skinner, J. F., and Stiddard, M. H. B., *J. chem. Soc. (A)* (1968), 38.
1262. Druce, D. M., Lappert, M. F., and Riley, P. N. K., *Chem. Commun.* (1967), 486.
1263. Chatt, J., *J. chem. Soc.* (1950), 2301.
1264. Adams, D. M., and Chandler, P. J., *J. chem. Soc. (A)* (1967), 1009.
1265. Hatfield, W. E., and Whyman, R., *Transit. Metal Chem.,* **5** (1969), 138.
1266. Dieck, H. T., and Brehm, H. P., *Chem. Ber.,* **102** (1969), 3577.
1267. Zelonka, R. A., and Baird, M. C., *Chem. Commun.* (1971), 780.
1268. Isslieb, K., and Frohlich, H. O., *Chem. Ber.,* **95** (1962), 375.
1269. Holtje, R., and Schlegel, H., *Z. anorg. allg. Chem.,* **243** (1940), 246.
1270. Lutsenko, I. F., Kazankova, M. A., and Malykbina, J. G., *Zh. obshch. Khim.,* **37** (1967), 2364.
1271. Dilts, J. A., and Shriver, D. F., *J. Am. chem. Soc.,* **91** (1969), 4088.
1272. Bezman, S. A., Churchill, M. R., Osborn, J. A., and Wormald, J., *J. Am. chem. Soc.,* **93** (1971), 2063.
1273. Davidson, J. M., *Chem. Ind.,* (1964), 2021.
1274. Cariati, F., and Naldini, L., *Gazz. chim. ital.,* **95** (1965), 3.
1275. Lippard, S. J., and Ucko, D. A., *Inorg. Chem.,* **7** (1968), 1051.
1276. Lippard, S. J., and Melmed, K. M., *J. Am. chem. Soc.,* **89** (1967), 3924.
1277. Cariati, F., and Naldini, L., *J. inorg. nucl. Chem.,* **28** (1966), 2243.
1278. Klanberg, F., Muetterties, E. L., and Guggenberger, L. J., *Inorg. Chem.,* **7** (1968), 2272.
1279. Brice, V. T., and Shore, S. G., *Chem. Commun.* (1970), 1312.
1280. Lippard, S. J., and Melmed, K. M., *Inorg. Chem.,* **8** (1969), 2755.
1281. Muetterties, E. L., and Alegranti, C. W., *J. Am. chem. Soc.,* **92** (1970), 4114.
1282. Layton, A. J., Nyholm, R. S., Pneumaticakis, G. A., and Tobe, M. L., *Nature, Lond.,* **214** (1967), 1109.
1283. Layton, A. J., Nyholm, R. S., Pneumaticakis, G. A., and Tobe, M. L., *Nature, Lond.,* **218** (1968), 950.

1284. Wells, A. F., *Z. Kristallogr. Kristallgeom.*, **94** (1936), 447.
1285. Mann, F. G., Purdie, D., and Wells, A. F., *J. chem. Soc.* (1936), 1503.
1286. Costa, G., Reisenhofer, E., and Stefani, L., *J. inorg. nucl. Chem.*, **27** (1965), 2581.
1287. Costa, G., Pellizer, G., and Rubessa, F., *J. inorg. nucl. Chem.*, **26** (1964), 961.
1288. Jardine, F. H., Rule, L., and Vohra, A. G., *J. chem. Soc. (A)* (1970), 238.
1289. Moers, F. G., and Veld, P. H. Op., *J. inorg. nucl. Chem.*, **22** (1970), 3225.
1290. Seidel, W., *Z. anorg. allg. Chem.*, **341** (1965), 70.
1291. Seidel, W., *Z. anorg. allg. Chem.*, **335** (1965), 316.
1292. Lippard, S. J., Stowers, D., and Ucko, D. A., unpublished work quoted in ref. 32.
1293. Lewis, D. F., Lippard, S. J., and Weckler, P. S., *J. Am. chem. Soc.*, **92** (1970), 3805.
1294. Cotton, F. A., and Goodgame, D. M. L., *J. chem. Soc.* (1960), 5267.
1295. Messmer, G. G., and Palenik, G. J., *Can. J. Chem.*, **47** (1969), 1440; *Inorg. Chem.*, **8** (1969), 2750.
1296. Anderson, W. A., Carty, A. J., Palenik, G. J., and Schreiber, G., *Can. J. Chem.*, **49** (1971), 761.
1297. Cooper, D., and Plane, R. A., *Inorg. Chem.*, **5** (1966), 2204.
1298. Ziolo, R. F., and Dori, Z., *J. Am. chem. Soc.*, **90** (1968), 6560.
1299. Ziolo, R. F., Gaughan, A. P., Dori, Z., Pierpont, C. G., and Eisenberg, R., *J. Am. chem. Soc.*, **92** (1970), 738; *Inorg. Chem.*, **10** (1971), 1289.
1300. Gaughan, A. P., Ziolo, R. F., and Dori, Z., *Inorg. Chim. Acta*, **4** (1970), 640.
1301. Reichle, W. T., *Inorg. nucl. Chem. Lett.*, **5** (1969), 981.
1302. Collier, J. W., Fox, A. R., Hinton, I. G., and Mann, F. G., *J. chem. Soc.* (1964), 1819.
1303. Ditts, J. A., and Johnson, M. P., *Inorg. Chem.*, **5** (1966), 2079.
1304. Bennett, M. A., Kneen, W. R., and Nyholm, R. S., *Inorg. Chem.*, **7** (1968), 552.
1305. Lippard, S. J., and Palenik, G. J., *Inorg. Chem.*, **10** (1971), 1322.
1306. Mann, F. G., Wells, A. F., and Purdie, D., *J. chem. Soc.* (1937), 1828.
1307. Dahl, O., and Larsen, O., *Acta. chem. scand.*, **22** (1968), 2037.
1308. Brooks, E. H., Glockling, F., and Hooton, K. A., *J. chem. Soc.* (1965), 4283.
1309. Schweckendiek, G., *Germ. Pat.*, 836 647; *Chem. Abstr.*, **52** (1958), 14679.
1310. Panattoni, G., and Frasson, E., *Acta. crystallogr.*, **16** (1963), 1258.
1311. Cariati, F., and Naldini, L., *Gazz. chim. ital.*, **95** (1965), 201.
1312. Welch, F. J., and Wu, C., *J. org. Chem.*, **30** (1965), 1229.
1313. Beck, W., and Schuierer, E., *Z. anorg. allg. Chem.*, **347** (1966), 304.
1314. Cahours, A., and Gal, H., *Justus Liebigs Annln Chem.*, **155** (1870), 355; **156** (1870), 302.
1315. Naldini, L., Malatesta, L., Cariati, F., and Simonetta, G., *Coord. Chem. Rev.*, **1** (1966), 255.
1316. Kowala, C., and Swan, J. M., *Aust. J. Chem.*, **19** (1966), 547.
1317. Meyer, J. M., and Allred, A. L., *J. inorg. nucl. Chem.*, **30** (1968), 1328.
1318. Bellon, P. L., Manassero, M., and Sansoni, M., *Ricerca scient.*, **39** (1969), 173.
1319. Nichols, D. I., and Charleston, A. S., *J. chem. Soc. (A)* (1969), 2581.
1320. Cariati, F., Galzzioli, D., and Naldini, L., *Chimica Ind., Milano,* **52** (1970), 995; *Chem. Abstr.*, **74** (1971), 18869c.
1321. Westland, A. D., *Can. J. Chem.*, **47** (1969), 4135.
1322. Lindet, L., *C. r. hebd. Séanc. Acad. Sci., Paris,* **98** (1884), 1382; **101** (1885), 164.
1323. Arbuzov, A. E., and Zoroastrova, V. M., *Dokl. Akad. Nauk, S.S.S.R.*, **84** (1963), 503.
1324. Arai, G. J., *Recl Trav. chim. Pays-Bas Belg.* **81** (1962), 307.
1325. Bower, L. M., and Stiddard, M. H. B., *J. chem. Soc. (A)* (1968), 706.
1326. Coffey, C. E., Lewis, J., and Nyholm, R. S., *J. chem. Soc.* (1964), 1741.
1327. Kasenally, A. S., Lewis, J., Manning, A. R., Millar, J. R., Nyholm, R. S., and Stiddard, M. H. B., *J. chem. Soc. (A)* (1967), 3407.
1328. Glockling, F., and Wilbey, M. D., *J. chem. Soc. (A)* (1968), 2168.
1329. Mason, R., Robinson, E. B., and Towl, A. D. C., *J. Am. chem. Soc.*, **93** (1971), 4591.
1330. Bradform, C. W., and Nyholm, R. S., *Chem. Commun.* (1967), 384.
1331. Dolcetti, G., Nicolini, M., Giustiniani, M., and Belluco, U., *J. chem. Soc. (A)* (1969), 1387.
1332. Reed, C. A., and Roper, W. R., *J. chem. Soc. (A)* (1970), 506.
1333. Mannon, Kh. A. I. F., *Acta. crystallogr.*, **23** (1967), 649.

1334. Blundell, T. L., and Powell, H. M., *J. chem. Soc. (A)* (1971), 1685.
1335. Malatesta, L., Naldini, L., Simonetta, G., and Cariati, F., *Chem. Commun.* (1965), 212.
1336. Cariati, F., Naldini, L., Simonetta, G., and Malatesta, L., *Inorg. Chim. Acta,* 1 (1967), 24.
1337. McParthin, M., Malatesta, L., and Mason, R., *Chem. Commun.* (1969), 334.
1338. Albano, V. G., Bellon, P. L., Manassero, M., and Sansoni, M., *Chem. Commun.* (1970), 1210.
1339. Perutz, M. F., and Weisz, O., *J. chem. Soc.* (1946), 438.
1340. Mason, R., Robinson, E. B., and Towl, A. D. C., *J. Am. chem. Soc.,* 93 (1971), 4592.
1341. Boschi, T., Crociani, B., Cattalini, L., and Marangoni, G., *J. chem. Soc. (A)* (1970), 2408.
1342. Nichols, D. I., *J. chem. Soc. (A)* (1970), 1216.
1343. Charlton, J. S., and Nichols, D. I., *J. chem. Soc. (A)* (1970), 1484.
1344. Coates, G. E., and Ridley, D., *J. chem. Soc.* (1964), 166.
1345. Deacon, G. B., and Green, J. H. S., *Chem. Ind.* (1965), 1031; *Chem. Commun.* (1966), 629.
1346. Deacon, G. B., and Green, J. H. S., *Spectrochim. Acta.,* A24 (1968), 845.
1347. Deacon, G. B., and Green, J. H. S., *Spectrochim. Acta.,* A24 (1968), 1921.
1348. Schmelz, M. J., Hill, M. A. G., and Curran, C., *J. phys. Chem., Ithaca,* 65 (1961), 1273.
1349. Hatfield, W. E., and Yoke, J. T., *Inorg. Chem.,* 1 (1962), 470.
1350. Evans, R. C., Mann, F. G., Peiser, H. S., and Purdie, D., *J. chem. Soc.* (1940), 1209.
1351. Deacon, G. B., *J. inorg. nucl. Chem.,* 24 (1962), 1221.
1352. Nöth, H., and Vetter, H. J., *Chem. Ber.,* 96 (1963), 1479.
1353. Coates, G. E., and Lauder, A., *J. chem. Soc.* (1965), 1875.
1354. Deacon, G. B., and West, B. O., *J. inorg. nucl. Chem.,* 24 (1962), 169.
1355. Mann, F. G., and Purdie, D., *J. chem. Soc.* (1940), 1230.
1356. Ewart, G., Payne, D. S., Porte, A. L., and Lane. A. P., *J. chem. Soc.* (1962), 3984.
1357. Deacon, G. B., and West, B. O., *J. chem. Soc.* (1961), 5127.
1358. Jain, S. C., and Rivest, R., *Inorg. Chim. Acta.,* 4 (1970), 291.
1359. Davis, A. R., Murphy, C. J., and Plane, R. A., *Inorg. Chem.,* 9 (1970), 423.

PART 3
Transition Metal Complexes Containing Monotertiary Arsines and Stibines

J. C. CLOYD JR and C. A. McAULIFFE

*Department of Chemistry, University of Manchester,
Institute of Science and Technology*

17 INTRODUCTION

The ability of tertiary arsines to coordinate with metallic salts was known long before the development of modern coordination chemistry. Thus, in 1870 Cahours and Gal[1] prepared the adducts of palladium, platinum and gold chlorides with triethylarsine, while in 1908 Dehn and Williams[2] isolated complexes of trimethylarsine with mercuric chloride.

Organoarsines and stibines form most stable complexes with class (b) acceptors, that is those metal ions that lie at the right of the periodic table[3,4]. In general, monodentate arsines and stibines do not form stable complexes with elements of the first transition series in their 'normal' oxidation state, although many stable adducts are known with these elements in low formal oxidation states. The most stable complexes are formed by the heavier transition metal ions, for example Pt and Au, and the current interest in the theory of the nature of metal–phosphorus bonds[5,6] obviously must be extended to metal–arsenic and metal–antimony bonds.

18 TITANIUM, ZIRCONIUM, HAFNIUM

Although $AsPh_3$ and $TiCl_4$ were known to form a complex as early as 1924[7], the nature of the adduct has been thoroughly investigated only recently. When the two reactants are mixed in dry benzene the red-black, 1:1 adduct, $[TiCl_4(AsPh_3)]$ (m.p. 126-8°C), is formed[8]. The complex was characterised by elemental analysis and by magnetic measurements which yielded a room temperature magnetic susceptibility of -0.26×10^{-3} c.g.s. units (calculated value -0.27×10^{-3} c.g.s. units). The report[9] of a 'deep red' $[TiCl_4(AsPh_3)_2]$ complex must be regarded with some suspicion in view of the above data. The lability of $[TiCl_4(AsPh_3)]$ in the equilibrium

$$TiCl_4 + AsPh_3 \rightleftharpoons [TiCl_4(AsPh_3)]$$

TABLE 18 CATALYST SYSTEMS CONTAINING GROUP IVA METAL COMPLEXES

	Components		Use†	Reference
Group IVA	Group VB	Other		
$TiCl_4$	R_3As	Al	P	14
$TiCl_3$	R_3As, R_3Sb	$RAlX_2$	P	15
$TiCl_3$	Ph_3As, Ph_3Sb	Et_3Al	P	16
$TiCl_3$	Ph_3As	CaH_2	P	17
$TiCl_3$	$(CH_2=CH)_3Sb$	Et_3Al	P	18
$TiCl_3$	$Et_3Sb . 2ZnCl_2$	Et_3Al	P	19
$TiCl_3$	$Et_3Sb . ZnCl_2$	Et_3Al	P	20
‡	R_3As, R_3Sb	none	X	21
$TiCl_3$	Et_3Sb, Et_3Bi	none	M	22
$TiCl_3$	$(C_6H_{13})_3As$	$NaAlEt_4$	P	23
π-allylic compound	R_3As	Lewis acid	P,O	24
$3TiCl_3-AlCl_3$	R_3As, R_3Sb	$RAlX_2$	P	25
$TiCl_4$	Et_3Sb	$LiAlH_4$	A	26
$ZrCl_3, TiCl_3$	Ph_3Sb	Na	P	27
π-allylic compound	R_3As, R_3Sb	R_2AlX	P,O,D	28
$Ti(OBu^n)_4$	Et_3As, Et_3Sb	$EtAlCl_2$	A	29
Ti halide	R_3As, R_3Sb	Na, Li, K, Mg, Zn	A	30
§	R_3As, R_3Sb	Na, Li, K, Al, Mg, Zn, Cu	P	31
$Ti(BH_4)_3$	Bu_3Sb	none	P	32
$TiCl_4$	$Ph_2BiH, (BuO)_2SbH$	none	P	33
$TiCl_3, (BuO)_2TiCl_2$	R_3As, R_3Sb	$RAlCl_2$	P	34

† P—monoolefin polymerisation; O—monoolefin oligomerisation; X—oxo process;
   D—alkadiene polymerisation; M—methyl methacrylate polymerisation;
   A—1-olefin polymerisation.
‡ complex unspecified.
§ halide, alkyl halide, or acetylacetonate of Ti or Zr.

is indicated by the failure of this solution to follow Beer's law[8]. At low concentrations the solvent (benzene) must be competing effectively with the $AsPh_3$ in complex formation. The corresponding bromide adduct has also been reported[10], but an analysis of the far-i.r. spectra of the bromide and chloride has failed to yield an unambiguous assignment of the geometry of $[TiX_4(AsPh_3)]$ [11].

When an excess of AsH_3 was condensed onto $TiCl_4$ at $-196°C$ and allowed to warm to $-112°C$, the bright yellow $[TiCl_4(AsH_3)]$ formed, which was stable at room temperature only under about 80 cm Hg (1.1 kN m^{-2}) pressure or at 'lower temperatures' at which it could be sublimed[12]. The Raman spectrum of this complex was recorded, and, mainly on the basis of a comparison with the spectrum of the corresponding PH_3 complex, $[TiCl_4(AsH_3)]$ was considered to possess a trigonal-bipyramidal structure with the AsH_3 occupying an equatorial position.

The reaction between $SbPh_3$ and $TiCl_4$ in dry benzene is reported[8] to give initially a deep violet coloured solution from which a gummy black solid is quickly deposited. No characterisable product could be isolated. The failure to isolate a $SbPh_3$ complex may be due to either the expected poorer donor ability of $SbPh_3$ compared with $AsPh_3$, or a reduction of $TiCl_4$ by the $SbPh_3$.

Whereas no other monotertiary arsine or stibine complexes with titanium (or zirconium or hafnium) compounds have been reported, a surprisingly large number of patents have dealt with catalyst systems employing these compounds (table 18). These catalyst systems often contain Lewis acids in addition to the Lewis bases. The function of the Lewis acid is often not clearly delineated, but its presence has as pronounced an effect on the catalytic activity as does the Lewis base[13,24,28].

19 VANADIUM, NIOBIUM, TANTALUM

19.1 VANADIUM

Vanadium halides exhibit a strong tendency to undergo reduction in the presence of arsines and this fact probably accounts for the paucity of information on this system. Vanadium tetrachloride is reported[35] to undergo reduction with $AsPh_3$ as indicated by analytical and magnetic measurements. Although e.s.r. measurements on a THF solution containing VCl_3 and $Bu^n_2AsCH=CH_2$ indicate the presence of a *bis* adduct[36], the existence of a V(III) species is thought to be unlikely. Perhaps a complement to the above observations is the one that no stable complex was isolated between $VOCl_2$ and $AsPh_3$, although an interaction was presumed to occur[37].

As was the case with titanium (chapter 18) there is a paucity of definitely characterised complexes, but a relatively active interest in catalysis systems containing vanadium and monotertiary arsines[21,24,28,31,33,34] exists.

The arsine and stibine-substituted carbonyls or mixed carbonyl–nitrosyls of vanadium might be expected to provide fruitful research in this area, but only the di-μ-arsenido-bridged compound, $[(OC)_4VAsPh_2]_2$[38], has been prepared. Patent literature has alluded to the existence of $[V(CO)_4(AsPh_3)_2]$[39,40] and $[V(CO)_4(SbPh_3)_2]$[39-42], the latter compound having been described as pyrophoric[39]. These compounds are easily prepared from, for instance, the ligand and $V(CO)_6$ in hexane at room temperature[39]. Other zero-valent vanadium compounds with arsines and stibines have been mentioned briefly[43].

19.2 NIOBIUM

Niobium pentachloride reacts with triphenylarsine, -stibine, and -bismuthine to give both the mono- and disubstituted complexes $NbCl_5.L$ and $NbCl_5.2L$[44]. Only the elemental analyses (C, H, Cl, Nb) and the preparations are reported for these compounds, and thus it is not possible to assign a structure, though other heavier members of the transition metal series can exhibit coordination numbers of seven.

A patent[41] has indicated the possibility of monotertiary arsine and stibine-substituted $Nb(CO)_6$.

No other monotertiary arsine or stibine-substituted compounds with the members of Group VA have been reported.

20 CHROMIUM, MOLYBDENUM, TUNGSTEN

20.1 CHROMIUM

20.1.1 Halide and other complexes

No simple adducts of monotertiary arsines or stibines with chromium halides have been reported, although interaction must occur, since mixtures of these compounds are reported to be efficient catalysis systems[21,24,28,31,33]. Triphenylarsine is reported to interact with CrO_2Cl_2 [45], but the coordination to the arsenic atom may be through oxygen, since hydrolysis of the adduct leads to $Ph_3As=O$ as one of the products.

The surprisingly air stable, red crystalline complex obtained from $CrCl_3$ and o-lithiobenzyldimethylarsine (CL), is an example of arsine bonded to Cr(III) [46]. The room temperature magnetic moment is 3.80 B.M. for this octahedral complex.

(CL)

20.1.2 Carbonyl complexes

Chromium hexacarbonyl will react with monotertiary arsines and stibines to form mono- and disubstituted complexes (table 19), although forcing conditions must be employed. For the monosubstituted derivative reaction in a high-boiling solvent (for example, diglyme) is required[47-53]; for the disubstituted derivatives reaction is best effected without solvent in a sealed tube at elevated temperatures[53]. The basicity of the ligand may govern the degree and ease of substitution[54].

A more facile method of preparing these compounds is by the u.v. irradiation of a solution (for example THF) of the reactants[55], or by reaction of the ligand with a suitable precursor such as $[(THF)Cr(CO)_5]$ [56] or $[(MeSCH_2CH_2SMe)Cr(CO)_4]$ [57]. The trisubstituted derivative fac-$[L_3Cr(CO)_3]$ can be obtained by this method from $[(olefin)Cr(CO)_3]$ [53,58] (where 'olefin' is, for example, a π-bonded aryl). The utility of these methods for preparing the monosubstituted derivatives is indicated by the preparation of $[(H_3As)Cr(CO)_5]$ [59]. An interesting variation[60] on these methods is illustrated by equation (1)

$$(OC)_5 CrS(CH_3)Sn(CH_3)_3 + (CH_3)_2 AsCl \rightarrow$$
$$\text{(a)}$$
$$(OC)_5 CrAs(CH_3)_2 SCH_3 + ClSn(CH_3)_3 \qquad (1)$$
$$\text{(b)}$$

Compound (a) contains chromium bonded to sulphur; on reaction with dimethylchloroarsine, trimethyltin chloride is split out, and the newly formed ligand, $(CH_3)_2AsSCH_3$, is bonded to chromium through the arsenic.

Although Booth[61] earlier referred to the existence of arsenido-bridged carbonyls of chromium, the reaction of $Cr(CO)_6$ with As_2Me_4 is reported[62] to give the diarsine-bridged (CLIa) and a polymeric material, $[Cr(CO)_4As_2Me_4]_m$ ($m \approx 13$). The di-μ-arsenido-bridged compound (CLIb) was not isolated so that its existence

TABLE 19 GROUP VIA METAL CARBONYL SUBSTITUTION PRODUCTS WITH MONODENTATE TERTIARY ARSINES AND STIBINES

Complex	Colour	M.p.(°C)†	Dipole moment (D)	Ref.
$[Cr(CO)_5AsMe_2(SMe)]$	pale yellow	53	–	61
$[Cr(CO)_5AsMe_3]$	colourless	51	–	61
$[Cr(CO)_5As(c\text{-}C_6H_{11})_3]$	bright yellow	150	–	59
$[Cr(CO)_5(AsMe_2Cl)]$	yellow	59	–	59
$[Cr(CO)_5AsH_3]$	yellow	69d.	3.99	57
$[Mo(CO)_5AsH_3]$	yellow	71d.	–	57
$[W(CO)_5AsH_3]$	bright yellow	104d.	–	57
$[W(CO)_5As(c\text{-}C_6H_{11})_3]$	colourless	155	–	59
cis-$[Cr(CO)_4(SbPh_3)_2]$	yellow	–	–	58
cis-$[Cr(CO)_4(AsMe_2Ph)_2]$	lemon yellow	93	6.7	54
cis-$[Mo(CO)_4(AsMe_2Ph)_2]$	off-white	82	6.8	54
cis-$[Mo(CO)_4(SbPh_3)_2]$	–	–	6.41	72
cis-$[W(CO)_4(AsMe_2Ph)_2]$	pale yellow	97	7.1	54
fac-$[Cr(CO)_3(AsMe_2Ph)_3]$	lemon yellow	145–150d.	7.4	54
fac-$[Mo(CO)_3(AsMe_2Ph)_3]$	off white	121	7.7	54
fac-$[Mo(CO)_3(SbPh_3)_3]$	–	–	6.93	72
fac-$[W(CO)_3(AsMe_2Ph)_3]$	off white	134	8.3	54
trans-$[Mo(CO)_2(SbPh_3)_4]$	yellow	–	0	72
cis-$[Mo(CO)_2(SbPh_3)_4]$	pale yellow	–	4.37	72
$[Mo_2(CO)_{10}(AsMe_2)_2]$	yellow	117	3.9	85
$[W_2(CO)_{10}(AsMe_2)_2]$	yellow	145	3.1	85
$[Mo(CO)_4AsMe_2]_2$	orange	310–315d.	1.1	85
$[W(CO)_4AsMe_2]_2$	orange	>350	1.1	85

† d. indicates decomposition.

is more doubtful than that of the corresponding phosphine complex[62]. The electrochemistry of some of these complexes has been investigated briefly elsewhere[63,64].

(CLI a)   (CLI b)

The pentacyclic arsine, $(AsCH_3)_5$, reacts with $Cr(CO)_6$ to give
$[(AsCH_3)_5\{Cr(CO)_5\}_2]$ in which the $(AsCH_3)_5$ ring is bonded to two $Cr(CO)_5$
moieties in a fashion similar to that in $(CLIa)$[65]. If an extended reflux is employed,
a compound whose elemental analysis and molecular weight indicates a
formulation $[Cr(CO)_4(AsCH_3)_5]_2$ is obtained. A possible structure is suggested
involving one broken (μ-arsenido) and one bridging $(AsCH_3)_5$ ring.

20.2 MOLYBDENUM AND TUNGSTEN

20.2.1 Halide complexes
The red, paramagnetic complexes $[MoCl_4(AsPh_3)_2]$ (μ_{eff} = 2.36 B.M.) [66] and
$[MoCl_4(AsPr^n_3)_2]$ (μ_{eff} = 2.40 B.M.) [67] are obtained by displacing alkyl cyanide
from $[MoCl_4(RCN)_2]$ by the appropriate arsine. The depressed magnetic moment
is presumed to be due in part to spin–orbit coupling[66]. Only two equivalents of
RCN are displaced from $[MoBr_3(RCN)_3]$ by $AsPh_3$ to give the orange-yellow
$[MoBr_3(AsPh_3)_2(RCN)]$ (μ_{eff} = 3.83 B.M.) [66]. If $MoCl_5$ is first added to ethanol,
and then $AsMe_2Ph$ added to the resulting brown oil, the blue-green, diamagnetic
complex $[MoOCl_2(AsMe_2Ph)_3]$ (ν(Mo—O) at 952 cm^{-1}) is obtained[68]. The
octahedral $[MoCl_4(AsR_3)_2]$ are presumed to possess a *trans* configuration; the
position of ν(Mo—O) in $[MoOCl_2(AsMe_2Ph)_3]$ suggests that the oxygen is *trans* to
one of the chlorine atoms.

20.2.2 Carbonyl complexes
In general, the substitution in $Mo(CO)_6$ and $W(CO)_6$ by monotertiary arsines and
stibines is analogous to the substitution in $Cr(CO)_6$ [47,48,53-7,59]. The *cis* configuration appears to be the most stable isomeric form for the disubstituted
complexes, since *trans*-$[Mo(CO)_4(AsEt_3)_2]$ isomerises[68] rapidly at 40°C in
pentane solution. The trisubstituted compounds prefer the *fac* configuration.
With certain ligands $(As(OMe)_3, AsEt_3)$, $[Mo(CO)_3L_3]$ disproportionates to give
$[Mo(CO)_4L_2]$ and $[Mo(CO)_5L]$ [68,69]. The fact that steric hindrance may partly
govern the degree of substitution of $M(CO)_6$ [54,70] is supported by the existence
of *cis*- and *trans*-$[Mo(CO)_2(SbPh_3)_4]$ [71]. Apparently the larger antimony atom
provides less steric crowding by holding the attached phenyl groups further apart
from each other. The monosubstituted complexes have been used in analyses of ν(C—O)
frequencies[72-9], and in mechanistic investigations[70,77-9]. Similar analyses (*vide infra*)
have been performed for *cis*-$[Mo(CO)_4(SbEt_3)_2]$ and *fac*-$[Mo(CO)_3(SbEt_3)_3]$ [80].
The kinetics of the substitution of $[Mo(CO)_4(cycloocta-1,5-diene)]$ by $AsPh_3$
and $SbPh_3$ [81] and the kinetics of substitution of *cis*-$[Mo(CO)_4L_2]$ by
α,α'-bipyridine [82] have been investigated. The substitution by $AsPh_3$ and $SbPh_3$
of $[(o\text{-phenanthroline})Mo(CO)_4]$ proceeds more readily than with $Mo(CO)_6$,
presumably owing to the presence of the two 'hard' nitrogen atoms[83,84].

The hexacarbonyls react with diarsines, As_2R_4, to give complexes[62,85] of
structures (CLIa) and (CLIb). The di-μ-arsenido bridged (CLIa) is obtained with Mo and

W in contrast to chromium, although the polymeric ($m \approx 20$) complex $[Mo(CO)_4As_2Me_4]_m$ was also obtained[62]. The electrochemistry of these compounds has also been investigated[63,64,86].

20.2.3 Carbonyl halide complexes

The carbonyl halide complexes (table 20) $[MX_2(CO)_3L_2]$ (M = Mo, W; X = halogen; L = AsR_3, SbR_3) have been prepared by halogen oxidation of

TABLE 20 MONODENTATE TERTIARY ARSINE AND STIBINE SUBSTITUTION PRODUCTS WITH GROUP VIA METAL HALIDE CARBONYLS AND HALIDE NITROSYLS

Complex	Colour	M.p.(°C)[†]	Carbonyl or nitrosyl stretching frequency (cm^{-1})	Ref.
$[MoCl_2(CO)_3(AsPh_3)_2]$	yellow		2038,1965,1925	91
$[MoCl_2(CO)_3(SbPh_3)_2]$	orange		2035,1965,1920	90
$[MoBr_2(CO)_3(AsPh_3)_2]$	orange		2020,1960,1915	90
$[MoBr_2(CO)_3(SbPh_3)_2]$	orange		2020,1960,1915	90
$[MoBr_2(CO)_3(AsMe_2Ph)_2]$	yellow	95–110d.		88
$[MoI_2(CO)_3(AsPh_3)_2]$	yellow			89
$[MoI_2(CO)_3(SbPh_3)_2]$	red			89
$[MoI_2(CO)_3(AsMe_2Ph)_2]$	light brown	99–110d.		88
$[WCl_2(CO)_3(AsPh_3)_2]$	bright yellow		2020,1940,1900	93
$[WCl_2(CO)_3(SbPh_3)_2]$	yellow		2015,1935,1900	92
$[WBr_2(CO)_3(AsPh_3)_2]$	orange		2015,1940,1900	92
$[WBr_2(CO)_3(SbPh_3)_2]$	orange		2015,1940,1905	92
$[WBr_2(CO)_3(AsMe_2Ph)_2]$	yellow	110–120d.		88
$[WI_2(CO)_3(AsPh_3)_2]$	yellow			89
$[WI_2(CO)_3(SbPh_3)_2]$	red			89
$[WI_2(CO)_3(AsMe_2Ph)_2]$	yellow	124–135d.		88
$[MoCl_2(NO)_2(AsPh_3)_2]$	green	220		97
$[MoBr_2(NO)_2(AsPh_3)_2]$	green	237		96
$[MoI_2(NO)_2(AsPh_3)_2]$	purple	242		96
$[WCl_2(NO)_2(AsPh_3)_2]$	green	278		97
$[WBr_2(NO)_2(AsPh_3)_2]$	green	>300		96

[†] d. indicates decomposition.

$[M(CO)_4L_2]$ [87], carbon monoxide displacement from $[M(CO)_4X_2]$ [88–91] or by the disproportionation reaction of the incomplete equation (2)[92].

$$[Me_6C_6W(CO)_3Cl]BPh_4 + L \xrightarrow{LiCl} [WCl_2(CO)_3L_2] + [Me_6C_6W(CO)_3] \quad (2)$$

The complexes are non-conducting, diamagnetic, seven-coordinate monomers. They differ from their phosphine analogues in that none of them lose CO to form $[MX_2(CO)_2L_2]$ [87,88]. The molybdenum compounds react with sodium salts of dialkyldithiocarbamates, R_2NCS_2Na (R = CH_3, C_2H_5; R_2N = pyrrolidyl), with expulsion of one equivalent of L and CO to form $[Mo(CO)_2L(R_2NCS_2)_2]$ [93]. The anionic complexes $[Mo(CO)_3LX_3]^-$ (L = $AsPh_3$, $SbPh_3$; X = Cl, Br) can be obtained from either $[Et_4N][Mo(CO)_4X_3]$ and L, or from $[MoX_2(CO)_3L_2]$ and

Et_4NX [94]. These complexes are isolated in two isomeric forms, distinguished by their characteristic carbonyl stretching frequencies.

20.2.4 Nitrosyl halide complexes

These complexes of general formula $[MX_2(NO)_2L_2]$ (M = Mo, W; X = halogen; L = $AsPh_3$, $SbPh_3$) can be prepared by either refluxing benzene solutions of $[M(NO)_2X_2]$ and L [95,96] or by passing NO through benzene solutions of the corresponding tricarbonyl $[MX_2(CO)_3L_2]$ [97]. The latter route is not stereo-specific, and it is possible to isolate all three *cis* NO isomers (CLIIa–c). The *cis* position for the nitrosyls was deduced from the appearance of two $\nu(N-O)$ in the i.r. spectra (one $\nu(N-O)$ would be expected for *trans* nitrosyls). These authors[97] suggest that the isomer produced by the former route is that of (CLIIa). A review of the halocarbonyls and nitrosyls of the Group VIA metals has appeared[98].

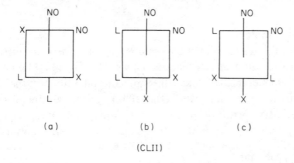

(a)   (b)   (c)

(CLII)

21 MANGANESE, TECHNETIUM, RHENIUM

21.1 MANGANESE

21.1.1 Halide complexes

Although these types of complexes are mentioned as catalysts[21,22,24,99], reports of isolated complexes are rare. Manganese(II) halides react with $AsPh_3$ in THF to yield $[MnX_2(AsPh_3)_2]$, which are high-spin and probably tetrahedral[100]. In acetone solution the double salts $[MnX_4]$ $[AsPh_4]_2$ are formed[101]. Manganese(III) salts are reduced by tertiary arsines[4].

21.1.2 Carbonyl complexes

The product from the reaction of $AsPh_3$ and $Mn_2(CO)_{10}$ depends critically on the solvent, temperature and mode of reaction (table 21). The reaction can be accomplished photochemically[103] and this appears to be the best method of obtaining the substituted dimer $[Mn(CO)_4AsPh_3]_2$, although the presence of a paramagnetic impurity (possibly the monomer $[Mn(CO)_5AsPh_3]$) is indicated by the slight paramagnetism (0.38 B.M.). Reactions in high-boiling solvents give either the monomer[102,107], the dimer[104] or the di-μ-arsenido-bridged dimer $[Mn(CO)_4AsPh_2]_2$ [103,105-7]. The situation does not yet seem to be clear, however, since the yields of the arsenido-bridged dimers depend on both solvent and temperature[105], and paramagnetic monomers may be present in the products from photochemically induced reactions[103]. Furthermore, the dimeric (solid state) arsenido-bridged $[Mn(CO)_4As(p\text{-}MeC_6H_4)_2]_2$ is monomeric in molten naphthalene[107]. Whereas the existence of the monomeric $[Mn(CO)_4AsR_2]$ has not been considered in the studies of table 21, the possible presence of this species may further complicate matters. Finally in a kinetic investigation of the substitution of $Mn_2(CO)_{10}$ by $AsPh_3$ in high-boiling solvents, evidence was found only for the presence of the substitution products of the dimeric $Mn_2(CO)_{10}$ and not for the radical $[Mn(CO)_4L]$ [108]. The dimers $[Mn(CO)_4AsPh_3]_2$ [109] and $[Mn(CO)_4(AsPh_2)]_2$ [107] have been prepared by independent methods; the i.r. spectra of the former[110] and the latter[111,112] have been investigated.

21.1.3 Substitution products of manganese carbonyl halides

The reaction of $[Mn(CO)_5X]$ with $AsPh_3$ or $SbPh_3$ to give $[MnX(CO)_4L]$ takes place by an S_N1 mechanism, with rates decreasing in the order $Cl > Br > I$ [113]. Substitution of a second CO ligand to form $[MnX(CO)_3L_2]$ is generally more difficult[114-16] than for monosubstitution[117]. The rate of CO exchange[118] in, and the rate of formation of, the disubstituted products have been investigated[119]; the

TABLE 21 THE REACTION OF Ph₃As WITH Mn₂(CO)₁₀

Compound[†]	Colour	M.p.(°C)[‡]	Analyses	Molecular calc.	Weight found	Reaction conditions	Ref.
[Mn(CO)$_4$(AsPh$_3$)]	bright orange	–	As,Mn	473.1	462§	16–18 h at 120° in p-xylene	102
[Mn(CO)$_4$(SbPh$_3$)]	bright yellow	–	Sb,Mn	620.0	631§	20 h at 90° in p-xylene	102
[Mn(CO)$_4$(AsPh$_3$)]$_2$ ‖	dark orange	155	C,H,Mn	946	840§	22 h u.v. irradiation in sealed tube	103
[Mn(CO)$_4$(AsPh$_3$)]$_2$		161–162	C,H	–	–	u.v. irradiation in c-C$_6$H$_{12}$ or 120° in xylene, both in sealed tube	104
[Mn(CO)$_4$(AsPh$_2$)]$_2$	yellow	255–260d.	C,H,As	–	–	5 h at 140° in evacuated tube	103
[Mn(CO)$_4$(AsPh$_2$)]$_2$	yellow	255–260d.	C,H,Mn,As	792	773¶	2 h reflux (139°) in xylene	105
[Mn(CO)$_4$(AsPh$_2$)]$_2$			–	–	–	from Ph$_2$AsH and Mn(CO)$_5$Cl in refluxing C$_3$H$_6$ (1 h) in presence of p-toluidine	106

† these are the formulations of the authors.
‡ d. indicates decomposition.
§ in benzene.
‖ this compound is slightly paramagnetic, μeff = 0.38 B.M.
¶ in naphthalene.

latter study showed that replacement with PPh_3 is easier than with either $AsPh_3$ or $SbPh_3$, even though $v(C-O)$ frequencies indicate little differences in π-bonding abilities in the ligands. The far-i.r. spectra of some of these complexes have been reported[120], and the rates of reaction of $AsPh_3$ and $SbPh_3$ with $[Et_4N][cis\text{-}Mn(CO)_4X_2]$ have been investigated[121]. In a general review of Lewis base–metal carbonyl complexes, Manuel[122] has included a section on manganese carbonyl halides.

Wojcicki and Farona have discussed[123,124] the $AsPh_3$ and $SbPh_3$ complexes, $[Mn(CO)_4(SCN)L]$, $cis\text{-}[Mn(CO)_3(SCN)L_2]$ and $trans\text{-}[Mn(CO)_3(NCS)L_2]$. In the monosubstituted and cis-disubstituted complexes the thiocyanate is $trans$ to CO and is S-bonded, whereas in the $trans$ complexes the isothiocyanate is $trans$ to L and is N-bonded. This bonding isomerism is considered to be a result of the different π-bonding abilities of the ligands in the $trans$ position.

The i.r. spectrum (solid state) of the dark yellow non-electrolyte, $[Mn(CO)_4(AsPh_3)NO_3]$, exhibits unusual nitrate absorptions which are taken to be indicative of nitrate bonded to Mn through the nitrogen[125]. The complex, $cis\text{-}[Mn(CO)_3(AsPh_3)(hfac)]$ (hfac = 1,1,1,5,5,5-hexafluoropentane-2,4-dione) has been reported to be an example of Mn(I) bonded to two 'hard' donor atoms[126].

The mono- and disubstituted carbonyl halides react with NO to give the more stable $[Mn(NO)_3L]$ and $[Mn(NO)_2L_2I]$ (L = $AsPh_3$, $SbPh_3$), respectively[127]. The compounds $[Mn(NO)_3AsPh_3]$ and $[Mn(NO)(CO)_3(AsPh_3)]$ have been the subject of a kinetic investigation[128].

These monosubstituted carbonyls also include compounds in which the anionic ligand is Ph_3Ge [129] and Ph_3Sn [130,131]. These compounds have been described as antiknock additives for petrol[132]. Heating $[HMn(CO)_4(SbPh_3)]$ with Et_2Zn . bipy produces $\{Mn(CO)_4(SbPh_3)\}_2Zn$. bipy [133]. Triphenylarsine will also add to $[Ph_3PAuMn(CO)_5]$ to give $[Ph_3PAuMn(CO)_4AsPh_3]$ [134].

21.1.4 Mixed bridged-carbonyl dimers

In addition to the di-μ-arsenidotetracarbonyldimanganese(I) complexes, a number of similarly bridged compounds have been prepared containing only one arsenido bridge and one other different bridging moiety. The sealed tube reaction (18 h at 120–130°C) of $Mn_2(CO)_{10}$ and $(CF_3)_2AsI$ produces the deep red arsenido-iodo-bridged $[Mn(CO)_4As(CF_3)_2I]_2$ [135]. Reaction of this dimer with HgX_2 (X = SCF_3, $SeCF_3$, SMe) gives similarly unsymmetrically bridged species; reaction with $LiPR_2$ (R = Me, Et) gives the mixed arsenido-phosphido-bridged complexes[136]. The structures of these compounds have been confirmed by spectroscopic studies[137].

Oddly, the reaction (cyclohexane; 50°C for 1 h) between Me_2AsI and $Me_3SnMn(CO)_5$ gives a polymeric compound, presumably $[(Me_2As)_2I_2Mn_4(CO)_{12}]$ [138]. The structure is suggested to consist of a $\{Mn(CO)_3\}_4$ tetrahedron, alternatively or consecutively bridged by Me_2As and I moieties.

21.2 TECHNETIUM

Despite the limited availability of this element, some monotertiary arsine and stibine complexes have been studied. The brown complex $[TcCl_4(AsPh_3)_2]$ is reported to be obtained from $TcCl_4$ and $AsPh_3$ in refluxing ethanol[139]. This complex is paramagnetic, μ_{eff} = 3.62 B.M., and its electronic spectrum has been interpreted in terms of octahedral Tc(IV).

The carbonyl chloride $Tc(CO)_4Cl$, reacts readily with $AsPh_3$ or $SbPh_3$ to give the colourless $[Tc(CO)_3L_2Cl]$, for which an octahedral structure is assumed[140]. The chlorocarbonyl also reacts with As_2Ph_4 in refluxing benzene to give the colourless $[Tc(CO)_3ClAsPh_2]_2$; the structure is considered to involve Ph_2As moieties bridging two $Tc(CO)_3Cl$ groups[141].

21.3 RHENIUM

21.3.1 Halide complexes

Rhenium halides (table 22) form complexes with monotertiary arsines and stibines, but the composition of the product is governed by the choice of rhenium reactant. For instance, $ReCl_4$ reacts with $AsPh_3$ in acetone at room temperature to yield the presumably binuclear $[ReCl_3AsPh_3]_2$ [142], although a small amount of contaminating $[ReOCl_3(AsPh_3)_2]$ is also produced. The reduction of $[ReOCl_3(AsMe_2Ph)_2]$ in refluxing benzene with excess $AsMe_2Ph$ gives $[ReCl_3(AsMe_2Ph)_3]$ [143], which has been shown by ^1H n.m.r. to possess the *mer* configuration[144]. This reduction

TABLE 22 RHENIUM COMPLEXES WITH TERTIARY ARSINES AND STIBINES

Complex	Colour	M.p.($^\circ$C)†	Comments	Ref.
$[Re(CO)_4(AsMe_2Ph)]_2$	pale yellow	128		155
$[Re_2(CO)_7(AsMe_2Ph)_3]$	white	150–154		155
cis-$[Re(CO)_3(AsMe_2Ph)_2]$	light yellow	89–92		155
cis-$[Re(CO)_4(AsMe_2Ph)Cl]$	white	77	also Br^-	155
fac-$[Re(CO)_3(AsMe_2Ph)_2Cl]$	white	116	also Br^-	155
$[Re(CO)_3(AsPh_2)Cl]_2$	colourless	259d.	also I^-	141
$[Re(CO)(NO)(AsPh_3)_2Cl_2]$	orange-yellow	270		159
$[ReOCl_3(AsMe_2Ph)_2]$	blue	170–175d.	$\nu(Re—O)$ = 984,989 cm^{-1}	141
$[ReOCl_3(AsPh_3)_2]$	green-yellow	–	$\nu(Re—O)$ = 967 cm^{-1}	150
$[ReOBr_3(AsPh_3)_2]$	yellow	–	$\nu(Re—O)$ = 974,980 cm^{-1}	150
$[ReOCl_3(SbPh_3)_2]$	green	–	$\nu(Re—O)$ = 976 cm^{-1}	150
$[ReOBr_3(SbPh_3)_2]$	green	–	$\nu(Re—O)$ = 969,973 cm^{-1}	150
$[ReOCl_3(AsEt_2Ph)_2]$	turquoise	120–122d.	dipole moment = 8.0 D.	149
$[ReCl_4(AsPh_3)_2]$	scarlet	–		148
$[ReCl_4(SbPh_3)_2]$	royal blue	–		148
$[ReBr_3(AsPh_3)]_n$	purple	–	$n = 3$ (?)	146
$[ReCl_3(AsPh_3)]_n$	yellow-green	–	$n = 2$ (?)	142

† d. indicates decomposition.

ability is generally applicable only to the more basic arsines. The reaction between $ReBr_3$ and $AsPh_3$ in acetone gives the trimeric $[ReBr_3AsPh_3]_3$ [145,146].

Rhenium(IV) chloride complexes can be obtained by reacting $AsPh_3$ or $SbPh_3$ with $[ReCl_4(MeCN)_2]$. The substitution proceeds reluctantly, however, and elevated temperatures are required to obtain scarlet $[ReCl_4(AsPh_3)_2]$ and royal blue $[ReCl_4(SbPh_3)_2]$ [147,148]. The conditions of the reaction are critical, since $[ReCl_4(MeCN)_2]$ tends to disproportionate at elevated temperatures [148].

21.3.2 Oxy-halide complexes

These complexes (table 22) of formulation $[ReOX_3L_2]$ (X = Cl, Br; L = AsR_3, SbR_3) can be prepared by boiling the perrhenate, ligand and the corresponding HX in ethanol [149]. However, this method often involves long periods of reflux and results in low yields. An improvement in yield is realised if the perrhenate and HX aqueous solution are added to glacial acetic acid before treatment with the ligand [150]. Two other more convenient preparations involve the exchange between alkylarsines (for example, $AsMe_2Ph$) and $[ReOCl_3(PPh_3)_2]$ [151], and the reaction of excess $AsPh_3$ with $[Bu^n_4N][ReBr_4O]$ [152].

The reduction of $[ReOCl_3(AsEt_2Ph)_2]$ with $LiAlH_4$ gave a crude product, which could not be obtained pure, but which was tentatively identified as $[ReH_7(AsEt_2Ph)_2]$ [153]. The isolation of $[ReH_5(AsPh_3)_3]$ from the reaction of $(Et_4N)_2ReH_9$ and $AsPh_3$ in refluxing isopropanol [154] tends to support this heptahydrido-formulation, since arsines are known to react with metal multi-hydrides to substitute H_2 for one ligand. The hydride $[Et_4N][ReH_8(AsPh_3)]$ has also been isolated [154].

21.3.3 Complexes containing carbonyl

The substitution of $Re_2(CO)_{10}$ by $AsMe_2Ph$ is accomplished by u.v. irradiation [155]. The product obtained depends on the length of irradiation in a situation somewhat similar to the substitution of $Mn_2(CO)_{10}$. Three products can be isolated: (1) $[Re(CO)_4L]_2$, (2) $[Re_2(CO)_7L_3]$ and (3) $[Re(CO)_3L_2]$ (table 22). There is a suggestion that (1) contains some paramagnetic species in CCl_4 solution, since the ^1H n.m.r. signal is a broad ill-defined singlet. Compound (2) is described as para-magnetic. These products react with either halogens or chlorinated hydrocarbon solvents (for example, CH_2Cl_2) to give the carbonyl halides $[Re(CO)_4LX]$ or $[Re(CO)_3L_2X]$ (X = Cl, Br) [155].

The reaction between $[Re(CO)_5X]$ (X = Cl, I) and As_2Ph_4 leads to the di-μ-arsenido-bridged dimer $[Re(CO)_3XAsPh_2]_2$ [141] analogous to the corresponding technetium dimer. If a 1:1 mole ratio of reactants is employed, there is spectro-scopic evidence for $[I(CO)_4RePh_2AsAsPh_2]$ [141]. Kinetic data has been recorded for the substitution of $[Re(CO)_4(AsPh_3)Cl]$ by other ligands [156].

Treatment of $[Re(CO)_4(AsMe_2Ph)Br]$ with hydrazine is reported [157] to give a molecular nitrogen complex, $[Re(CO)_3(NH_2)(N_2)(AsMe_2Ph)]$. The compound has not been fully characterised, and its precise nature must await further investigation.

The monosubstitution products of $[Ph_3MRe(CO)_5]$ (M = Ge, Sn) with $AsPh_3$ and $SbPh_3$ may be obtained directly by heating the reactants together at elevated ($> 200°C$) temperatures[158]. The reaction of $AsPh_3$ with $[Re(CO)(NO)(cyclooctene)Cl_2]_2$ breaks the chloride bridges and displaces the olefin to give $[Re(CO)(NO)Cl_2L_2]$ [159].

22 IRON, RUTHENIUM, OSMIUM

22.1 IRON

22.1.1 Complexes that do not contain CO or NO

Compounds of iron with monotertiary arsines and stibines have been reported as catalysts[21,24,28]. It is probable that one such system contains an $Fe(0)$–$SbPh_3$ complex as the catalytic agent[160]. It may also be mentioned here that an 'Fe(III)–triethanolamine–EDTA' complex is reported to remove AsH_3 from the air[161]. The process must be considered to be catalytic since on complexation with AsH_3 the Fe(III) is reduced to Fe(II), and the Fe(III) can then be regenerated with oxygen.

Other Fe(II) complexes containing $AsPh_3$ or $SbPh_3$ are $[FeL_5(EPh_3)](ClO_4)_2$ (L = p-tolylisocyanide; E = As, Sb) formed by the displacement of the weakly coordinated perchlorate in $[FeL_5(OClO_3)]ClO_4$ [162] and $Na_3[Fe(CN)_5(EPh_3)]$ produced by ligand substitution with $Na_3[Fe(CN)_5(NH_3)]$ [163]. The paramagnetic Fe(III) complexes $Na_2[Fe(CN)_5(EPh_3)]$ (E = As, μ_{eff} = 2.44 B.M.; E = Sb, μ_{eff} = 2.12 B.M.) can be obtained from the corresponding Fe(II) complexes by bromine oxidation[163]. The Mössbauer spectra of $[Fe(CN)_5(EPh_3)]^{n-}$ (n = 2, 3) have been studied[164,165], and the thermal decomposition of $[Fe(CN)_3(EPh_3)]^{3-}$ has been reported[166]; these complexes react with HCl to give $H_3[Fe(CN)_5(EPh_3)]$ [167].

Ferric chloride with methyldiphenyl- or dimethyltolylarsine yields complexes that dissociate so readily in solution that their constitution is not certain; $2FeCl_3 . 4L$ (L = arsine) may be formulated as $[FeCl_3L_2]_2$ or $[FeCl_2L_4][FeCl_4]$ [168]. In ethyl acetate triphenylarsine forms a complex of probable formula $[FeCl_3(AsPh_3)_2]_2$ [169]. The dark yellow $[FeCl_3(AsPh_3)]$ can be obtained quite readily from refluxing solutions of $[Fe(CO)_4(AsPh_3)]$ in chlorinated solvents (for example $CHCl_3$) [170]. It is suggested that this complex is spin-free and tetrahedral on the basis of magnetic measurements, and these conclusions have been supported by Mössbauer studies[171].

Arsine and stibine ligands will cleave the dimeric $[Fe(S_2C_2R_2)_2]_2^{n-}$ (n = 0, 1, 2) compounds to give the monomeric five-coordinate species $[LFe(S_2C_2R_2)_2]^{m-}$ (m = 0, 1). When n = 0 (R = CF_3) and L = $AsPh_3$ or $SbPh_3$, the monomer with m = 0 is obtained directly[172]; when n = 1 (R = CF_3) and L = $AsPh_3$, the monomer (m = 0) can also be obtained from the reaction according to equation (3)[172]

$$2[Fe(S_2C_2R_2)_2]_2^- + AsPh_3 \rightarrow [Fe(S_2C_2R_2)_2]_2^{2-} + [(AsPh_3)Fe(S_2C_2R_2)_2]$$

$$(3)$$

For n = 2 and L = $AsPr^n_3$ the monomer with m = 1 is obtained[173]. Although in this last reaction no mention is made about exclusion of oxygen from the system, the same reaction repeated in the presence of air is said to give the monomer with $m = 1$ and the ligand in the final form as $Ph_3As=O$ [174].

22.1.2 Carbonyl complexes

The direct substitution of iron carbonyls by monotertiary arsines or stibines is rather difficult[61], but substitution (table 23) to afford either $[Fe(CO)_4L]$ or $[Fe(CO)_3L_2]$ can be readily accomplished by either irradiation of mixtures of the reactants[175-7], or by refluxing the reactants in a suitable solvent (for example, THF)[178,179]. In the latter procedure either $Fe(CO)_5$ or $Fe_3(CO)_{12}$ may be used as the iron carbonyl[179]. The mono- and disubstituted compounds can also be obtained from the reaction (0°C to room temperature) between $H_2Fe(CO)_4$ and $AsPh_3$ [180]. These compounds have been studied spectroscopically[72,76].

Products of substitution from iron pentacarbonyl and triphenylstibine were first described by Reppe and Schweckendiek[181,182] in catalytic studies. Other early work included the preparation of $[Fe(CO)_3(SbCl_3)_2]$ [183].

The mono- and disubstituted $AsPh_3$ complexes $[Fe(CO)_4L]$ and $[Fe(CO)_3L_2]$ will dissolve in 98 per cent H_2SO_4 to give slightly stable (the solutions decompose within 10 min) protonated species, identified by their ^1H n.m.r. spectra[184]. The protonated complex derived from $[Fe(CO)_3(SbPh_3)_2]$ decomposes even more rapidly (within a minute). The adduct formed between HgX_2 (X = hal) and $[Fe(CO)_3L_2]$ (L = $AsPh_3$, $SbPh_3$) is presumed to be a metal–donor type of complex, that is $L_2(CO)_3Fe \rightarrow HgX_2$ [185]. These adducts are said to be more stable than the parent carbonyls $(CO)_5Fe \rightarrow HgX_2$.

The substitution of the carbonyl bromide $[Fe(CO)_4Br_2]$ by $AsPh_3$ to give $[Fe(CO)_3Br_2(AsPh_3)]$ is thought to proceed through an intermediate which involves the linkage[186]

$$ Fe \underset{Br}{\overset{Br}{<}} \hspace{-6pt} > As $$

A similar type of intermediate was postulated in the reaction of $H_2Fe(CO)_4$ with $AsPh_3$ [180]. The preparation of the disubstituted carbonyl halides $[Fe(CO)_2X_2L_2]$ (L = $AsPh_3$, $SbPh_3$) has also been described[187].

The di-μ-arsenido-bridged iron carbonyls $[Fe(CO)_3AsR_2]_2$ are obtained from the reaction of $Fe_3(CO)_{12}$ and $As_2(C_6F_5)_4$ in refluxing toluene[188]. Curiously, the only product from $Fe(CO)_5$ and As_2Me_4 (180°C for 20h) was $[(OC)_4FeMe_2AsAsMe_2Fe(CO)_4]$ [85]. A similar arsenido-bridged compound is obtained from the reaction between $Fe(CO)_5$ and $(CF_3)_2AsI$ [189]. The product $[Fe(CO)_3IAs(CF_3)_2]_2$ is assumed to contain non-bridging iodine atoms, which instead occupy a sixth coordination position about the Fe atoms. In view of the difficulties experienced in substituting iron carbonyls with arsines directly, it is not surprising that only one $AsPh_3$ (or $SbPh_3$) will undergo substitution with $[Fe(CO)_3(SEt)]_2$ in a refluxing hydrocarbon to give the unsymmetrically substituted compound $[Fe_2(CO)_5(AsPh_3)(SEt)_2]$ [190].

TABLE 23 IRON COMPLEXES WITH TERTIARY ARSINES AND STIBINES

Compound	Colour	M.p.(°C)†	Comments	Ref.
[Fe(CO)₄AsPh₃]	pale yellow	160-161,175-176d.	ν(C—O) = 2041,1961,1923 cm⁻¹	175,178,
	yellow	178	ν(C—O) = 2063,1985,1946 cm⁻¹	179,184
[Fe(CO)₃(AsPh₃)₂]	yellow	170-172,193-195d.	ν(C—O) = 1876 cm⁻¹	175,178,
	golden yellow	225-226d.	ν(C—O) = 1894 cm⁻¹	179,184
[Fe(CO)₄SbPh₃]	yellow-brown	136	ν(C—O) = 2058,1985,1946 cm⁻¹	178
[Fe(CO)₃(SbPh₃)₂]	yellow-brown	196d.	ν(C—O) = 1873 cm⁻¹	178,184
[Fe(CO)₄AsMe₃]	colourless	124		176
[Fe(CO)₄AsEt₃]	yellow	68		176
[Fe(CO)₄As(c-C₆H₁₁)₃]	bright yellow	176		176
[Fe(CO)₃{As(c-C₆H₁₁)₃}₂]	bright yellow	259		176
[Fe(CO)₃(AsPh₃)₂(HgBr₂)]	yellow	96-98d.		185
[Fe(CO)₃(SbPh₃)₂(HgBr₂)]		118-125d.		185
[Fe(CO)₃(AsMe₂)]₂	orange	>350		185
[Fe(CO)₃{As(C₆F₅)₂}₂]₂		252		188
Hg[Fe(CO)₂(NO)AsPh₃]₂		170d.		212
Hg[Fe(CO)₂(NO)SbPh₃]₂		175d.		212
[Fe(NO)₂(AsPh₃)Br]		140-143		208
[Fe(NO)₂(AsPh₃)I]		96-98		208
[Fe(NO)₂(SbPh₃)I]		74-76		208
[Fe(NO)₂(AsPh₃)CN]		110d.		209
[Fe(NO)₂(AsMe₂)]₂	dark red	227-230	ν(N—O) = 1757,1727 cm⁻¹	206
[Fe(NO)₂(AsPh₂)]₂	purple-red	252d.		207
[Fe(NO)₂(AsPh₂)]₂	black	125-127		207
Bu₄N[FeS₄C₄(AsPrⁿ₃)(CN)₄]	green-black	150-180		173
[FeS₄C₄(AsPh₃)(CF₃)₄]	black	158-159.5		172
[FeS₄C₄(SbPh₃)(CF₃)₄]	black	171-173		172

† d. indicates decomposition.

The compound $[Fe_3(CO)_9X_2]$ (X = S, Se, Te) will substitute one or two moles of $AsPh_3$ quite readily to give $[Fe_3(CO)_8X_2(AsPh_3)]$ or $[Fe_3(CO)_7X_2(AsPh_3)_2]$[191]. The tetragonal-pyramidal cluster Fe_3X_2 is presumed to remain intact. In addition, the $[Fe_3(CO)_9(AsPh_3)Te_2]$ complex can be isolated, but it is likely that the arsine is coordinated to one of the tellurium atoms.

An unexpected bridging arsenido moiety is found in the degradation products of $[ffarsFe_3(CO)_{10}]$ (ffars = $Me_2As\overline{C=C(AsMe_2)CF_2CF_2}$)[192]. Refluxing for 1h in cyclohexane yields '$[ffarsFe_3(CO)_9]$', and crystallographic examination has shown that ffars has been degraded and the resulting complex $[\overline{(AsMe_2C=CCF_2CF_2})(AsMe_2)Fe_3(CO)_9]$ has a structure which may be described in terms of three $Fe(CO)_3$ groups and an $AsMe_2$ group being linked together at the corners of a tetrahedrally distorted square-planar arrangement. The rearranged ligand, $AsMe_2\overline{C=CCF_2CF_2}$, is bonded to the first atom by an arsenic linkage, to the second iron atom by a π-bond from the cyclobutene ring, and to the third iron atom by a σ-bond from the cyclobutene ring[193,194]. The formation of this complex involves the breakage of an As—C bond and the formation of Fe—As and Fe—C σ-bonds. Cullen *et al.*[192] have pointed out that such fragmentations are more common than had been originally expected[195,196].

Cyclic pentaarsines react with $Fe(CO)_5$ to give compounds of empirical formula $[\{Fe(CO)_3\}_2(AsR_4)]$, the mass spectra of which do not show individual $AsR_4{}^+$ species[197]. The structure of $[\{Fe(CO)_3\}_2(AsMe_4)]$ has been determined, establishing the presence of two arsenido bridges in the molecule and the absence of a cyclic arsine ligand [198].

22.1.3 Complexes containing NO

Although triphenylphosphine[199] and triphenylarsine will displace both CO groups from $[Fe(NO)_2(CO)_2]$, triphenylstibine forms only the monosubstitution product $[Fe(NO)_2(CO)(SbPh_3)]$ [200,201]. Both $[Fe(NO)_2(CO)(AsPh_3)]$ and $[Fe(NO)_2(AsPh_3)_2]$ can be obtained from the sealed tube reaction (16 h at 85°C) between $AsPh_3$ and $[Fe(NO)_2(CO)_2]$ [201]. The disubstituted $[Fe(NO)_2(AsPh_3)_2]$ can also be obtained[202] through a disproportionation reaction (equation 4)

$$Na[Fe(NO)(CO)_3] + 2AsPh_3 + 2CF_3COOH \xrightarrow{-65°C} [Fe(NO)_2(AsPh_3)_2]$$
$$+ H_2Fe(CO)_4 \qquad (4)$$

The substitution of $[Fe(NO)_2(CO)_2]$ by $AsPh_3$ has also been investigated mechanistically[203,204]; it has been found that the rate is solvent dependent and is also affected by certain catalysts (for example Bu_4NX, X = hal). An i.r. study of NO and CO absorptions in these compounds has been carried out[205].

Tetrasubstituted diarsines react with either $Hg\{Fe(NO)(CO)_3\}_2$, $[Fe(NO)_2(CO)_2]$ [206] or $[Fe(NO)_2X]_2$ (X = Br, I) [207] under forcing conditions to yield the diamagnetic arsenido-bridged compounds $[Fe(NO)_2(AsR_2)]_2$. Under milder conditions $[Fe(NO)_2X(AsR_2)]_2$ is obtained[207], with structure similar to

$[Fe(CO)_3IAs(CF_3)_2]_2$ (section 22.1.2). The dimeric nitrosyl halides will also react with $AsPh_3$ or $SbPh_3$ to give monomeric paramagnetic $[Fe(NO)_2LX]$ [208], from which the corresponding cyanides can be obtained by metathesis[209].

The reaction between $Hg\{Fe(NO)(CO)_3\}$ and molten $AsPh_3$ or $SbPh_3$ gives $Hg\{Fe(NO)(CO)_2L\}$ [210], in contrast to the above reaction with tetrasubstituted diarsines[206]. Alternatively, the anion $[Fe(NO)(CO)_2L]^-$ can be used[211]. These methods have been used to yield similar compounds with tin- [210–12], mercury- [210,212], lead- [211] and cadmium-iron[210] bonds. The structures are assumed[211,212] to involve trigonal-bipyramidal coordination about the iron atom, with a linear M—Fe—L grouping.

22.2 RUTHENIUM AND OSMIUM

22.2.1 Halide and derived complexes
Unlike ferric chloride, ruthenium(III) halides form monomeric complexes with $AsPh_2Me$. Hypophosphorous acid reduces these $[RuX_3(AsPh_2Me)_3]$ complexes to Ru(II) derivatives $[RuX_2(AsPh_2Me)_4]$. This same ligand will reduce Ru(IV) halides to Ru(III) initially[213]. In a similar fashion the $[OsX_3(AsPh_2Me)_3]$ complexes, derived from K_2OsX_6, may be reduced to $[OsX_2(AsPh_2Me)_4]$ [214].

Treatment of Ru and Os halides with $AsPh_3$ in high-boiling alcohols produces complexes originally formulated as $[MX(AsPh_3)_3]$ [215,216]. The work of Chatt and Shaw with similar phosphine derivatives implicated the alcohol in a process known as 'reductive carbonylation', and the complexes actually formed are $[MHX(CO)(PR_3)_3]$ [217]. It is highly likely that Vaska's arsine complexes may be similarly formulated. Reference to tables 24 and 25 illustrates how reaction conditions can affect the products.

The 'reductive carbonylation' reaction for osmium derivatives is best obtained by heating $AsPh_3$ with $(NH_4)_2OsX_6$ (X = Cl, Br) in mixtures of ethylene glycol-ethanol for 1–2 h at 100–140°C, which yields $[OsHX(CO)L_3]$ [221]. If these solutions are acidified with HX, then the complexes $[OsX_3L_3]$ are obtained[143] via the step-wise reduction

$$OsO_4 \rightarrow OsX_6^{2-} \rightarrow OsX_4L_2 \rightarrow OsX_3L_3$$

The isolation of $[OsX_4L_2]$ in this process depends on the nature of L, but the chlorides can be obtained by oxidation of $[OsCl_3L_3]$ with chlorine in the presence of light[222]. The $SbPh_3$ ligand is also known, and $[OsCl_3(SbPh_3)_3]$ undergoes reduction with excess $SbPh_3$ to yield $[OsCl_2(SbPh_3)_4]$ [223,224].

If $RuCl_3 . 3H_2O$ and $AsPhEt_2$ are refluxed in 2-methoxyethanol (1 h), $[RuCl_3L_3]$ is obtained[225]; extended (17 h) reflux in methanol results in *cis*- and *trans*-$[RuCl_2L_4]$ [225], whereas shorter reflux periods (2 h) in the same solvent gives $[RuCl_3(AsPh_3)MeOH]$ [226]. The 'reductive carbonylation' product $[RuHCl(CO)L_3]$ appears to be obtainable only from the reaction between

TABLE 24 RUTHENIUM COMPLEXES WITH TERTIARY ARSINES AND STIBINES

Compound	Colour	M.p.(°C)†	Dipole moment (D) and μ_{eff} (B.M.)	Comments	Ref.
cis-[RuCl₂(AsPhMe₂)₄]	orange	179–190d.	7.75 D		225
trans-[RuCl₂(AsPhMe₂)₄]	red	156–160d.	1.15 D		225
[RuCl₂(SbPh₃)₄]		240–242			228
[RuCl₂(SbPh₃)₄]	red	230–233			226
[RuCl₃(AsPhEt₂)₃]	red-brown	173–176d.			225
[RuCl₃(AsPh₃)₂CH₃OH]	green	149	1.79 B.M.		226
[RuCl₃(AsPh₃)₂CH₃NO₂]	green	198–199d.	2.00 B.M.		233
[RuBr₃(AsPh₃)₂CH₃OH]	dark brown	166–167d.	1.74 B.M.		226
Et₄N[RuCl₄(AsPh₃)₂]	orange-brown	218d.	2.10 B.M.		233
Me₄N[RuBr₄(AsPh₃)₂]	purple-black	230–231d.	1.84 B.M.		233
[RuCl₂(CO)(AsPh₃)₃]	fawn	202			226
[RuCl₂(CO)(AsPhMe₂)₃]	yellow	181–189			227
[RuCl₂(CO)(SbPh₃)₃]	orange-brown	234–240			226
[RuCl₂(CO)₂(AsPh₃)₂]	colourless	292–295		config CLIII (see text)	236
[RuCl₂(CO)₂(AsPhEt₂)₂]	white	142–145	4.0 D		225
[RuCl₂(CO)₂(SbPh₃)₂]	white, colourless	248–265	4.8 D	config CLIII (see text)	228
[RuBr₂(CO)₂(AsPh₃)₂]	almost colourless	268,300	4.3 D	config CLIII (see text)	236
[RuBr₂(CO)₂(AsEt₃)₂]	colourless	106		config CLIII (see text)	236
[RuBr₂(CO)₂(AsEt₃)₂]	deep yellow	86	6.2 D	config CLIV (see text)	236
[RuBr₂(CO)₂(SbPh₃)₂]	bright yellow	280	3.3 D	config CLIII (see text)	236
[RuI₂(CO)₂(AsEt₃)₂]		207	4.1 D	config CLIII (see text)	239
[RuI₂(CO)₂(AsEt₃)₂]		207	6.4 D	config CLIV (see text)	239
[RuI₂(CO)₂(AsPh₂Me)₂]		220			220
[RuHCl(CO)(AsPhMe₂)₃]	cream	91–92		ν(C—O) = 1908 cm⁻¹	227
[Ru(CO)₃{As(C₆F₅)₃}₂]₂		231			188
[RuCl₃(NO)(AsPh₃)₂]	yellow-orange	>320d.		ν(N—O) = 1872 cm⁻¹	245
[RuCl₃(NO)(AsPh₂Me)₂]	light orange	209–211		ν(N—O) = 1878 cm⁻¹	245
[RuCl₃(NO)(AsPhMe₂)₂]	brown	174–175		ν(N—O) = 1852 cm⁻¹	243
[RuCl₃(NO)(AsEt₃)₂]	brown	135–146.5	2.55 D	ν(N—O) = 1823 (vs) cm⁻¹	243
[RuCl₃(NO)(SbPh₃)₂]	yellow	220		ν(N—O) = 1833 cm⁻¹	245
[RuCl₃(NO)(SbEt₃)₂]	orange-red	109–110	1.9 D	ν(N—O) = 1829,1975 cm⁻¹	243
[RuBr₃(NO)(AsEt₃)₂]	brown	142–143.5		ν(N—O) = 1830 cm⁻¹	243
[RuI₃(NO)(AsEt₃)₂]	red-brown	131.5–132.5		ν(N—O) = 1819 cm⁻¹	243
[RuI₂(NO)(AsPhMe₂)₂]	dark brown	168		ν(N—O) = 1835 cm⁻¹ dimeric?	244

† d. indicates decomposition.

TABLE 25 OSMIUM COMPLEXES WITH TERTIARY ARSINES AND STIBINES

Compound	Colour	M.p.(°C)†	Comments	Ref.
mer-[OsCl₃(AsPhEt₂)₃]	red	185–194d.		222
fac-[OsCl₃(AsPhEt₂)₃]	orange	173–177		222
fac-[OsCl₃(AsPhMe₂)₃]	yellow	195–197d.		222
[OsCl₃(AsPhMe₂)₃]	red-brown	185–186	μ_{eff} = 1.9 B.M.	143
[OsCl₃(AsPrⁿ₃)₃]	orange	159–160	μ_{eff} = 1.9 B.M.	143
[OsCl₃(SbPh₃)₃]		169–170		223
[OsBr₃(AsPrⁿ₃)₃]	purple	185–188		143
mer-[OsBr₃(SbPh₃)₃]				227
[OsCl₂(SbPh₃)₄]				223
trans-[OsCl₄(AsPhEt₂)₂]	brown	163–165		222
[OsCl₄(AsPrⁿ₃)₂]	green-brown	155–156	μ_{eff} = 1.6 B.M.	143
[OsCl₄(AsPhMe₂)₂]	brown	177–178	μ_{eff} = 1.5 B.M.	143
[OsH₄(AsPhMe₂)₂]	colourless	oil		222
[OsCl₂(CO)(SbPh₃)₃]	pale green			223
[OsCl₂(CO)₂(SbPh₃)₂]	yellow, white (isomers)		ν(C—O) = 2032,1964 cm⁻¹	223,237
[OsI₂(CO)₂(AsPh₃)₂]	white		ν(C—O) = 2037,1971 cm⁻¹	237
[OsI₂(CO)₂(SbPh₃)₂]			ν(C—O) = 2031,1966 cm⁻¹	237
[OsCl₂(N₂)(AsPhMe₂)₃]	cream	150–151d.	ν(N—N) = 2070 cm⁻¹, benzene solution	247
[OsCl₂(N₂)(AsPhEt₂)₃]	fawn	154–156d.	ν(N—N) = 2059 cm⁻¹, benzene solution	247
[OsCl₃(NO)(AsPh₃)₂]	red-orange	ca.300	ν(N—O) = 1840 cm⁻¹	241
[OsBr₃(NO)(AsPh₃)₂]	red-orange	ca.280	ν(N—O) = 1840 cm⁻¹	241
[OsI₃(NO)(AsPh₃)₂]	red-orange	ca.300	ν(N—O) = 1835 cm⁻¹	241
[OsCl₃(NO)(SbPh₃)₂]	red	ca.245	ν(N—O) = 1805 cm⁻¹; dipole moment = 2.5 D	241
[OsBr₃(NO)(SbPh₃)₂]	red	ca.240	ν(N—O) = 1810 cm⁻¹	241
[OsI₃(NO)(SbPh₃)₂]	red	ca.235	ν(N—O) = 1830 cm⁻¹; dipole moment = 2.9 D	241

† d. indicates decomposition.

$RuCl_2(CO)L_3$ and KOH in ethanol[227]. The reaction between K_2RuCl_6 and $SbPh_3$ gives $[RuCl_2(SbPh_3)_4]$ [228].

Spectroscopic investigations have deduced the meridional configuration for $[MX_3L_3]$ (M = Ru, Os; L = arsine) [229]. The ^1H n.m.r. spectrum of $[OsCl_4(AsPhMe_2)_2]$ is consistent with the *trans* configuration[230]. An interesting identification of the metal oxidation state in these complexes has been obtained from their charge transfer spectra[231].

These halide complexes undergo a number of unusual reactions. The complexes $[RuX_3(AsPh_3)_2(MeOH)]$ (X = Cl, Br) react readily with R_4NX to yield anionic derivatives, $(R_4N)[RuX_4(AsPh_3)_2]$ [232,233]. The addition of CO to refluxing solutions of the metal trihalide followed by addition of the ligand, or passing CO through solutions of the halide and ligand, produce either $[MX_2(CO)L_3]$ or $[MX_2(CO)_2L_2]$ (M = Ru, Os; X = Cl, Br), depending upon the amount of ligand used[223,226-8,234]. The $[RuCl_2(CO)_2(AsPh_3)_2]$ can also be obtained from refluxing $CHCl_3$ solutions of $Ru_3(CO)_{12}$ and $AsPh_3$ [235]. Both the ruthenium [236] and osmium[237] complexes can be obtained by treating the respective carbonyl halide with monotertiary arsine or stibine. The i.r. spectra of the osmium complexes have been briefly mentioned[238], and the isolation and identification of two stereoisomers† of $[RuI_2(CO)_2(AsEt_3)_2]$ (CLIII and CLIV) has been accomplished[239].

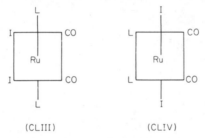

(CLIII)   (CLIV)

Reduction of $[RuCl_2(CO)_2(AsPh_3)_2]$ by zinc dust in hot DMF under moderate (410 kN m^{-2}) pressure yields the five-coordinate (C_{3v}) $[Ru(CO)_3(AsPh_3)_2]$ [240]. The parent carbonyl shows no tendency to trimerise and readily undergoes oxidative addition reactions with the loss of one CO molecule.

The nitrosyl halide complexes $[MX_3(NO)L_2]$ (M = Ru, Os) (tables 24 and 25) can be obtained in a way analogous to the preparation of the carbonyl halide complexes[223,228,241-3], or directly from reaction of the metal nitrosyl halides and the ligand[244,245]. The position of the $\nu(N-O)$ absorption[244,246] and the diamagnetism of these complexes indicates a dimeric structure[244].

When the $[OsCl_3L_3]$ (L = AsPhMe_2, AsPhEt_2) compounds are reduced with zinc under nitrogen (100 atm), the stable dinitrogen complexes $[OsCl_2(N_2)L_3]$ are formed[247]. Under different reducing conditions (namely, BH_4^-), these complexes form the hydrido derivatives $[OsH_4L_3]$ [248,249]. These hydrides will react with a further mole of arsine or CO to give $[OsH_2L_4]$ or $[OsH_2(CO)L_3]$ [250].

Triphenylarsine reacts with the dimeric carbonyl carboxylates $[M(CO)_3(RCO_2)]_2$

† Five stereoisomers are possible for $MA_2B_2C_2$, of which one can exist as optical isomers.

(M = Ru, Os) eliminating two moles of CO to give the dimeric $[M(CO)_2(RCO_2)(AsPh_3)]_2$ [251]. The structure is presumed from spectroscopic data to involve two bridging carboxylates and a metal–metal bond.

22.2.2 Carbonyl complexes

The reaction between $Ru_3(CO)_{12}$ and $AsPh_3$ in refluxing $CHCl_3$ to give $[Ru(CO)_2Cl_2(AsPh_3)_2]$ has already been mentioned. However, if the same reactants are refluxed in hexane, $[Ru_3(CO)_{10}(AsPh_3)_2]$ is obtained [235]. The Ru_3 cluster is presumed to remain intact, and, in view of the product obtained in $CHCl_3$, both $AsPh_3$ are assumed to coordinate to the same Ru atom. The kinetics of this substitution have been discussed [252]. Polymeric ruthenium carbonyl iodide $[RuI_2(CO)_2]_n$ reacts readily with $AsPhMe_2$ [220], $AsPh_3$ and $SbPh_3$ [253] forming $[RuI_2(CO)_2L_2]$. The di-μ-arsenido-bridged dimer $[Ru(CO)_3As(C_6F_5)_2]_2$ is obtained from the reaction between $Ru_3(CO)_{12}$ and the diarsine [189]. The unusual compound $[Ru_6(CO)_{17}C]$ will react with $AsPh_3$ expelling one mole of CO to form $[Ru_6(CO)_{16}C(AsPh_3)]$ [254].

Finally, the use of $[OsHCl_2(AsPh_3)_3]$ [255], $[RuCl_3(AsPh_3)_3]$ [256] and other ruthenium complexes [257-60] as catalysts has been noted.

23 COBALT, RHODIUM, IRIDIUM

23.1 COBALT

23.1.1 Halide and derived complexes

Unlike their phosphine analogues, monotertiary arsines do not readily complex with cobalt halides. Under non-competing conditions (anhydrous cobalt(II) halide and arsine *in vacuo*) the $CoX_2(AsEt_3)_2$ complexes are formed, but they lose one mole of arsine on standing[261]. The halide used must also influence the stability of the products, since $CoI_2(AsPh_3)_2$ may be isolated from nitromethane solution[262]. The work of Ercolani and coworkers on cationic ligands has included the remarkable stabilisation of cobalt–arsine bonds in the complexes $[CoX_3(L^+)]$ (X = Br, I; $L^+ = Ph_2AsCH_2CH_2As^+(CH_2Ph)Ph_2$) [263]. The presence of the positive charge on the ligand is undoubtedly important in stabilising these complexes. Triphenylarsine reacts with anhydrous cobalt(II) nitrate, but the product obtained is $(Ph_3AsO)_2Co(NO_3)_2$ [264]. The anion $[Co(CN)_4]^{2-}$ can accommodate two triphenylarsine ligands to form $[Co(CN)_4(AsPh_3)_2]$ [265].

Cobalt–bisdithiolene complexes react with $AsPh_3$ and $SbPh_3$ in a fashion similar to the reactions with iron complexes (section 22.1.1). The derivatives $[LCo(S–S)_2]^{n-}$, where S–S is a dithiolene ligand, L is a monotertiary arsine or stibine and $n = 0$[172] or 1[173,266], are known. White and Farona have prepared a fascinating series of adducts of $[Co(sacsac)_2]$ (sacsacH = dithioacetylacetone) of the form $[Co(sacsac)_2L]$ (L = $AsPh_3$, $SbPh_3$, $BiPh_3$) [267]. These complexes are rare examples of the coordination of heavy Group VB donor atoms to cobalt in the formal +2 oxidation state. The complexes have been assigned a five-coordinate, square-pyramidal structure, in which L occupies the apical position, presumably.

The 'cobaloxime' compounds, $Co(dmg)_2$ (dmg is the anion of dimethylglyoxime), have attracted interest because of their similarities to vitamin B_{12}. The cobaloximes will coordinate[268–71] with $AsPh_3$ and $SbPh_3$ to form 1:1 or 1:2 adducts. The 1:1 adducts can also act as (incompletely) reversible oxygen carriers[272]. Furthermore, treatment of $[(Ph_3As)Co(dmg)_2Cl]$ with potassium in THF followed by $ClBPh_2$ results in the formation of a compound in which the $LCo(dmg)_2$ unit is intact, but a Co–B bond is formed and $Ph_2B{<}^O_O$ linkages have replaced the hydrogen-bonding between the dmg groups[273].

23.1.2 Carbonyl complexes

Dicobalt octacarbonyl $Co_2(CO)_8$ will react with $AsPh_3$ and $SbPh_3$ to give both mono- and disubstituted products[274–7] (table 26). The reaction favours the formation of $[Co_2(CO)_6L_2]$, and the monosubstituted compound exists only in solution under CO at atmospheric pressure[274]. The latter is best prepared from

TABLE 26 COBALT COMPLEXES WITH TERTIARY ARSINES AND STIBINES

Compound	M.p.(°C)†	Colour	Comments	Ref.
[Co(CO)$_3$AsEt$_3$]$_2$	135d.		ν(C—O) = 1953 cm^{-1}	278
[Co(CO)$_3$AsPh$_3$]$_2$	195d.		ν(C—O) = 1951,1970 cm^{-1}	278
[Hg{Co(CO)$_3$AsPh$_3$}$_2$]	198–199d.		ν(C—O) = 1944,1955 cm^{-1}	284
[Hg{Co(CO)$_3$AsEt$_3$}$_2$]	146–147		ν(C—O) = 1946,1990 cm^{-1}	284
[Hg{Co(CO)$_3$SbPh$_3$}$_2$]	185–187d.		ν(C—O) = 1951,1995 cm^{-1}	284
[Hg{Co(CO)$_3$SbMe$_3$}$_2$]	145–147		ν(C—O) = 1951,1994 cm^{-1}	284
[Cl$_2$Sn{Co(CO)$_3$AsPh$_3$}$_2$]	180	orange-red		285
[Cl$_2$Sn{Co(CO)$_3$SbPh$_3$}$_2$]	188	red		285
[Co(NO)(CO)$_2$AsPh$_3$]	111–113d.		ν(N—O) = 1765 cm^{-1}	291
[Co(NO)(CO)(AsPh$_3$)$_2$]	125d.	red-brown	ν(N—O) = 1722 cm^{-1}	291
[Co(NO)(CO)(SbPh$_3$)$_2$]	119d.	deep red-brown	ν(N—O) = 1719 cm^{-1}	291
[Co(NO)$_2$(AsPh$_3$)Br]	132–133	black	dipole moment = 6.40 D	300
[Co(NO)$_2$(SbPh$_3$)Br]	126–128	black	dipole moment = 5.26 D	300
[Co(NO)$_2$(AsPh$_3$)CN]	127			209
[Co(NO)$_2$(AsPh$_3$)NCS]	121–123d.	black		301
[Co(NO)$_2$(AsPh$_3$)SEt]	55–56d.	black-brown		301
[Co(NO)$_2$(SbPh$_3$)SEt]	51–52d.	black-brown		301
trans-[Co(NO)(AsPh$_3$)SEt]$_2$	110d.	dark brown	ν(N—O) = 1691 cm^{-1}	298
cis-[Co(NO)(AsPh$_3$)SPh]$_2$	137–140d.	black-brown	ν(N—O) = 1718,1692 cm^{-1}	298
Bu$_4$N[CoS$_4$C$_4$(CN)$_4$(AsPr^n$_3$)]	127–128	orange-brown		173
Bu$_4$N[CoS$_4$C$_4$(CN)$_4$(SbPh$_3$)]	158–159	red-brown		173
[CoS$_4$C$_4$(CF$_3$)$_4$(AsPh$_3$)]	186.5–188	brown	μ_{eff} = 1.73 B.M.	172
[CoS$_4$C$_4$(CF$_3$)$_4$(SbPh$_3$)]	211–212	brown	μ_{eff} = 1.79 B.M.	172

† d. indicates decomposition.

equimolar quantities of $Co_2(CO)_8$ and $Co_2(CO)_6L_2$ [276]. The i.r. spectrum of the disubstituted dimer suggests the following structure[278]

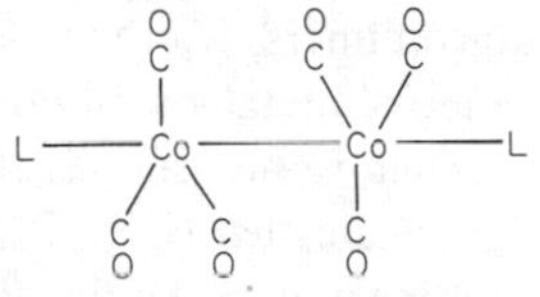

(CLV)

although there is also evidence for the presence of small amounts of carbonyl-bridged isomers in solution[279]. These dimers have been mentioned as effective low-pressure hydroformylation catalysts[280].

The tetrameric carbonyl $Co_4(CO)_{12}$ also reacts with $AsPh_3$ and $SbPh_3$, but only one CO is substituted[281]. In contrast, the novel $[RCCo_3(CO)_9]$ (R = Ph, Bu) will substitute one or two moles of CO for arsine[281]. However, if R=Cl then $[Co_3(CO)_9C]_2$ and R_3AsCl_2 are the products[282]. More recent work on $[RCCo_3(CO_9)]$ (R = Ph, Me, halide) has shown that one or two CO groups can be replaced by boiling in hexane or at room temperature under u.v. irradiation. A novel route employed PtL_4 with the enneacarbonyl at $20°C$ in benzene to yield the monosubstituted product[283].

Complexes of formula $M\{Co(CO)_3L\}_2$ (M = Hg [284] or Cl_2Sn[285]), can be obtained by refluxing the parent carbonyl with monotertiary arsines or stibines in benzene. The structure of the mercury cobaltate is similar to (CLV) but with a Co—Hg—Co linkage[284].

Attempts to prepare di-μ-arsenido-bridged cobalt dimers result in the formation of presumably polymeric $[R_2AsCo(CO)_3]_x$ [286,287]. The reaction between Me_2AsH and $HCo(CO)_4$ is said[286] to give the unstable $[Me_2AsCo(CO)_4]$ intermediate, which rapidly loses CO to form the polymer. The PF_3 analogue of $HCo(CO)_4$ will undergo substitution with $AsPh_3$ or $SbPh_3$ to yield the monosubstituted $[HCo(PF_3)_3L]$ [288].

Triphenylarsine will displace the olefins from (π-cyclooctenyl)-π-cyclootta-1,5-dienecobalt, $Co(C_8H_{13})(C_8H_{12})$, under hydrogen at atmospheric pressure to form $[HCo(AsPh_3)_4]$ [289]. The reaction proceeds through the intermediate $HCoL_3$, and by suitable adjustments of reactants $[CoH_3L_3]$ and $[CoH(N_2)L_3]$ can also be obtained. It may be noted here that, whereas $AsEt_3$ or $SbPh_3$ do not displace PPh_3 from $[Co(N_2)(PPh_3)_3]$, an excess of each ligand causes a lowering of the $\nu(N{\equiv}N)$ frequency[290].

23.1.3 Nitrosyl complexes

The substitution of one or two moles of CO from $Co(NO)(CO)_3$ by $AsPh_3$ or $SbPh_3$ (table 26) depends upon the reaction temperature[200,291]. No trisubstitution product $[Co(NO)L_3]$ was obtained even under extreme conditions. The kinetics of monosubstitution have been extensively studied[291-5], and involve a dissociation; CO is lost to form the transient $Co(NO)(CO)_2$, which rapidly associates to form

$[Co(NO)(CO)_2L]$. The i.r. spectra[205,296] and polarographic behaviour[297] of these complexes have been studied.

The monosubstituted $[Co(NO)(CO)_2(AsPh_3)]$ reacts with thiols with expulsion of CO to form the sulphido-bridged dimers, $[Co(NO)(AsPh_3)SR]_2$ [298]. Physical measurements indicate the presence of metal–metal bonding and admit to the presence of *cis* and *trans* isomers; such isomerism has also been identified by i.r. techniques in the halogeno-bridged analogues $[Co(NO)(AsR_3)X]_2$ [299].

Triphenylarsine or triphenylstibine will cleave the dimeric nitrosyl halides $[Co(NO)_2X]_2$ to form the monomeric $[Co(NO)_2LX]$, which are said to be diamagnetic and tetrahedral[300]. The corresponding cyanides can be made by metathesis[209]. Triphenylarsine will also cleave the sulphido-bridged dimers (X = SCN, SR) to yield the monomers $[Co(NO)_2L(SR)]$; the thiocyanate monomer is bonded through the nitrogen[301]. The halogeno dimers react with Ph_4As_2 resulting in the cleavage of the halogeno bridges to give the diamagnetic diarsine-bridged $[Cl(NO)_2Co(Ph_2AsAsPh_2)Co(NO)_2Cl]$ [207].

23.2 RHODIUM

23.2.1 Halide and carbonyl halide complexes

Monotertiary arsines react with $RhCl_3 . 3H_2O$ in alcohols to give *mer-* and *fac-* $[RhL_3Cl_3]$, and, according to the amount of arsine employed, the chloro-bridged $[Rh_2L_4Cl_6]$ or $[Rh_2L_3Cl_6]$[302] (table 27). The methyldiphenylarsine complexes $[RhL_3X_3]$ [303,304] are reduced by hypophosphorous acid, and the original products were thought to be Rh(II) derivatives, $[Rh(AsPh_2Me)_3X_2]_2$ [305], but on being shown to be diamagnetic were considered to be mixed Rh(I) and Rh(III) salts[306]. Further work showed these products to be $[RhHX_2(AsPh_2Me)_3]$ [304]. Further characterisation of the reported $[RhX_2(AsPhMe_2)_4]$ [307] complexes is obviously necessary. The isomers of $[RhL_3Cl_3]$ have been distinguished by their dipole moments[302] and by spectroscopic studies[299,308]. Metathetical reactions with Br^- or I^- indicate one chloride is more labile than the other two since $[RhL_3XCl_2]$ can be isolated[309]; with $SbPh_3$, Hill and McAuliffe have obtained a series $[Rh(SbPh_3)_3XCl_2]$ (X = Br, I, SCN, $SnCl_2$) [310]. The hydrido complexes $[RhHL_3X_2]$ [61] still appear to be best obtained by reduction (H_3PO_2 or Zn) of $[RhL_3X_3]$ [311].

The Rh(I) chlorocarbonyl complexes *trans*-$[Rh(CO)L_2Cl]$ are obtained when CO is introduced to the alcoholic solutions of $RhCl_3 . 3H_2O$ and arsine[312]. Similar addition of CO to the chloro-bridged complexes $[Rh_2L_4Cl_6]$ provide the Rh(III) complexes $[Rh(CO)L_2Cl_3]$ [312]. The Rh(I) complexes can also be conveniently prepared by adding formaldehyde to the $RhCl_3$–arsine solutions[313], or from $[Rh_2(CO)_4Cl_2]$ [314]. Vallarino reacted $[Rh_2(CO)_4Cl_2]$ with Ph_3L (L = P, As, Sb) and obtained a series $[Rh(CO)(LPh_3)_2Cl]$ [315], but could not prepare a corresponding bismuthine [316]. By a similar reaction Hieber and coworkers claim to have produced the dicarbonyl $[Rh(CO)_2(SbPh_3)_3Cl]$ [317,318]. Vallarino oxidised

$[Rh(CO)(LPh_3)_2X]_2$ with halogens and obtained $[Rh(CO)(LPh_3)_2X_3]$ [316]. The four-coordinate $[Rh(CO)L_2Cl]$ are all diamagnetic, and the dipole moments of about 3 Debye units suggest a *trans* arrangement of the ligands[319]. The addition of $AsPh_3$ or $SbPh_3$ to carbonylated ethanolic solutions of rhodium iodide gives the Rh(III) complex $[Rh(CO)L_2I_3]$ [320] when the ligand is dissolved in acetone. If, however, the ligand is added to this solution in ethanol, there is some evidence for the formation of $[Rh(CO)L_2I]$ [320]. In view of the ready reaction of arsines with the above rhodium halides it is interesting to note that of the three ligands, $(C_6F_5)_{3-n}AsPh_n$ (n = 0-2), only one (n = 2) would react with $Rh_2(CO)_4Cl_2$ to give $[Rh(CO)L_2Cl]$. This ligand, and the other ligands, did not complex with $RhCl_3 . 3H_2O$ or $Rh_2Cl_2(C_2H_4)_4$ [321,322].

One of the most stimulating areas of interest in rhodium chemistry is the use of complexes to catalyse organic reactions. Wilkinson's group demonstrated that $Rh(PPh_3)_3X$ (X = Cl, Br, I) are very good olefin hydrogenation catalysts, but $Rh(AsPh_3)_3Cl$ is much less efficient[323].

The potentially bidentate ligand, *o*-dimethylarsinoanisole combines with $RhCl_3 . 3H_2O$ to yield a complex of formulation $[RhL_2Cl_3]$ [323]. A crystallographic examination[324,325] has shown that the coordination about rhodium is octahedral, consisting of three chlorides, one bidentate ligand and one ligand bound only through the arsenic atom. A similar situation exists when the ligand is *o*-dimethyl-arsinodimethylaniline[326], and also presumably with the corresponding stibine ligands[327].

The arsines, *cis*- and *trans*-β-styryldiphenylarsine, react with $[Rh(CO)_2Cl]_2$ to produce $[Rh(CO)L_2Cl]$, for which there is no evidence for interaction between the metal and the C=C bond[328]. However, a related ligand, tris(*o*-styryl)arsine, reacts with the cycloocta-1,5-diene–rhodium halide dimer to produce compounds of formula $[RhLX]$ [329]. Preliminary X-ray crystallographic and other data support a monomeric five-coordinate structure in which all three olefinic bonds are coordinated.

Tetraphenyldiarsine will also react with rhodium chloride carbonyl dimer, but the product obtained depends upon the solvent[330]. Reaction in petroleum ether affords $[Rh(CO)(AsPh_2)_2Cl]$, whereas in hot benzene *cis* and *trans* isomers of the di-μ-arsenido compound $[Rh(CO)(AsPh_2)Cl]_2$, are obtained.

Reference has already been made to the reaction between $[Rh(CO)_2Cl]_2$ with $SbPh_3$, which has involved some controversy[61,331,332]. A more recent investigation[333] has shown that the product obtained from benzene solution is $[Rh(CO)L_3Cl] . C_6H_6$. The clathrated benzene is held quite tightly and its presence was shown both spectroscopically and by preparing the unsolvated complex in diethyl ether. These authors point out that the confusion concerning the formulation arose in part because of the similar carbon, hydrogen and chlorine content of $[Rh(CO)L_4Cl]$, $[Rh(CO)L_3Cl]$ and $[Rh(CO)L_3Cl] . C_6H_6$. This is a point that should be care-fully attended to in these systems. Finally, the authors point out that $[Rh(CO)L_2Cl]$ can be isolated, but the reaction must be conducted in petroleum ether at low temperatures or with less than the required amount of $SbPh_3$.

TABLE 27 RHODIUM COMPLEXES WITH TERTIARY ARSINES AND STIBINES

Compound	Colour	M.p.(°C)†	Comments	Ref.
fac-[RhCl₃(AsEt₃)₃]	yellow	166–173		302
mer-[RhCl₃(AsEt₃)₃]	red	111–112.5		302
fac-[RhCl₃(AsPhMe₂)₃]	yellow	198–206d.		302
mer-[RhCl₃(AsPhMe₂)₃]	orange-red	190–194d.	dipole moment = 7.25 D	302
[RhCl₃(AsPh₂Me)₃]	orange	208		362
[Rh₂Cl₆(AsEt₃)₄]	orange-brown	240–256d.	dipole moment = 11.3 D	302
[Rh₂Cl₆(AsEt₃)₃]	brown	260–265d.	dipole moment = 8.1 D	302
mer-[RhBr₃(AsPhMe₂)₃]	red	192–200d.		309
mer-[RhI₃(AsPhMe₂)₃]	red-brown	192–201d.		309
mer-[RhBrCl₂(AsPhMe₂)₃]	orange	191–195d.		309
mer-[RhICl₂(AsPhMe₂)₃]	blood-red	167–173d.		309
[RhHCl₂(AsPh₂Me)₃]	yellow	172–175}	isomers (?)	362
[RhHCl₂(AsPh₂Me)₃]	yellow-orange	157–159}		375
[RhHBr₂(AsPh₂Me)₃]	orange	160–165	ν(Rh–H) = 2090 cm⁻¹	375
[RhHI₂(AsPh₂Me)₃]	brown	150–154	ν(Rh–H) = 2075 cm⁻¹	375
[RhCl(CO){AsPh₂(C₆F₅)}]	yellow	138–144		321
trans-[RhCl(CO)(AsEt₃)₂]	yellow	56–60		311
trans-[RhCl(CO)(AsPhEt₂)₂]	yellow	94–96		311
trans-[RhCl(CO)(trans-vdiars)₂]⁺ ‡	yellow	150–152		328
[RhCl(CO)(As₂Ph₄)₂]	yellow-brown	175d.		330
trans-[RhBr(CO)(AsPh₂)]₂	brown-black	208d.		330
cis-[RhCl(CO)(AsPh₂)]₂	black	235d.		330
trans-[Rh(NCS)(CO)(AsPh₃)₂]	yellow	188–190	ν(C–N) = 2149,2093 cm⁻¹	356
[RhCl(CO)(SbPh₃)₃] . C₆H₆	magenta-red	154–155		333
trans-[Rh(NCS)(CO)(SbPh₃)₂]	bright orange	181–183	ν(C–N) = 2109 cm⁻¹	356

$[RhBr(CO)(SbPh_3)_3]$	violet-red	163–164	dipole moment = 3.15 D	333
trans-$[RhCl_3(CO)(AsEt_2Ph)_2]$	orange-yellow	145–153		311
$[RhI_3(CO)(AsPh_3)_2]$	red			320
$[RhI_3(CO)(SbPh_3)_2]$	red			
$[RhHCl_2(CO)(AsPh_3)_2]$	yellow	198–199	$\nu(Rh-H) = 2087$ cm^{-1}	365
$[RhHCl_2(CO)(SbPh_3)_2]$	yellow	159	$\nu(Rh-H) = 2035$ cm^{-1}	365
$[Rh(CO)_2(AsPh_3)_3]ClO_4$	orange-brown	261–265d.		364
$[Rh(CO)_2(SbPh_3)_3]PF_6$	yellow	205		365
$[RhCl(CO)(AsPh_3)_2 . BBr_3]$	yellow-orange	187		358
$[RhCl_2(HgCl)(AsPh_2Me)_3]$	brown	150–155d.		362
$[RhBr_2(HgBr)(AsPh_2Me)_3]$	bright purple	170d.		362
$[RhCl(AsPh_3)_3]$	yellow			334
$[RhCl(SbPh_3)_3]$	yellow			334
$[RhCl(As \{o\text{-styryl})_3\}]$		239d.	also Br and I	329
$[RhCl(H_2)(AsPh_3)_2]$			$\nu(Rh-H)$, 2022, (benzene solution)	334
$[RhCl(O_2)(AsPh_3)_2]$	brown		$\nu(O-O) = 890$ cm^{-1}	334
$[RhHCl(SiCl_3)(AsPh_3)_2]$	yellow	178–182d.	$\nu(Rh-H) = 2080$ cm^{-1}	359
$[RhHCl(SiCl_3)(SbPh_3)_2]$	orange-yellow	183–185d.	$\nu(Rh-H) = 2080$ cm^{-1}	359
$[Rh(NO)(AsPh_3)_3]$	dark-red	ca. 137d.	$\nu(N-O) = 1630$ cm^{-1}	373
$[Rh(NO)(SbPh_3)_3]$		110d.	$\nu(N-O) = 1629$ cm^{-1}	373
$[RhCl_2(NO)(AsPh_3)_2]$	orange-brown	249–250	$\nu(N-O) = 1626$ cm^{-1}	372
$[RhBr_2(NO)(AsPh_3)_2]$	red-brown	259–260	$\nu(N-O) = 1656,1629$ cm^{-1}	372
$[RhI_2(NO)(AsPh_3)_2]$	violet-brown	249–250		372
$[RhCl(NO)(NO_2)(AsPh_3)_2]$	olive-brown	211–213d.	$\nu(N-O) = 1406,1300,815$ cm^{-1}	357

† d. indicates decomposition.

‡ trans-vdiars = trans-$Ph_2AsCH=CHAsPh_2$.

The complexes $[RhL_3Cl]$ ($L = AsPh_3$, $SbPh_3$) are prepared from the ligand and $[(C_2H_4)_2RhCl]_2$ in methanol[334]. The arsine complex will add hydrogen and oxygen to form $[Rh(H_2)L_3Cl]$ and $[Rh(O_2)L_3Cl]$. The complex $[Rh(H_2)L_3Cl]$ was shown by n.m.r. to be octahedral in solution with a solvent molecule occupying the sixth position.

As well as $[Rh(AsPh_3)_3X]$ other rhodium complexes have elicited interest as catalysts[255,335-52]. The catalysts have been described as complexes of the type $[Rh(CO)L_2X]$ [335-7] $[RhL_3X_3]$ [339-46] or as mixtures of arsines and stibines with various rhodium compounds[255,347-52]. These catalysts have been used for a variety of purposes (for example, hydrogenation, polymerisation, hydroformylation), and, in general, they are less active than the phosphine analogues. This generalisation must be approached with caution, since for a specific process the arsine (or stibine)-containing catalyst can be more active.

23.2.2 Reactions of Rh(I) and Rh(III) complexes
The Rh(I) complexes $[Rh(CO)L_2X]$ undergo the 'oxidative addition' reaction to give Rh(III) complexes[353-4]. The *cis* addition of halogens has been demonstrated by spectroscopic methods[309], and the reader is referred to the previous review for a more lengthy discussion of this type of reaction[355].

The thiocyanate complexes $[Rh(CO)(NCS)L_2]$ (L = arsine or stibine) are prepared by metathesis from the corresponding chloride complexes[356]. These complexes possess the *trans* structure and the thiocyanate is N-bonded. However, in solution, the arsine complexes form bridged ($-NCS-$) dimers, $[RhL(CO)(NCS)]_2$.

Nitric oxide reacts with either $[Rh(CO)(AsPh_3)_2Cl]$ or $[Rh(AsPh_3)_3Cl]$ to give $[Rh(NO)(NO_2)L_2Cl]$ [357]. The presence of N_2O in the reaction indicates that disproportionation of the NO has taken place; consequently, the reaction may be viewed as the oxidative addition of N_2O_3 (that is, $NO^+ NO_2^-$).

Two other reactions illustrate the utility of these Rh(I) complexes. The Lewis acid BBr_3 will form the stable adduct $[Br_3B \cdot Rh(CO)(AsPh_3)_2Cl]$ from $[Rh(CO)(AsPh_3)_2Cl]$ [358]. The addition of silanes $HSiR_3$ can be accomplished with $[RhL_3Cl]$ ($L = AsPh_3$ or $SbPh_3$) to give $[L_2RhH(SiR_3)Cl]$, although reaction occurs only under fairly extreme conditions[359].

The hydrido complexes $[RhHX_2(AsMePh_2)_3]$ (X = Cl, Br) undergo reaction with HgY_2 (Y = F, Cl, Br, I, OAc) to give the complexes $[L_3X_2RhHgY]$ [360-2]. Physical measurements and chemical reactions[362] confirm the existence of the Rh—Hg covalent bond.

Solutions of phenyl(cycloocta-1,5-diene)triphenylphosphinerhodium(I) containing an excess of $AsPh_3$ yield the unusual compounds $[HRh(PPh_3)(AsPh_3)_3]$ and $[Rh(PPh_3)(AsPh_3)_2(O_2)]$ when hydrogen and oxygen respectively are bubbled through them[363]. The oxygen-containing compound is probably a dimer, and is explosive.

Finally, if $AsPh_3$ is added to carbonylated solutions of rhodium perchlorate, the complex $[Rh(CO)_2L_3]ClO_4$ is isolated[364]. The corresponding $SbPh_3$ complex is obtained by adding $AlCl_3$ (or $FeCl_3$) to solutions of $[Rh(CO)L_3Cl]$ [365].

23.2.3 Chelated rhodium complexes

The complex $[(CO)_2Rh(chelate)]$ (chelate is 8-oxyquinolate) will add one equivalent of $AsPh_3$ to form $[L(CO)Rh(chelate)]$ [366].

However, the dimeric $[Rh(CO)_2(S_2PF_2)]_2$ reacts with $AsPh_3$ or $SbPh_3$ to give the presumably polymeric compounds of empirical formula $[Rh(CO)L(S_2PF_2)]$ [367]. Solutions of $RhCl_3 . 3H_2O$ and dimethylglyoxime, when treated with $AsPh_3$ or $SbPh_3$, give $[Rh(dmg)_2LCl]$[368] (dmg is the monoanion of dimethylglyoxime), whereas the addition of $AsEt_3$ to rhodium(II) acetate yields the presumably dimeric $[Rh(OAc)_2L]_2$ [369].

23.2.4 Carbonyl complexes

Triphenylarsine will substitute one or two moles of carbon monoxide from $Rh_4(CO)_{12}$ [370], with the products probably retaining the structure of the parent carbonyl. In contrast, reaction between $AsPh_3$ and $Rh_6(CO)_{16}$ provides $[Rh_6(CO)_6L_9]$ [371]. Again, the rhodium cluster is presumed to remain intact, but there is complete absence of the characteristic $\nu(C{-}O)$ stretching modes in the i.r. spectrum.

23.2.5 Nitrosyl complexes

The dimeric nitrosyl chloride $[Rh(NO)_2Cl]_2$ reacts with $AsPh_3$ to give the monomeric $[Rh(NO)L_2Cl_2]$ [372-4], or $[Rh(NO)L_3]$ [373] (table 27). In a similar reaction with $SbPh_3$ $[Rh(NO)L_2Cl_2]$ could not be isolated in a crystalline form, but $[Rh(NO)L_3]$ was readily obtained[373].

23.3 IRIDIUM

23.3.1 Halide and hydride complexes

The reaction between $IrCl_3$ (or chloroiridate(III)) and monotertiary arsines leads to $[IrL_3Cl_3]$ [374,375], which can also be prepared by treatment of $[IrH_3L_3]$ with Cl_2 or HCl [376]. Spectroscopic evidence[377] and exchange of AsR_3 for PR_3 in $[RhCl_3(R_3P)_3]$ [378] support the *mer* configuration. Mixed phosphine–arsine complexes have been synthesised[379] by treating *mer*-$[Ir(AsR_3)_3Cl_3]$ (R_3 = $PhEt_2$, Et_3) with excess $PPhMe_2$ to produce *mer*-$[Ir(AsR_3)(PPhMe_2)_2Cl_3]$. Similar reactions between arsines and bromoiridates produce the hydrido complexes $[IrHL_3Br_2]$ and $[IrH_2L_3Br]$ [380]. The chloro-hydride $[IrHL_3Cl_2]$ can be obtained by reduction of $[IrCl_3L_3]$ [375,381] or by addition of HCl to $[IrL_3Cl]$ [382], although it is also formed in small amounts in the reaction between arsine and chloro-iridate[375,380]. Two isomers of $[IrHL_3X_2]$ are obtained from the reduction of $[IrL_3X_3]$, according to the reducing agent employed (Zn–HX or KOH in alcohol)[375]. These isomers differ in the ligand (X or L) *trans* to hydrogen. The trihydride $[IrH_3L_3]$ [380], obtained by reduction of $[IrHL_3Br_2]$ with $NaBH_4$, is also reported to exist in two isomeric forms. Hydride[383,384] or chloro-hydride[385,386] containing

TABLE 28 IRIDIUM COMPLEXES WITH TERTIARY ARSINES AND STIBINES

Compound	Colour	M.p.(°C)†	Comments	Ref.
mer-[IrCl₃(AsEt₃)₃]	orange	110–111.5	dipole moment = 6.7 D	374
mer-[IrCl₃(AsPhEt₂)₃]	orange			374
fac-[IrCl₃(SbPh₃)₃]	orange-yellow	223		388
mer-[IrCl₃(SbPh₃)₃]	orange	253		388
[IrCl₃(AsPh₂Me)₃]	yellow	138–140		375
[Ir₂Cl₆(SbPh₃)₄]	maroon	170–171		388
[IrCl₄(AsPr$^n_3$)₂]	violet	144–145d.	μ_{eff} = 1.9 B.M.	143
trans-[IrCl₄(AsPh₃)₂]		165d.	μ_{eff} = 1.75 B.M.	382
H[IrCl₄(SbPh₃)₂]	hazel	156–157		388
[IrCl(AsPh₃)₃]	orange-red			382
[IrCl(SbPh₃)₃]	red			382
[IrBr₃(AsPh₂Me)₃]	orange	223–226		375
mer-[IrBr₃(AsMePh₂)₃]	orange	206–208		377
[IrBr₃(SbPh₃)₃]	red	233–235		389
[IrBr₃(SbPh₃)₂]	brown	198		389
[IrI₃(AsPh₂Me)₃]	orange-brown	227–228		375
[IrHCl₂(AsMePh₂)₃]	white } (isomers) yellow }	225–227, 217–219	ν(Ir–H) = 2210 cm⁻¹ ν(Ir–H) = 2095 cm⁻¹ ν(Ir–H) = 2185 cm⁻¹	375
[IrHCl₂(AsPh₃)₃]	pale yellow	195d.	ν(Ir–H) = 2098 cm⁻¹	382
[IrHCl₂(AsEt₃)₃]		72.5–73.5		381
[IrHCl₂(SbPh₃)₃]	pale yellow } (isomers)	197	ν(Ir–H) = 2210 cm⁻¹	382
[IrHBr₂(AsPh₂Me)₃]	pale yellow } (isomers) orange }	228–231, 237–239	ν(Ir–H) = 2095 cm⁻¹	375
[IrHBr₂(AsPh₃)₃]	—	244d.		380
[IrHBr₂(SbPh₃)₃]	yellow	226		389
[IrHI₂(AsPh₂Me)₃]	pale orange } (isomers) orange-red }	236–239, 217–221	ν(Ir–H) = 2195 cm⁻¹ ν(Ir–H) = 2095 cm⁻¹	375
[IrH₂Cl(AsPh₃)₃]	colourless	183d.	ν(Ir–H) = 2160,2120 cm⁻¹	382
[IrH₂Cl(SbPh₃)₃]	off white	145d.	ν(Ir–H) = 2090 cm⁻¹	382
[IrH₂Br(AsPh₃)₃]	orange-yellow	234d.		380
fac-[IrH₃(AsPh₃)₃]				380
mer-[IrH₃(AsPh₃)₃]		223d.		380
[IrH₃(AsPh₃)₂]				380
[IrH₂(AsPh₃)₃]ClO₄				380

Compound	M.p. (°C)[†]	Colour		Ref.
[IrH₂(AsPh₃)₄]ClO₄	203d.			380
trans-[IrCl(CO)(AsPH₂)₂]₂	111–115	orange		330
trans-[IrCl(CO)(AsPhMe₂)₂]	236	yellow		396
trans-[IrBr(CO)(AsPh₃)₂]	105–114	yellow		394
trans-[IrBr(CO)(AsPhEt₂)₂]	95–101	yellow	dipole moment = 2.05 D	395
[IrCl₃(CO)(AsEt₃)₂]	140–143	pale yellow ⎫ (isomers)	dipole moment = 9.65 D, confign CLVII (see text)	399
[IrCl₃(CO)(AsPhEt₂)₂]	149–153	yellow ⎭	dipole moment = 2.95 D, confign CLVI (see text)	399
[IrCl₃(CO)(AsPhMe₂)₂]	192–193	yellow ⎫ (isomers)	dipole moment = 9.8 D, confign CLVII (see text)	400
	227–235	colourless ⎭	dipole moment = 3.1 D, confign CLVI (see text)	
[IrCl₃(CO)(SbPh₃)₂]	253	light yellow ⎫ (isomers)	confign CLVIII (see text)	388
	205	yellow ⎭		
[IrCl₃(CO)(SbPr^{n}₃)₂]	102–105	yellow	dipole moment = 2.4 D, confign CLVI (see text)	399
[IrBr₃(CO)(AsPh₃)₂]	304d.	yellow-green		394
[IrBr₃(CO)(AsPhMe₂)₂]	206–208	yellow	dipole moment = 3.05 D, confign CLVI (see text)	400
[IrBr₃(CO)(SbPh₃)₂]	220	yellow		389
[IrI₃(CO)(AsPhMe₂)₂]	246–249	orange	dipole moment = 3.1 D, confign CLVI (see text)	400
[IrI₃(CO)(AsPh₃)₂]	292	yellow-orange		394
[IrHCl₂(CO)(AsPh₃)₂]	295d.	ivory-white	ν(Ir–H) = 2200 cm⁻¹	365,394
[IrHCl₂(CO)(AsPhMe₂)₂]	132–136	white		396
[IrHCl₂(CO)(SbPh₃)₂]	163d.	bright yellow	ν(Ir–H) = 2157 cm⁻¹; ν(C–O) = 2016 cm⁻¹	365
[IrHBr₂(CO)(AsPh₃)₂]	309d.	light yellow		394
[IrH₂Br(CO)(AsPh₃)₂]	270d.	light yellow		394
[IrH₃(CO)(AsPh₃)₂]	156d.	white		394
[IrH(CO)(AsPh₃)₃]	175	yellow		394
[IrH(CO)₂(AsPh₃)₂]	135–136	white	ν(Ir–H) = 2080 cm⁻¹; ν(C–O) = 1970,1925 cm⁻¹	407
[IrH₂(CO)(AsPh₃)₂]ClO₄	250d.	white		394
[Ir(CO)₃(AsPhMe₂)₂]BPh₄	125–140d.	white	ν(C–O) = 1996,1976 cm⁻¹	406
[Ir(CO)₂(AsPhMe₂)₃]BPh₄	174d.	bright yellow	ν(C–O) = 1996,1940 cm⁻¹	406
[Ir(CO)₂(SbPh₃)₃]PF₆	126–134	pale yellow	Λ_M = 140 cm²/mho mole	365
[Ir(CO)(AsPhMe₂)₄]BPh₄	220–222	yellow	ν(C–O) = 1912 cm⁻¹	406
[IrCl₂(NO)(AsPh₃)₂]	249–253	light brown	ν(N–O) = 1545 cm⁻¹	372
[IrBr₂(NO)(AsPh₃)₂]	260–262	brown	ν(N–O) = 1564 sh, 1558 cm⁻¹	372
[IrI₂(NO)(AsPh₃)₂]	156d.	pink	ν(N–O) = 1563 sh, 1557 cm⁻¹	372
[IrBr(CO)(O₂)(AsPh₃)₂]	300d.	lime-green		394
[IrCl(SO₄)(CO)(AsPh₃)₂]	200d.	buff		401
[IrCl(CO)(NO₃)₂(AsPh₃)₂]				401

† d. indicates decomposition.

tertiary arsines or stibines and other ligands have been reported (table 28). When IrH$_5$(PR$_3$)$_2$ are treated with SbPh$_3$, [IrH$_3$(PR$_3$)$_2$(SbPh$_3$)] are obtained, and the *trans* phosphine configuration predominates in the products[387].

The AsPh$_3$ and SbPh$_3$ complexes [IrL$_3$Cl] are prepared from [IrCl(cyclooctene)$_2$]$_2$ and L in petroleum ether at room temperature[382]. The addition of HCl to the arsine complex has already been mentioned, and, in addition, these complexes will add hydrogen (irreversibly) to give [IrH$_2$ClL$_3$], and the arsine complex will add chlorine to give *trans*-[IrCl$_4$L$_2$] [143,229-31]. The Ir(I) complexes are extremely sensitive to oxygen, so much so that the isolation of pure [IrL$_3$Cl] is practically precluded.

The reaction of haloiridates with SbPh$_3$ is reported to proceed somewhat differently than with AsPh$_3$. The reaction[388] between the stibine and Na$_3$IrCl$_6$ yields Na[IrL$_2$Cl$_4$], [IrHL$_3$Cl$_2$] and *cis*- and *trans*-[IrL$_3$Cl$_3$], whereas the reaction with K$_3$IrBr$_6$ [389] in the presence of KBr affords K[IrL$_2$Br$_4$], [IrL$_3$Br] and [IrHL$_3$Br$_2$]. The complexes M[IrX$_4$L$_2$] lose MX to yield [IrL$_2$X$_3$], and the corresponding acids H[IrL$_2$X$_4$] can be obtained. Chatt and coworkers have prepared the Ir(IV) complex [Ir(AsPr^n$_3$)$_2$Cl$_4$] [390].

The two ortho-substituted ligands, *o*-dimethylarsinoanisole[391] and *o*-dimethylarsinoaniline[392] also (section 23.2.1) react with K$_3$IrCl$_6$, although presumably these complexes differ from the corresponding rhodium complexes.

23.3.2 Carbonyl complexes

The addition of AsPh$_3$ to the polymeric carbonyl halides [Ir(CO)$_3$X] produces *trans*-[Ir(CO)L$_2$X], X = Cl [393] and Br [394], although SbPh$_3$ again shows anomalous behaviour in giving [Ir(CO)L$_3$Cl] [394]. Similar reaction with As$_2$Ph$_4$ gives the di-μ-arsenido-bridged dimer [IrCl(CO)AsPh$_2$]$_2$ [330]. The *trans*-[Ir(CO)L$_2$X] complexes can also be obtained from the carbonylation of sodium bromoiridate with ligand[395], although similar treatment of chloroiridic acid yields [IrHCl$_2$(CO)L$_2$] [396]. This complex can be dehydrohalogenated with NaOMe or KOH to give *trans*-[Ir(CO)L$_2$Cl].

These square-planar complexes will undergo a number of oxidative-addition reactions[397]. The reaction with HBr in dry benzene gives a *cis* addition product, whereas in wet benzene or benzene–methanol a mixture of *cis* and *trans* addition products is formed[398]. The product obtained from halogen addition[394], [Ir(CO)L$_2$X$_3$], is also obtained from the carbonylation of [IrX$_3$L$_3$] [399,400]. Configurations (CLVI), (CLVII) and (CLVIII) are obtained, depending on the particular arsine employed (table 28).

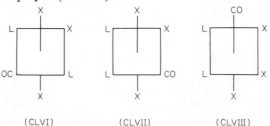

(CLVI)   (CLVII)   (CLVIII)

Some unusual reactions[401] are observed in these systems including the addition of carboxylic acids[402] and the preparation of a complex containing a carboxy group bonded to iridium[403]. Vaska and Chen[404] have studied the rate of oxygen uptake by *trans*-[Ir(CO)(AsPh$_3$)$_2$Cl], and found that it is greater than the analogous triphenylphosphine complex.

The isolation of [Ir(CO)$_3$(AsPh$_3$)$_2$]BPh$_4$ can be effected[405] by bubbling CO through solutions of *trans*-[Ir(CO)(AsPh$_3$)$_2$Cl] and NaBPh$_4$. This complex will add hydrogen to give [IrH$_2$(CO)$_2$(AsPh$_3$)$_2$]BPh$_4$. If additional arsine is added to these solutions, the complexes [Ir(CO)$_2$(AsPh$_3$)$_3$]BPh$_4$ and [Ir(CO)(AsPh$_3$)$_4$]BPh$_4$ can also be isolated[406]. The use of NaBH$_4$ in the CO-saturated solution of *trans*-[IrCl(CO)L$_2$] permits the isolation of [IrH(CO)$_2$L$_2$] [407]. Without the benefit of extra CO the complex [IrH$_3$(CO)L$_2$] is reported[394].

The only mention of nitrosyl complexes are the complexes [Ir(NO)(AsPh$_3$)$_2$X$_2$] (table 28), obtained from [IrCl$_2$(NO)(C$_8$H$_{12}$)] and AsPh$_3$ [372]. Quite recently Robinson and Uttley[408] have prepared [M(NO)(AsPh$_3$)$_2$X$_2$] (M = Rh, Ir; X = Cl, Br, I). This was accomplished by mixing the metal halide and triphenylarsine with N-methyl-N-nitroso-*p*-toluene-sulphonamide in a boiling alcoholic solvent. This is a convenient method of producing nitrosyls, obviating the use of gaseous NO or NOX.

Some of the complexes mentioned above have been employed as catalysts, and whereas the range of examples is not as large as with rhodium the use of tertiary arsine and stibine complexes of formulations [IrL$_3$X$_3$] [409], [IrHL$_3$X$_2$] and [IrH$_2$L$_3$X] [254,410,411], [IrH$_3$L$_3$] [412,413] and [Ir(CO)L$_2$X] [410] have been noted.

24 NICKEL, PALLADIUM, PLATINUM

24.1 NICKEL

24.1.1 Halide complexes

Salts of Ni(II) have been noted previously[61] to exhibit very little tendency to form stable complexes incorporating monotertiary arsines or stibines except with arsines of unusual donor strength[414]. By this criterion the formation of Ni(II) complexes with tertiary stibines would appear to be entirely unlikely. However, such complexes may be obtainable if particular attention is paid to the experimental conditions[415] (for example, the preparation of $[Co(AsEt_3)_2X_2]$, section 23.1.1). Also, Quagliano et al.'s unusual technique to obtain Ni–As coordination in such complexes as $[Ni(L^+)X_3]$ (L is $[Ph_2AsCh_2CH_2As(CH_2Ph)Ph_2]^+$) [263] might result in more instances of AsR_3 complexed with Ni(II).

24.1.2 Complexes of Ni(0)

The stepwise substitution of $Ni(CO)_4$ by tertiary arsines and stibines apparently does not proceed past the disubstituted stage[61,68,416-20]. The Raman and i.r. spectra of these complexes have been reported[76,421,422]. Curiously enough only monosubstitution occurs[69] with $BiEt_3$, whereas no substitution occurs with $BiPh_3$ or $Sb(OEt)_3$. The compound $[Ni(CO)_2(PPh_3)(SbPh_3)]$ [416] is an interesting example of a complex containing PPh_3 and $SbPh_3$ coordinated to the same nickel atom. Monosubstitution products are obtained with the arsines $Me_3SnAsMe_2$ [423,424], $Me_3SiAsMe_2$ [424] and $(Me_2N)_3As$ [425]. The first two ligands appear to be better donors than their completely aliphatic analogues[424].

The compound $Ni(PF_3)_4$ behaves in a fashion similar to $Ni(CO)_4$ in that both mono- and disubstitution products can be obtained with $AsPh_3$ and $SbPh_3$ [426]. Attempts to promote further substitution led to decomposition with the evolution of PF_3 and deposition of nickel.

The $SbPh_3$ and $AsPh_3$ Ni(0) complexes, $[NiL_4]$, can be obtained from the sealed tube reaction between the ligand and $K_4[Ni(CN)_4]$ [427]. No intermediate substitution products, that is, $K_{4-n}[Ni(CN)_{4-n}L_n]$ ($n = 1$-3), were isolated, but in a closely related system $[Ni(CNPr^i)_3(AsPh_3)]$ has been noted[428]. Methyldibromo-arsine also forms a Ni(0) complex $[Ni(AsBr_2Me)_4]$ [429].

These Ni(0) arsine and stibine compounds described above have seen use as catalysts, mainly in oligomerisations. The effects of the $Ni(CO)_4$ substituted derivatives have been noted[430-2], and the action of the complexes $[NiL_4]$ [432-8] deserves special mention. In the cyclooligomerisation of butadiene these complexes are presumed[438] to form first a π-complex with the diolefin with activation of the butadiene followed by subsequent carbon–carbon linkage. The new olefin remains on the nickel until it is displaced by further butadiene, reinitiating the cycle. This proposal has been supported by the isolation of one of the presumed intermediates[438]. Finally, monotertiary arsines are reported to poison Raney nickel catalysts[439]; this may be due to the hydrogenolysis of the C—As bond of AsR_3 to form R_2 as noted elsewhere[440].

24.1.3 Nitrosyl complexes

The $AsPh_3$ and $SbPh_3$ complexes, $[Ni(NO)L_2X]$ (X = Br, I) were obtained directly from $[Ni(NO)X]_4$, with no evidence for the formation of the intermediate $[Ni(NO)LX]_2$ [441]. The corresponding reaction between PPh_3 and $[Ni(NO)I]_4$, however, leads to the dimeric $[Ni(NO)(PPh_3)I]_2$ [442]. Similar reaction with As_2Ph_4 yields the dimeric $[Ni(NO)(AsPh_2)X]_2$ (X = Br, I), whose magnetic moment in THF solution (about 1.45 B.M.) is consistent with partial Ni–Ni interaction[207].

24.2 PALLADIUM

24.2.1 Halide complexes

The square-planar complexes $[PdX_2L_2]$ formed from palladium(II) salts and tertiary arsines and stibines have been used as derivatives in the characterisation of the ligands, especially by Mann and coworkers, and consequently a large number of these complexes are known[1,61,411,443-50]. The stability of these complexes is somewhat less than the corresponding phosphine complexes (with the stibine complexes being the least stable), yet $SbPh_3$ and $AsPh_3$ have been used in the quantitative determination of Pd(II) [451-3].

These $[PdL_2X_2]$ complexes exhibit *cis-trans* isomerism. In general, the *trans* isomer is the more stable, and although *cis* isomers may be isolated their existence is often noted only in solution in equilibrium with the *trans* isomer. The isomers have been investigated by a variety of techniques as a means of distinguishing them, including i.r.[454-6], n.m.r.[457] and (for X = Cl) ^{35}Cl nuclear quadrupole resonance[458-9] spectra. Shobatake and Nakamoto have recently carried out an extremely thorough i.r. and Raman study of $[Pd(AsPh_3)_2X_2]$ (X = Cl, Br), including assignment of $v(Pd–As)$ at 180 cm^{-1} [460]. The isomers differ in behaviour according to the following points: (1) melting point, (2) dipole moment, (3) *trans* isomer more intensely coloured than *cis* isomer, (4) *trans* isomer more soluble than *cis* isomer and (5) for $v(Pd–X)$ stretching frequencies, the *cis* isomer exhibits two or more absorptions and $v(Pd–X)$ varies considerably with changing L, whereas the *trans* isomer exhibits a single absorption and $v(Pd–X)$ shows little variation with L. Generally the tendency for the *trans* isomer to be formed appears to be greater for arsines than stibines; solutions of stibine complexes have been shown to contain up to 40 per cent of the *cis* isomer, which often precipitates from solution[444]. In points (1–4) the identification of either isomer practically requires the presence of both isomers for the purposes of comparison. For instance, the dipole moments or solubilities may exhibit wide variances, depending on the ligand involved. Point (5) may be generally applicable, although the region in which $\nu(Pd-X)$ occurs can be obscured by ligand absorptions.

The complex *trans*-$[Pd(AsPh_3)_2(CNS)_2]$ is unusual in that both the N-bonded isothiocyanato and S-bonded thiocyanato complexes can be obtained[461,462]. The S-bonded isomer is obtained from $K_2[Pd(SCN)_4]$ and $AsPh_3$ in EtOH at $0°C$, whereas reaction at $25°C$ leads to the N-bonded isomer. Furthermore, heating the S-bonded isomer causes it to isomerise in the solid state[462]. Much attention has been directed towards the identification of these isomers by studying characteristic regions in their i.r. spectra. The positions of $\nu(Pd-S)$ (about 300 cm^{-1}) and $\nu(Pd-N)$ (about 265 cm^{-1}) has been suggested[463] as a means of differentiation, but this assignment can be complicated by overlapping ligand bands[464]. The relative position of $\nu(C-S)$ has been used in identifying the isomers[465], but this assignment can be confused by the first overtone of the NCS deformation mode which occurs in this region and is of comparable intensity to $\nu(C-S)$ [466]. The most reliable means of identifying these isomers would appear to be the integrated absorption intensities of $\nu(C\equiv N)$, which for S-bonded isomers is $0.8–2.3 \times 10^4 \mu^{-1}$ cm^{-2}, and for N-bonded isomers is $9–12 \times 10^4 \mu^{-1}$ cm^{-2} [467].

The observance of ligand isomerism in $[PdL_2(CNS)_2]$ has prompted similar investigations into related systems. However, with NO_2^- [468], NCO^-[469,470] and $NCSe$ [471] as the possibly ambidentate ligands only N-bonded isomers were observed.

The synthesis of planar Pd(II) and Pt(II) complexes $[ML_2X_2]$ (L = 10-hydroxy- or 10-methyl-5,10-dihydrophenarsazine; X = Cl, Br, I) has recently been described[472]. ^{1}H n.m.r. and far-i.r. spectra indicate that the ligands are monodentate and co-ordinate via the arsenic atom. Moreover, a *cis* configuration has been assigned, which is quite surprising in view of the bulkiness of the ligand (CLIX).

R = OH, Me

(CLIX)

Triphenylarsine reacts with $[Pd(RNC)_2X_2]$ (R = *c*-C_6H_{11} or Ph; X = Cl, Br) displacing one mole of isocyanide and giving *cis*-$[Pd(RNC)(AsPh_3)X_2]$ [473]. A related complex is $[Pd(RNC)(AsPh_3)(MCl_3)Cl]$ where M is Sn or Ge [474]. The ligand *o*-allylphenyldimethylarsine forms the complexes $[PdX_2L]$ (X = Cl, Br), in which the double bond is coordinated to palladium, and on the basis of n.m.r. evidence the olefin is presumed to have isomerised[475].

Complexes of the type $[PdX(AsMe_3)_3]^+$ (X = halogen) and $[Pd(AsMe_3)_4]^{2+}$ can be prepared from $[PdL_2X_2]$ and one or two equivalents of $[LAgNO_3]$ respectively[476]. The complexes were isolated as their tetrafluoroborate salts (table 29). This method is probably successful because of the increased donor power of the trialkylated arsine, which complexes tenaciously. Other attempts to add further arsine ligands and increase the coordination number of $[PdL_2X_2]$ to form $[PdL_3X_2]$ have had only limited success. Thus, $[Pd(AsPhMe_2)_3X_2]$ complexes are known[450], but they readily dissociate to lose one molecule of ligand. In solution in the presence of excess ligand there is evidence for the $[PdL_3X]X$ species[61,450]. Malatesta and Angoletta have reported the preparation of $[Pd(AsPh_3)_4]$ [477].

The carbonyl complex $[Pd(CO)(AsPh_3)Cl_2]$ can be obtained by adding triphenylarsine to carbonylated solutions of $PdCl_2$ in 2-methoxyethanol [478]. Addition of more ligand results in $[Pd(AsPh_3)_2Cl_2]$.

Finally, the complexes $[PdL_2X_2]$ have been mentioned as catalysts for

TABLE 29 SOME PALLADIUM(II) COMPLEXES WITH TERTIARY ARSINES AND STIBINES

Compound	Colour	M.p.($^\circ$C)†	Ref.
cis-$[PdCl_2(AsMe_3)_2]$	yellow	236	455
trans-$[PdBr_2(AsMe_3)_2]$	orange	230–231	454
trans-$[PdI_2(AsMe_3)_2]$	red	178–179	454
cis-$[PdCl_2(SbMe_3)_2]$	green	150d.	455
$[PdCl_2(SbMe_3)_2]$		150–165d.	‡
trans-$[Pd(SCN)_2(AsPh_3)_2]$	yellow-orange	§	465
trans-$[Pd(NCS)_2(AsPh_3)_2]$	bright yellow	195d.	465
trans-$[Pd(NCO)_2(AsPh_3)_2]$	yellow	–	469
$Pr^n_4N[PdCl_3(AsMe_3)]$	deep orange	122	476
$[PdCl(AsMe_3)_3]BF_4$	pale yellow	208	476
$[Pd(AsMe_3)_4](BF_4)_2$	yellow	220	476
$[Pd_2Br_4(AsEt_3)_2]$	red-brown	180–182	488
$[Pd_2Br_2(AsMe_3)_4](BF_4)_2$	yellow	225d.	476
$[Pd_2Br_2(AsPh_3)_2(SPh)_2]$	orange	262d.	493
$[Pd(O_2CMe)_2AsPh_3]_2$	orange-red	–	494

† d. indicates decomposition.
‡ taken from ref. 61.
§ becomes bright yellow at about 130° and melts at 195°d.

polymerisation[479,480] and hydrogenation[481]. It is interesting to note that the activity of these complexes is enhanced by the addition of $SnCl_2$ [482] or similar halides[483].

24.2.2 Bridged dipalladium complexes

The preparation of the binuclear complexes can be effected by treating $[PdL_2X_2]$ with additional $PdCl_2$, or $PdCl_2$ with a defect of ligand. Treatment of a solution of the $[PdL_2Cl_2]$ complex with ammonium chloropalladite yields the more deeply coloured $[Pd_2L_2Cl_4]$ complex[448,484]. Similar bridged complexes $[Pd(AsMe_3)_4X_2](BF_4)_2$ have been obtained by treating $[PdL_2X_2]$ with an equimolar amount of $AgBF_4$ [478]. The $[Pd_2L_2X_4]$ complexes will react with tetraalkylammonium halides to give $[NR_4][PdLX_3]$ [476,485], but evidently these cannot be obtained by adding arsine to $[Pd_2X_6]^{2-}$, since $[PdL_2X_2]$ complexes result[486]. It has been shown by Wells[487] that $[Pd_2(AsMe_3)_2Br_4]$ has the *trans* planar symmetrical structure (CLX). The preponderance of *cis–trans* isomerism is not as evident for $[Pd_2L_2X_4]$ as for $[PdLX_2]$ complexes, and physical evidence[454,457,480,489] tends to indicate symmetrical *trans* isomers only. Complexes such as (CLXI) containing two different metal atoms have also been isolated[490].

$$Me_3As \diagdown Pd \diagup {}^{Br}\diagdown Pd \diagup {}^{Br}\diagdown AsMe_3 \qquad Pr_3As \diagdown Pd \diagup {}^{Br}\diagdown Hg \diagup {}^{Br}\diagdown AsPr_3$$

(CLX)   (CLXI)

The dinuclear complexes containing stibines are less stable than their arsine analogues[491]. When halogeno-bridged arsine complexes react with amines, stable $[Pd(AsR_3)(am)X_2]$ complexes are formed[484,492].

Complexes containing bridging groups other than halogens have been isolated[61]. The reaction between polymeric $[Pd(SPh)X]_n$ and $AsPh_3$ leads to a complex with –SPh bridges[493]. Similarly, reaction between trimeric palladium(II) acetate and ligand gives a complex containing both bridging and monodentate acetates[494], although if an excess of $AsPh_3$ is used $[Pd(AsPh_3)_2(O_2CCH_3)_2]$ is obtained containing only monodentate acetate[495]. The reaction between $[Pd(S_2PF_2)_2]$ and $AsPh_3$ also results in a mononuclear complex containing one bidentate and one monodentate $F_2PS_2{}^-$[496]. A novel bridged complex, $[Pd_2Cl_4(DATB)](DATB = 1,4$-bis-($o$-aminothiophenoxy)*trans*-but-2-ene), will react with one equivalent of $AsPh_3$ to form $[Pd_2Cl_4(DATB)(AsPh_3)]$, in which the arsine appears to have taken the place of the coordinated C=C bond[497]. Further reaction with $AsPh_3$ leads to $[Pd(AsPh_3)_2Cl_2]$.

24.3 PLATINUM

24.3.1 Halide complexes

The tendency to form square-planar complexes in this group decreases in the order Pt > Pd > Ni, although, in fact, platinum(II) salts form very similar $[ML_2X_2]$ complexes to palladium(II). However, the tendency to produce *cis*-$[ML_2X_2]$ complexes is much more pronounced for platinum. In most preparations of stibine complexes *cis–trans* mixtures are obtained; the very soluble *trans* isomers usually remain in solution in equilibrium with the less soluble *cis* isomers. Isomerisation studies have been carried out by Chatt and Wilkins[498–501], and it has also been shown that isomerisation is photochemically effected[502]. It has been pointed out[61] that these isomers differ in three ways: (1) *trans* isomers are more deeply coloured; (2) *cis* isomers are much less soluble in organic solvents and tend to have higher melting points; (3) *cis* isomers have large dipole moments (9–13 D). Quite a large number of $[PtL_2X_2]$ (L = various AsR_3, SbR_3; X = halide, pseudohalide) have been prepared[443,445,449,450,486,503–7], and these complexes have provided suitable examples for elucidation of reaction mechanisms[508–10] and for the study of the so-called *trans* effect[511-14]. The energy of $\nu(C-As)$ and $\nu(C-Sb)$ vibrations has also been assigned[515].

Whereas most arsines coordinate readily with Pt(II) salts, only the arsine (and none of the stibines) with $n = 2$ of $(C_6F_5)_{3-n}Ph_nAs$ complex with K_2PtCl_4 or PtI_2 [322,323]. It is also interesting to note here that complexes may be formed with arsines AsR_3 of large steric bulk (R is, for example, 2,4,6-trimethylphenyl) [516,517].

The isomerism of the thiocyanate noted in $[Pd(AsPh_3)_2(CNS)_2]$ (section 24.1.1) was not found in the corresponding platinum complexes[465], or with the $AsEt_3$ complex[518]. These complexes contain N-bonded thiocyanate; the $SbPh_3$ complex, however, contained S-bonded thiocyanate[465]. Similarly, only N-bonding was observed in the nitro[468] and cyanato[469] complexes.

The $[PtL_2X_2]$ complexes have been used to obtain compounds in which platinum is bonded to lead[519], tin[520] and silicon[521]. An oxalato complex $[Pt(AsPh_3)_2C_2O_4]$ has been obtained from oxalic acid and the acetate complex[522]. This complex evolves CO_2 under u.v. irradiation, but the nature of the platinum-containing products was uncertain.

The anionic complexes $[PtLX_3]^-$ could not be obtained from $[Pt_2X_6]^{2-}$ and arsine[486], but were produced from the reaction between the bridged dimer $[Pt_2L_2X_4]$ and R_4NX [476,485]. The cationic complexes $[PtL_3X]^+$ are prepared from either $[PtL_2X_2]$ and $LAgNO_3$ [476], or solutions of $[PtL_2X_2]$ and $NaClO_4$ in the presence of ligand[573]. This latter method has been used to obtain $[Pt(CO)L_2X]ClO_4$ by introducing CO to the solutions instead of L. The carbonyl complex can also be obtained from the action of CO on $[Pt_2Cl_2L_4]^{2+}$ [524]. Complexes $[Pt(AsMe_3)_4]^{2+}$ are obtained from $[PtL_2X_2]$ and $LAgNO_3$ [476]. The corresponding $SbPh_3$ complex may be five-coordinate by bonding with one of the nitrates, but this is not certain[525].

The novel complex *cis*-$[PtCl_2(Ph_2AsOH)(PEt_3)]$ containing Ph_2AsOH bonded to platinum through trivalent arsenic has been prepared by hydrolysis of the corresponding Ph_2AsCl complex[526]. These complexes lose HCl to form binuclear arsenato-bridged complexes. The corresponding Ph_2AsOMe complexes have also been isolated[527].

The separation of the optical isomers of AsEtMePh has been effected with the aid of a Pt(II) complex[528]. The process involves reaction of $[Pt_2Cl_4(\pm)L_2]$ with $(-)$stilbenediamine to give $[PtCl(\pm)L(-)L']$ Cl. The complexes containing the optical ligand sets $(+)(-)$ and $(-)(-)$ possessed sufficiently different solubilities to allow separation by crystallisation. The optically pure ligand was then recovered by treating the complex with cyanide.

A number of potentially chelating ligands containing one arsine and one olefin form complexes with Pt(II), in which the double bond may or may not be coordinated. The ligands included are (a) $CH_2=CH(CH_2)_3AsMe_2$ [529], (b) *o*-$CH_2=CHCH_2C_6H_4AsMe_2$ [475] and (c) *o*-, *m*- and *p*-$CH_2=CHC_6H_4AsMe_2$ [530,531]. With (a) and (b) both $[PtLX_2]$ and $[PtL_2X_2]$ can be isolated, depending on the amount of ligand used, although only X = NCS gives $[PtL_2X_2]$ with (b). As might be expected the *para* and *meta* isomers of (c) give only $[PtL_2X_2]$, whereas the *ortho* isomer gives both complexes as with (a), depending on the amount of ligand used. Hall and Nyholm, using the ligands tris(*o*-vinylphenyl)arsine and -stibine, have recently prepared [PtLBr$_2$] [532]. The complexes are four-coordinate with the three double bonds rapidly equilibrating in solution. The stibine complex reacts with bromine with resultant decomposition, but the arsine complex reacts with two moles of bromine to give a new complex in which two of the double bonds have been brominated. No Pt(IV) complex was isolated from the bromination reactions.

[Pt(AsPh$_3$)$_2$C$_2$O$_4$] (C$_2$O$_4$ = oxalate) is formed by reaction of oxalic acid with the carbonate complex[533]. The photochemical reaction

has been found to occur. The reaction between alkyl- and aryl-platinum(II)–olefin complexes with AsPh$_3$ and SbPh$_3$ has been found to give [L$_2$PtR$_2$] (R = σ-bonded alkyl or aryl)[534].

Finally, a number of the [PtL$_2$X$_2$] complexes have been used as catalysts[255,482,483,535-8]. The catalytic activity is enhanced in hydrogenations by the addition of SnCl$_2$ [482,483,535-8].

24.3.2 Bridged diplatinum complexes

The preparations of the bridged diplatinum complexes are very similar to those used in the preparation of the palladium analogues. A slight adaptation of the reaction between [PtL$_2$X$_2$] and PtCl$_2$ involving heating the reactants in a high-boiling hydrocarbon has resulted in improved yields of [Pt$_2$L$_2$Cl$_4$] [539]. A recent crystallographic examination of [Pt$_2$(AsMe$_3$)$_2$Cl$_4$] (CLXII) has confirmed the

(CLXII)

trans configuration, and has demonstrated the presence of three disparate Pt–Cl bond lengths[540-1]. Several workers[454,488,489,542] have measured ν(Pt–Cl), and, in general, have anticipated[542] the results of the crystallographic analysis.

The bridged cationic complexes [PtL$_4$Cl$_2$]$^{2+}$ (L = arsine or stibine)[476,524] have been obtained by reacting [PtCl$_2$L$_2$] with AgBF$_4$. These complexes give [Pt(CO)L$_2$Cl]$^+$ with CO; they react either with water to give the hydride *trans*-[PtHClL$_2$] or with alcohols to give the unusual [PtCl(COOR)L$_2$] (R = Me, Et)[524].

The bridging halide can be replaced by other bridging moieties in [Pt$_2$X$_4$L$_2$]. Reaction of [Pt$_2$(PPr^n$_3$)$_2$Cl$_4$] with Ph$_2$AsH in the presence of NaOEt yields the di-μ-arsenido-bridged *trans*-[Pt$_2$Cl$_2$(AsPh$_2$)$_2$(PPr^n$_3$)$_2$] [543]. The loss of HCl from *cis*-[PtCl$_2$(Ph$_2$AsOH)(PEt$_3$)] to form arsenato-bridged dimers has already been mentioned[526]. The bridged dimer will react with NaSPh to give a di-μ-sulphido-bridged dimer $[Pt_2(SPh)_2(Ph_2AsO)_2(PEt_3)_2]$, in which the arsenato group is bonded in the terminal position. The corresponding Ph_2PO^--bridged complexes *trans*-$[Pt_2Cl_2(Ph_2PO)_2(AsEt_3)_2]$ were also isolated[520], but if $(PhO)_2PO^-$ is used, only mononuclear square-planar complexes were obtained[544,545]. The halogeno-bridged carbonyl complexes $[Pt(CO)_2X_4]$ react with $AsPh_2Me$ to give the unstable impure $[Pt(CO)(AsPh_2Me)X_2]$ [546].

24.3.3 Complexes of Pt(IV)

The platinum(IV) complexes $[PtL_2X_4]$ with tertiary arsines are obtained by oxidation of $[PtL_2X_2]$ with halogens[506]. Both *cis* and *trans* isomers can be isolated depending on the isomer of $[PtL_2X_2]$ used, although chloride seems to be better than bromide and iodide in stabilising the *cis* complexes[547].

Reaction between *trans*-$[Pt(AsEt_3)_2X_4]$ (X = Cl, Br) and iodide[548], thiocyanate or selenocyanate[549], result in reduction to $[PtL_2X_2]$, and finally substitution of X for the other anion. This suggests that metathetical reactions between $[PtL_2X_4]$ and X' must be approached with a precautionary list of oxidation–reduction potentials.

The oxidation of $[PtL_2Br_2]$ [530] (L = o-CH_2=$CHC_6H_4AsMe_2$; section 24.3.1) also gives a platinum(IV) complex $[PtL_2Br_4]$, but the compound contains only three Pt–Br bonds, with the remaining coordination site being occupied by a Pt–C bond (CLXIII). Similar oxidation of the $[PtBr_2L]$ complexes (L is o-vinyl- and o-allylphenyldimethylarsine)[550] results in dimers $[PtBr_4L]_2$ containing bromide bridges and a similar Pt–C bond. The rearrangement of the allyl ligand has been confirmed by a preliminary crystallographic investigation[551].

(CLXIII)

24.3.4 Hydrido complexes

The hydride complex $[PtHClL_2]$ can be obtained either from the addition of HCl to $Pt(AsPh_3)_4$ [552-3] or the hydrazine reduction of *cis*-$[PtCl_2(AsEt_3)_2]$ [554]. The *trans* configuration is most likely on the basis of dipole moment measurements[554]. Solutions of $[PtHXL_2]$ and L' (L' is L, CO, phosphine, arsine or stibine) will give $[PtHL'L_2]^+$ only in the presence of a weakly coordinating anion such as perchlorate[523,555]. Other *trans*-$[PtHX(AsEt_3)_2]$ complexes can be prepared by metathesis from the chloride, and n.m.r. studies indicate that the thiocyanate complex exists in solution as a 2:1 mixture of N-bonded and S-bonded isomers[556].

25 COPPER, SILVER, GOLD

25.1 COPPER

Since this subject was last reviewed[61] not a great deal more information has become available. One development is the characterisation of apparently genuine Cu(II) derivatives containing arsines. Cu(II) is readily reduced to Cu(I) by tertiary arsines[557], but using the ligands o-$HO_2CC_6H_4AsR_2$ (R = Ph, p-MeC_6H_4) Sandhu and Parmar[558] have obtained $[CuL_2]$, which contain Cu(II) bonded to As(III). These complexes exhibit effective magnetic moments of 1.64 B.M. (R = Ph) and 1.91 B.M. (R = p-MeC_6H_4), and are probably polymeric with a distorted octahedral environment. With Cu(II) complexes it is necessary to exert great care in the interpretation of results. Thus, the complex once formulated as a Cu(I)–Cu(II) complex, $[Cu_2Cl_3(AsPh_2Me)_3]$[559,560], was later shown to be $[Cu(OAsPh_2Me)_4][CuCl_4]$[561].

Copper(I) is notable for its tendency to form complexes containing 1–4 equivalents of ligand per copper atom, although the Cu–L ratio varies according to the nature of the ligand. Perhaps one observation that is valid for halide complexes is that the copper remains four-coordinate even in the tetrameric 1:1 complexes[562]. Diphenylmethylarsine forms complexes with copper(I) in ligand–metal ratios of 4:1, 3:1, 2:1 and 1:1; Nyholm[563] formulated these as $[CuL_4]X$, $[CuL_3X]$, $[CuL_4][CuX_2]$ and $[CuLX]_4$, respectively. The 1:1 complexes are the most common, and four-coordination has been shown to be present by X-ray analysis of $[Cu(AsEt_3)I]$[564]. This complex has a structure with four four-coordinate copper atoms at the apices of a regular tetrahedron. Cuprous iodide forms $[CuLI]$ complexes very readily with arsines[565].

The recent work of Jardine and Young[566] on the Cu(I) complexes of $AsPh_3$ and $SbPh_3$ produced the expected tetrameric $[CuLX]_4$ complexes. The $[CuL_3X]$ and $[Cu_2L_3X_2]$ (which is postulated to contain halide bridges linking four- and three-coordinate copper atoms) were also isolated. The complexes are generally rather unstable in solution so that unambiguous structural assessment was not possible, but the complexes were thought to be analogous to the corresponding phosphine complexes. However, a three-coordinate monomer $[Cu(SbPh_3)_2Br]$ was believed to have been isolated, though little unequivocal evidence is available. The steric bulk of the $SbPh_3$ may assist in the formation of complexes of C.N. = 3. It may well be worth reexamining the $[Cu\{P(OPh)_3\}(AsPh_3)Br]$ complex[567] with regard to this question of three-coordination. Four-coordination has been proposed for the $[Cu(AsPh_3)_2CN]$ complex[568].

The addition of CsB_3H_8 to $[L_nCuCl]$ solutions gave the stable $[L_nCuB_3H_8]$ (L = $AsPh_3$, n = 2; L = $SbPh_3$, n = 3) complexes[569]. In the arsine complex the copper is four-coordinate and bonded to two arsenic atoms and two hydrogens of the $B_3H_8^-$ moiety[570]. In view of this it would be interesting to determine whether Cu(I) is five-coordinate in the stibine complex.

A stable 1:1 adduct is formed between $AsPh_3$ and $Cu(hfac)_2$ (hfac = hexa-fluoroacetylacetone); e.s.r. data has been presented[571]. The diamagnetic complexes $[Cu(AsR_3)_nCl]$ (n = 1, 2) and $[Cu_2(AsR_3)_3Cl_2]$ (R = p-MeC$_6$H$_4$) were prepared by treatment of $CuCl_2 \cdot 2H_2O$ with AsR_3 in the appropriate molar ratios[572].

The olefin-arsine o-CH_2=$CHCH_2C_6H_4AsMe_2$ forms $[CuL_2I]$ in which the olefinic bond is not coordinated[573].

25.2 SILVER

The coordination chemistry of Ag(I) resembles that of Cu(I) much more so than it does Au(I). The 1:1 complexes of trialkylarsines $[Ag(AsR_3)I]$ are assumed to be four-coordinate, since $[Ag(AsEt_3)I]$ is isomorphous with $[Cu(AsEt_3)I]$ [574]. The seemingly unique $Ag(BiPh_3)ClO_4$ needs to be reinvestigated in order to determine whether bismuth is coordinated or not[575].

A number of 2:1 complexes are known, for example $[Ag(AsPhMe_2)_2X]$ and $[Ag(AsPh_2Me)_2X]$ [576], and these may well be three-coordinate. There appear to be no examples of 3:1 complexes, but a number of 4:1 complexes, for example $[Ag(AsPh_3)_4]BF_4$ [577], are known.

Complexes of o-CH_2=$CHCH_2C_6H_4AsMe_2$ with Ag(I) salts could only be isolated with $AgNO_3$ [573]. The $\nu(C=C)$ in both the solid state and in solution indicated coordination of the olefinic double bond. The nitrate is probably bidentate and so the complex is four-coordinate. The question of the coordination of $AgNO_3$ with tertiary arsines is an interesting one, since the 1:1 halide complexes like their Cu(I) analogues are tetrameric. However, solutions of $AgNO_3$ and $AsPh_3$ observed photometrically[578] show only 1:1 adduct formation even with an excess of arsine, whereas $[Ag(AsPh_3)_4]^+$ can be isolated in the presence of BF_4^- [577]. Another spectrophotometric investigation[579] has attempted to deduce the characteristic coordination types for the Cu(I), Ag(I) and Au(I) salts. One result showed that 1 and 3 are the preferred coordination numbers of Ag(I).

25.3 GOLD

A recent report[580] has confirmed the earlier observation[61] that the preferred coordination number of Au(I), unlike its Cu(I) and Ag(I) analogues, is two, and the 1:1 complexes only rarely further coordinate; for example only $[AuLX]$ (L = $AsPh_3$, $SbPh_3$) could be obtained by Westland[580]. The preference for two-coordination is maintained even in $[Au(AsPh_3)(acac)]$, in which the anion of acetylacetone is not bidentate, but coordinated through C-3 [581].

Cahours and Gal[1,582] prepared $[Au(AsEt_3)Cl]$, and Mann and coworkers have further studied the trialkylarsine complexes[574]. With $AsPh_2Me$ and $AsPhMe_2$

gold(I) halides form 1:1 complexes, but AuCN behaves anomalously, probably because of polymerisation[583], even though [Au(PEt$_3$)CN] is known[584].

The i.r. spectra of some of these complexes have been examined, and, as expected, ν(Au–X) occurs over a range of frequencies in response to changes in the arsine[456,476].

The preparation of [AuLX] can be effected by treatment of AuX$_4$⁻ with the ligand; however, the reaction between HAuCl$_4$ and Ph$_3$As=S gives the pale yellow gold(III) complex [Au(AsPh$_3$)Cl$_3$] [585].

Gold complexes containing metal–metal heterobonds, [LAuMn(CO)$_5$] (L = AsPh$_3$, SbPh$_3$), have been briefly described[134,586]. Triphenylarsinegold(I) trimethylsilanolate, [Ph$_3$AsAuOSiMe$_3$], has recently been reported[587]. This complex is stable as a solid but in solution it is labile.

REFERENCES

1. Cahours, A., and Gal, H., *C. r. hebd. Séanc. Acad. Sci., Paris,* **71** (1870), 208.
2. Delm, W. M., and Williams, E., *Am. chem. J.,* **40** (1908), 103
3. Ahrland, S., Chatt, J., and Davies, N. R., *Q. Rev. chem. Soc.,* **12** (1958), 265.
4. Booth, G., *Adv. inorg. Chem. Radiochem.,* **6** (1964), 2.
5. Venanzi, L. M., *Chem. Br.,* **4** (1968), 162.
6. Pidcock, A., part 1 of this book.
7. Challenger, F., and Pritchard, F., *J. chem. Soc.* (1924), 864.
8. Westland, A. D., and Westland, L., *Can. J. Chem.,* **43** (1965), 426.
9. Anagnostopoulos, A. K., *Hém. Hron.* 31 (1966), 141; *Chem. Abstr.,* **66** (1967), 82009j.
10. Fowles, G. W. A., and Walton, R. A., *J. chem. Soc.* (1964), 4330.
11. Beattie, I. R., and Webster, M., *J. chem. Soc.* (1964), 3507.
12. Drake, J. E., and Riddle, C., *Inorg. nucl. Chem. Lett.,* **5** (1969), 665.
13. Takashi, Y., *Bull. chem. Soc. Japan,* **40** (1967), 1201.
14. Coover, H. W. Jr, *US Pat.* 3 026 309 (1962); *Chem. Abstr.,* **57** (1962), 24061.
15. Coover, H. W. Jr, and Shearer, N. H. Jr, *US Pat.* 3 081 287 (1963); *Chem. Abstr.,* **58** (1963), 12701e.
16. Coover, H. W. Jr, and Shearer, N. H. Jr, *US Pat.* 3 072 629 (1963); *Chem. Abstr.,* **58** (1963), 6940f.
17. Coover, H. W. Jr, and Shearer, N. H. Jr, *US Pat.* 3 058 969 (1962); *Chem. Abstr.,* **58** (1963), 1549b.
18. Asahi Chemical Industry Co. Ltd, Takashi, Y., Aijima, I., Kobayashi, Y., and Tsunoda, Y., *Jap. Pat.* 24 596 (1963); *Chem. Abstr.,* **60** (1964), 6949d.
19. Asahi Chemical Industry Co. Ltd, Takashi, Y., Aijima, I., Kobayashi, Y., and Tsunoda, Y., *Jap. Pat.* 24 735 (1963); *Chem. Abstr.,* **60** (1964), 5662h.
20. Asahi Chemical Industry Co. Ltd, Takashi, Y., Aijima, I., and Kobayashi, Y., *Jap. Pat.* 17 053 (1964); *Chem. Abstr.,* **62** (1965), 2847c.
21. Shell Internationale Research Maatschappij, N. V., Cannell, L. G., Slaugh, L. H., and Mullineaux, R. D., *Germ. Pat.* 1 186 455 (1965); *Chem. Abstr.,* **62** (1965), 16054f.
22. Kutner, A., *US Pat.* 3 193 540 (1965); *Chem. Abstr.,* **63** (1965), 16501b.
23. Coover, H. W. Jr, and Shearer, N. H. Jr, *US Pat.* 3 196 140 (1965); *Chem. Abstr.,* **63** (1965), 11724e.
24. Studiengesellschaft Kohle m.b.h., *Neth. Pat. Appl.* 6 409 179 (1965); *Chem. Abstr.,* **63** (1965), 5770h.
25. Phillips Petroleum Co., *Neth. Pat. Appl.* 6 507 150 (1965); *Chem. Abstr.,* **64** (1966), 16010d.
26. Coover, H. W. Jr, and Shearer, N. H. Jr, *US Pat.* 3 216 988 (1965); *Chem. Abstr.,* **64** (1966), 3714h.
27. Tokuya Soda Co. Ltd, Machida, K., *Jap. Pat.* 21 793 (1965); *Chem. Abstr.,* **64** (1966), 2192g.
28. Studiengesellschaft Kohle m.b.h., *Belg. Pat.* 651 596 (1965); *Chem. Abstr.,* **64** (1966), 9928c.
29. Coover, H. W. Jr, *US Pat.* 3 222 337 (1965); *Chem. Abstr.,* **64** (1966), 5227d.
30. Coover, H. W. Jr, and Shearer, N. H. Jr, *US Pat.* 3 220 997 (1965); *Chem. Abstr.,* **64** (1966), 5227e.
31. Eastman Kodak Co., *Fr. Pat.* 1 401 841 (1965); *Chem. Abstr.,* **65** (1966), 5554a.
32. Mirviss, S. B., Dougherty, H. W. Jr, and Looney, R. W., *US Pat.* 3 310 547 (1967); *Chem. Abstr.,* **66** (1967), 116115h.
33. Aftandilian, V. D., *US Pat.* 3 285 890 (1966); *Chem. Abstr.,* **66** (1967), 11289q.
34. Coover, H. W. Jr, and Newton, H. S. Jr, *US Pat.* 3 284 427 (1966); *Chem. Abstr.,* **66** (1967), 11292k.
35. Bridgland, B. E., Fowles, G. W. A., and Walton, R. A., *J. inorg. nucl. Chem.,* **27** (1965), 383.

36. Henrici-olive, G., and Olive, S., *Chem. Commun.* (1969), 596.
37. Du Preez, J. G. H., and Gibson, M. L., *Jl. S. Afr. chem. Inst.*, **23** (1970), 184; *Chem. Abstr.*, **74** (1971), 49201h.
38. Hieber, W., and Kummer, R., *Z. Naturf.*, **20b** (1965), 271.
39. Werner, R. P. M., *US Pat.* 3 247 233 (1966); *Chem. Abstr.*, **65** (1966), 5489c.
40. Werner, R. P. M., *US Pat.* 3 236 755 (1966); *Chem. Abstr.*, **64** (1966), 15395f.
41. Ethyl Corp., Werner, R. P. M., and Podall, H. E., *Fr. Pat.* 1 322 168 (1963); *Chem. Abstr.*, **60** (1964), 3015c.
42. Werner, R. P. M., *US Pat.* 3 254 953 (1966); *Chem. Abstr.*, **65** (1966), 8962a.
43. Studiengesellschaft Kohle m.b.h., Wilke, G., Mueller, E. W., Kroener, M., Heimbach, P., and Breil, H., *Germ. Pat.* 1 191 375 (1965); *Chem. Abstr.*, **63** (1965), 7045d.
44. Desnoyers, J., and Rivest, R., *Can. J. Chem.*, **43** (1965), 1879.
45. Makhija, R. C., and Stairs, R. A., *Can. J. Chem.*, **47** (1969), 2293.
46. Tzschach, A., and Nindel, H., *J. organometal. Chem.*, **24** (1970), 159.
47. Vidali, M., Ros, R., and Rizzardi, G., *Gazz. chim. ital.*, **98** (1968), 1240.
48. Rizzardi, G., Ros, R., and Sindellari, L., *Gazz. chim. ital.*, **98** (1968), 1231.
49. Matthews, C. N., *US Pat.* 3 117 983 (1964); *Chem. Abstr.*, **60** (1964), 6870g.
50. Diamond Alkali, Co., Matthews, C. N., *Br. Pat.* 946 674 (1964); *Chem. Abstr.*, **61** (1964), 3149a.
51. Magee, T. A., Matthews, C. N., Wang, T. S., and Wotiz, J. H., *J. Am. chem. Soc.*, **83** (1961), 3200.
52. Matthews, C. N., Magee, T. A., and Wotiz, J. H., *J. Am. chem. Soc.*, **81** (1959), 2273.
53. Jenkins, J. M., Moss, J. R., and Shaw, B. L., *J. chem. Soc. (A)* (1969), 2796.
54. Werner, H., and Prinz, R., *Chem. Ber.*, **99** (1966), 3582.
55. Schumann, H., and Breunig, H. J., *J. organometal. Chem.*, **27** (1971), c28.
56. Strohmeier, W., and Mueller, F. J., *Chem. Ber.*, **102** (1969), 3608.
57. Dobson, G. R., and Houk, L. W., *Inorg. Chim. Acta*, **1** (1967), 287.
58. Whiting, M. C., *Br. Pat.* 941 061 (1963); *Chem. Abstr.*, **60** (1964), 3006d.
59. Fischer, E. O., Bathelt, W., and Mueller, J., *Chem. Ber.*, **103** (1970), 1815.
60. Ehrl, W. O., and Vahrenkamp, H., *Chem. Ber.*, **103** (1970), 3563.
61. Booth, G., *Adv. inorg. Chem. Radiochem.*, **6** (1964), 1.
62. Hayter, R. G., *Inorg. Chem.*, **3** (1964), 711.
63. Dessy, R. E., and Wieczorek, L., *J. Am. chem. Soc.*, **91** (1969), 4963.
64. Dessy, R. E., Kornmann, R., Smith, C., and Hayter, R., *J. Am. chem. Soc.*, **90** (1968), 2001.
65. Elmes, P. S., and West, B. O., *Aust. J. Chem.*, **23** (1970), 2247.
66. Allen, E. A., Feenan, K., and Fowles, G. W. A., *J. chem. Soc.* (1965), 1636.
67. Butcher, A. V., and Chatt, J., *J. chem. Soc. (A)* (1970), 2652.
68. Bouquet, G., and Bigorgne, M., *Bull. Soc. chim. Fr.* (1962), 433.
69. Benlian, D., and Bigorgne, M., *Bull. Soc. chim. Fr.* (1963), 1583.
70. Zingales, F., Faraone, F., Uguagliati, P., and Belluco, U., *Inorg. Chem.*, **7** (1968), 1653.
71. Barbeau, C., and Turcotte, J., *Can. J. Chem.*, **48** (1970), 3583.
72. Singh, S., Singh, P. P., and Rivest, R., *Inorg. Chem.*, **7** (1968), 1236.
73. Dalton, J., Paul, I., Smith, J. G., and Stone, F. G. A., *J. chem. Soc. (A)* (1968), 1195.
74. Graham, W. A. G., *Inorg. Chem.*, **7** (1968), 315.
75. Darensbourg, D. J., and Brown, T. L., *Inorg. Chem.*, **7** (1968), 959.
76. Bigorgne, M., and Benard, J., *Rev. Chim. miner.*, **3** (1966), 831; *Chem. Abstr.*, **67** (1967), 59211f.
77. Graham, J. R., and Angelici, R. J., *Inorg. Chem.*, **6** (1967), 2082.
78. Angelici, R. J., and Ingemanson, C. M., *Inorg. Chem.*, **8** (1969), 83.
79. Ingemanson, C. M., and Angelici, R. J., *Inorg. Chem.*, **7** (1968), 2646.
80. Benlian, D., and Bigorgne, M., *Bull. Soc. chim. Fr.* (1967), 4106.
81. Zingales, F., Graziani, M., and Belluco, U., *J. Am. chem. Soc.*, **89** (1967), 256.
82. Graziani, M., Zingales, F., and Belluco, U., *Inorg. Chem.*, **6** (1967), 1582.
83. Houk, L. W., and Dobson, G. R., *Inorg. Chem.*, **5** (1966), 2119.
84. Dalton, J., Paul, I., Smith, J. G., Stone, F. G. A., *J. chem. Soc. (A)* (1968), 1208.
85. Chatt, J., and Thornton, D. A., *J. chem. Soc.* (1964), 1005.
86. Dessy, R. E., and Weissman, P. M., *J. Am. chem. Soc.*, **88** (1966), 5117.

87. Moss, J. R., and Shaw, B. L., *J. chem. Soc. (A)* (1970), 595.
88. Colton, R., and Rix, C. J., *Aust. J. Chem.,* 22 (1969), 305.
89. Colton, R., and Tomkins, I. B., *Aust. J. Chem.,* 19 (1966), 1519.
90. Colton, R., and Tomkins, I. B., *Aust. J. Chem.,* 19 (1966), 1143.
91. Anker, M. W., Colton, R., and Tomkins, I. B., *Aust. J. Chem.,* 20 (1967), 9.
92. Stiddard, M. H. B., and Townsend, R. E., *J. chem. Soc. (A)* (1969), 2355.
93. Colton, R., and Rose, G. G., *Aust. J. Chem.,* 23 (1970), 1111.
94. Bowden, J. A., and Colton, R., *Aust. J. Chem.,* 22 (1969), 905.
95. Johnson, B. F. G., *J. chem. Soc. (A)* (1967), 475.
96. Cotton, F. A., and Johnson, B. F. G., *Inorg. Chem.,* 3 (1964), 1609.
97. Anker, M. W., Colton, R., and Tomkins, I. B., *Aust. J. Chem.,* 21 (1968), 1149.
98. Anker, M. W., Colton, R., and Tomkins, I. B., *Rev. pure appl. Chem.,* 18 (1968), 23.
99. Pengilly, B. W., *US Pat.* 3 062 786 (1962); *Chem. Abstr.,* 58 (1963), 1560h.
100. Naldini, L., *Gazz. chim. ital.,* 90 (1960), 1337.
101. Naldini, L., and Sacco, A., *Gazz. chim. ital.,* 89 (1959), 2258,
102. Hieber, W., and Freyer, W., *Chem. Ber.,* 92 (1959), 1765.
103. Osborne, A. G., and Stiddard, M. H. B., *J. chem. Soc.* (1964), 634.
104. Lewis, J., Manning, A. R., and Miller, J. R., *J. chem. Soc. (A)* (1966), 845.
105. Lambert, R. F., *Chem. Ind.* (1961), 830.
106. Hayter, R. G., *J. Am. chem. Soc.,* 86 (1964), 823.
107(a). Ugo, R., and Bonati, F., *J. organometal. Chem.,* 8 (1967), 189.
107(b). Lambert, R. F., *US Pat.* 3 080 407 (1963); *Chem. Abstr.,* 59 (1963), 3960b.
108. Wawersik, H., and Basolo, F., *Inorg. Chim. Acta,* 3 (1969), 113.
109. Johnson, B. F. G., Johnston, R. D., Lewis, J., and Robinson, B. H., *J. organometal. Chem.,* 10 (1967), 105.
110. Parker, D. J., and Stiddard, M. H. B., *J. chem. Soc. (A)* (1966), 695.
111. Abel, E. W., Dalton, J., Paul, I., Smith, J. G., and Stone, F. G. A., *J. chem. Soc. (A)* (1968), 1203.
112. Stone, F. G. A., US Clearinghouse for Federal Scientific and Technical Information 1967; *Chem. Abstr.,* 71 (1969), 17235u.
113. Angelici, R. J., and Basolo, F., *J. Am. chem. Soc.,* 84 (1962), 2495.
114. Wilkinson, G., *US Pat.* 3 092 646 (1963); *Chem. Abstr.,* 62 (1965), 10277a.
115. Hieber, W., and Schropp, W. Jr, *Z. Naturf.,* 14b (1959), 460.
116. Abel, E. W., and Wilkinson, G., *J. chem. Soc.* (1959), 1501.
117. Hieber, W., Beck, W., and Tengler, H., *Z. Naturf.,* 15b (1960), 411.
118. Hieber, W., and Wollmann, K., *Chem. Ber.,* 95 (1962), 1552.
119. Angelici, R. J., and Basolo, F., *Inorg. Chem.,* 2 (1963), 728.
120. Bennett, M. A., and Clark, R. J. H., *J. chem. Soc., Suppl. 1* (1964), 5560.
121. Smith, F. E., and Butler, I. S., *Can. J. Chem.,* 47 (1969), 1311.
122. Manuel, T. A., *Adv. organometal. Chem.,* 3 (1965), 181.
123. Wojcicki, A., and Farona, M. F., *Inorg. Chem.,* 3 (1964), 151.
124. Farona, M. F., and Wojcicki, A., *Inorg. Chem.,* 4 (1965), 1402.
125. Addison, C. C., and Kilner, M., *J. chem. Soc. (A)* (1968), 1539.
126. Hartman, F. A., Kilner, M., and Wojcicki, A., *Inorg. Chem.,* 6 (1967), 34.
127. Hieber, W., and Tengler, H., *Z. anorg. allg. Chem.,* 318 (1962), 136.
128. Wawersik, H., and Basolo, F., *J. Am. chem. Soc.,* 89 (1967), 4626.
129. Nesmeyanov, A. N., Anisimov, K. N., Kolobova, N. E., and Antonova, A. B., *Izv. Akad. Nauk SSSR Ser. Khim.* (1966), 160; *Chem. Abstr.,* 64 (1966), 12720h.
130. Gorsich, R. D., *J. Am. chem. Soc.,* 84 (1962), 2486.
131. Gorsich, R. D., *US Pat.* 3 030 397 (1962); *Chem. Abstr.,* 57 (1962), 7309g.
132. Gorsich, R. D., *US Pat.,* 3 030 396 (1962); *Chem. Abstr.,* 57 (1962), 7309e.
133. Carey, N. A. D., and Noltes, J. G., *Chem. Commun.* (1968), 1471.
134. Kasenally, A. S., Lewis, J., Manning, A. R., Miller, J. R., Nyholm, R. S., and Stiddard, M. H. B., *J. chem. Soc.* (1965), 3407.
135. Grobe, J., *Z. anorg. allg. Chem.,* 331 (1964), 63.
136. Grobe, J., Helgerud, J. E., and Stierand, H., *Z. anorg. allg. Chem.,* 371 (1969), 123.
137. Grobe, J., and Sheppard, N., *Z. Naturf.,* 23b (1968), 901.
138. Abel, E. W., and Hutson, G. V., *J. inorg. nucl. Chem.,* 30 (1968), 2339.

139. Fergusson, J. E., and Hickford, J. H., *J. inorg. nucl. Chem.*, **28** (1966), 2293.
140. Hieber, W., Lux, F., and Herget, C., *Z. Naturf.*, **20b** (1965), 1159.
141. Hieber, W., and Opavsky, W., *Chem. Ber.*, **101** (1968), 2966.
142. Cotton, F. A., Robinson, W. R., and Walton, R. A., *Inorg. Chem.*, **6** (1967), 223.
143. Chatt, J., Leigh, G. J., Mingos, D. M. P., and Paske, R. J., *J. chem. Soc. (A)* (1968), 2636.
144. Randall, E. W., and Shaw, D., *J. chem. Soc. (A)* (1969), 2867.
145. Cotton, F. A., and Lippard, S. J., *J. Am. chem. Soc.*, **86** (1964), 4497.
146. Cotton, F. A., Lippard, S. J., and Mague, J. T., *Inorg. Chem.*, **4** (1965), 508.
147. Rouschias, G., and Wilkinson, G., *Chem. Commun.* (1967), 442.
148. Rouschias, G., and Wilkinson, G., *J. chem. Soc. (A)* (1968), 489.
149. Chatt, J., and Rowe, G. A., *J. Chem. Soc.* (1962), 4019.
150. Johnson, N. P., Lock, C. J. L., and Wilkinson, G., *J. chem. Soc.* (1964), 1054.
151. Chatt, J., Garforth, J. D., and Johnson, N. P., *J. chem. Soc.* (1964), 601.
152. Cotton, F. A., and Lippard, S. J., *Inorg. Chem.*, **5** (1966), 9.
153. Chatt, J., and Coffey, R. S., *J. chem. Soc. (A)* (1969), 1963.
154. Ginsberg, A. P., *Chem. Commun.* (1968), 857.
155. Singleton, E., Moelwyn-Hughes, J. T., and Garner, A. W. B., *J. organometal. Chem.*, **21** (1970), 449.
156. Zingales, F., and Trovati, A., *Rc. Ist. lomb. Sci. Lett., A*, **101** (1967), 527; *Chem. Abstr.*, **69** (1968), 30540t.
157. Moelwyn-Hughes, J. T., and Garner, A. W. B., *Chem. Commun.* (1969), 1309.
158. Nesmeyanov, A. N., Kolobova, N. E., Anisimov, K. N., and Khandozhko, V. N., *Izv. Akad. Nauk SSSR, Ser. Khim.* (1966), 163; *Chem. Abstr.*, **64** (1966), 12718d.
159. Uguagliati, P., Trovati, A., and Zingales, F., *Inorg. Chem.*, **10** (1971), 851.
160. B. F. Goodrich, Co., *Neth. Pat. Appl.* 6 609 478 (1967); *Chem. Abstr.*, **67** (1967), 32323g.
161. Drahorad, J., and Kucera, J., *Czech. Pat.* 117 293 (1966); *Chem. Abstr.*, **65** (1966), 16549e.
162. Bonati, F., Minghetti, G., and Leoni, R., *J. organometal. Chem.*, **25** (1970), 223.
163. Nast, R., and Krueger, K. W., *Z. anorg. allg. Chem.*, **341** (1965), 189.
164. Burger, K., Korecz, L., Papp, S., and Mohai, B., *Radiochem. radioanal. Lett.*, **2** (1969), 153; *Chem. Abstr.*, **72** (1970), 26884c.
165. Fluck, E., and Kuhn, P., *Z. anorg. allg. Chem.*, **350** (1967), 263.
166. Papp, S., *Proc. 3rd. Symp. co-ord. Chem.*, **1** (1970), 507; *Chem. Abstr.*, **74** (1971), 60389w.
167. Papp, S., and Schoenweitz, T., *J. inorg. nucl. Chem.*, **32** (1970), 697.
168. Nyholm, R. S., *J. Proc. R. Soc. N.S.W.*, **78** (1944), 229; *Chem. Abstr.*, **40** (1946), 3996.
169. Naldini, L., *Gazz. chim. ital.*, **90** (1960), 1231.
170. Singh, P. P., and Rivest, R., *Can. J. Chem.*, **46** (1968), 1773.
171. Birchall, T., *Can. J. Chem.*, **47** (1969), 1351.
172. Balch, A. L., *Inorg. Chem.*, **6** (1967), 2158.
173. McCleverty, J. A., Atherton, N. M., Connelly, N. G., and Winscom, C. J., *J. chem. Soc. (A)* (1969), 2242.
174. Balch, A. L., *Inorg. Chem.*, **10** (1971), 276.
175. Lewis, J., Nyholm, R. S., Sandhu, S. S., and Stiddard, M. H. B., *J. chem. Soc.* (1964), 2825.
176. Strohmeier, W., and Mueller, F. J., *Chem. Ber.*, **102** (1969), 3613.
177. Lewis, J., Nyholm, R. S., Osborne, A. G., Sandhu, S. S., and Stiddard, M. H. B., *Chem. Ind.* (1963), 1398.
178. Clifford, A. F., and Mukherjee, A. K., *Inorg. Chem.*, **2** (1963), 151.
179. Clifford, A. F., and Mukherjee, A. K., *Inorg. Synth.*, **8** (1966), 185.
180. Farmery, K., and Kilner, M., *J. chem. Soc. (A)* (1970), 634.
181. B.A.S.F., and Schweckendiek, W., *Germ. Pat.* 834 991 (1952); *Chem. Abstr.*, **52** (1958), 9192.
182. Reppe, W., and Schweckendiek, W. J., *Justus Liebigs Annln Chem.*, **560** (1948), 104.
183. Wilkinson, G., *J. Am. chem. Soc.*, **73** (1951), 5502.
184. Davison, A., McFarlane, W., Pratt, L., and Wilkinson, G., *J. chem. Soc.* (1962), 3653.
185. Adams, D. M., Cook, D. J., and Kemmitt, R. D. W., *J. chem. Soc. (A)* (1968), 1067.

186. Cohen, I. A., and Basolo, F., *J. inorg. nucl. Chem.*, **28** (1966), 511.
187. Hieber, W., and Thalhofer, A., *Angew. Chem.*, **68** (1956), 679.
188. Cooke, M., Green, M., and Kirkpatrick, D., *J. chem. Soc. (A)* (1968), 1507.
189. Grobe, J., *Z. anorg. allg. Chem.*, **361** (1968), 32.
190. Hieber, W., and Zeidler, A., *Z. anorg. allg. Chem.*, **329** (1964), 92.
191. Cetini, G., Stanghellini, P. L., Rossetti, R., and Gambino, O., *J. organometal. Chem.*, **15** (1968), 373.
192. Cullen, W. R., Harbourne, D. A., Liengme, B. V., and Sams, J. R., *Inorg. Chem.*, **9** (1970), 702.
193. Einstein, F. W. B., and Svensson, A. M., *J. Am. chem. Soc.*, **91** (1969), 3663.
194. Einstein, F. W. B., Pilatti, A. M., and Restivo, R., *Inorg. Chem.*, **10** (1971), 1947.
195. Bosnich, B., Nyholm, R. S., Pauling, P. J., and Tobe, M. L., *J. Am. chem. Soc.*, **90** (1968), 4741.
196. Cullen, W. R., and Harbourne, D. A., *Can. J. Chem.*, **47** (1969), 3371.
197. Elmes, P. S., and West, B. O., *Coord. Chem. Rev.*, **3** (1968), 279.
198. Gatehouse, B. M., *Chem. Commun.* (1969), 948.
199. Reed, H. W. B., *J. chem. Soc.* (1954), 1931.
200. Malatesta, L., and Araneo, A., *J. chem. Soc.* (1957), 3803.
201. McBride, D. W., Stafford, S. L., and Stone, F. G. A., *Inorg. Chem.*, **1** (1962), 386.
202. Hieber, W., Klingshirn, W., and Beck, W., *Chem. Ber.*, **98** (1965), 307.
203. Morris, D. E., and Basolo, F., *J. Am. chem. Soc.*, **90** (1968), 2536.
204. Morris, D. E., and Basolo, F., *J. Am. chem. Soc.*, **90** (1968), 2531.

205. Beck, W., and Lottes, K., *Chem. Ber.*, **98** (1965), 2657.
206. Hayter, R. G., and Williams, L. F., *Inorg. Chem.*, **3** (1964), 717.
207. Hieber, W., and Kummer, R., *Z. anorg. allg. Chem.*, **344** (1966), 292.
208. Hieber, W., and Kramolosky, R., *Z. anorg. allg. Chem.*, **321** (1963), 94.
209. Hieber, W., and Fuehrling, H., *Z. anorg. allg. Chem.*, **373** (1970), 48.
210. Hieber, W., and Klingshirn, W., *Z. anorg. allg. Chem.*, **323** (1963), 292.
211. Manning, A. R., and Casey, M., *J. chem. Soc. (A)* (1971), 256.
212. Casey, M., and Manning, A. R., *J. chem. Soc. (A)* (1970), 2258.
213. Dwyer, F. P., Humpoletz, J. E., and Nyholm, R. S., *J. Proc. R. Soc. N.S.W.*, **80** (1946), 217; *Chem. Abstr.*, **42** (1948), 2885.
214. Dwyer, F. P., Nyholm, R. S., and Tyson, B. T., *J. Proc. R. Soc. N.S.W.*, **81** (1947), 272.
215. Vaska, L., *Z. Naturf.*, **15b** (1960), 56.
216. Vaska, L., and Sloane, E. M., *J. Am. chem. Soc.*, **82** (1960), 1263.
217. Chatt, J., and Shaw, B. L., *Chem. Ind.* (1961), 290.
218. Dwyer, F. P., Humpoletz, J. E., and Nyholm, R. S., *J. Proc. R. Soc. N.S.W.*, **80** (1946), 217; *Chem. Abstr.*, **42** (1948), 2885.
219. Vaska, L., *Chem. Ind.*, (1961), 1402.
220. Irving, R. J., *J. chem. Soc.* (1956), 2879.
221. Vaska, L., *J. Am. chem. Soc.*, **86** (1964), 1943.
222. Douglas, P. G., and Shaw, B. L., *J. chem. Soc. (A)* (1970), 334.
223. Araneo, A., and Bianchi, C., *Gazz. chim. ital.*, **97** (1967), 885; *Chem. Abstr.*, **67** (1967), 104686c.
224. Valenti, V., Sgamellotti, A., Cariati, F., and Araneo, A., *Gazz. chim. ital.*, **98** (1968), 983; *Chem. Abstr.*, **70** (1969), 7499g.
225. Chatt, J., Shaw, B. L., and Field, A. E., *J. chem. Soc.* (1964), 3466.
226. Stephenson, T. A., and Wilkinson, G., *J. inorg. nucl. Chem.*, **28** (1966), 945.
227. Lupin, M. S., and Shaw, B. L., *J. chem. Soc. (A)* (1968), 741.
228. Araneo, A., and Martinengo, S., *Rc. Ist. lomb. Sci. Lett.*, **A99** (1965), 829; *Chem. Abstr.*, **65** (1966), 1748d.
229. Chatt, J., Leigh, G. J., and Mingos, D. M. P., *J. chem. Soc. (A)* (1969), 1674.
230. Chatt, J., Leigh, G. J., Mingos, D. M. P., Randall, E. W., and Shaw, D., *Chem. Commun.* (1968), 419.
231. Leigh, G. J., and Mingos, D. M., *J. chem. Soc. (A)* (1970), 587.
232. Stephenson, T. A., *Inorg. nucl. Chem. Lett.*, **4** (1968), 687.
233. Stephenson, T. A., *J. chem. Soc. (A)* (1970), 889.
234. Bruce, M. I., and Stone, F. G. A., *J. chem. Soc. (A)* (1967), 1238.

235. Bruce, M. I., Gibbs, C. W., and Stone, F. G. A., *Z. Naturf.,* **23b** (1968), 1543.
236. Hieber, W., and John, P., *Chem. Ber.,* **103** (1970), 2161.
237. Hales, L. A. W., and Irving, R. J., *J. chem. Soc. (A)* (1967), 1932.
238. Hales, L. A. W., and Irving, R. J., *Spectrochim. Acta,* **A23** (1967), 2981.
239. John, P., *Chem. Ber.,* **103** (1970), 2178.
240. Collman, J. P., and Roper, W. R., *J. Am. chem. Soc.,* **87** (1965), 4008.
241. Araneo, A., Valenti, V., and Cariati, F., *J. inorg. nucl. Chem.,* **32** (1970), 1877.
242. Araneo, A., *Gazz. chim. ital.,* **96** (1966), 1560; *Chem. Abstr.,* **66** (1967), 61380h.
243. Chatt, J., and Shaw, B. L., *J. chem. Soc. (A)* (1966), 1811.
244. Irving, R. J., and Laye, P. G., *J. chem. Soc.* (1966), 161.
245. Fairey, M. B., and Irving, R. J., *J. chem. Soc. (A)* (1966), 475.
246. Fairey, M. B., and Irving, R. J., *Spectrochim. Acta,* **A20** (1964), 1757.
247. Chatt, J., Leigh, G. J., and Richards, R. L., *J. chem. Soc. (A)* (1970), 2243.
248. Douglas, P. G., and Shaw, B. L., *Chem. Commun.* (1969), 624.
249. Leigh, G. J., Levinson, J. J., and Robinson, S. D., *Chem. Commun.* (1969), 705.
250. Bell, B., Chatt, J., and Leigh, G. J., *Chem. Commun.* (1970), 576.
251. Crooks, G. R., Johnson, B. F. G., Lewis, J., Williams, I. G., and Gamlen, G., *J. chem. Soc. (A)* (1969), 2761.
252. Candlin, J. P., and Shortland, A. C., *J. organometal. Chem.,* **16** (1969), 289.
253. Hieber, W., and Heusinger, H., *J. inorg. nucl. Chem.,* **4** (1957), 179.
254. Johnson, B. F. G., Lewis, J., and Williams, I. G., *J. chem. Soc. (A)* (1970), 901.
255. Fotis, P., Jr., and McCollum, J. D., *US Pat.* 3 324 018 (1967); *Chem. Abstr.,* **67** (1967), 53616v.
256. Wilkinson, G., *Fr. Pat.* 1 459 643 (1966); *Chem. Abstr.,* **67** (1967), 53652d.
257. Shell Internationale Research Maatschappij N. V., Slaugh, L. H., *Belg. Pat.* 621 662 (1963); *Chem. Abstr.,* **59** (1963), 11268d.
258. Chabardes, P., and Colevray, L., *Fr. Pat.* 1 526 197 (1968); *Chem. Abstr.,* **71** (1969), 5006g.
259. Hiraki, K., and Hirai, H., *J. polym. Sci.,* **B7** (1969), 449.
260. Araneo, A., Valenti, V., Cariati, F., and Sgamellotti, A., *Annali Chim.,* **60** (1970), 768; *Chem. Abstr.,* **74** (1971), 134271a.
261. Hatfield, W. E., and Yoke, J. T., III, *Inorg. Chem.,* **1** (1962), 475.
262. Goodgame, D. M. L., Goodgame, M., and Cotton, F. A., *Inorg. Chem.,* **1** (1962), 239.
263. Ercolani, C., Quagliano, J. V., and Vallarino, L. M., *Chem. Commun.* (1969), 1094.
264. Fereday, R. J., Logan, N., and Sutton, N., *J. chem. Soc. (A)* (1969), 2699.
265. Maki, N., and Ohshima, K., *Bull. chem. Soc. Japan,* **43** (1970), 3970.
266. Langford, C. H., Billig, E., Shupack, S. I., and Gray, H. B., *J. Am. chem. Soc.,* **86** (1964), 2958.
267. White, J. F., and Farona, M. F., *Inorg. Chem.,* **10** (1971), 1080.
268. Schrauzer, G. N., and Lee, L., *J. Am. chem. Soc.,* **90** (1968), 6541.
269. Schrauzer, G. N., and Windgassen, R. J., *Chem. Ber.,* **99** (1966), 602.
270. Schrauzer, G. N., and Kratel, G., *Chem. Ber.,* **102** (1969), 2392.
271. Schrauzer, G. N., and Kratel, G., *Angew. Chem.,* **77** (1965), 130.
272. Schrauzer, G. N., and Lee, L. P., *J. Am. chem. Soc.,* **92** (1970), 1551.
273. Schmid, G., Powell, P., and Nöth, H., *Chem. Ber.,* **101** (1968), 1205.
274. Bor, G., *1st. Proc. Int. Symp. New Aspects Chem. Metal Carbonyls Deriv.* (1968); *Chem. Abstr.,* **71** (1969), 119101q.
275. Bor, G., and Markó, L., *Chem. Ind.* (1963), 912.
276. Szabo, P., Fekete, L., Bor, G., Nagy-magos, Z., and Markó, L., *J. organometal. Chem.,* **12** (1968), 245.
277. Hieber, W., and Freyer, W., *Chem. Ber.,* **93** (1960), 462.
278. Manning, A. R., *J. chem. Soc. (A)* (1968), 1135.
279. Manning, A. R., *J. chem. Soc. (A)* (1968), 1665.
280. Slaugh, L. H., and Mullineaux, R. D., *J. organometal. Chem.,* **13** (1968), 469.
281. Cetini, G., Gambino, O., Rossetti, R., and Stanghellini, P. L., *Inorg. Chem.,* **7** (1968), 609.
282. Robinson, B. H., and Tham, W. S., *J. organometal. Chem.,* **16** (1969), P45.
283. Matheson, T. W., Robinson, B. H., and Tham, W. S., *J. chem. Soc. (A)* (1971), 1457.
284. Manning, A. R., *J. chem. Soc. (A)* (1968), 1018.

285. Bonati, F., Cenini, S., Morelli, D., and Ugo, R., *J. chem. Soc. (A)* (1966), 1052.
286. Baay, Y. L., and MacDiarmid, A. G., *Inorg. nucl. Chem. Lett.,* **3** (1967), 159.
287. Baay, Y. L., and MacDiarmid, A. G., *Inorg. Chem.,* **8** (1969), 986.
288. Kruck, T., Lang, W., Derner, N., and Stadler, M., *Chem. Ber.,* **101** (1968), 3816.
289. Rossi, M., and Sacco, A., *Chem. Commun.* (1969), 471.
290. Otvos, I., Speier, G., and Markó, L., *Acta chim. hung.,* **66** (1970), 27; *Chem. Abstr.,* **74** (1971), 27611q.
291. Hieber, W., and Ellerman, J., *Chem. Ber.,* **96** (1963), 1643.
292. Thorsteinson, E. M., and Basolo, F., *J. Am. chem. Soc.,* **88** (1966), 3929.
293. Cardaci, G., Foffani, A., Distefano, G., and Innorta, G., *Inorg. Chim. Acta,* **1** (1967), 340.
294. Cardaci, G., and Foffani, A., *Inorg. Chim. Acta,* **2** (1968), 252.
295. Thorsteinson, E. M., and Basolo, F., *Inorg. Chem.,* **5** (1966), 1691.
296. Horrocks, W. D., Jr, and Taylor, R. A., *Inorg. Chem.,* **2** (1963), 273.
297. Piazza, G., Foffani, A., and Paliani, G., *Z. phys. Chem. Frankf. Ausg.* **60** (1968), 177; *Chem. Abstr.,* **70** (1969), 8463u.
298. Hieber, W., and Ellerman, J., *Chem. Ber.,* **96** (1963), 1650.
299. Beck, W., and Lottes, K., *Z. anorg. allg. Chem.,* **335** (1965), 258.
300. Hieber, W., and Heinicke, K., *Z. anorg. allg. Chem.,* **316** (1962), 305.
301. Hieber, W., Bauer, I., and Neumair, G., *Z. anorg. allg. Chem.,* **335** (1965), 250.
302. Chatt, J., Johnson, N. P., and Shaw, B. L., *J. chem. Soc.* (1964), 2508.
303. Dwyer, F. P., and Nyholm, R. S., *J. Proc. R. Soc. N.S.W.,* **75** (1941), 140.
304. Lewis, J., Nyholm, R. S., and Reddy, G. K. N., *Chem. Ind.* (1960), 1386.
305. Dwyer, F. P., and Nyholm, R. S., *J. Proc. R. Soc. N.S.W.,* **75** (1941), 127.
306. Nyholm, R. S., *Inst. int. Chim. Solvay, Conseil Chim.* (1956), p. 225.
307. Dwyer, F. P., and Nyholm, R. S., *Proc. R. Soc. N.S.W.,* **76** (1942), 133.
308. Bennett, M. A., Clark, R. J. H., and Milner, D. L., *Inorg. Chem.,* **6** (1967), 1647.
309. Brookes, P. R., and Shaw, B. L., *J. chem. Soc. (A)* (1967), 1079.
310. Hill, W. E., and McAuliffe, C. A., unpublished observation.
311. Reddy, G. K. N., and Leelamani, E. G., *Indian J. Chem.,* **7** (1969), 929.
312. Chatt, J., and Shaw, B. L., *J. chem. Soc. (A)* (1966), 1437.
313. Evans, D., Osborn, J. A., and Wilkinson, G., *Inorg. Synth.,* **11** (1968), 99.
314. McCleverty, J. A., and Wilkinson, G., *Inorg. Synth.,* **8** (1966), 214.
315. Vallarino, L. M., *J. chem. Soc.* (1957), 2287.
316. Vallarino, L. M., *Proceedings of the 1st International Conference on Coordination Chemistry* (1957, Rome), *J. inorg. nucl. Chem.,* **7** (1958), 288.
317. Hieber, W., and Heusinger, H., *Angew. Chem.,* **68** (1956), 678.
318. Hieber, W., Heusinger, H., and Vohler, O., *Chem. Ber.,* **90** (1957), 2425.
319. Hieber, W., and Floss, J. G., *Z. anorg. allg. Chem.,* **291** (1957), 314.
320. Kingston, J. V., and Scollary, G. R., *J. inorg. nucl. Chem.,* **31** (1969), 2557.
321. Kemmitt, R. D. W., Nichols, D. I., and Peacock, R. D., *J. chem. Soc. (A)* (1968), 2149.
322. Nichols, D. I., *J. chem. Soc. (A)* (1969), 1471.
323(a). Osborne, J. A., Jardine, F. H. J., Young, J. F., and Wilkinson, G., *J. chem. Soc. (A)* (1966), 1711.
323(b). Panattoni, C., Volponi, L., Bombieri, G., and Graziani, R., *Gazz. chim. ital.,* 97 (1967), 1006; *Chem. Abstr.,* **67** (1967), 104724p.
324. Graziani, R., Bombieri, G., Volponi, L., and Panattoni, C., *Chem. Commun.* (1967), 1284.
325. Graziani, R., Bombieri, G., Volponi, L., Panattoni, C., and Clark, R. J. H., *J. chem. Soc. (A)* (1969), 1236.
326. Bombieri, G., Graziani, R., Panattoni, L., and Volponi, L., *Chem. Commun.* (1967), 977.
327. Volponi, L., De Paoli, G., and Zarli, B., *Inorg. nucl. Chem. Lett.,* **5** (1969), 947.
328. Mague, J. T., and Mitchener, J. P., *Inorg. Chem.,* **8** (1969), 119.
329. Hall, D. I., and Nyholm, R. S., *Chem. Commun.* (1970), 488.
330. Hieber, W., and Kummer, R., *Chem. Ber.,* **100** (1967), 148.
331. Ugo, R., Bonati, F., and Cenini, S., *Rc. Ist lomb. Sci. Lett.,* **A98** (1964), 627; *Chem. Abstr.,* **66** (1967), 43269z.
332. Hieber, W., and Volker, F., *Chem. Ber.,* **99** (1966), 2614.

333. Ugo, R., Bonati, F., and Cenini, S., *Inorg. Chim. Acta*, **3** (1969), 220.
334. Mague, J. T., and Wilkinson, G., *J. chem. Soc. (A)* (1966), 1736.
335. Imperial Chemical Industries, Ltd, *Neth. Pat. Appl.* 6 603 612 (1966); *Chem. Abstr.*, **66** (1967), 28511d.
336. Evans, D., Osborn, J. A., and Wilkinson, G., *J. chem. Soc. (A)* (1968), 3133.
337. Paulik, F. E., Roth, J. F., and Robinson, K. K., *Fr. Pat.* 1 560 961 (1969); *Chem. Abstr.*, **72** (1970), 54762e.
338. Dewhirst, K. C., *US Pat.* 3 489 786 (1969); *Chem. Abstr.*, **72** (1970), 89833f.
339. Blum, J., Becker, J. Y., Rosenman, H., and Bergmann, E. D., *J. chem. Soc. (B)* (1969), 1000.
340. Blum, J., Rosenman, H., and Bergmann, E. D., *Tetrahedron Lett.* (1967), 3665.
341. Blum, J., and Biger, S., *Tetrahedron Lett.* (1970), 1825.
342. Shell Internationale Research Maatschappij N.V., *Neth. Pat. Appl.* 6 601 668 (1966); *Chem. Abstr.*, **66** (1967), 11791x.
343. O'Connor, C., and Wilkinson, G., *Tetrahedron Lett.* (1969), 1375.
344. Hussey, A. S., and Takeuchi, Y., *J. org. Chem.*, **35** (1970), 643.
345. Coffey, R. S., *Br. Pat.* 1 121 643 (1968); *Chem. Abstr.*, **69** (1968), 60537q.
346. Imperial Chemical Industries, Ltd, *Neth. Pat. Appl.* 6 602 062 (1966); *Chem. Abstr.*, **66** (1967), 10556y.
347. Pruett, R. L., and Smith, J. A., *S. Afr. Pat.* 68 04 937 (1968); *Chem. Abstr.*, **71** (1969), 90819s.
348. Lawrenson, M. J., and Foster, G., *Germ. Pat.* 1 812 504 (1969); *Chem. Abstr.*, **71** (1969), 101313a.
349. Japan Synthetic Rubber Co., Ltd, *Br. Pat.* 1 184 751 (1970); *Chem. Abstr.*, **72** (1970), 112035h.
350. Hung, P. N., and Lefevre, G., *C. r. hebd. Séanc. Acad. Sci., Paris, Ser. C*, **265** (1967), 519; *Chem. Abstr.*, **68** (1968), 12350f.
351. Robinson, K. K., Paulik, F. E., Hershman, A., and Roth, J. F., *J. Catal.*, **15** (1969), 245.
352. Paulik, F. E., Hershman, A., Roth, J. F., Craddock, J. H., Knox, W. R., and Schultz, R. G., *S. Afr. Pat.* 68 02 174 (1968); *Chem. Abstr.*, **71** (1969), 12573t.
353. Deeming, A. J., and Shaw, B. L., *J. chem. Soc. (A)* (1969), 597.
354. Douek, I. C., and Wilkinson, G., *J. chem. Soc. (A)* (1969), 2604.
355. Chow, K. K., Levason, W., and McAuliffe, C. A., part 2 of this book.
356. Jennings, M. A., and Wojcicki, A., *Inorg. Chem.*, **6** (1967), 1854.
357. Hughes, W. B., *Chem. Commun.* (1969), 1126.
358. Powell, P., and Nöth, H., *Chem. Commun.* (1966), 637.
359. Haszeldine, R. N., Parish, R. V., and Parry, D. J., *J. chem. Soc. (A)* (1969), 683.
360. Nyholm, R. S., and Vrieze, K., *Proc. chem. Soc.* (1963), 138.
361. Vrieze, K., *Proceedings of the 8th International Conference on Coordination Chemistry* (ed. V. Gutman), Springer, Vienna (1964), p. 153; *Chem. Abstr.*, **66** (1967), 121640e.
362. Nyholm, R. S., and Vrieze, K., *J. chem. Soc.* (1965), 5331.
363. Takesada, M., Tamazaki, H., and Hagihara, N., *Bull. chem. Soc. Japan*, **41** (1968), 270.
364. Reddy, G. K. N., and Susheelamma, C. H., *Chem. Commun.* (1970), 54.
365. Hieber, W., and Frey, V., *Chem. Ber.*, **99** (1966), 2614.
366. Ugo, R., La Monica, G., Cenini, S., and Bonati, F., *J. organometal. Chem.*, **11** (1968), 159.
367. Hartman, F. A., and Lustig, M., *Inorg. Chem.*, **7** (1968), 2669.
368. Powell, P., *J. chem. Soc. (A)* (1969), 2418.
369. Kitchens, J., and Bear, J. L., *J. inorg. nucl. Chem.*, **31** (1969), 2415.
370. Whyman, R., *Chem. Commun.* (1970), 230.
371. Johnson, B. F. G., Lewis, J., and Robinson, P. W., *J. chem. Soc. (A)* (1970), 1100.
372. Crooks, G. R., and Johnson, B. F. G., *J. chem. Soc. (A)* (1970), 1662.
373. Hieber, W., and Heinicke, K., *Z. anorg. allg. Chem.*, **316** (1962), 321.
374. Chatt, J., Field, A. E., and Shaw, B. L., *J. chem. Soc.* (1963), 3371.
375. Reddy, G. K. N., and Leelamani, E. G., *Z. anorg. allg. Chem.*, **362** (1968), 318.
376. Canziani, F., Sartorelli, U., and Zingales, F., *Rc. Ist lomb. Sci. Lett.*, **A99** (1965), 21; *Chem. Abstr.*, **65** (1966), 3297e.
377. Jenkins, J. M., and Shaw, B. L., *J. chem. Soc.* (1965), 6789.
378. Powell, J., and Shaw, B. L., *J. chem. Soc. (A)* (1968), 617.

379. Angoletta, M., and Caglio, G., *Gazz. chim. ital.*, 99 (1969), 46.
380. Canziani, F., and Zingales, F., *Rc. Ist lomb. Sci. Lett.*, A96 (1962), 513; *Chem. Abstr.*, 62 (1965), 11401b.
381. Imperial Chemical Industries, Ltd, Chatt, J., and Shaw, B. L., *Br. Pat.* 965 811 (1964); *Chem. Abstr.*, 61 (1964), 11654b.
382. Bennett, M. A., and Milner, D. L., *J. Am. chem. Soc.*, 91 (1969), 6983.
383. Araneo, A., Bonati, F., and Minghetti, G., *Inorg. Chim. Acta*, 4 (1970), 61.
384. Mann, B. E., Masters, C., and Shaw, B. L., *Chem. Commun.* (1970), 846.
385. Araneo, A., and Martinengo, S., *Gazz. chim. ital.*, 95 (1965), 61; *Chem. Abstr.*, 63 (1965), 6585f.
386. Blake, D. M., and Kubota, M., *J. Am. chem. Soc.*, 92 (1970), 2578.
387. Mann, B. E., Masters, C., and Shaw, B. L., *Chem. Commun.* (1968), 703; (1970), 846; *J. inorg. nucl. Chem.*, 33 (1971), 2195.
388. Araneo, A., Martinengo, S., Pasquale, P., and Zingales, F., *Gazz. chim. ital.*, 95 (1965), 1435; *Chem. Abstr.*, 67 (1967), 28855q.
389. Araneo, A., and Martinengo, S., *Gazz. chim. ital.*, 95 (1965), 825; *Chem. Abstr.*, 63 (1965), 10981h.
390. Chatt, J., Leigh, G. J., Mingos, D. M. P., and Paske, R. J., *J. chem. Soc. (A)* (1968), 2636.
391. Bombieri, G., Volponi, L., and Sindellari, L., *Ricera scient.*, 36 (1966), 1226; *Chem. Abstr.*, 66 (1967), 101196m.
392. Panattoni, C., Sindellari, L., and Volponi, L., *Ricera scient. Ser. 2, Pt.2 Rend., Sez. A*, 8 (1965), 1149; *Chem. Abstr.*, 64 (1966), 1373a.
393. Hieber, W., and Frey, V., *Chem. Ber.*, 99 (1966), 2607.
394. Canziani, F., Sartorelli, U., and Zingales, F., *Chimica Ind., Milano*, 49 (1967), 469; *Chem. Abstr.*, 67 (1967), 49972r.
395. Chatt, J., Johnson, N. P., and Shaw, B. L., *J. chem. Soc. (A)* (1967), 604.
396. Deeming, A. J., and Shaw, B. L., *J. chem. Soc. (A)* (1968), 1887.
397. Deeming, A. J., and Shaw, B. L., *J. chem. Soc. (A)* (1969), 1128.
398. Blake, D. M., and Kubota, M., *Inorg. Chem.*, 9 (1970), 989.
399. Chatt, J., Johnson, N. P., and Shaw, B. L., *J. chem. Soc.* (1964), 1625.
400. Shaw, B. L., and Smithies, A. C., *J. chem. Soc. (A)* (1968), 2784.
401. Levison, J. J., and Robinson, S. D., *Inorg. nucl. Chem. Lett.*, 4 (1968), 407.
402. Deeming, A. J., and Shaw, B. L., *J. chem. Soc. (A)* (1969), 1802.
403. Deeming, A. J., and Shaw, B. L., *J. chem. Soc. (A)* (1969), 443.
404. Vaska, L., and Chen, L. S., *Chem. Commun.* (1971), 1080.
405. Church, M. J., Mays, M. J., Simpson, R. N. F., and Stefanini, F. P., *J. chem. Soc. (A)* (1970), 2909.
406. Deeming, A. J., and Shaw, B. L., *J. chem. Soc. (A)* (1970), 2705.
407. Yagupsky, G., and Wilkinson, G., *J. chem. Soc. (A)* (1969), 725.
408. Robinson, S. D., and Uttley, M. F., *J. chem. Soc. (A)* (1971), 1254.
409. Imperial Chemical Industries, Ltd, *Fr. Pat.* 1 523 527 (1968); *Chem. Abstr.*, 71 (1969), 80696k.
410. Lyons, J. E., *Chem. Commun.* (1969), 564.
411. Yamaguchi, M., *J. Soc. chem. Ind. Japan*, 70 (1967), 675; *Chem. Abstr.*, 67 (1967), 99542w.
412. *Neth. Pat. Appl.* 6 608 122 (1966); *Chem. Abstr.*, 67 (1967). 26248v.
413. Coffey, R. S., *Br. Pat.* 1 130 743 (1968); *Chem. Abstr.*, 70 (1969), 59417q.
414. Cass, R. C., Coates, G. E., and Hayter, R. G., *J. chem. Soc.* (1955), 4007.
415. Tayim, H. A., Bouldoukian, A., and Awad, F., *J. inorg. nucl. Chem.*, 32 (1970), 3799.
416. B.A.S.F., Reppe, W., Schweckendiek, W., Magin, A., and Klager, K., *Germ. Pat.* 805 642 (1951); *Chem. Abstr.*, 47 (1953), 602.
417. B.A.S.F. and Schweckendiek, W., *Germ. Pat.* 834 991 (1952); *Chem. Abstr.*, 52 (1958), 9192.
418. Bouquet, G., and Bigorgne, M., *Bull. Soc. chim. Fr.* (1962), 433.
419. Reppe, W., and Schwekendiek, W., *Justus Liebigs Annln Chem.*, 560 (1948), 104.
420. Wilkinson, G., *Z. Naturf.*, 9b (1954), 446.
421. Bouquet, G., Loutellier, A., and Bigorgne, M., *J. mol. Struct.*, 1 (1968), 211.

422. Bigorgne, M., and Bouquet, G., *C. r. hebd. Séanc. Acad. Sci., Paris, Ser. C,* **264** (1967), 1485; *Chem. Abstr.,* **67** (1967), 59210e.
423. Abel, E. W., Crow, J. P., and Illingworth, S. M., *Chem. Commun.* (1968), 817.
424. Abel, E. W., Crow, J. P., and Illingworth, S. M., *J. chem. Soc. (A)* (1969), 1631.
425. Vetter, H. J., and Nöth, H., *Z. anorg. allg. Chem.,* **330** (1964), 233.
426. Kruck, T., Baur, K., Glinka, K., and Stadler, M., *Z. Naturf.,* **23b** (1968), 1147.
427. Behrens, H., and Mueller, A., *Z. anorg. allg. Chem.,* **341** (1965), 124.
428. Nast, R., and Schulz, H., *Chem. Ber.,* **103** (1970), 785.
429. Maier, L., *Angew. Chem.,* **71** (1959), 574.
430. Feldman, J., Saffer, B. A., and Frampton, O. D., *US Pat.* 3 480 685 (1969); *Chem. Abstr.,* **72** (1970), 66350g.
431. Imperial Chemical Industries, Ltd, *Neth. Pat. Appl.* 6 514 643 (1966); *Chem. Abstr.,* **65** (1966), 13841f.
432. Cities Service Research and Development Co., Jackson, H. G., Clark, R. F., and Storrs, C. D *Belg. Pat.* 633 646 (1963); *Chem. Abstr.,* **61** (1965), 593e.
433. Studiengesellschaft Kohle m.b.H., Wilke, G., Mueller, E. W., Kroner, M., Heimbach, P., and Breil, H., *Fr. Pat.* 1 320 729 (1963); *Chem. Abstr.,* **59** (1963), 14026h.
434. Shell Internationale Research Maatschappij N.V., Zoche, G., Muller, E. W., and Korte, F., *Br. Pat.* 1 019 968 (1966); *Chem. Abstr.,* **64** (1966), 15743h.
435. Feldman, J., Frampton, O., Saffer, B., and Thomas, M., *Am. chem. Soc., Div. petrol. Chem., Preprints,* **9** (1964), A55; *Chem. Abstr.,* **64** (1966), 14076f.
436. Scholven-Chemie A.-G., *Neth. Pat. Appl.* 6 612 339 (1967); *Chem. Abstr.,* **67** (1967), 63689k.
437. Heimbach, P., and Wilke, G., *Justus Liebigs Annln Chem.,* **727** (1969), 183.
438. Wilke, G., Bogdanovic, B., Heimbach, P., Kroener, M., and Mueller, E. W., *Adv. Chem. Ser.,* **34** (1962), 137; *Chem. Abstr.,* **58** (1963), 5333a.
439. Horner, L., Reuter, H., and Herrmann, E., *Justus Liebigs Annln Chem.* **660** (1962), 1.
440. Schoenberg, A., Brosowski, K. H., Singer, E., *Chem. Ber.,* **95** (1962), 2984.
441. Hieber, W., and Bauer, I., *Z. anorg. allg. Chem.,* **321** (1963), 107.
442. Hieber, W., and Bauer, I., *Z. Naturf.,* **16b** (1961), 556.
443. Chatt, J., Duncanson, L. A., Gatehouse, B. M., Lewis, J., Nyholm, R. S., Tobe, M. L., Todd, P. F., and Venanzi, L. M., *J. chem. Soc.* (1959), 4073.
444. Chatt, J., and Wilkins, R. G., *J. chem. Soc.* (1953), 70.
445. Jensen, K. A., *Z. anorg. allg. Chem.,* **229** (1936), 225.
446. Lyon, D. R., Mann, F. G., and Cookson, G. H., *J. chem. Soc.* (1947), 662.
447. Mann, F. G., and Purdie, D., *J. chem. Soc.* (1935), 1549.
448. Mann, F. G., and Wells, A. F., *J. chem. Soc.* (1938), 702.
449. Morgan, G. T., and Yarsley, V. E., *J. chem. Soc.,* **127** (1925), 184.
450. Nyholm, R. S., *J. chem. Soc.* (1950), 848.
451. Senise, P., and Levi, F., *Analytica chim. Acta,* **30** (1964), 509.
452. Senise, P., *Analyt. Chem., Proc. int. Symp.,* Birmingham Univ. (1962), 171; *Chem. Abstr.,* **59** (1963), 9298b.
453. Senise, P., and Levi, F., *Analytica chim. Acta,* **30** (1964), 422.
454. Goodfellow, R. J., Evans, J. G., Goggin, P. L., and Duddell, D. A., *J. chem. Soc. (A)* (1968), 1604.
455. Park, P. J. D., and Hendra, P. J., *Spectrochim. Acta,* **A25** (1969), 227.
456. Coates, G. E., and Parkin, C., *J. chem. Soc.* (1963), 421.
457. Duddell, D. A., Evans, J. G., Goggin, P. L., Goodfellow, R. J., Rest, A. J., and Smith, J. G., *J. chem. Soc. (A)* (1969), 2134.
458. Fryer, C. W., and Smith, J. A. S., *J. organometal. Chem.,* **18** (1969), 35.
459. Fryer, C. W., and Smith, J. A. S., *J. chem. Soc. (A)* (1970), 1029.
460. Shobatake, K., and Nakamoto, K., *J. Am. chem. Soc.,* **92** (1970), 3332.
461. Basolo, F., Burmeister, J. L., and Poe, A. J., *J. Am. chem. Soc.,* **85** (1963), 1700.
462. Burmeister, J. L., and Basolo, F., *Inorg. Synth.,* **12** (1970), 218.
463. Keller, R. N., Johnson, N. B., and Westmoreland, L. L., *J. Am. chem. Soc.,* **90** (1968), 2729.
464. Goodgame, D. M. L., and Malerbi, B. W., *Spectrochim. Acta,* **A24** (1968), 1254.
465. Burmeister, J. L., and Basolo, F., *Inorg. Chem.,* **3** (1964), 1587.

466. Sabatini, A., and Bertini, I., *Inorg. Chem.*, **4** (1965), 1665.
467. Larsson, R., and Miezis, A., *Acta chem. scand.*, **23** (1969), 37.
468. Burmeister, J. L., and Timmer, R. C., *J. inorg. nucl. Chem.*, **28** (1966), 1973.
469. Norbury, A. H., and Sinha, A. I. P., *J. chem. Soc. (A)* (1968), 1598.
470. Norbury, A. H., and Sinha, A. I. P., *Inorg. nucl. Chem. Lett.*, **3** (1967), 355.
471. Burmeister, J. L., and Gysling, H. J., *Inorg. Chim. Acta*, **1** (1967), 100.
472. Allen, E. A., and Nixon, L. A., *J. inorg. nucl. Chem.*, **31** (1969), 1467.
473. Crociani, B., Boschi, T., and Belluco, U., *Inorg. Chem.*, **9** (1970), 2021.
474. Crociani, B., Boschi, T., and Nicolini, M., *Inorg. Chim. Acta*, **4** (1970), 577.
475. Bennett, M. A., Kneen, W. R., and Nyholm, R. S., *Inorg. Chem.*, **7** (1968), 556.
476. Duddell, D. A., Goggin, P. L., Goodfellow, R. J., Norton, M. G., and Smith, J. G., *J. chem. Soc. (A)* (1970), 545.
477. Malatesta, L., and Angoletta, M., *J. chem. Soc.* (1957), 1186.
478. Kingston, J. V., and Scollary, G. R., *Chem. Commun.* (1969), 455.
479. Shier, G. D., *US Pat.* 3 442 883 (1969); *Chem. Abstr.*, **71** (1969), 13522n.
480. Shell Internationale Research Maatschappij N.V., *Belg. Pat.* 638 289 (1964); *Chem. Abstr.*, **62** (1965), 11937d.
481. Itatani, H., and Bailar, J. C. Jr, *J. Am. Oil Chem. Soc.*, **44** (1967), 147.
482. Tayim, H. A., and Bailar, J. C. Jr, *J. Am. chem. Soc.*, **89** (1967), 4330.
483. Bailar, J. C. Jr, and Itatani, H., *J. Am. chem. Soc.*, **89** (1967), 1592.
484. Mann, F. G., and Purdie, D., *J. chem. Soc.* (1936), 873.
485. Goodfellow, R. J., Goggin, P. L., and Duddell, D. A., *J. chem. Soc. (A)* (1968), 504.
486. Livingstone, S. E., and Whitley, A., *Aust. J. Chem.*, **15** (1962), 175.
487. Wells, A. F., *Proc. R. Soc.*, **A167** (1938), 169.
488. Adams, D. M., and Chandler, P. J., *J. chem. Soc. (A)* (1969), 588.
489. Goodfellow, R. J., Goggin, P. L., and Venanzi, L. M., *J. chem. Soc. (A)* (1967), 1897.
490. Mann, F. G., and Purdie, D., *J. chem. Soc.* (1940), 1230.
491. Chatt, J., and Venanzi, L. M., *J. chem. Soc.* (1957), 2351.
492. Chatt, J., and Venanzi, L. M., *J. chem. Soc.* (1957), 2445.
493. Boschi, T., Crociani, B., Toniolo, L., and Belluco, U., *Inorg. Chem.*, **9** (1970), 532.
494. Stephenson, T. A., and Wilkinson, G., *J. inorg. nucl. Chem.*, **29** (1967), 2122.
495. Stephenson, T. A., Morehouse, S. M., Powell, A. R., Heffer, J. P., and Wilkinson, G., *J. chem. Soc. (A)* (1965), 3632.
496. Tebbe, F. N., and Muetterties, E. L., *Inorg. Chem.*, **9** (1970), 629.
497. Goodall, D. C., *J. chem. Soc. (A)* (1966), 1562.
498. Chatt, J., and Wilkins, R. G., *J. chem. Soc.* (1951), 2532.
499. Chatt, J., and Wilkins, R. G., *J. chem. Soc.* (1952), 273.
500. Chatt, J., and Wilkins, R. G., *J. chem. Soc.* (1952), 4300.
501. Chatt, J., and Wilkins, R. G., *J. chem. Soc.* (1956), 525.
502. Haake, P., and Hylton, T. A., *J. Am. chem. Soc.*, **84** (1962), 3774.
503. Adams, D. M., Chatt, J., Gerratt, J., and Westland, A. D., *J. chem. Soc.* (1958), 276.
504. Burrows, G. J., and Parker, R. H. J., *J. Proc. R. Soc. N.S.W.*, **68** (1934), 39; *Chem. Abstr.*, **29** (1935), 3620.
505. Malatesta, L., and Cariello, C., *J. chem. Soc.* (1958), 2323; *J. inorg. nucl. Chem.*, **8** (1958), 561.
506. Nyholm, R. S., *J. chem. Soc.* (1950), 843.
507. Razumova, Z., *Zh. neorg. Khim.*, **3** (1958), 1126; *Chem. Abstr.*, **53** (1959), 3970.
508. Cattalini, I., Belluco, U., Martelli, M., and Ettorre, R., *Gazz. chim. ital.*, **95** (1965), 567; *Chem. Abstr.*, **63** (1965), 12655a.
509. Belluco, U., Cattalini, L., Basolo, F., and Pearson, R. G., *J. Am. chem. Soc.*, **87** (1965), 241.
510. Pearson, R. G., Sobel, H., and Songstad, J., *J. Am. chem. Soc.*, **90** (1968), 319.
511. Cheeseman, T. P., Odell, A. L., and Raethel, H. A., *Chem. Commun.* (1968), 1496.
512. Adams, D. M., Chatt, J., Gerratt, J., and Westland, A. D., *J. chem. Soc.* (1964), 734.
513. Chatt, J., and Westland, A. D., *J. chem. Soc. (A)* (1968), 88.
514. Pidcock, A., Richards, R. E., and Venanzi, L. M., *J. chem. Soc. (A)* (1968), 1970.
515. Jensen, K. A., and Nielsen, P. H., *Acta chem. scand.*, **17** (1963), 1875.

516. Negoiu, D., and Serban, V., *An. Univ. Bucuresti, Ser. Stiint. Natur. Chim.,* 18 (1969), 55; *Chem. Abstr.,* 73 (1970), 62127n.
517. Negoiu, D., and Serban, V., *An. Univ. Bucuresti, Ser. Stiint. Natur. Chim.,* 18 (1969), 113; *Chem. Abstr.,* 74 (1971), 106695y.
518. Pecile, C., *Inorg. Chem.,* 5 (1966), 210.
519. Deganello, G., Carturan, G., and Belluco, U., *J. chem. Soc. (A)* (1968), 2873.
520. Young, J. F., Gillard, R. D., and Wilkinson, G., *J. chem. Soc.* (1964), 5176.
521. Chatt, J., Eaborn, C., and Kapoor, P. N., *J. organometal. Chem.,* 23 (1970), 109.
522. Blake, D. M., and Nyman, C. J., *Chem. Commun.* (1969), 483.
523. Church, M. J., and Mays, M. J., *J. chem. Soc. (A)* (1970), 1938.
524. Cherwinski, W. J., and Clark, H. C., *Can. J. Chem.,* 47 (1969), 2665.
525. Westland, A. D., *J. chem. Soc.* (1965), 3060.
526. Chatt, J., and Heaton, B. T., *J. chem. Soc. (A)* (1968), 2745.
527. Chatt, J., and Heaton, B. T., *Spectrochim. Acta,* A23 (1967), 2220.
528. Bosnich, B., and Wild, S. B., *J. Am. chem. Soc.,* 92 (1970), 459.
529. Bennett, M. A., Kouwenhoven, H. W., Lewis, J., and Nyholm, R. S., *J. chem. Soc.* (1964), 4570.
530. Bennett, M. A., Chatt, J., Erskine, G. J., Lewis, J., Long, R. F., and Nyholm, R. S., *J. chem. Soc. (A)* (1967), 501.
531. Bennett, M. A., Kneen, W. R., and Nyholm, R. S., *J. organometal. Chem.,* 26 (1971), 293.
532. Hall, D. I., and Nyholm, R. S., *J. chem. Soc. (A)* (1971), 1491.
533. Blake, D. M., and Nyman, C. J., *Chem. Commun.* (1969), 483.
534. Kistner, C. R., Blackman, J. D., and Harris, W. C., *Inorg. Chem.,* 8 (1969), 2165.
535. Bamford, W. R., Lovie, J. V., and Watt, J. A. C., *J. chem. Soc. (C)* (1966), 1137.
536. Imperial Chemical Industries, Ltd, Lovie, J. C., and Watt, J. A. C., *Br. Pat.* 1 023 797 (1966); *Chem. Abstr.,* 64 (1966), 19824e.
537. Frankel, E. N., Emken, E. A., Itatani, H., and Bailar, J. C. Jr, *J. org. Chem.,* 35 (1967), 1447.
538. Adams, R. W., Batley, G. E., and Bailar, J. C. Jr, *J. Am. chem. Soc.,* 90 (1968), 6051.
539. Goodfellow, R. J., and Venanzi, L. M., *J. chem. Soc.* (1965), 7533.
540. Watkins, S. F., *Chem. Commun.* (1968), 504.
541. Watkins, S. F., *J. chem. Soc. (A)* (1970), 168.
542. Taylor, M. J., Odell, A. L., and Raethel, H. A., *Spectrochim. Acta,* A24 (1968), 1855.
543. Chatt, J., and Davidson, J. M., *J. chem. Soc.* (1964), 2433.
544. Pidcock, A., and Waterhouse, C. R., *Inorg. nucl. Chem. Lett.,* 3 (1967), 487.
545. Pidcock, A., and Waterhouse, C. R., *J. chem. Soc. (A)* (1970), 2080.
546. Irving, R. J., and Magnusson, E. A., *J. chem. Soc.* (1956), 1860.
547. Adams, D. M., and Chandler, P. J., *J. chem. Soc. (A)* (1967), 1009.
548. Peloso, A., and Dolcetti, G., *J. chem. Soc. (A)* (1967), 1944.
549. Peloso, A., Ettorre, R., and Dolcetti, G., *Inorg. Chim. Acta,* 1 (1967), 307.
550. Bennett, M. A., Erskine, G. J., and Nyholm, R. S., *J. chem. Soc. (A)* (1967), 1260.
551. Bennett, M. A., Erskine, G. J., Lewis, J., Mason, R., Nyholm, R. S., Robertson, G. B., and Towl, A. D. C., *Chem. Commun.* (1966), 395.
552. Cariati, F., Ugo, R., and Bonati, F., *Chem. Ind.* (1964), 1714.
553. Cariati, F., Ugo, R., and Bonati, F., *Inorg. Chem.,* 5 (1966), 1128.
554. Chatt, J., and Shaw, B. L., *J. chem. Soc.* (1962), 5075.
555. Toniolo, L., Giustiniani, M., and Belluco, U., *J. chem. Soc. (A)* (1969), 2666.
556. Powell, J., and Shaw, B. L., *J. chem. Soc.* (1965), 3879.
557. Hatfield, W. E., and Whyman, R., *Transit. Metal Chem.,* 5 (1969), 47.
558. Sandhu, S. S., and Parmar, S. S., *Chem. Commun.* (1968), 1335.
559. Mellor, D. P., Burrows, G. J., and Morris, B. S., *Nature, Lond.,* 141 (1938), 414.
560. Mellor, D. P., and Craig, D. P., *J. Proc. R. Soc. N.S.W.,* 75 (1941), 27; *Chem. Abstr.,* 35 (1941), 7310.
561. Nyholm, R. S., *J. chem. Soc.* (1951), 1767.
562. Kettle, S. F. A., *Theor. Chim. Acta,* 4 (1966), 150.
563. Nyholm, R. S., *J. chem. Soc.* (1952), 1257.
564. Wells, A. F., *Z. Kristallogr. Kristallgeon.,* 94 (1936), 447; *Chem. Abstr.,* 31 (1937), 590.
565. Mann, F. G., Purdie, D., and Wells, A. F., *J. chem. Soc.* (1936), 1503.

566. Jardine, F. H., and Young, F. J., *J. chem. Soc. (A)* (1971), 2444.
567. Azanovskaya, M. M., *Zh. obshch. Khim.*, 27 (1957), 1363; *Chem. Abstr.*, 52 (1958), 2738.
568. Cooper, D., and Plane, R. A., *Inorg. Chem.*, 6 (1966), 2209.
569. Lippard, S. J., and Ucko, D. A., *Inorg. Chem.*, 7 (1968), 1051.
570. Lippard, S. J., and Melmed, K. M., *Inorg. Chem.*, 8 (1969), 2755.
571. Baird, M. C., and Zelonka, R. A., *Chem. Commun.* (1971), 780.
572. Negoiu, D., Negoiu, M., and Ilinca, F., *An. Univ. Bucuresti, Ser. Stiint. Natur. Chim.*, 18 (1969), 139.
573. Bennett, M. A., Kneen, W. R., and Nyholm, R. S., *Inorg. Chem.*, 7 (1968), 552.
574. Mann, F. G., Wells, A. F., and Purdie, D., *J. chem. Soc.* (1937), 1828.
575. Nuttall, R. H., Roberts, E. R., and Sharp, D. W. A., *J. chem. Soc.* (1962), 2854.
576. Burrows, G. J., and Parker, R. H., *J. Am. chem. Soc.*, 55 (1933), 4133.
577. Wittig, G., and Hellwinkel, D., *Chem. Ber.*, 97 (1964), 769.
578. Olson, D. C., and Bjerrum, J., *Acta chem. scand.*, 20 (1966), 143.
579. Bjerrum, J., George, R. S., Hawkins, C. J., and Olson, D. C., *Proc. Symp. Coord. Chem., Tihany, Hung.* (1964), 187; *Chem. Abstr.*, 64 (1966), 18940f.
580. Westland, A. D., *Can. J. Chem.*, 47 (1969), 4135.
581. Gibson, D., Johnson, B. F. G., and Lewis, J., *J. chem. Soc. (A)* (1970), 367.
582. Cahours, A., and Gal, H., *Justus Liebigs Annln Chem.*, 156 (1870), 302; *C. r. hebd. Séanc. Acad. Sci., Paris*, 70 (1870), 1380.
583. Dwyer, F. P., and Stewart, D. M., *J. Proc. R. Soc. N.S.W.*, 83 (1949), 177; *Chem. Abstr.*, 57 (1962), 13806.
584. Shell Development Co., Van Peski, A., and Van Melsen, J. A., *US Pat.* 2 150 349 (1938).
585. King, M. G., and McQuillan, G. P., *J. chem. Soc. (A)* (1967), 898.
586. Brier, P. N., Chalmers, A. A., Lewis, J., and Wild, S. B., *J. chem. Soc. (A)* (1967), 1889.
587. Shioteni, A., and Schmidbaur, H., *J. Am. chem. Soc.*, 92 (1970), 7003.

PART 4
The Chemistry of Multidentate Ligands Containing Heavy Group VB Donors

B. CHISWELL

*Department of Chemistry, University of Queensland,
St Lucia, Queensland, Australia*

26 INTRODUCTION

26.1 SCOPE OF REVIEW

This review aims to cover the field of multidentate ligands (not including bidentates) containing at least one phosphorus, arsenic, antimony or bismuth donor atom.

The nearest approach to a similar review was compiled by Booth[1] in 1964. Although some reference will be made for continuity of argument to papers mentioned by Booth, this review will concentrate in the main on the field that has grown so rapidly since 1965.

The greater bulk of the work published in the area under review has stemmed from four research schools and their offshoots, and the majority of papers are associated with the names of Venanzi, Nyholm, Meek and Sacconi.

The review will concentrate upon tri- and tetradentate ligands. A large volume of work has now been published on such ligands, and a substantial body of speculative interpretation has developed (chapter 29.1). On the other hand ligands of a higher denticity containing heavy Group VB donor atoms are rare, although Barclay et al.[2] have reported that the hexadentate arsine, tetrakis-(3-dimethylarsinopropyl)-o-phenylenediarsine, gives low-spin octahedral compounds with iron(II), cobalt(II) and nickel(II) metal atoms.

A stimulating review[3] has recently been written on the synthesis of pentadentate ligands, but reference was made chiefly to nitrogen and oxygen donor ligands. Pentadentate ligands containing heavy Group VB donors of the type shown in (CLXIV) (as yet unknown), would be able to fill all five coordination sites on a trigonal-bipyramidal metal atom.

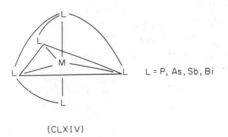

L = P, As, Sb, Bi

(CLXIV)

Other types of ligands, also as yet unknown, that will be of interest in the field of this review are macrocyclic ligands containing the heavier Group VB donor atoms of similar type to those described for nitrogen by Busch and other workers[4], and ligands that are designed to force a square-pyramidal shape upon a divalent metal atom in the same manner that the now famous QAS (tris-[o-diphenyl-arsinophenyl-]arsine) and QP (tris-[o-diphenylphosphinophenyl-]phosphine) were designed by Venanzi[5] to yield trigonal-bipyramidal complexes. A possible example[6] is shown in (CLXV).

$$[Ph_2P(CH_2)_2]_2-C=C-[(CH_2)_2PPh_2]_2$$
(CLXV)

26.2 LIGAND TYPES

26.2.1 Linear and tripod ligands

The term *tripod* ligand has been used to describe ligands that possess an apical Group VB donor atom, or an apical tetrahedral carbon atom, with three molecular chains attached, each of which contains a potential donor atom. Such ligands can act as tetradentates (CLXVI) or tridentates (CLXVII) respectively. The origin of the name tripod is obvious and it is easy to distinguish such ligands from those in which the donors occur in the unbifurcated molecular chain of the ligand.

However, the use of the term tripod to describe tridentate ligands of the type shown in (CLXVIII) is confusing. Such ligands are not strictly tripods, and are structurally no different to long-chain tridentates in which the donor atoms are in the unbifurcated ligand chain. Thus, although these ligands are structurally likely to place the two D groups on *cis*-corners of a square pyramid, octahedron or trigonal bipyramid as does $MeAs[(CH_2)_3AsMe_2]_2$ in the square-pyramidal complex $[Ni(ligand)Br_2]$ [7], the same ligand may span three corners of a square plane with the two D groups occupying *trans*-positions. Thus palladium(II) and platinum(II) complexes of the type $[M(ligand)Br]^+$ with the above ligand have a square-planar stereochemistry[8].

As this review will show, the types of tripod tetradentates known have expanded rapidly since Venanzi *et al.* reported[9] the first examples. The types of molecular chains that have been used to connect the L group of (CLXVI) with its three donor atoms are:

(i) phenyl groups with the further potential donor atoms in the *ortho*-positions. Such ligands have five-membered chelate rings.
(ii) ethylene chains. Again five-membered chelate rings are formed.
(iii) propylene chains. Six-membered chelate rings are obtained.

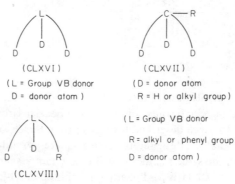

(CLXVI)
(L = Group VB donor
D = donor atom)

(CLXVII)
(D = donor atom
R = H or alkyl group)

(L = Group VB donor

R = alkyl or phenyl group
D = donor atom)

(CLXVIII)

The shape of these tetradentate tripod ligands places fairly rigid limits on the stereochemistry available to the metal atom, particularly in ligands using the above type (i) molecular chains. The author still remembers making a molecular model of $P[o\text{-}C_6H_4PPh_2]_3$ in 1961, and noting with pleasure how beautifully a trigonal bipyrimidal atom dropped into the outstretched ligand claws!

26.2.2 Hybrid ligands

Much recent work has concentrated on multidentate ligands possessing at least two different types of donor atoms. In many cases such ligands possess both *hard* and *soft* donors[10]. Such ligands have been called *hybrid* ligands to indicate that the overall donor behaviour of the ligand exhibits a cross between typical hard and soft behaviour. Sacconi[11] in particular has looked closely at hybrid ligands of many types, and discussed the factors governing the spin-multiplicity of five-coordinate cobalt(II) and nickel(II) compounds (chapter 29).

Purely for the sake of classification in this review the donors nitrogen, oxygen and sulphur are taken as hard donors, while all the other Group VB donors and selenium are classified as soft donors.

26.2.3 Donor atom sets and sequences

A donor atom set denotes the donors present without reference to their sequence in the ligand. Thus the linear quadridentate, ethylene-bis-(oxyethylene)-bis-(diphenylphosphine) $[Ph_2P(CH_2)_2O(CH_2)_2O(CH_2)_2PPh_2]$, has the donor set P_2O_2, but the donor sequence POOP. The donor sequence in tripod ligands will be shown in the review as L(ABC), where L = the apical ligand and A, B and C the molecular chain donors.

In some instances multidentates do not use all of the potential donor atoms. As this part of the review is organised on the basis of types of ligand, the potential dentate action has been the rule in classification. Thus potential quadridentates, for example, regardless of their coordinating ability in specific cases, are discussed in chapter 28.

26.2.4 Nomenclature and abbreviations

The very complexity of the multidentates discussed in this review, has led workers to use abbreviations for their ligands. Although in specific papers such abbreviations are readily comprehended, in a review article it is impossible to use any of the systems of abbreviation advanced by different groups of workers as ambiguities must thereby arise. I have therefore chosen to use

(i) full names for ligands
(ii) only widely accepted abbreviations, for example Me for methyl, Ph for phenyl, $o\text{-}C_6H_4$ for *ortho*-substituted benzene.

Thus, for example, tris-(o-diphenylphosphinophenyl)arsine will be shown as $As[o\text{-}C_6H_4PPh_2]_3$.

27 TRIDENTATE LIGANDS

27.1 TRIDENTATES WITH SOFT DONOR ATOMS ONLY

27.1.1 Tridentates with donor set P_3

Ligands with this donor set are of both tripod and linear types shown in (CLXVII) and (CLXVIII) respectively (L = D = phosphorus). Work on both types originated with Hart[22] in 1960 and Chatt and Watson[12] in 1961. These workers found that the three ligands $PhP[o\text{-}C_6H_4PEt_2]_2$, $PhP[(CH_2)_2PPh_2]_2$ and $MeC[CH_2PPh_2]_3$ gave *cis*-compounds of the type [M(CO)$_3$ ligand] when reacted with the Group VIA metal hexacarbonyls. This work was later extended by other workers[13], and the interaction of the phosphine ligands with nickel tetracarbonyl was also studied[14]. In this latter case the triphosphines were shown to be capable of spanning three of the tetrahedral apices about the Ni(0) atom to yield [Ni(CO)ligand].

Recent work on $MeC[CH_2PPh_2]_3$ has dealt with its rhenium[15], and cobalt and nickel[16] complexes. With rhenium(V) the ligand yields:

(i) blue six-coordinate compounds of formula [ReO(ligand)Cl$_3$] in which the ligand exhibits bidentate action;
(ii) green seven-coordinate compounds of the same formula in which the ligand acts as a tridentate.

Reduction of the Re(V) compounds yields octahedral Re(III) complexes of the type [Re(ligand)Cl$_3$]. Similarly, in the tetrahedral cobalt(II) salts of formula [Co(ligand)X$_2$] (X = Cl or Br), the triphosphine acts as a bidentate, but where X = ClO$_4$ the four-coordinate complex [Co(ligand)ClO$_4$]ClO$_4$ is obtained with a tridentate ligand group. On the other hand, the compounds [Co(ligand)(NCS)$_2$] and [Ni(ligand)X$_2$] (X = Cl, Br or NCS), are all five-coordinate.

Venanzi and coworkers have published a number of papers dealing with the chemistry of the ligand $PhP[o\text{-}C_6H_4PPh_2]_2$ [9]. With palladium(II) and platinum(II) four-coordinate compounds of the type [M(ligand)I] I are obtained, in which the ligand occupies three of the four corners of a square plane[17]. Although it was reported[18(a)] that with chromium(III) chloride and bromide the ligand reacted to give octahedral [Cr(ligand)X$_3$], a later paper[18(b)] claimed that the compound [Cr(ligand)(CO)$_3$], obtained by interaction of Cr(CO)$_6$ and the triphosphine, did not yield isolable Cr(III) compounds upon oxidation, as did the corresponding tetraphosphine $P[o\text{-}C_6H_4PPh_2]_3$. With manganese(I) pentacarbonyl halides the triphosphine yields the octahedral cation [Mn(CO)$_3$(ligand)]$^+$ [19], but with ruthenium(0) five-coordinate compounds of the type [Ru(CO)$_2$(ligand)] were obtained[20].

Somewhat surprisingly the linear tridentate 2,4-diethyl-1,5-bis(diphenyl)-3-phenyl-1,3,5-triphospha-2,4-diazapentane [$Ph_2P.N(Et).P(Ph).N(Et).PPh_2$] has been shown by X-ray work[21] to coordinate all three phosphorus atoms in the *cis* positions of an octahedron in $[M(CO)_2(ligand)]$.

27.1.2 Tridentates with donor set As$_3$

The first example of a tridentate As$_3$ ligand was the compound bis(3-dimethylarsinopropyl)methylarsine, whose transition metal compounds, M(ligand)X$_2$, were briefly described in 1953[23]. The four-coordinate palladium(II) and platinum(II) compounds, of types [M(ligand)Br] Br and $[M(ligand)_2](ClO_4)_2$, were more fully described much later[8]. In 1960 it was demonstrated that in nickel(II) halide compounds of the triarsine the metal atom was five-coordinate[7]. The X-ray structure of [Ni(ligand)Br$_2$] was shown to be a distorted square pyramid with the three arsenic atoms and the nickel atom in the plane of the square, one bromine normal to this plane above the nickel atom, and the second bromine depressed 20° below the square plane. It was suggested at the time that the arsine methyl groups pushed this bromine atom down, a contention supported by the X-ray analysis of [Pd(ligand)Br] Br, in which the palladium and arsenic atoms in the square plane with the bromine depressed 10° from this plane.

The interaction of various metal halides with the triarsine $MeAs[o\text{-}C_6H_4AsMe_2]_2$ have been reported. With cobalt(II) unstable low-spin five-coordinate compounds of type $[Co(ligand)_2]^{2+}$ have been isolated. These readily oxidise to the cobalt(III) compounds [Co(ligand)X$_3$] (where X = Cl, Br, I, NCS, NO$_2$ or NO$_3$)[30]. Other d^6 metals such as Rh(III) and Ir(III) also give compounds of the same formulation, as well as the bis-compounds $[M(ligand)_2](ClO_4)_3$[31]. The *fac*- and *mer*-isomers of some of the compounds [Rh(ligand)X$_3$] were isolated and distinguished on the basis of their i.r. spectra in the Rh–X stretch region. Nanda and Tobe[32] have measured the rate of displacement of chloride ion from the [Co(ligand)Cl$_3$] complex.

An interesting synthesis of the ligand $MeAs[o\text{-}C_6H_4AsMe_2]_2$ has been revealed in the study of the supposedly octahedral low-spin tris(o-phenylenebisdimethylarsine)nickel(II) perchlorate[44]. This anomalous compound has been found to be the five-coordinate $[Ni(MeAs[o\text{-}C_6H_4AsMe_2]_2)(Me_2As[o\text{-}C_6H_4AsMe_2])](ClO_4)_2$, possessing both a tridentate and a bidentate arsine moiety, the former being synthesised from the latter during complex formation[42]. Electronic spectral[43] and X-ray work show the compound to be essentially square pyramidal in shape. As the electronic spectra of the compounds $[Ni(triarsine)_2](ClO_4)_2$ (where triarsine = $MeAs[o\text{-}C_6H_4AsMe_2]_2$, $MeAs[(CH_2)_3AsMe_2]_2$ or $MeC[CH_2AsMe_2]_3$) are identical with the spectrum of this $[Ni(As)_5](ClO_4)_2$ compound, it is suggested that all three bis-triarsine compounds are also five-coordinate[42].

The titanium(IV) halide adducts of $MeAs[(CH_2)_3AsMe_2]_2$ have been reported to be [TiF$_4$]$_3$. triarsine, [TiCl$_4$]$_2$. triarsine, [TiBr$_4$]. triarsine and [TiI$_3$triarsine]I[33].

Nyholm and coworkers have studied the interaction of the tripod triarsine $MeC[CH_2AsMe_2]_3$[34], and the two triarsines, $MeAs[o\text{-}C_6H_4AsMe_2]_2$[35] and $MeAs[(CH_2)_3AsMe_2]_2$[36], with Group VIB metal hexacarbonyls. With all three hexacarbonyls the compounds *cis*-[M(ligand)(CO)$_3$] (where M = Cr, Mo or W) are obtained with a stability order $W > Mo > Cr$ [35]. With $MeC[CH_2AsMe_2]_3$, the complex $[W(ligand)(CO)_4]$, in which the ligand acts as a bidentate, was also prepared[34]. Oxidation reactions of these tricarbonyls with halogens have been studied. Both the molybdenum and tungsten compounds yield seven-coordinate complexes $[M(ligand)(CO)_3X]X$ (where $X = Br$ or I) [34-6]. Heating of these compounds removes one carbonyl group to yield the uncharged compound $[M(ligand)(CO)_2X_2]$ [35,37]. The halogen oxidation of $[Cr(MeAs\{o-C_6H_4AsMe_2\}_2)(CO)_3]$ yields the corresponding seven-coordinate $[Cr(ligand)(CO)_3X]X$ compound, which upon heating loses all its carbon monoxide to yield $[Cr(ligand)X_3]$ [35]. The other linear triarsine compound $[Cr(MeAs\{(CH_2)_3AsMe_2\}_2(CO)_3]$ is also readily oxidised to $[Cr(ligand)(CO)_2I]^+$, but the tripod ligand complex $[Cr(MeC\{CH_2AsMe_2\}_3)(CO)_3]$ yields $[Cr(ligand)(CO)_2I]^+$ upon reaction with iodine[37]. With Group VIIA carbonyls, octahedral compounds of the types $[Re(CO)_3(ligand)Cl]$ (ligand acting as a bidentate) and $[M(CO)_3ligand]$ (where $M = Mn$ or Re) are formed where the ligand is $MeC[CH_2AsMe_2]_3$ [34].

The monovalent tetrahedral compounds $[M(ligand)(halide)]$ (where $M = Cu$ or Ag), in which the tridentate ligand is either the tripod $MeC[CH_2AsMe_2]_3$ or $MeAs[o-C_6H_4AsMe_2]_2$, have been used in the preparation of metal–metal bond compounds of the type $[(triarsine)M'–M''(CO)_x]$ (where $M' = Cu$ or Ag; $M'' = Mn$, $x = 5$; $M'' = Fe$ or $Co, x = 4$) [38-40]. The main study here has been concerned with the effect of the metal–metal bond upon the carbonyl i.r. bands, and the ligand has been chosen to give a stable monovalent copper or silver atom possessing only one halogen atom. X-ray analyses of these complexes indicate that the copper and silver atoms both possess distorted tetrahedral stereochemistry[40,41].

In the early 1960s Venanzi and coworkers produced a series of papers and reviews dealing with the transition metal complexes of bis(o-diphenylarsino)-phenylarsine, $PhAs[o-C_6H_4AsPh_2]_2$. With platinum(II) this ligand forms four-coordinate complexes of type $[Pt(ligand)I]X$ (where $X = I$ or ClO_4) [24], but both four-coordinate $[Pd(ligand)I]ClO_4$ and five-coordinate $[Pd(ligand)I_2]$ could be obtained[25]. Rhenium(II) compounds of the triarsine are five-coordinate $[Re(ligand)X_2]$ (where $X = Cl$, Br or I), while reaction of the ligand with $ReCl_3$ or $ReBr_3$ yields the rhenium(V) compounds $[ReO(ligand)X_3]$ [26]. The compound $[Rh(triarsine)Cl_3]$ is obtained by the reaction of the ligand with $[Rh(Ph_3P)_3Cl_3]$ [27].

Much of the above work on $PhAs[o-C_6H_4AsPh_2]_2$ is collected in a review[5]. More recent work on this ligand has resulted in the preparation of $[Cr(triarsine)(CO)_3]$ by ligand interaction with $Cr(CO)_6$ [18], five-coordinate $[Re(ligand)(CO)_2]$ [20] and square-pyramidal nickel(II) compounds of types $[Ni(ligand)X_2]$ (where $X = Br$ or I) and $[Ni_2(ligand)_3H_2O](ClO_4)_4$ [28]. In this latter compound, it is postulated

(CLXIX)

that the ligand bridges as shown in (CLXIX) (L = As). Attempts to prepare cobalt(II) compounds with the ligand were unsuccessful[28]. It has been suggested that this lack of complexing ability can be attributed to the poor σ-donor ability of the ligand −AsPh$_2$ groups[29].

27.1.3 Tridentates with donor set As$_2$P

The ligand bis(*o*-diphenylarsinophenyl)phenylphosphine, PhP[*o*-C$_6$H$_4$AsPh$_2$]$_2$, has been reported[28] to give similar compounds with nickel(II) as the corresponding triarsine (P = As, section 27.2.2), namely square-pyramidal [Ni(ligand)X$_2$] and [Ni$_2$(ligand)$_3$H$_2$O](ClO$_4$)$_4$ (structure (CLXIX), L = P). Unlike the triarsine this AsPAs ligand yielded square-pyramidal green cobalt(II) complexes of formulation [Co(ligand)X$_2$] (where X = Br or NCS).

27.2 HYBRID TRIDENTATES WITH HARD AND SOFT DONOR ATOMS

27.2.1 Tridentates with donor sets PN$_2$ and P$_2$N

The nuclear magnetic resonance work of Venanzi *et al.*[45] on the ligands Ph$_{3-n}$P[*o*-C$_6$H$_4$NMe$_2$]$_n$ (where *n* = 2 or 3) revealed that in both instances the phosphorus and only one nitrogen atom were coordinated in the complexes [M(ligand)X$_2$] (where M = Pd or Pt; X = Cl, Br or I). At room temperature there is very rapid exchange of the single coordinated nitrogen between the two or three potential donors, but a structure in which one definite nitrogen was coordinated could be 'frozen out' at low temperatures. The complexes with nickel(II) and cobalt(II) halides with this ligand are tetrahedral with the phosphorus and one nitrogen donor coordinated[74]. Dobson *et al.*[46] also found that when PhP[(CH$_2$)$_2$NEt$_2$]$_2$ reacted with Mo(CO)$_6$ to yield octahedral [Mo(CO)$_4$(ligand)], only one of the two nitrogen donors was coordinated. On the other hand, if the similar ligand with the PNP donor sequence (EtN[(CH$_2$)$_2$PPh$_2$]$_2$) was reacted with molybdenum hexacarbonyl, tridentate action ensued and octahedral [Mo(CO)$_3$(ligand)] was obtained[46].

Sacconi and Morassi have studied the interaction of the tridentates RN[(CH$_2$)$_2$PPh$_2$]$_2$ [47](donor sequence PNP) and [Ph$_2$P(CH$_2$)$_2$]N(R)- [(CH$_2$)$_2$NEt$_2$] (PNN)[48], with cobalt(II) and nickel(II) halides. They found that for the ligand with donor sequence PNP the compounds [Ni(ligand)X$_2$] changed from low-spin five-coordinate to square-planar [Ni(ligand)X]$^+$ as the R group on the central nitrogen donor atom became more bulky, that is as R varied from H to Me to cyclohexyl. With the same changes in the bulkiness of the R group for the five-coordinate compounds [Co(ligand)X$_2$], tetrahedral [Co(ligand)X]$^+$ were obtained[47]. The ligand with donor sequence PNN yielded high-spin five-coordinate compounds of type [Co(ligand)X$_2$] (when X = Cl, Br and I, R on central N atom = H or Me; when X = NCS, R = Me). However, when X = NCS and R = H another example of singlet–triplet equilibrium for this d^7 ion was established[48]. Critical comparison of the effect of the bulkiness of the central nitrogen R group in these two ligands is not possible as the tridentate (cyclohexyl)N[(CH$_2$)$_2$PPh$_2$]-[(CH$_2$)$_2$NEt$_2$] was not reported.

The X-ray crystal structure of dibromo(bis[2-diphenylphosphinoethyl]amine)-nickel(II), [Ni(HN[(CH$_2$)$_2$PPh$_2$]$_2$)Br$_2$], has been reported[49]. The structure is a distorted square pyramid, with one bromine at the apical position, the two phosphorus, the other bromine and nitrogen atoms roughly in the basal plane, and the nickel atom 16 pm above this plane. The length of the apical Ni–Br bond is longer by 37 pm than the Ni–Br distance in the basal plane. This can be justified in terms of the low-spin nature of the compound, which requires that there are two electrons in the $d_{z^2}^*$ anti-bonding orbital (normal to the basal plane), and none in the $d_{x^2-y^2}^*$ orbital (in the basal plane). The bromine in the basal plane is thus repelled less than that in the apical position of the square pyramid.

In a lengthy paper dealing with hybrid tridentates prepared by condensing *o*-substituted benzaldehydes and 3-aminodiphenylphosphine (or arsine) to yield a number of different donor atom sequences, Sacconi *et al.*[50] describe the five-coordinate nickel(II) complexes of the linear tridentate ligand shown in structure (CLXX) (L = P). In the compounds [Ni(ligand)X$_2$], where X = Br a high-spin compound is formed, but where X = I the compound is low spin.

(CLXX)

A series of recent papers has dealt with tridentates with the donor sequence PNP, in which the nitrogen donor is a pyridine nitrogen atom (CLXXI). Some of the five-coordinate compounds formed by these two ligands with both cobalt(II) and nickel(II) of type [M(ligand)X$_2$] exhibit spin equilibrium. Thus when M = Co, n = 2 and X = Cl, the complex is high spin, but when X = Br or I, the complexes exhibit anomalous magnetic behaviour over a temperature range[51,53]; however, for

(CLXXI)

nickel when n = 2 and X = Cl, the magnetic moment of the complex varies from 3.16 B.M. at 294 K to 1.32 B.M. at 99K, while when X = Br and I, both compounds are diamagnetic[52,53]. Somewhat surprisingly the five-coordinate complexes of iron(II), [Fe(PNP)I$_2$] and [Fe(PNP)I(NCS)], with ligand (CLXXI) (when n = 2) also exhibit spin equilibrium[53]. These five-coordinate compounds of cobalt, nickel and iron are postulated to be distorted trigonal bipyramidal in stereochemistry[53]. The PNP ligand, where n = 1 in (CLXXI), also yields five-coordinate (distorted square pyramidal) compounds of type [M(ligand)X$_2$] in which M = Fe, Co or Ni and X = halide or NCS. The iron compounds are high spin, but the cobalt and nickel complexes exhibit spin equilibrium depending upon the nature of the X group[54].

27.2.2 Tridentates with donor sets AsN_2 and As_2N

Tridentates with the donor sequence AsNAs have been little studied. The interaction of $HN[(CH_2)_2AsPh_2]_2$ and nickel(II) iodide yields diamagnetic $[Ni(ligand)I]^+$, but with nickel(II) bromide the compound $[Ni(ligand)Br_2]$ is obtained[55]. This compound has an anomalous magnetic moment of 2.25 B.M., and this value is unchanged over a temperature range. It is suggested from this and a study of the solid spectrum of the compound that there is a 50/50 mixture of high-spin octahedral and diamagnetic five-coordinate nickel atoms in the structure of the compound.

The donor atom sequence NNAs has now been well studied. The ligand shown in (CLXX) (L = As) forms the high-spin five-coordinate compounds $[Ni(ligand)X_2]$ (where X = Br or I) [50]. High-spin five-coordinate complexes $[M(ligand)X_2]$ (where M = Co, Ni or Mn; X = halogen) and octahedral complexes of types $[M(ligand)X_2]$ (high spin where M = Co or Ni; X = NCS or NO_3) and $[M(ligand)_2]^{2+}$ (where M = Co, Ni or Mn) have been prepared[56] from a series of NNAs ligands (see

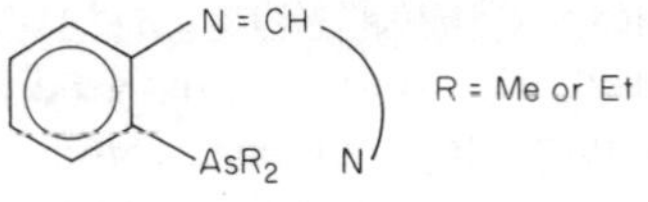

(CLXXII)

structure (CLXXII)) derived from the interaction of both *o*-dimethylarsinoaniline[64] and *o*-diethylarsinoaniline with the aldehydes:

(i)   pyridine-2-aldehyde
(ii)   6-methylpyridine-2-aldehyde
(iii) *o*-methylaminobenzaldehyde
(iv)   *o*-dimethylaminobenzaldehyde
(v)   2-pyrrolealdehyde

27.2.3 Tridentates with donor sets P_2O, As_2O, PS_2, P_2S and As_2S

Meek and coworkers have been responsible for a series of papers on the complexes of the SPS ligand, bis(*o*-methylthiophenyl)phenylphosphine(PhP[*o*-C_6H_4SMe]$_2$). In square-planar $[Pd(ligand)Cl_2]$ the ligand uses the phosphorus and only one sulphur donor atom, but in $[Pd(ligand)Cl]ClO_4$ all three potential donors are coordinated[57,113]. With nickel(II) salts five-coordinate diamagnetic compounds of types $[Ni(ligand)X_2]$, $[Ni(ligand)_2](ClO_4)_2$ and $[Ni(ligand)(bidentate)](ClO_4)$, are obtained[58]. The X-ray crystal structure of $[Ni(ligand)I_2]$ consists of discrete square-pyramidal molecules with the phosphorus, sulphur and two iodine atoms all in the basal plane, and the nickel only slightly displaced above this plane towards the apical sulphur atom[59]. It is argued that the ligand field stabilisation energy favours the square-pyramidal over the trigonal-bipyramidal structure for the five-coordinate d^8 configuration, and that the Ni → P π back-bonding keeps the nickel atom approximately in the basal plane[59]. The point is made[58] that five-coordinate nickel(II) is a product of the type of donor set rather than of specific demands of the ligand, since the compounds [Ni(bidentate)Cl$_2$] and [Ni(bidentate)$_2$Cl$_2$] (where bidentate means both ligands shown in structure (CLXXIII)) all have spectra similar to [Ni(PhP[o-C$_6$H$_4$SMe]$_2$)Cl$_2$] and are thus also five-coordinate.

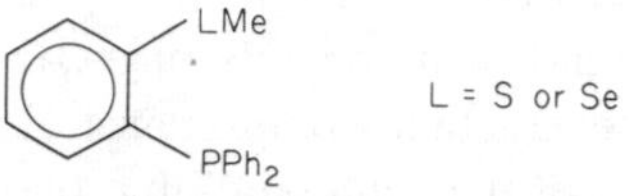

$L = S$ or Se

(CLXXIII)

Degischer and Schwarzenbach have also studied sulphur–phosphorus tridentates. The ligand S[(CH$_2$)$_2$PPh$_2$]$_2$ with a PSP donor sequence yields both four- and five-coordinate complexes with nickel(II) and palladium(II) halides[60], while the interesting series of ligands R$_{3-x}$P[(CH$_2$)$_2$SH]$_x$ (where R = H, Et or Ph; x = 1, 2 or 3), give diamagnetic square-planar rather than five- or six-coordinate complexes with the d^8 metal ions Ni(II), Pd(II), Pt(II) and Au(III) [61].

The ligand S[(CH$_2$)$_2$AsPh$_2$]$_2$ (AsSAs) has been shown to yield low-spin five-coordinate nickel(II) complexes of type [Ni(ligand)X$_2$] [55]. The similar potential tridentate in which the central sulphur atom is replaced by an oxygen donor, O[(CH$_2$)$_2$AsPh$_2$]$_2$, gives compounds of the same formulation with nickel(II) halides, but in this latter case spectral studies indicate that the structure is tetrahedral; the oxygen atom is uncoordinated and the bidentate diarsine ligand apparently forms an eight-membered chelate ring[55,62]. The contention that the oxygen atom is not coordinated is supported by the X-ray analysis of the somewhat similar compound, dichloro[oxydiethylenebis(diphenylphosphine)]nickel(II), [Ni(O[(CH$_2$)$_2$PPh$_2$]$_2$)Cl$_2$], which has been shown[63] to possess a distorted tetrahedral stereochemistry with the shortest nickel–oxygen distance of 364 pm.

27.2.4 Tridentates with three different donor atoms; PNO, PNS, AsNO and AsNS

Sacconi *et al.*[50] have described the nickel(II) halide complexes of the tridentates ONP and ONAs shown in (CLXXIV) with Y = O. These are high-spin five-coordinate

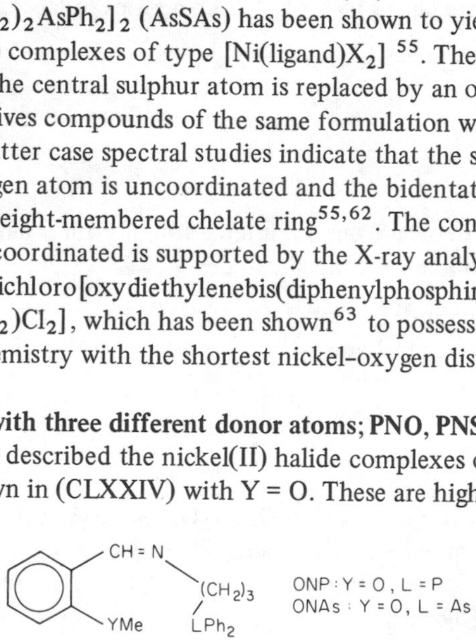

ONP : Y = O, L = P
ONAs : Y = O, L = As

(CLXXIV)

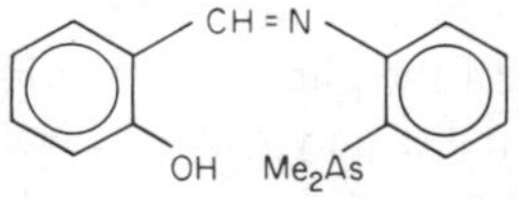

(CLXXV)

compounds of the type [Ni(ligand)X$_2$]. On the other hand when the oxygen donor is anionic, as in the case of the ONAs ligand (CLXXV), high-spin octahedral [Ni(ONAs)$_2$] and polymeric [Ni(ONAs)X] are formed[56]. With cobalt(II) halides this ligand yields diamagnetic cobalt(III) compounds [Co(ONAs)$_2$]$^+$ [56].

The donor set SNP places the five-coordinate nickel(II) atom at approximately the high–low spin crossover point. Thus, with the tridentate SNP of structure (CLXXIV) with Y = S and L = P, the complex five-coordinate [Ni(SNP)Br$_2$] is high spin, but the corresponding iodo-compound is low spin[50]. On the other hand, the donor set SNAs of the ligand shown in (CLXXIV) with Y = S and L = As yields high-spin five-coordinate [Ni(SNAs)X$_2$] compounds[50].

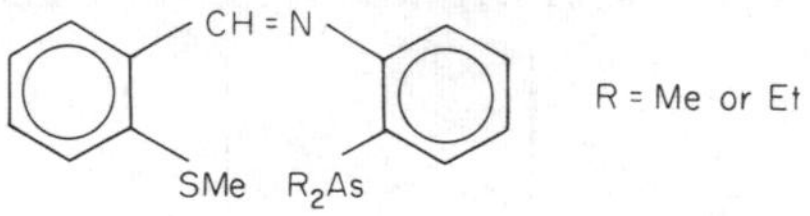

R = Me or Et

(CLXXVI)

The ligands (CLXXVI) with the donor sequence SNAs give high-spin five-coordinate compounds [M(ligand)X$_2$] (where M = Co or Ni, X = halogen), high-spin octahedral bis-complexes, [Ni(ligand)$_2$]$^{2+}$, but a mixture of low- and high-spin [Co(ligand)$_2$]$^{2+}$ compounds[56].

When the sulphur donor is included in a thiophene ring, none of the four ligands (CLXXVII) shows a tendency to act as a tridentate. With cobalt(II) and nickel(II) salts the sulphur atom remains uncoordinated[56].

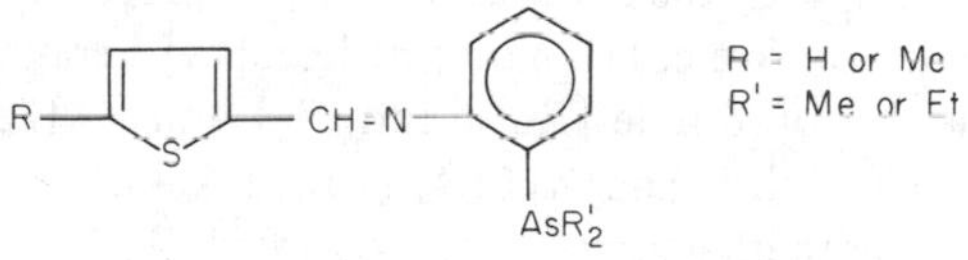

R = H or Me
R' = Me or Et

(CLXXVII)

28 TETRADENTATE LIGANDS

28.1 TETRADENTATES WITH SOFT DONOR ATOMS ONLY

28.1.1 Tetradentates with donor set P_4

To review the subject of P_4 tetradentates is to review the chemistry of one ligand—the tripod tetraphosphine of Venanzi—tris(o-diphenylphosphinophenyl)-phosphine (see structure (CLXXVIII), where Y = Z = P and R = Ph). Together with

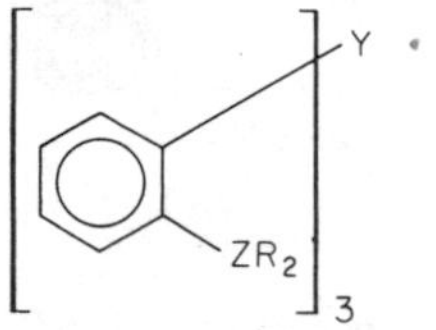

(CLXXVIII)

the corresponding tetraarsine (structure (CLXXVIII) where Y = Z = As and R = Ph), this tetraphosphine was the forerunner of a large number of tripod tetradentates (structure (CLXXVIII) where Y = Group VB donor, Z = Group VB or VIB donor and R = alkyl or Ph group), which have been principally used for the systematic study of five-coordinate metal complexes.

The first paper dealing with the complexes of $P[o\text{-}C_6H_4PPh_2]_3$ was published in 1963[17]. It dealt with the five-coordinate [M(ligand)X]$^+$ compounds, where M = Pd or Pt, and the six-coordinate [Ru(ligand)Cl$_2$] and [Os(ligand)Cl$_2$] compounds. The nickel(II) complexes, [Ni(ligand)X] ClO$_4$ (where X = Cl, NO$_3$ or ClO$_4$), were shown to be five-coordinate diamagnetic compounds[65]. Assignments of electronic spectral peaks were made on the basis of a trigonal-bipyramidal structure, and notwithstanding the large extinction coefficient values, d–d transitions were considered to give rise to the peaks. This assignment instituted a still-continuing discussion upon the nature of the electronic transitions giving rise to the peaks in the spectra of complexes of this P_4 and the corresponding As$_4$ ligand (see chapter 29).

Chromium(III) chloride and bromide have been reported to interact with the tetraphosphine to yield [Cr(ligand)X$_3$], in which the ligand acts as a tridentate[18(a)]. The octahedral Cr(0), Cr(I) and Cr(III) compounds in which the ligand acts as a tetradentate have been prepared[18(b)] by the following reactions

$$P_4 + Cr(CO)_6 \rightarrow [Cr(CO)_3P_4] \tag{i}$$

$$[Cr(CO)_3P_4] \xrightarrow{\text{heat}} [Cr(CO)_2P_4] \tag{ii}$$

$$[Cr(CO)_2P_4] \xrightarrow{\text{oxidn}} [Cr(CO)_2P_4]^+ \tag{iii}$$

$$[Cr(CO)_2P_4]^+ + I_2 \rightarrow [CrI_2P_4]^+ \tag{iv}$$

The manganese(I) compounds of $P[o\text{-}C_6H_4PPh_2]_3$ have been reported[19]. The ligand acts as a bidentate in $[Mn(CO)_3(\text{ligand})X]$, as a tridentate in $[Mn(CO)_3(\text{ligand})]^+$ and as a tetradentate in $[Mn(CO)_2(\text{ligand})]^+$. Attempts to oxidise this last compound to a manganese(II) complex were unsuccessful.

With the three iron(II) halides the tetraphosphine yields trigonal-bipyramidal $[Fe(\text{ligand})X]^+$, with magnetic moments in the region of 3 B.M. Diamagnetic octahedral complexes of type $[Fe(\text{ligand})X_2]$ (where X = CN or NCS) were also prepared[66]. The Mössbauer spectra of these iron compounds have been reported[77].

Cobalt(I), (II) and (III) complexes of $P[o\text{-}C_6H_4PPh_2]_3$ have been noted[67]. In each case the ligand acts as a tetradentate to give low-spin compounds of types

$$[Co^I(CO)\text{ligand}]^+ \quad\text{--trigonal bipyramid} \tag{i}$$
$$[Co^{II}(\text{ligand})X]^+ \quad\text{--trigonal bipyramid} \tag{ii}$$
$$[Co^{III}(\text{ligand})X_2]^+ \text{--octahedral} \tag{iii}$$

The structure of the compound of type (ii), $[Co(\text{ligand})Cl]BPh_4$, has been determined by crystallographic analysis[68]. This compound has a very distorted trigonal bipyramidal shape, and it is suggested that a Jahn–Teller effect is operative. In a low spin d^7 system the degeneracy of the d_{xy} and $d_{x^2-y^2}$ orbitals must be removed by a distortion of the three-fold symmetry. The problem of the geometry of complexes derived from tripod quadridentates has been recently discussed in a very interesting paper[69] (see chapter 29).

An early review[70] of the metal complexes of $P[o\text{-}C_6H_4PPh_2]_3$ lists the metal ions that form trigonal-bipyramidal compounds as Fe(II), Co(I), Co(II), Ni(II), Pd(II) and Pt(II), while octahedral complexes are formed by Fe(II), Ru(II) Os(II) and Co(III). In the foregoing cases the ligand acts as a tetradentate, but in the compounds $[Cr(\text{ligand})X_3]$ it exhibits tridentate behaviour, and acts as a bidentate in the tetrahedral complexes $[Hg(\text{ligand})X_2]$ [70,71].

Recent work on the tetraphosphine has been described in two papers dealing with the nickel(II)[72] and ruthenium(0) and (II)[20] complexes of a large number of different tetradentate tripod ligands. As the nickel work is basically concerned with comparative geometry of different ligands, it will be discussed in chapter 29. The six-coordinate complexes $[Ru(\text{ligand})X_2]$ (where X = Cl, Br, I, NCS or CN) have been prepared[20]. Reductive carbonylation of the compounds yields the five-coordinate $[Ru(CO)(\text{ligand})]$, which can be readily reoxidised by halogen to $[Ru(\text{ligand})X_2]$ (where X = Cl, Br or I).

28.1.2 Tetradentates with donor set As_4

Whereas the chemistry of the P_4 donor set was restricted to the one tripod tetraphosphine, the ligands with the As_4 set are of three types

$$\text{tripod } As[o\text{-}C_6H_4AsR_2]_3 \quad (R = Ph \text{ or } Me) \tag{i}$$
$$\text{tripod } As[(CH_2)_3AsMe_2]_3 \tag{ii}$$
$$\text{linear } As_4 \text{ (CLXXIX)} \quad (L = As) \tag{iii}$$

Although complexes of both tripod tetraarsines were described in 1961, the great bulk of published work deals with the ligand $As[o\text{-}C_6H_4AsPh_2]_3$. The preparation of the ligand was reported in a brief note[9]. This was quickly followed by a paper dealing with the five-coordinate platinum(II) compounds, $[Pt(ligand)X]Y$ (where X = Cl, Br, I or SCN; Y = Cl, Br, I, SCN, ClO_4 or BPh_4)[24]. The trigonal-bipyramidal structure of the $[Pt(ligand)I]BPh_4$ complex was demonstrated by X-ray crystal analysis[73]. The apical arsenic of the ligand occupies one apex of the bipyramid, and the iodine atom the other apex. The tripod legs of the quadridentate come down from the apical arsenic atom to allow the three terminal arsenics to form a trigonal plane about the platinum atom. Palladium(II) compounds of the ligand appear to be similar in type and structure to the platinum(II) compounds[25]. Octahedral palladium(IV) and platinum(IV) compounds, $[M(ligand)Cl_2]Cl_2$, have also been reported[17].

(CLXXIX)

Diamagnetic ruthenium(II) complexes of $As[o\text{-}C_6H_4AsPh_2]_3$ of the type $[Ru(ligand)X_2]$ (where X = Cl, Br, I, NCS or NO_3) have been reported[75], and their octahedral stereochemistry confirmed by the X-ray crystallographic analysis of the dibromocompound[76]. Osmium(II) compounds are of a similar type[17]. Recent work[20] on ruthenium has resulted in the preparation of the five-coordinate compounds $[Ru(CO)(ligand)]$ by reductive carbonylation of $[RuCl_2(ligand)]$. Rhenium(II) complexes are also of the octahedral type $[Re(ligand)X_2]$ (where X = Cl or Br), but the reaction of rhenium trichloride and the tetraarsine yields $[Re(As_3AsO)Cl_2]$ in which one of the arsenic atoms of the ligand has been oxidised to its oxide[26].

The complexes of palladium, platinum, ruthenium and rhenium with $As[o\text{-}C_6H_4AsPh_2]_3$ have been reviewed by Venanzi[5].

On interaction of $As[o\text{-}C_6H_4AsPh_2]_3$ with $[Rh_2X_2(cyclooctadiene)_2]$ (X = Cl, Br or I), $[Rh_2Cl_2(CO)_4]$ or $[RhCl(CO)(Ph_3P)_2]$, the corresponding halide complex $[RhX(ligand)]$ is obtained in each case[27]. These rhodium(I) compounds are diamagnetic and five-coordinate, and can be oxidised by the corresponding halogen to yield octahedral $[RhX_2(ligand)]X$. No rhodium(II) intermediates were obtained.

The trigonal-bipyramidal compounds of nickel(II), $[Ni(ligand)X]^+$ (where X = Cl, Br, I, NCS, CN, NO_3 or ClO_4; ligand = $As[o\text{-}C_6H_4AsPh_2]_3$), have been noted[65], and the phosphorus-31 nuclear magnetic resonance spectra of the nickel(II) compounds and the corresponding palladium(II) and platinum(II) compounds with this and a large number of similar tripod tetradentates, have been discussed[78].

A 1964 review[70] listed the known five-coordinate compounds, $[M(ligand)X]^+$, of $As[o\text{-}C_6H_4AsPh_2]_3$ for the metal ions Ni(II), Rh(I), Pt(II) and Pd(II); and the known six-coordinate compounds, $[M(ligand)X_2]^{n+}$ for the metal ions Re(II), Ru(II), Os(II) ($n = 0$ in each case), Rh(III) ($n = 1$), Pd(IV) and Pt(IV) ($n = 2$). Tetrahedral Hg(II) compounds ($n = 0$) were also reported[70,71].

The complex cobalt(I) cation, $[Co(CO) ligand]^+$ (ligand $= As[o\text{-}C_6H_4AsPh_2]_3$) has been reported[67], but compounds of cobalt with higher oxidation numbers do not appear to have been prepared with this ligand.

Recent work has seen the preparation of the diamagnetic five-coordinate complexes $[M(ligand)X]^{+}$[81] (where M = Ni and X = Cl, Br, I, NCS, NO_2, N_3, NO_3 or CH_3COO; M = Pd or Pt and X = Cl, Br, I or NCS), and the compound $[Ru(ligand)Cl_2]$[20], with the new ligand $As[o\text{-}C_6H_4AsMe_2]_3$. Studies of the electronic spectra of these compounds indicate that the intensities of peaks attributed to d–d transitions are lower than those of the analogous $As[o\text{-}C_6H_4AsPh_2]_3$ compounds, and it is concluded that the ligand field bands in complexes of this latter ligand 'borrow' intensity from $\pi \to \pi^*$ transitions of the phenyl groups (see chapter 29).

The divalent iron and nickel complexes, formed by the tripod tetraarsine $As[(CH_2)_3AsMe_2]_3$, of the type $[M(ligand)X_2]$, as well as the trivalent iron, cobalt and nickel compounds, $[M(ligand)X_2]^+$, were reported by Barclay and Barnard[79] to be all octahedral low-spin compounds. Later work by Benner and Meek[80] demonstrated that the supposedly nickel(III) complex was in fact the trigonal-bipyramidal complex $[Ni(ligand)X] X$. The spectra of the above compounds have been compared with those of similar compounds derived from other tetradentate tripod ligands[82] (see chapter 29).

An X-ray analysis of the complex $[Pd(tetraarsine)Cl] ClO_4 . C_6H_6$ (tetraarsine = the ligand shown in (CLXXIXa)) has revealed that the compound possesses a square-pyramidal stereochemistry[83]. The palladium atom is coplanar with three arsenic and one chlorine atoms, while the apical arsenic (which is one of the two not at the end of the ligand chain) is pulled back from the vertical position, above the palladium atom, by the ligand chain; the apical arsenic–Pd–basal plane angle is $81°$.

Recently the linear As_4 ligand shown in (CLXXIXb) (L = As) has been shown[84] to yield four-coordinate $[NiAs_4]^{2+}$, five-coordinate $[NiAs_4X] ClO_4$ (where X = Cl, Br or I), and five-coordinate $[CoAs_4X] ClO_4$ (where X = Cl or Br) species.

28.1.3 Tripod tetradentates with donor sets of type $A(B)_3$ (A = apical atom)

In an early paper in 1961[9] describing the preparation of $As[o\text{-}C_6H_4AsPh_2]_3$, Venanzi *et al.* also described the mixed donor ligand tris(o-diphenylarsinophenyl)-phosphine $P[o\text{-}C_6H_4AsPh_2]_3$ (structure (CLXXVIII), Y = P, Z = As; R = Ph). However, it was some ten years before complexes of this ligand were described. A recent paper[20] dealing with the ruthenium complexes of an array of ligands of the type shown in (CLXXVIII) (where Y = P, As or Sb; Z = P or As), has reported the preparation of $[Ru(ligand)Cl_2]$. Another paper[72], dealing with the spectra of the five-coordinate [Ni(ligand)X] BPh$_4$ compounds, compares a large range of tripod tetradentate ligand complexes and will be discussed in chapter 29.

The complexes derived from the similar mixed PAs$_3$ ligand, tris(3-dimethyl-arsinopropyl)phosphine (P[(CH$_2$)$_3$AsMe$_2$]$_3$) have been more widely studied. The diamagnetic five-coordinate [Ni(ligand)X]$^+$ compounds of this ligand have been reported[85], and were assigned a trigonal-bipyramidal stereochemistry on the basis of spectral similarities to the [M{As(o-C$_6$H$_4$AsPh$_2$)$_3$}X]$^+$ compounds. This stereochemistry has been confirmed by the X-ray analysis[86] of [Ni(ligand)(CN)]ClO$_4$. The trigonal-bypyramid has an apical phosphorus and three arsenic atoms in the equatorial plane. The nickel atom is displaced 19 pm up from the equatorial plane towards the phosphorus. This contrasts with the structure of [Pt(As[o-C$_6$H$_4$AsPh$_2$]$_3$)I]BPh$_4$, in which the platinum is displaced down from the arsenic equatorial plane towards the iodine atom[73]. The spectra of these five-coordinate nickel compounds have been studied[82] and compared with spectra of similar compounds derived from ligands of the general type shown in (CLXXVIII) (see chapter 29).

Other tripod tetradentates with an apical phosphorus atom that have been described are tris(o-methylselenophenyl)phosphine (P[o-C$_6$H$_4$SeMe]$_3$) and tris(o-vinylphenyl)phosphine (P[o-C$_6$H$_4$CH=CH$_2$]$_3$). The former ligand yields five-coordinate diamagnetic nickel(II) compounds [Ni(ligand)X] ClO$_4$ [87]. Comparison of similar nickel compounds with three equatorial sulphur, arsenic or phosphorus atoms shows a spectrochemical series R$_2$Se $<$ R$_2$S $<$ R$_3$As $<$ R$_3$P. Transitions from the equatorial d$_{x^2-y^2}$ or d$_{xy}$ orbitals to the d$_{z^2}$ orbital are more energetic for selenium than sulphur, but less than for arsenic or phosphorus. This indicates an increasing electron delocalisation in the order S $<$ Se $<$ As $<$ P between the equatorial metal d$_{x^2-y^2}$ and d$_{xy}$ orbitals and orbitals of e symmetry on the donor atoms in the equatorial plane. The ligand P[o-C$_6$H$_4$CH=CH$_2$]$_3$ acts as a bidentate in [Pt(ligand)Br$_2$], with only one of the vinyl double bonds coordinated[88].

Work on mixed donor tripod tetradentates with an apical arsenic atom has only developed recently. The ruthenium complex of As[o-C$_6$H$_4$PPh$_2$]$_3$, [Ru(ligand)Cl$_2$], has been described[20], while comparative spectral work on the [Ni(ligand)X] BPh$_4$ complexes has been undertaken[72] (see chapter 29). As in the case of the corresponding phosphine, the ligand As[o-C$_6$H$_4$CH=CH$_2$]$_3$ only acts as a bidentate in [Pt(ligand)Br$_2$] [88].

Much recent work has concentrated upon tripod tetradentates possessing an antimony atom at the apex of the tripod. The ruthenium(II) compounds [Ru(ligand)Cl$_2$] of Sb[o-C$_6$H$_4$PPh$_2$]$_3$ and Sb[o-C$_6$H$_4$AsPh$_2$]$_3$ have been described[20], and comparative spectral studies have been undertaken on the [Ni(ligand)X] BPh$_4$ complexes of both ligands, as well as of those of a number of similar tripod tetradentates[72]. As noted for the corresponding phosphine and arsine ligands, Sb[o-C$_6$H$_4$CH=CH$_2$]$_3$ operates only as a bidentate in [Pt(ligand)Br$_2$] [88].

McAuliffe *et al.* have studied the nickel(II) and palladium(II) compounds of the Sb(As)$_3$ ligand Sb[o-C$_6$H$_4$AsMe$_2$]$_3$ recently. Five-coordinate compounds of types

[Ni(ligand)X] X and [Ni(ligand)X] BPh$_4$ (where X = Cl, Br, I, NCS or NO$_3$) have been prepared[89]. The ligand has also been found to yield the unusual [Ni(ligand)$_2$]X$_2$ (where X = ClO$_4$ or BPh$_4$) complexes, in which it is proposed that there is one bidentate SbAs ligand and one tridentate AsSbAs ligand[90]. The [Pd(ligand)X] Y (where X = Cl, Br, I or SCN and Y = Cl, CNS or BPh$_4$) complexes are trigonal bipyramidal in structure[91]. The electronic spectra of the compounds show a shift of the ligand field bands to higher energy in the order I$^-$ > Br$^-$ > Cl$^-$, and this nephelauxetic behaviour is discussed in relation to the compression of the Pd–Sb linkage upon chelation. The increase in size of the apical atom in a trigonal-bipyramidal structure increases the length of the metal–equatorial donor distances; this lifts the metal out of the trigonal equatorial plane and 'compresses' the metal–apical donor atom bond[69].

The complexes of the tripod tetradentate Sb[(CH$_2$)$_x$AsMe$_2$]$_3$ with nickel(II) have been shown to be five-coordinate [Ni(ligand)X] BPh$_4$ (where x = 3 and X = Cl, Br, I, NCS, NO$_3$ or CN) [82], while the complexes [Ni(ligand)X] Y (where x = 2 and X and Y = halogen) have been used as new internal pressure calibrants[92].

28.1.4 Tetradentates with donor set P$_2$As$_2$

The new P$_2$As$_2$ ligand (structure (CLXXIXb), L = P) has been shown[84] to yield four-coordinate [Ni ligand] (ClO$_4$)$_2$, five-coordinate [Ni(ligand)X] BPh$_4$ (where X = Cl, Br or I) and five-coordinate [Co(ligand)X] BPh$_4$ (where X = Cl or Br) complexes.

28.2 HYBRID TETRADENTATES WITH HARD AND SOFT DONOR ATOMS

The greater part of the work in the field of all hybrid tetradentates containing both hard and soft donors is connectéd with the school of Sacconi. Although the possible number of tripod tetradentates of the type shown in (CLXXX) is very large, this group of workers has already studied a large number of the possible ligands. Much less work has been reported on the similar ligands in which the three terminal donors are joined to the apical donor by a propylene chain or the edge of a benzene ring.

$$W \Bigg\langle \begin{array}{l} (CH_2)_2 - XR_n \\ (CH_2)_2 - YR_n \\ (CH_2)_2 - ZR_n \end{array}$$

where W = Group VB donor
X, Y and/or Z = Group VB donor and n = 2
X, Y and/or Z = Group VIB donor and n = 1
R = alkyl or phenyl group.
(CLXXX)

28.2.1 Tripod tetradentates with donor sets of type $A(B)_3$ (A = apical atom)

Sacconi and Bertini[93,94] have noted that the $N(P)_3$ ligand $N[(CH_2)_2PPh_2]_3$ gives deep-coloured diamagnetic complexes with nickel(II) halides of type $[Ni(ligand)X]^+$. The regular trigonal-bipyramidal stereochemistry of the nickel atom has been confirmed by X-ray analysis[95]. The nitrogen atom is at one apex with the iodine at the other; the three phosphorus atoms are in the trigonal plane, with the nickel only 12 pm out of this plane towards the iodine atom. The cobalt(II) compounds $[Co(ligand)X]X$ of the ligand are high spin if X = Cl or Br, but low spin when X = I or NCS [94]. X-ray crystallographic work has been undertaken on examples of both the low- and the high-spin compounds. $[Co(ligand)I]I$ has been found to possess a square-pyramidal structure[96]. The nitrogen, two of the three phosphorus atoms and the iodine form the square basal plane. The cobalt atom is 34 pm above the base of the plane towards the third phosphorus donor. The X-ray structure of the high-spin compound $[Co(ligand)Cl]PF_6$ is very interesting[97]. The three phosphorus atoms and the chlorine create a very distorted trigonal-bipyramidal arrangement about the cobalt atom. The ligand acts as a tripod with the nitrogen at the apex and the three phosphorus atoms creating the trigonal basal plane. However, the cobalt is well removed down below this plane towards the chlorine atom giving a very long cobalt–nitrogen bond length of 267.5 pm. Thus, the three phosphorus–cobalt–chlorine bond angles approach a tetrahedral angle, and the molecule as a whole can be considered as a distorted tetrahedron with a P_3Cl chromophore. It is suggested[97], that the square-pyramidal arrangement of the $N(P)_3$ ligand favours a low-spin state.

Five-coordinate chromium(II) complexes of the $N(P)_3$ ligand $N[(CH_2)_2PPh_2]_3$ of the type $[Co(ligand)X]BPh_4$ have been described[98].

Venanzi *et al.*[45] reported the first attempts to produce complexes with a $P(N)_3$ tetradentate. These workers prepared the complexes $[M(ligand)X_2]$ (where M = Pd or Pt; ligand = $P[o\text{-}C_6H_4NMe_2]_3$; X = Cl, Br or I), and demonstrated by nuclear magnetic resonance that, although the ligand operated only as a PN bidentate, the nitrogen actually bonded exchanged rapidly at room temperature between the three potential donors. This ligand was also shown[74] to operate as a bidentate in the tetrahedral complexes $[M(ligand)X_2]$ (where M = Co or Ni; X = halogen).

Tris(α-picolyl)phosphine has been reported to yield five-coordinate Mn(II), Co(II) and Ni(II) chlorocompounds[99].

Schwarzenbach[61] has described the compounds of the d^8 transition metals Ni(II), Pd(II), Pt(II) and Au(III) with the $P(S)_3$ ligand $P[(CH_2)_2SH]_3$. Such compounds are diamagnetic and square planar.

The ligand tris(o-methylthiophenyl)phosphine ($P[o\text{-}C_6H_4SMe]_3$) has been shown to yield highly coloured trigonal-bipyramidal nickel(II) compounds of types $[Ni(ligand)X]^+$ (where X = halogen), and $[Ni(ligand)L]^{2+}$ (where L = mono-dentate ligand)[100]. The two lowest energy bands in these complexes were assigned to $^1A_1 \rightarrow {}^1E(D)$ electron transitions. It is of interest that the corresponding arsine with the donor set $As(S)_3$ was reported not to react with nickel(II) salts under the conditions used to prepare the above nickel $P(S)_3$ complexes[100]. The structure of these $[NiP(S)_3X]^+$ compounds has been confirmed by the X-ray analysis of $[Ni\{P(o\text{-}C_6H_4SMe)_3\}Cl]\,ClO_4$ [101]. This compound has been found to be very nearly a regular trigonal bipyramid. The three sulphur atoms are in the equatorial plane, with the phosphorus and chlorine atoms filling the apical positions. The nickel atom is only slightly displaced from this plane by 6.1 pm towards the chlorine. The average nickel–sulphur bond length of 226.7 pm is claimed to indicate only slight Ni–S π-bonding.

The palladium(II) compounds of $P[o\text{-}C_6H_4SMe]_3$ have been reported[57] as $[Pd(ligand)Cl_2]$ and $[Pd_2(ligand)Cl_4]$. In the former case the ligand acts as a PS bidentate, but it is claimed that it acts as a bridging tetradentate in the latter compound.

The $[NiN(As)_3X]^+$ compounds (where $N(As)_3 = N[(CH_2)_2AsPh_2]_3$) are claimed to be diamagnetic and trigonal bipyramidal in structure[55,93].

Although the ligand $As[o\text{-}C_6H_4SMe]_3$ does not appear to form nickel(II) compounds[100], this ligand has been reported to act as an SAs bidentate in $[Pd(ligand)Cl_2]$, and as a bridging tetradentate in $[Pd_2(ligand)Cl_4]$ [57].

28.2.2 Hybrid tetradentates with donor sets of type $N(B_2C)$ (N = apical nitrogen donor)

In a paper published in 1968 Sacconi and Morassi[102] listed the cobalt(II) and nickel(II) complexes of a series of tripod hybrid tetradentates of the type shown in (CLXXXI), where V, W and Z were Group VB and VIB donor atoms. In a series of papers since this preliminary note Sacconi *et al.* have published this work more fully.

$$N \Big\langle \begin{matrix} (CH_2)_2 - V \\ (CH_2)_2 - W \\ (CH_2)_2 - Z \end{matrix}$$

(CLXXXI)

The $N(N_2P)$ ligand (structure (CLXXXI); $V = W = NEt_2$ and $Z = PPh_2$) yields Ni(II) or Co(II) compounds of formula $[M(ligand)X]\,Y$ (where $X = Cl$, Br, I, CN or NCS and $Y = I$ or BPh_4) [103]. The cobalt compounds are all high-spin five-coordinate, but the nickel compounds are either high-spin five-coordinate, or low-spin probably square planar, depending upon the nature of X. In the compound $[NiN(N_2P)(NCS)_2]$, the phosphorus atom is uncoordinated[103]. This same ligand also yields the five-coordinate chromium(II) complex $[Cr(ligand)Br]\,BPh_4$ [98].

When the donor set of the above $N(N_2P)$ ligand is changed to $N(NP_2)$ to give the ligand of structure (CLXXXI) ($V = NEt_2$ and $W = Z = PPh_2$) the nickel(II) complexes, $[NiN(NP_2)X]^+$, are all five-coordinate and low spin. On the other hand the cobalt(II) compounds of this formulation with this $N(NP_2)$ ligand are still high spin[103].

The $N(OP_2)$ ligand (structure (CLXXXI); $V = OMe$ and $W = Z = PPh_2$) has been shown[104,105] to yield complexes of the types $[M(ligand)X]\,Y$ (where $M = Co$ or

Ni), [Co(ligand)X]$_2$$^{2+}$ and [Co(ligand)X$_2$] from its interaction with cobalt(II) and nickel(II) salts. The nickel compounds are low-spin square-pyramidal structures, and the X-ray crystal structure of the compound [NiN(OP$_2$)NCS] PF$_6$ indicates a distorted square-pyramidal structure[106] (CLXXXII). The two phosphorus, the nitrogen and the iodine atoms create the basal plane, and the nickel is only 14 pm above this plane towards the oxygen atom. The Ni—O bond length is long (248.0 pm) and the N–CH$_2$–CH$_2$–OMe chain appears too short to allow the oxygen atom to occupy a position normal to the basal plane above the nickel atom, the basal plane–$\widehat{\text{Ni}}$–O angle being 78.6°. In d^8 C$_{4v}$ symmetry the d$_{x^2-y^2}$ orbital is empty, but the d$_{z^2}$ orbital pointing towards the oxygen atom possesses two electrons. The nickel–oxygen bond is therefore destabilised and the compound approaches a square-planar structure as indicated by its electronic spectrum[104]. The similar [NiN(OAs$_2$)I] BPh$_4$ is accorded the same type of elongated structure[109]. The [Co(ligand)X]$^+$ (where X = halogen) compounds are high-spin, distorted trigonal-bipyramidal in structure, but when X = NCS the compound is low-spin elongated square-pyramidal with a structure similar to that in (CLXXXII)[104]. In the [Co(ligand)X$_2$] compounds the N(OP$_2$) ligand acts as a tridentate (NP$_2$) to give the low-spin five-coordinate iodide, or as a bidentate (P$_2$) to yield the high-spin tetrahedral chloride or bromide[104].

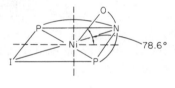

(CLXXXII)

The interaction of the N(N$_2$As) ligand (structure (CLXXXI); V = W = NEt$_2$ and Z = AsPh$_2$) with cobalt(II) halide yields the high-spin five-coordinate complexes [Co(ligand)X] Y (where X = Cl, Br, I, CN or NCS and Y = BPh$_4$ or I) [103]. The nickel(II) halide compounds, [Ni(ligand)X]$^+$, are either high-spin five-coordinate, or low-spin square planar[103]. However, the [Ni(ligand)(NCS)$_2$] complex has been shown by X-ray analysis to be square-pyramidal in shape, with the arsenic donor unbonded[107,108].

The N(NAs$_2$) ligand (structure (CLXXXI); V = NEt$_2$ and W = Z = AsPh$_2$) has also been reported to yield five-coordinate cobalt(II) and nickel(II) complexes[103].

28.2.3 Hybrid tetradentates with donor sets of type N(N'BC) (N = apical nitrogen donor)

The N(NOP) and N(NSP) ligands (structure (CLXXXI); V = NEt$_2$, W = OMe or SMe respectively and Z = PPh$_2$) both yield the high-spin trigonal-bipyramidal cobalt(II) compounds [Co(ligand)X] BPh$_4$ [109]. Of the nickel(II) compounds of structure [Ni(ligand)X] BPh$_4$, only the complex in which X = I and ligand = N(NOP) has low-spin square-pyramidal structure. All other complexes where X = Cl, Br or I and ligand = N(NOP) or N(NSP) are diamagnetic square-planar compounds in which the ligand has a tridentate N_2P donor set[109].

It is clear from the foregoing results of studies on potential hybrid tripod tetradentates containing oxygen or sulphur donors that in many cases these Group VIB donors bond only weakly, or in some instances not at all, to the metal ion.

28.2.4 Hybrid tetradentates with linear donor sets A_2B_2

The donor atom sequence POOP occurs in the ligand $Ph_2P(CH_2)_2O(CH_2)_2O(CH_2)_2PPh_2$. In the diamagnetic compound $[Ni(POOP)I_2]$, the ligand acts as a P_2 bidentate to yield a four-coordinate nickel atom with a stereochemistry somewhere between that of a tetrahedron and a square plane[100]. The two oxygen atoms are at 320 and 316 pm respectively from the nickel atom, and cannot therefore be considered to coordinate[110].

Two ligands with the donor atom sequence PSSP have been described with the general formula $Ph_2P(CH_2)_nS(CH_2)_nS(CH_2)_nPPh_2$, where $n = 2$ [60] and $n = 3$ [111]. Both ligands give diamagnetic Ni(II), Pd(II) and Pt(II) compounds of the type $[M(ligand)](ClO_4)_2$, which can add a halide ion to form five-coordinate complexes, for example $[M(ligand)I]^+$. For the ligand with $n = 3$ the nickel complex is trigonal-bipyramidal, while the palladium and platinum complexes are square-pyramidal[111]. This same ligand forms low-spin square-pyramidal $[Co(ligand)X]^+$ compounds[111].

The AsOOAs tetradentate $(o\text{-}C_6H_4AsMe_2)O(CH_2)_2O(o\text{-}C_6H_4AsMe_2)$, like the POOP ligand mentioned above, fails to coordinate its oxygen atoms in the complexes $[Pd(ligand)X_2]$ (where X = Cl, Br, I or SCN), and thus operates as a bidentate arsine ligand[112]. The similar AsSSAs ligand $(o\text{-}C_6H_4AsMe_2)S(CH_2)_2S(o\text{-}C_6H_4AsMe_2)$ also operates as a diarsine in the $[Pd(ligand)X_2]$ (X = Cl, Br, I or SCN) complexes. It is suggested that the two arsenic atoms span *trans*-positions on the palladium metal atom in these compounds[112].

Palladium(II) complexes of the linear SAsAsS tetradentates, $(o\text{-}C_6H_4SMe)As(Ph)(CH_2)_nAs(Ph)(o\text{-}C_6H_4SMe)$ have been described for the values of $n = 2$, 3 and 4 [113,114]. Three types of complex palladium(II) compounds have been isolated:

(i) $[Pd(ligand)X_2]$ ($n = 2$ or 3) with the ligand exhibiting bidentate (As_2) action.
(ii) $[Pd_2(ligand)X_4]$ ($n = 2$, 3 or 4) in which the ligand bridges between the two palladium atoms using all four donors.
(iii) $[Pd_2(ligand)_2X_2]^{2+}$ ($n = 2$, 3 or 4), which are dimeric with two arsenic and one exchanging sulphur atom coordinated to each palladium together with a halogen atom.

The mode of action of these ligands is somewhat complex, but again one sees the tendency for the sulphur donors (this time the terminal donor groups) to fail to coordinate in these multidentate ligands.

29 GEOMETRICAL AND SPECTRAL CONCLUSIONS

The foregoing chapters in part 4 dealing with tri- and tetradentates containing the heavier Group VB donor atoms illustrates very clearly that the main topic of concern has centred on the ability of these ligands to yield five-coordinate transition metal complexes. The schools of Venanzi, Meek and Sacconi have concentrated upon the subject of five-coordination almost exclusively; and although many of the compounds that they describe may have octahedral, tetrahedral or square-planar stereochemistry, there is little doubt that both the design of ligands and the methods of preparation of their complexes indicate that five-coordination was the dominant theme in the work described. Thus, Sacconi's $N(OP_2)$ [104-6] and $N(OAs_2)$ [103] tripod multidentates, although tending to yield square-planar nickel(II) compounds, were basically studied to see if they would act as tetradentates and yield five-coordinate compounds with the week OMe donor group bonded to the central metal atom.

The only major body of work described in chapters 27 and 28 that has not been in the main devoted to the study of five-coordination has emanated from the Nyholm school. This group has tended to use particular arsine multidentates for the very specific task of stabilising low-valence state compounds and metal carbonyl complexes. The value of mono- and bidentate arsines in offering such stability by ready back-donation from the metal to the ligand to yield π-bonding systems has been known for some time, and it is not surprising that these workers should use multidentate arsine systems to obtain very stable complexes. However, as the work of Nyholm *et al.* has been basically concerned with the stabilisation of metal carbonyl complexes in both unusual oxidation states and coordination numbers, and with the i.r. spectra of such compounds, the structures of the coordinated multidentate ligands and the electronic spectra of the complexes have not been broadly analysed.

The study of five-coordinate metal complexes assumes wider proportions each year, and much new work has been described since the excellent review of Muetterties and Schunn[115] in 1966, and the more recent reviews of Sacconi[116] in 1968 and Ciampolini[117] in 1969.

The two topics of geometry and spectra of five-coordinate metal complexes are inextricably interwoven. Thus the stereochemistry of nearly all the compounds discussed in chapters 27 and 28 are inferred from their electronic spectra. It is, therefore, not feasible to attempt a separation of these two topics under separate headings. However, in the following discussion, papers dealing mainly with geometrical considerations will be discussed before the papers which are mainly spectroscopic in content.

The two idealised five-coordinate geometries are those of the trigonal bipyramid and the square pyramid. The closeness of the relationships between the various four-,

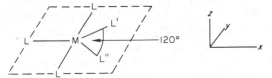

Figure 22

five- and six-coordinate geometries is something that is often not fully recognised, and much talk of discrete geometries is so much verbiage when the distorted geometries of a large number of complexes are considered. Thus the distinction between the trigonal bipyramid and the square pyramid is something that is more obvious on paper than in actual structures. Figure 22 illustrates the former structure. The three M—L bonds are in the xy plane, and the M—L$'$ and M—L$''$ bonds are 60° above and below this plane respectively in the xy plane. A movement upward of only 60° and 30° for the M—L$'$ and M—L$''$ bonds respectively will produce a regular square-pyramidal situation in which L$'$ is the apical ligand. It is obvious that relatively modest distortion of one of the idealised geometries of five-coordination can produce systems with structures between the two paradigms. Furthermore, when one realises that much of the work on five-coordination has been carried out with complexes of multidentate ligands (themselves sterically restrictive) it can be seen that idealised geometries do not readily occur.

Other distortions of the idealised five-coordinate geometry can take place. A small distortion of the metal atom out of the trigonal plane can rapidly lead to a distorted quasi-tetrahedral structure (see figure 23). On the other hand tetragonal distortion of the square-pyramidal geometry can quickly yield an essentially square-planar system[105].

The splitting of the 3d atomic orbitals in a D_{3h} (trigonal-bipyramidal) and a C_{4v} (square-pyramidal) ligand field is shown in figure 24. Energetically, the splittings are roughly to scale.

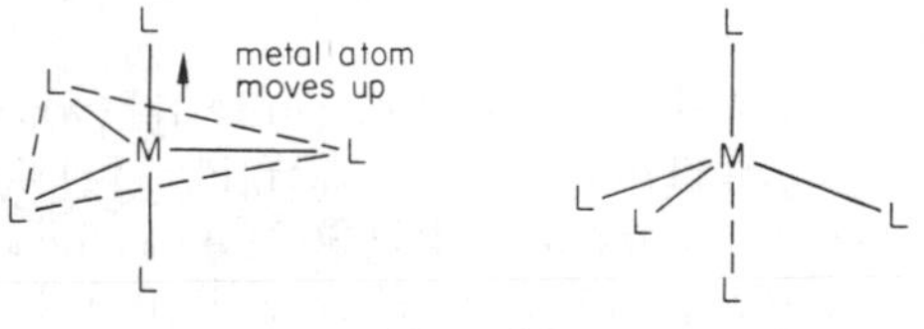

Figure 23

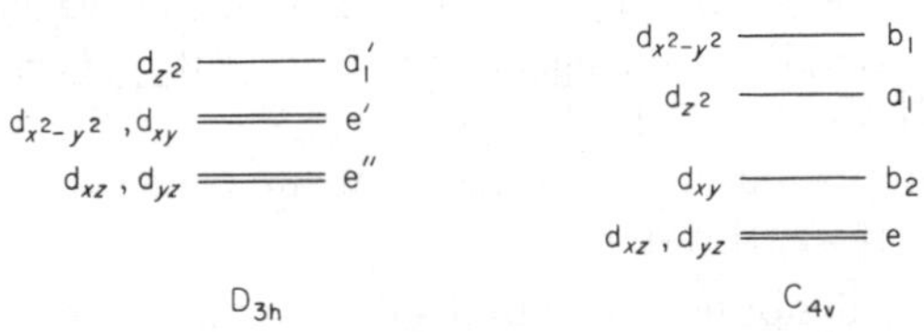

$$
\begin{array}{ll}
d_{z^2} & a_1' \\
d_{x^2-y^2}, d_{xy} & e' \\
d_{xz}, d_{yz} & e''
\end{array}
\qquad
\begin{array}{ll}
d_{x^2-y^2} & b_1 \\
d_{z^2} & a_1 \\
d_{xy} & b_2 \\
d_{xz}, d_{yz} & e
\end{array}
$$

D_{3h} C_{4v}

Figure 24

To Venanzi in 1964[5] the ligand $As[o\text{-}C_6H_4AsPh_2]_3$ had shown conclusively that ligands could be designed that demonstrated a marked preference for forcing a trigonal-bipyramidal stereochemistry upon a transition metal ion. In a review at a later date Wood[118] discussed the X-ray structures of five (and other)-coordinate compounds, but offered no further comment than the fact that both of the idealised geometries had been shown to exist in actual compounds.

A most important discussion upon the geometry of tripod ligands was published by Venanzi *et al.*[69] in 1970. In this work the tetradentate ligands discussed are of types:

(i) $X[o\text{-}C_6H_4Y]_3$ (where X = P, and Y = PPh_2, $AsPh_2$, NMe_2, SMe or SeMe; X = As, and Y = PPh_2, $AsPh_2$ or $AsMe_2$; X = Sb, and Y = PPh_2 or $AsPh_2$);
(ii) $X[(CH_2)_2Y]_3$ (where X = N and Y = PPh_2 or $AsPh_2$);
(iii) $X[(CH_2)_3Y]_3$ (where X = P and Y = $AsMe_2$).

As some of the structural features and physical properties of the complexes formed by these ligands could not be accounted for on the basis of the electronic properties of the donor atoms alone, calculations were undertaken to determine what geometrical restraints the various ligands placed upon the stereochemistry of the metal atom.

The main problems to be accounted for were listed as:

(i) The X-ray analyses of $[Co(P[o\text{-}C_6H_4PPh_2]_3)Cl]BPh_4$ [68], $[Pt(As[o\text{-}C_6H_4AsPh_2]_3)I]BPh_4$ [73], $[Ni(P[o\text{-}C_6H_4SMe]_3)Cl]ClO_4$ [101] and $[Ni(N[(CH_2)_2PPh_2]_3)I]I$ [95(b)], show that the metal atom is below the trigonal equatorial plane towards the halogen atom, whereas in $[Ni(P[(CH_2)_3AsMe_2]_3)CN]ClO_4$ [86], the nickel lies above the trigonal plane of the three arsenic atoms towards the apical phosphorus.
(ii) The ligand $P[o\text{-}C_6H_4NMe_2]_3$ acts only as a PN bidentate with Co(II), Ni(II), Pd(II) and Pt(II) [45,74].
(iii) The ligand $Sb[o\text{-}C_6H_4PPh_2]_3$ forms more than one type of palladium(II) and platinum(II) complex[69].
(iv) The electronic spectra of the cations $[Ni(ligand)Cl]^+$ (where ligand = $P[o\text{-}C_6H_4PPh_2]_3$, $As[o\text{-}C_6H_4PPh_2]_3$ or $Sb[o\text{-}C_6H_4PPh_2]_3$) show that the $^1A_1 \rightarrow a^1E$ transition decreases in energy in the order P > Sb > As as the apical atom is changed[72], in contrast to the spectrochemical order $R_3P > R_3As > R_3Sb$ derived for monodentate ligands[119].

Calculations were carried out on the positions of the donor metal atoms, using bond lengths and angles derived from X-ray data on the above and other relevant compounds. It was found that the ligands $X[o\text{-}C_6H_4YR_2]_3$ (where X = P, As or Sb, and Y = N, P or As), and $P[o\text{-}C_6H_4ZMe]_3$ (where Z = S or Se), do not form idealised trigonal bipyramids in which the metal and equatorial trigonal donor atoms are coplanar. It was shown that:

(i) The M–equatorial donor bond length has to be longer than the usual length in a M–monodentate donor bond of the same type.

(ii) The metal atom is below the equatorial plane towards the halogen atom.

The results of the calculations offer explanations for the four problems raised above:

(i) The results are, of course, supported by the X-ray structures mentioned above. The only case in which the metal atom is above the equatorial plane towards the tripod apical donor is in the compound in which the chelate rings are six rather than five-membered; that is, three bridging methylene groups between the apical and terminal donors in the ligand.

(ii) The ligand $P[o\text{-}C_6H_4NMe_2]_3$ is calculated to have particularly long M–equatorial donor bond lengths; and the metal atom is a long way below the equatorial plane. Thus, although the bidentate action of this ligand was originally attributed[45] to the effect of the two methyl groups in reducing the nitrogen coordinating power, the calculations indicate that this action is more likely due to the steric requirements of the ligand.

(iii) The ligand $Sb[o\text{-}C_6H_4PPh_2]_3$ is also calculated to have a relatively poor fit for the Ni(II), Pd(II) and Pt(II) ions, and it is therefore not surprising that mono-meric trigonal-bipyramidal complexes are not obtained[89,91].

(iv) The anomalous order of the spectrochemical series for the apical atoms of these tripod ligands (namely P > Sb > As) shown by the nickel complexes, is explained on the basis of the steric requirements of the multidentates, which prevent the donors taking up bond length positions similar to those with mono-dentate ligands with the same donor atoms. The effective ligand-field strength of the apical donor is thus changed and the anomalous spectrochemical series is obtained[72]. This argument has been reinforced by the study of the electronic spectra of the trigonal-bipyramidal complexes [Ni(ligand)X] BPh_4 (where X = Cl, Br, I, NCS, NO_3 or CN) of the three flexible tetradentate ligands $D[(CH_2)_3AsMe_2]_3$ (where D = P, As or Sb)[82]. The replacement in this study of the sterically rigid *o*-phenylene groups by flexible propylene chains allows the ligand groups to accommodate themselves to the trigonal-bipyramidal steric requirements of the metal atom. The electronic spectra thus show the normal spectrochemical order, P > As > Sb.

Recently Baracco and McAuliffe[91] have studied the electronic spectra of the $[Pd(Sb[o\text{-}C_6H_4AsMe_2]_3X]^+$ compounds where X = Cl, Br, I or SCN. They found, as did Venanzi *et al.*[81] for the similar complexes with $As[o\text{-}C_6H_4AsMe_2]_3$, that the iodide ion exerts an apparently greater ligand-field strength than the bromide ion. This is the reverse of the normal spectrochemical series. On the basis of the calculations of Venanzi *et al.*[69], which showed that for trigonal-bipyramidal complexes:

(a) the M–axial ligand bond length is shorter in tetradentate tripod complexes than it is in monodentate complexes;

(b) the M–equatorial ligand bonds are longer in tetradentate tripod complexes than they are in monodentate complexes;

Baracco and McAuliffe[91] have argued that the axial Pd—Sb bond is 'compressed' and thus that the normally polarisable antimony atom has become less polarisable. The palladium ion now exerts a greater polarising effect on the halide ion present in the complex and there is a changeover from spectrochemical to nephelauxetic behaviour for the halide donors, with a concomitant reversal of the apparent spectrochemical order.

Venanzi *et al.*[69] also undertook calculations on the geometry of octahedral compounds and were able to offer an explanation for the failure to obtain the octahedral compound $[Ru(Sb[o\text{-}C_6H_4PPh_2]_3)Cl_2]$.

Sacconi has undertaken a broad study of the five-coordinate cobalt(II) and nickel(II) complexes derived from hybrid tridentate and tripod tetradentate ligands. Much of his work on nickel(II) compounds is summarised in a review[120]. A recent article has discussed in detail the factors governing the spin multiplicity of such compounds[11].

For five-coordinate cobalt(II) and nickel(II) complexes, high- or low-spin nature will depend upon the energy separation between the d_{z^2} and $d_{x^2-y^2}$ orbitals (see figure 24). With typically soft donors, such as phosphorus or arsenic, low-spin complexes may be expected, while hard donors such as nitrogen or oxygen will yield high-spin compounds. The donor atoms sulphur and selenium, the halogens and thiocyanate will be found in both high- and low-spin situations. Sacconi[11] has studied the donor sets present at the crossover point when hybrid multidentates are involved.

Energetically speaking, the square-pyramidal structure is the most favoured for low-spin cobalt(II) compounds, while the trigonal-bipyramidal stereochemistry is favoured by low-spin nickel(II) compounds. The energy level diagrams of figure 24 are roughly to scale, and indicate that low spin d^7 is favoured by C_{4v}, and low spin d^8 by D_{3h} symmetry.

Having noted that the capacity for spin-pairing increases with the overall nucleophilic activity, and decreases with the overall electronegativity of the donor atoms, Sacconi[11] placed different multidentate donor sets in order according to their values of:

(i) the sum of the donor atom electronegativities $(\Sigma\chi)$;
(ii) the sum of the nucleophilic reactivity constants of the donor groups $(\Sigma n°)$.

He was then able to locate the crossover point between high- and low-spin cobalt(II) and nickel(II) in terms of $\Sigma\chi$ and $\Sigma n°$.

For five-coordinate complexes with hybrid ligands of the types [M (tridentate) (halogen)$_2$] and [M (tetradentate) (halogen)]$^+$ (where M = Co or Ni), Sacconi showed that the limiting values for $\Sigma\chi$, above which values high-spin complexes are obtained, were as shown in table 30.

The overall nucleophilic reactivity constants $(\Sigma n°)$ can be taken as a measure of the softness of the donor atoms. Sacconi[11] found that where $\Sigma n°$ is greater than a certain value, low-spin complexes are obtained. Table 31 lists the limiting values above which low-spin compounds occur.

TABLE 30 VALUES OF $\Sigma\chi$ ABOVE WHICH HIGH-SPIN FIVE-COORDINATE COMPLEXES OCCUR

Compound type	$\Sigma\chi$	Donor set with approx. this $\Sigma\chi$
[Ni (tridentate) (halogen)$_2$]	12.76	N_2AsI_2
[Co (tridentate) (halogen)$_2$]	12.65	N_2PI_2
[Ni (tetradentate) (halogen)]$^+$	13.2	NS_3Cl
[Co (tetradentate) (halogen)]$^+$	11.46	NP_3I

Recently Nelson *et al.*[53] have studied five-coordinate iron(II) compounds with a linear hybrid tridentate possessing a PNP donor sequence of the type [Fe(PNP)(halogen)$_2$]. They have found that the crossover from low to high spin occurs at approximately $\Sigma\chi > 11.6$ and $\Sigma n° < 31.5$ with the donor set NP_2I_2. These values would indicate that iron(II) compounds of type [Fe (tridentate) X_2] have a smaller tendency to yield low-spin complexes than either similar cobalt(II) or nickel(II) complexes.

Sacconi's work is a worthy attempt to offer generalisations on a large volume of data. Undoubtedly many future papers will use these concepts. However, as with many such generalisations of this nature, there are certain inherent problems. Firstly, as Sacconi himself points out[11], the actual stereochemistry of the complex will have a large effect upon the spin multiplicity of the complex and, as shown by Venanzi *et al.*[69], the actual geometry of multidentate complexes may be greatly distorted. Both $\Sigma\chi$ and $\Sigma n°$ are measurements on the donor atoms as individual entities, or simple monodentate ligands, and the applicability of such values in sterically hindered complexes is debatable. Secondly, the spin crossover point for nickel(II) varies from $\Sigma\chi = 12.76$ for the donor set N_2AsI_2 to $\Sigma\chi = 13.2$ for the set N_2S_2I, depending on whether the ligand is tri- or tetradentate. Although only a variation of approximately 0.5 units, this variation actually represents a change of 10 per cent on the range of $\Sigma\chi$ values quoted; the lowest $\Sigma\chi$ given is 10.45 for P_4I, while the highest is 15.97 for N_2O_2Cl, giving a range of $\Sigma\chi$ values of roughly five units. The variation in $\Sigma\chi$ for the crossover point in five-coordinate cobalt(II) compounds is even greater. For [Co (tridentate) (halogen)$_2$] $\Sigma\chi = 12.67$, while for [Co (tetradentate) (halogen)]$^+$ $\Sigma\chi = 11.46$ at the crossover point. This represents a variation of over one unit, or 20 per cent of the range

TABLE 31 VALUES OF $\Sigma n°$ ABOVE WHICH LOW-SPIN FIVE-COORDINATE COMPLEXES OCCUR

Compound type	$\Sigma n°$	Donor set with approx. this $\Sigma n°$
[Ni (tridentate) (halogen)$_2$]	25.5	$SNAsI_2$
[Co (tridentate) (halogen)$_2$]	29.0	N_2PBr_2
[Ni (tetradentate) (halogen)]$^+$	22.3–26	N_2S_2NCS (22.3)
[Co (tetradentate) (halogen)]$^+$	34.89	NP_3I

of values quoted. The obvious conclusion would appear to be that changes in ligand steric requirements are particularly important in determining the spin state of the complex. A similar situation arises in the $\Sigma n°$ values for the cross-over spin point for cobalt(II) complexes. There is a variation of almost six units in the overall nucleophilic reactivity constants, between the value for the crossover in [Co (tridentate) (halogen)$_2$] and in [Co (tetradentate) (halogen)]$^+$. This represents a variation of over 25 per cent of the range of values quoted (from 14.04 for N_2O_2Cl to 41.18 for P_4I).

The theoretical applications of ligand-field theory to five-coordinate metal complexes, with particular reference to electronic spectra, have been reviewed by Wood[121] and Furlani[122], and the generalisations made have been specifically applied in Sacconi's papers and reviews[11,116,120]. However, the major basic work in the area springs from papers of Ciampolini[123,132] upon a ligand-field model for high-spin five-coordinate cobalt(II) complexes, and a paper by Venanzi *et al.*[124] on a ligand-field model for trigonal-bipyramidal complexes of d^6, d^7 and d^8 metal ions. Since the publication of this paper in 1967, Venanzi and coworkers have produced a number of papers and a review dealing with the topic. The schools of Sacconi and Meek in particular have used this theoretical interpretation of spectra to great purpose in the elucidation of the stereochemistry of complexes by the study of their electronic spectra (for example, see references 120 and 117 and references quoted therein, as well as any papers in the references bearing the names of these workers).

Although in even the earliest papers dealing with the complexes of the tetra-arsine, $As[o\text{-}C_6H_4AsPh_2]_3$, Venanzi[70] attributed the very intense extinction coefficients ($\epsilon_{max} \approx 5000$) of the highly coloured five-coordinate complexes of type [M(ligand)X]$^+$ to d–d transitions, and in 1965 he offered a detailed analysis of the electronic spectra of the platinum and palladium compounds with a number of X groups[125] on this basis, Jørgensen[126] in 1966 claimed that this interpretation was not very plausible. He postulated that the first absorption bands of these tetraarsine complexes were due to ligand charge-transfer phenomena[126,130].

Using two-parameter ligand-field calculations and tensor operator theory, Venanzi *et al.*[124] obtained energy-level diagrams for the trigonal-bipyramidal complexes of d^6, d^7 and d^8 metal ions. The spectra of the trigonal-bipyramidal complexes of type [M(tetradentate)X]$^+$ (where M = Fe(II), Co(II) or Ni(II); tetradentate = $P[o\text{-}C_6H_4PPh_2]_3$ or $As[o\text{-}C_6H_4AsPh_2]_3$; X = halogen) were analysed in terms of the energy-level diagrams produced. The importance of this work is best seen by reference to the review article[127] published by these authors in 1967.

As evidence that the low-energy band of the electronic spectra of the trigonal-bipyramidal complexes with very high molar extinction coefficients ranging from 500 to 8000 are predominantly ligand-field rather than charge-transfer bands, the authors[127] list five observations:

(i) High-intensity, low-energy bands are observed in complexes of phosphorus and arsenic ligands with octahedral structure of the type [M^{n+}(tetradentate)X$_2$]$^{(n-2)+}$ [70].

(ii) The spectra of all the trigonal-bipyramidal complexes can be satisfactorily treated by this ligand-field model.

(iii) It is not likely that the complexes $[M(P[o\text{-}C_6H_4PPh_2]_3)X]^+$ (where M = Fe or Co) have charge-transfer bands with wavelengths as high as 1000 nm.

(iv) Changes in the nature of the X group in the complexes $[Ni(As[o\text{-}C_6H_4AsPh_2]_3)X]^+$ cause large changes in the position of the lowest energy band.

(v) The lowest energy bands in the complexes $[Ni(P[o\text{-}C_6H_4LMe]_3)Br]^+$ (where L = S or Se) are at approximately the same wavelength as in $[Ni(P[o\text{-}C_6H_4PPh_2]_3)Br]^+$, but are of much lower intensity.

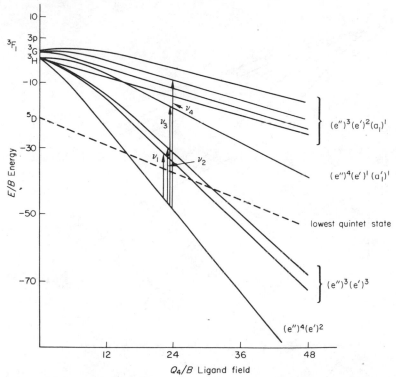

Figure 25 Energy-level diagram for trigonal-bipyramidal, low-spin complexes of iron(II), d^6

Simplified versions of the energy-level diagrams for trigonal-bipyramidal low-spin complexes of d^6, d^7 and d^8 metal ions are shown in figures 25, 26 and 27 respectively. In these diagrams, the parameter Q_4 is given by

$$Q_4 = \frac{ge^2 \langle r^4 \rangle}{R^5}$$

where

g = effective ligand charge
e = electronic charge

300 *The Chemistry of Multidentate Ligands*

R = metal–donor bond length
$\langle r^4 \rangle$ is an operator related to the radial part of the 3d wave function, and
B is the Racah parameter.

It should be noted that these are essentially diagrams of the Tanabe and Sugano type[131], but that the energies expressed in units of B have zero points which refer to the term of the free ion rather than the ground state of the complex. It should also be kept in mind that in a regular trigonal-bipyramidal (D_{3h}) ligand field, the

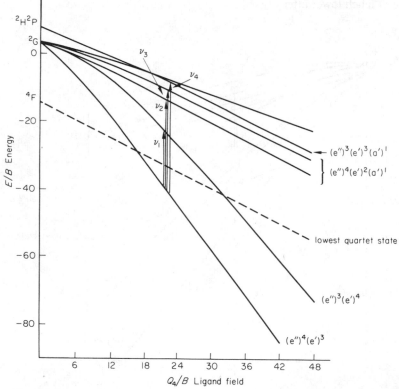

Figure 26 Energy-level diagram for trigonal-bipyramidal, low-spin complexes of cobalt(II), d^7

splitting of the 3d atomic orbitals of the metal ion is, in order of increasing orbital energy, $d_{xz} = d_{yz}$ (e'') < $d_{x^2-y^2} = d_{xy}$ (e') < d_{z^2} (a$_1'$) (see figure 24).

Table 32 demonstrates the correlation that has been obtained between electronic spectra and the calculated energy-level diagrams[124,127]. Although three specific compounds are quoted as examples in table 32, it is claimed that the spectra of a number of trigonal-bipyramidal complexes can be fitted to the relevant energy-level diagram.

The problem about the intensity of these ligand-field bands is as yet not completely resolved. In 1967 Venanzi *et al.*[124,127] had no doubt that 'the presence of benzene

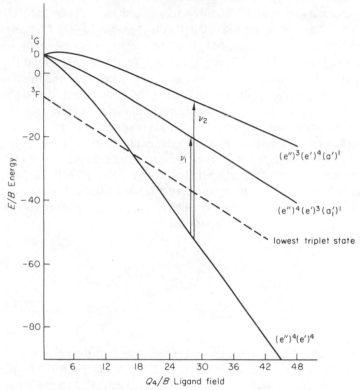

Figure 27 Energy-level diagram for trigonal-bipyramidal, low-spin complexes of nickel(II), d^8

TABLE 32 ELECTRONIC SPECTRAL TRANSITIONS FOR TRIGONAL-BIPYRAMIDAL d^6, d^7 AND d^8 METAL COMPLEXES

Metal ion	Electronic transition	Name in fig. 25 to 27	Approximate wavelength (nm)
d^6 Fe(II)[†] $(e'')^4(e')^2 \rightarrow (e'')^3(e')^3$		ν_1	1100
$\rightarrow (e'')^3(e')^3$		ν_2	910
$\rightarrow (e'')^4(e')^2(a_1')^1$		ν_3	555
$\rightarrow (e'')^3(e')^2(a_1')^1$		ν_4	400
d^7 Co(II)[†] $(e'')^4(e')^3 \rightarrow (e'')^3(e')^4$		ν_1	100
$\rightarrow (e'')^4(e')^2(a_1')^1$		ν_2	625
$\rightarrow (e'')^4(e')^2(a_1')^1$		ν_3	555
$\rightarrow (e'')^3(e')^3(a_1')^1$		ν_4	500
d^8 Ni(II)[†] $(e'')^4(e')^4 \rightarrow (e'')^4(e')^3(a_1')^1$		ν_1	590
$\rightarrow (e'')^3(e')^4(a_1')^1$		ν_2	385

[†] The approximate wavelength values were obtained from the spectra of the compounds:
(i) $[Fe(P[o\text{-}C_6H_4PPh_2]_3)NO_3]^+$
(ii) $[Co(P[o\text{-}C_6H_4PPh_2]_3)NO_3]^+$
(iii) $[Ni(P[o\text{-}C_6H_4PPh_2]_3)Cl]^+$

rings is not necessary to achieve high intensities', illustrating the point by reference to the spectra of trigonal-bipyramidal $[Ni\{P[(CH_2)_3AsMe_2]_3\}X]^+$ [85]. However, in a later paper[81] comparing the trigonal-bipyramidal nickel(II) complexes of the tetra-arsines, $As[o\text{-}C_6H_4AsPh_2]_3$ and $As[o\text{-}C_6H_4AsMe_2]_3$, it was pointed out that the $[Ni(ligand)X]^+$ compounds of the former ligand have much higher extinction coefficient values than the similar compounds of the latter ligand. It was concluded that in the case of the former ligand, the ligand-field bands 'borrow' intensity from the $\pi \rightarrow \pi^*$ transitions in the many phenyl groups present.

Subsequent to the above calculations[124], the determination of the X-ray structure of $[Co(P[o\text{-}C_6H_4PPh_2]_3)Cl]BPh_4$ [68] revealed that this compound deviated significantly from D_{3h} symmetry, with the P–Co–P angles of the three terminal phosphorus donors forming the very distorted trigonal base of structure (CLXXXIII). With this in mind Norgett and Venanzi[129] have performed new calculations using the Angular Overlap model.

(CLXXXIII)

The distortion found in this compound could arise from the operation of a Jahn–Teller effect, since d^7 ions in either D_{3h} or C_{3v} symmetry give rise to orbitally degenerate ground states. Such distortion has been suggested[67] to explain the observed magnetic properties of the compounds $[Co(P[o\text{-}C_6H_4PPh_2]_3)X]^+$. It would be expected that as the odd electron will be in the antibonding e' orbitals (d_{xy} and $d_{x^2-y^2}$), the splitting of these orbitals will be large (figure 28) and an appreciable Jahn–Teller distortion will occur.

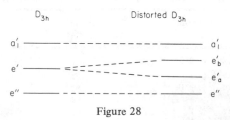

Figure 28

The calculated energy level diagram is shown in figure 29. The three main bands in the spectrum at 1040, 606 and 510 nm are respectively assigned to the one-electron transitions

$$(e'')^4(e'_a)^2(e'_b)^1 \rightarrow (e'')^4(e'_a)^1(e'_b)^2 \text{ (that is, } e'_a \rightarrow e'_b) \tag{i}$$
$$\rightarrow (e'')^4(e'_a)^2(a'_1)^1 \text{ (that is, } e'_b \rightarrow a'_1) \tag{ii}$$
$$\rightarrow (e'')^3(e'_a)^2(e'_b)^2 \text{ (that is, } e'' \rightarrow e'_b) \tag{iii}$$

Comparison of figures 26 and 29 indicates a significant difference in interpretation (see also table 32). The most important feature of this more recent paper[129] lies in its last paragraph. As the parameters used to construct the energy-level diagrams are based on best-fit analysis of the position and absorption maxima of the spectral bands, it is important that these values be accurately determined. Venanzi *et al.* [129] point out that, for example, the highest energy band in the five-coordinate nickel tetraphosphine complexes occur as shoulders, and are thus difficult to describe

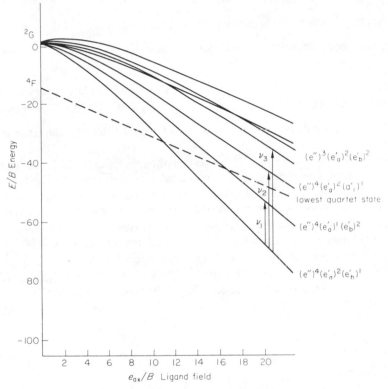

Figure 29 Energy-level diagram for distorted trigonal-bipyramidal complexes of d^7 metal ions calculated using the Angular Overlap model

accurately. Furthermore, as the calculated absorption maxima are fairly insensitive to changes in e'_{eq}/e'_{ax} [†] (the ratio that must be given a numerical value to allow a spectral fit), it is reasonable to suppose that all the spectra could be adequately described by a single value of this ratio. This prevents any realistic comparison of the bonding in the trigonal-bipyramidal tetraphosphine complexes of Fe(II), Co(II) and Ni(II).

† Where e' = electron charge$/4\pi$; e'_{eq} is in the equatorial, and e'_{ax} in the axial plane; together these two parameters replace the parameters Q_2 and Q_4 used in the earlier calculations[124].

Recently Venanzi *et al.*[128] have shown that the ligand-field spectra of five-coordinate complexes derived from a wide range of tripod tetradentates and nickel(II), palladium(II) and platinum(II) salts, change dramatically on cooling to temperatures in the vicinity of 100 K to 77 K. In D_{3h} symmetry the lowest energy band corresponds to the transition $^1A'_1 \rightarrow {}^1E'$, while in a C_{3v} environment it corresponds to the $^1A_1 \rightarrow a^1E$ transition. This lowest energy peak is usually asymmetric, and may even split into a double peak at room temperature[80,85]. On cooling it is found that this asymmetry (and splitting if it occurs) tends to disappear. The authors suggest two possible reasons:

(i) a temperature-dependent static distortion of the ground-state of the type

   regular trigonal bipyramidal $\rightleftharpoons$ distorted trigonal bipyramidal

occurs.

(ii) a Jahn–Teller distortion of the doubly degenerate excited state occurs in the complexes, which are essentially trigonal bipyramidal in the solutions being studied.

Analysis of the spectral results leads the authors to suggest that it is the latter effect taking place.

As noted above Ciampolini[123,132] has also applied the ligand-field model to the calculation of energy-level diagrams for five-coordinate complexes. He has been particularly interested in high-spin five-coordinate cobalt(II) compounds, and the discussion offered does not refer to multidentate ligands possessing heavy Group VB donors. However, the papers are of general interest to studies in five-coordination, as energy-level diagrams are calculated for both regular and various types of distorted trigonal-bipyramidal and square-pyramidal high-spin d^7 systems. Thus the energy-level diagrams for the point groups C_{4v} and C_{3v} are reported together with the correlation diagrams between C_{4v} and D_{3h} via a series of distorted C_{2v} intermediate structures.

REFERENCES

1. Booth, G., *Adv. inorg. Chem. Radiochem.*, **6** (1964), 1.
2. Barclay, G. A., Harris, C. M., and Kingston, J. V., *Chem. Commun.* (1968), 965.
3. Lions, F., *Rev. Pure appl. Chem.*, **19** (1969), 177.
4. Curtis, N. F., *Coord. Chem. Rev.*, **3** (1968), 3.
5. Venanzi, L. M., *Ligand Shape and its Stereochemical Consequences*, Consiglio Nazionale Delle Ricerche, Rome (1964).
6. Chiswell, B., unpublished work.
7. Mair, G. A., Powell, H. M., and Henn, D. E., *Proc. chem. Soc.* (1960), 415.
8. Barclay, G. A., Nyholm, R. S., and Parrish, R. V., *J. chem. Soc.* (1961), 4433.
9. Howell, T. E. W., Pratt, S. A. J., and Venanzi, L. M., *J. chem. Soc.* (1961), 3167.
10. Basolo, F., and Pearson, R. G., *Mechanisms of Inorganic Reactions*, Wiley, New York (1967), pp. 23-5.
11. Sacconi, L., *J. chem. Soc. (A)* (1970), 248.
12. Chatt, J., and Watson, H. R., *J. chem. Soc.* (1961), 4980.
13. Massey, A. G., and Nyholm, R. S., *J. Polym. Sci.*, **62** (1962), 5126.
14. Chatt, J., and Hart, F. A., *J. chem. Soc.* (1965), 812.
15. Davis, R., and Ferguson, J. E., *Inorg. Chim. Acta*, **4** (1970), 16.
16. Davis, R., and Ferguson, J. E., *Inorg. Chim. Acta*, **4** (1970), 23.
17. Hartley, J. G., Venanzi, L. M., and Goodall, D. C., *J. chem. Soc.* (1963), 3930.
18. (a) Howell, I. V., Venanzi, L. M., and Goodall, D. C., *J. chem. Soc. (A)* (1967), 395; (b) Howell, I. V., and Venanzi, L. M., *J. chem. Soc. (A)* (1967), 1007.
19. Chiswell, B., and Venanzi, L. M., *J. Chem. Soc. (A)* (1966), 417.
20. Halfpenny, M. T., and Venanzi, L. M., *Inorg. Chem. Acta*, **5** (1971), 91.
21. Cheung, K. K., Lai, T. F., and Lam, S. Y., *J. chem. Soc. (A)* (1970), 3345.
22. Hart, F. A., *J. chem. Soc.* (1960), 3324.
23. Barclay, G. A., and Nyholm, R. S., *Chem. Ind.* (1953), 378.
24. Brewster, J. A., Savage, C. A., and Venanzi, L. M., *J. chem. Soc.* (1961), 3699.
25. Savage, C. A., and Venanzi, L. M., *J. chem. Soc.* (1962), 1548.
26. Mawby, R. J., and Venanzi, L. M., *J. chem. Soc.* (1962), 4447.
27. Mawby, R. J., and Venanzi, L. M., *Essays in Coord. Chem.*, *Expl Suppl. IX* (1964), p. 240.
28. Hill, W. E., and McAuliffe, C. A., personal communication.
29. Workman, M. O., McAuliffe, C. A., and Meek, D., *Inorg. nucl. Chem. Lett.*, **5** (1969), 147.
30. Cunninghame, R. G., Nyholm, R. S., and Tobe, M. L., *J. chem. Soc.* (1964), 5800.
31. Cunninghame, R. G., Nyholm, R. S., and Tobe, M. L., *J. chem. Soc. (A)* (1971), 227.
32. Nanda, R. K., and Tobe, M. L., *J. chem. Soc. (A)* (1966), 1740.
33. Barclay, G. A., Gregor, I. K., and Wild, S. B., *Chem. Ind.* (1964), 1710.
34. Nyholm, R. S., Snow, M. R., and Stiddard, M. H. B., *J. chem. Soc.* (1965), 6570.
35. Cook, C. D., Nyholm, R. S., and Tobe, M. L., *J. chem. Soc.* (1965), 4194.
36. Masek, J., Nyholm, R. S., and Stiddard, M. H. B., *Colln Czech. chem. Commun. Engl. Edn* **29** (1964), 1714.
37. Nyholm, R. S., Snow, M. R., and Stiddard, M. H. B., *J. chem. Soc.* (1965), 6570.
38. Kasenally, A. S., Nyholm, R. S., and Stiddard, M. H. B., *J. Am. chem. Soc.*, **86** (1964), 1884.
39. Nyholm, R. S., Snow, M. R., and Stiddard, M. H. B., *J. chem. Soc.* (1965), 5343.
40. Kilbourn, B. T., Blundell, T. L., and Powell, H. M., *Chem. Commun.* (1965), 444.
41. Blundell, T. L., and Powell, H. M., *J. chem. Soc. (A)* (1971), 1685.
42. Bosnich, B., Nyholm, R. S., Pauling, P. J., and Tobe, M. L., *J. Am. chem. Soc.*, **90** (1968), 4741.
43. Preer, J. R., and Gray, H. B., *J. Am. chem. Soc.*, **92** (1971), 7306.
44. Nyholm, R. S., *J. chem. Soc.* (1950), 2061.

45. Fritz, H. P., Gordon, I. R., Schwarzhaus, K. E., and Venanzi, L. M., *J. chem. Soc.* (1965), 5210.
46. Dobson, G. R., Taylor, R. C., and Walsh, T. D., *Chem. Commun.* (1966), 281.
47. Sacconi, L., and Morassi, R., *J. chem. Soc. (A)* (1968), 2997.
48. Morassi, R., and Sacconi, L., *J. Am. chem. Soc.*, 92 (1970), 5241.
49. (a) Orioli, P. L., and Sacconi, L., *Chem. Commun.* (1968), 1310;
    (b) Orioli, P. L., and Ghilardi, C. A., *J. chem. Soc. (A)* (1970), 1511.
50. Sacconi, L., Speroni, G. P., and Morassi, R., *Inorg. Chem.*, 7 (1968), 1521.
51. Nelson, S. M., and Kélly, W. S. J., *Chem. Commun.* (1969), 94.
52. Nelson, S. M., and Kelly, W. S. J., *Chem. Commun.* (1968), 436.
53. Kelly, W. S. J., Ford, G. H., and Nelson, S. M., *J. chem. Soc. (A)* (1971), 388.
54. Dahlhoff, W. V., and Nelson, S. M., *J. chem. Soc. (A)*, (1971), 2184.
55. Sacconi, L., Bertini, I., and Mani, F., *Inorg. Chem.*, 7 (1968), 1417.
56. Lee, K. W., Ph.D. Thesis, Univ. of Queensland (1971).
57. Dyer, G., Workman, M. O., and Meek, D. W., *Inorg. Chem.*, 6 (1967), 1404.
58. Workman, M. O., Dyer, G., and Meek, D. W., *Inorg. Chem.*, 6 (1967), 1543.
59. Meek, D. W., and Ibers, J. A., *Inorg. Chem.*, 8 (1969), 1915.
60. Degischer, C., and Schwarzenbach, G., *Helv. chim. Acta*, 49 (1966), 1927.
61. Schwarzenbach, G., *Chemické Zvesti*, 19 (1965), 200.
62. Sacconi, L., and Gelsomini, J., *Inorg. Chem.*, 7 (1963), 291.
63. Greene, P. T., and Sacconi, L., *J. chem. Soc. (A)* (1970), 866.
64. Chiswell, B., and Verrall, K. A., *J. prakt. Chem.*, 312 (1970), 751.
65. Dyer, G., Hartley, J. G., and Venanzi, L. M., *J. chem. Soc.* (1965), 1293.
66. Halfpenny, M. T., Hartley, J. G., and Venanzi, L. M., *J. chem. Soc. (A)* (1967), 627.
67. Hartley, J. G., Kerfoot, D. G. E., and Venanzi, L. M., *Inorg. chim. Acta*, 1 (1967), 145.
68. Blundell, T. L., Powell, H. M., and Venanzi, L. M., *Chem. Commun.* (1967), 763.
69. Dawson, J. W., Lane, B. C., Mynott, R. J., and Venanzi, L. M., *Inorg. chim. Acta*, 5 (1971), 25.
70. Venanzi, L. M., *Angew. Chem. int. Edn*, 3 (1964), 453.
71. Dyer, G., Goodall, D. C., and Venanzi, L. M., unpublished observations.
72. Higginson, B. R., McAuliffe, C. A., and Venanzi, L. M., *Inorg. chim. Acta*, 5 (1971), 37.
73. Mair, G. A., Powell, H. M., and Venanzi, L. M., *Proc. chem. Soc.* (1961), 170.
74. Christopher, R. E., Gordon, I. R., and Venanzi, L. M., *J. chem. Soc. (A)* (1968), 205.
75. Hartley, J. G., and Venanzi, L. M., *J. chem. Soc.* (1962), 182.
76. Mais, R. H. B., Powell, H. M., and Venanzi, L. M., *Chem. Ind.* (1963), 1204.
77. Fluck, E., and Brauch, K. F., *Z. anorg. allg. Chem.*, 364 (1969), 107.
78. Dawson, J. W., and Venanzi, L. M., *J. Am. chem. Soc.*, 90 (1968), 7229.
79. Barclay, G. A., and Barnard, A. K., *J. chem. Soc.* (1961), 4269.
80. Benner, G. S., and Meek, D. W., *Inorg. Chem.*, 6 (1967), 1399.
81. Headley, O. St. C., Nyholm, R. S., McAuliffe, C. A., Sindellari, L., Tobe, M. L., and Venanzi, L. M., *Inorg. chim. Acta*, 4 (1970), 93.
82. McAuliffe, C. A., and Meek, D. W., *Inorg. chim. Acta*, 5 (1971), 270.
83. Blundell, T. L., and Powell, H. M., *J. chem. Soc. (A)* (1967), 1650.
84. Gee, D. R., Halfpenny, M. T., and McAuliffe, C. A., personal communication.
85. Benner, G. S., Hatfield, W. E., and Meek, D. W., *Inorg. Chem.*, 3 (1964), 1544.
86. Stevenson, D. L., and Dahl, L. F., *J. Am. chem. Soc.*, 89 (1967), 3424.
87. Dyer, G., and Meek, D. W., *Inorg. Chem.*, 6 (1967), 149.
88. Hall, D. I., and Nyholm, R. S., *J. chem. Soc. (A)* (1971), 1491.
89. Baracco, L., and McAuliffe, C. A., *J. chem. Soc. (Dalton)* (1972), 948.
90. Baracco, L., Halfpenny, M. T., and McAuliffe, C. A., *Chem. Commun.* (1971), 1502.
91. Baracco, L., and McAuliffe, C. A., personal communication.
92. Ferraro, J. A., *Inorg. nucl. Chem. Lett.*, 6 (1970), 823.
93. Sacconi, L., and Bertini, I., *J. Am. chem. Soc.*, 89 (1967), 2235.
94. Sacconi, L., and Bertini, I., *J. Am. chem. Soc.*, 90 (1968), 5443.
95. (a) Dapporto, P., and Sacconi, L., *J. chem. Soc. (A)* (1970), 1804;
    (b) *Chem. Commun.* (1969), 1091.
96. Orioli, P. L., and Sacconi, L., *Chem. Commun.* (1969), 1012.
97. Sacconi, L., Di Vaira, M., and Bianchi, A., *J. Am. chem. Soc.*, 92 (1970), 4465.

98. Mani, F., and Sacconi, L., *Inorg. chim. Acta*, **4** (1970), 365.
99. Chiswell, B., *Aust. J. Chem.*, **20** (1967), 2533.
100. Dyer, G., and Meek, D. W., *Inorg. Chem.*, **4** (1965), 1398.
101. Haugen, L. P., and Eisenberg, R., *Inorg. Chem.*, **8** (1969), 1072.
102. Sacconi, L., and Morassi, R., *Inorg. nucl. Chem. Lett.*, **2** (1968), 449
103. Sacconi, L., and Morassi, R., *J. chem. Soc. (A)* (1969), 2904.
104. Morassi, R., and Sacconi, L., *J. chem. Soc. (A)* (1971), 492.
105. Dapporto, P., Morassi, R., and Sacconi, L. *J. chem. Soc. (A)* (1970), 1298.
106. Bianchi, A., and Ghilardi, C. A., *J. chem. Soc. (A)* (1971), 1096.
107. Di Vaira, M., and Sacconi, L., *Chem. Commun.* (1969), 10.
108. Di Vaira, M., *J. chem. Soc. (A)* (1971), 148.
109. Morassi, R., and Sacconi, L., *J. chem. Soc. (A)* (1971), 1487.
110. Dapporto, P., and Sacconi, L., *J. chem. Soc. (A)* (1971), 1914.
111. Du Bois, T. D., and Meek, D. W., *Inorg. Chem.*, **8** (1969), 146.
112. Cannon, R. D., Chiswell, B., and Venanzi, L. M., *J. chem. Soc. (A)* (1967), 1277.
113. Dutta, R. L., Meek, D. W., and Busch, D. H., *Inorg. Chem.*, **9** (1970), 1215.
114. Dutta, R. L., Meek, D. W., and Busch, D. H., *Inorg. Chem.*, **9** (1970), 2098.
115. Muetterties, E. L., and Schunn, R. A., *Q. Rev. chem. Soc.*, **20** (1966), 245.
116. Sacconi, L., *Pure appl. Chem.*, **17** (1968), 95.
117. Ciampolini, M., *Struct. and Bond.*, **6** (1969), 52.
118. Wood, J. S., *Coord. Chem. Rev.*, **2** (1967), 403.
119. Goggin, P. L., Knight, R. J., Sindellari, L., and Venanzi, L. M., *Inorg. chem. Acta*, **5** (1971), 62.
120. Sacconi, L., *Transit. Metal Chem.*, **4** (1968), 230.
121. Wood, J. S., *Inorg. Chem.*, **7** (1968), 852.
122. Furlani, C., *Coord. Chem. Rev.*, **3** (1968), 141.
123. Ciampolini, M., and Bertini, I., *J. chem. Soc. (A)* (1968), 2241.
124. Norgett, M. J., Thornley, J. H. M., and Venanzi, L. M., *J. chem. Soc. (A)* (1967), 540.
125. Dyer, G., and Venanzi, L. M., *J. chem. Soc.* (1965), 2771.
126. Jørgensen, C. K., *Coord. Chem. Rev.*, **1** (1966), 164.
127. Norgett, M. J., Thornley, J. H. M., and Venanzi, L. M., *Coord. Chem. Rev.*, **2** (1967), 99.
128. Dawson, J. W., Venanzi, L. M., Preer, J. R., Hix, J. E., and Gray, H. B., *J. Am. chem. Soc.*, **93** (1971), 778.
129. Norgett, M. J., and Venanzi, L. M., *Inorg. chim. Acta*, **2** (1968), 107.
130. Jensen, K. A., and Jørgensen, C. K., *Acta, chem. scand.*, **19** (1965), 451.
131. Tanabe, Y., and Sugano, S., *J. phys. Soc. Japan*, **9** (1954), 753.
132. (a) Ciampolini, M., *Inorg. Chem.*, **5** (1966), 35;
     (b) Ciampolini, M., Nardi, N., and Speroni, G. P., *Coord. Chem. Rev.*, **1** (1966), 222.

PART 5
Metal Complexes with Ditertiary Arsines

ELMER C. ALYEA

University of Guelph, Guelph, Ontario

30 INTRODUCTION †

In 1939 Chatt and Mann[1] reported the isolation of 1:1 and 1:2 complexes of the types $[PdCl_2(L-L')]$ and $[PdCl_2(L-L')_2]$ with the ditertiary arsines represented by structural formulae (CLXXXIV) (R = Me, Bu) and (CLXXXV) (R = Bu, Ph; R' = Bu, Ph). Since the initial investigation of ditertiary arsines as chelating agents, however, only the coordinating ability of the o-phenylenebis(dimethylarsine) ligand of (CLXXXIV) (R = Me), hereafter referred to as diars, has been extensively studied.

(CLXXXIV)   (CLXXXV)

Primarily as a result of the efforts of Nyholm and coworkers since 1950, complexes of diars are now known with most of the transition metals, with the exception of the lanthanides and actinides‡, as well as with several main group metals and metalloids. The metal atom in these complexes often has an uncommon oxidation state[2] and may also attain an unusually high coordination number. Although the versatility of diars as a chelating ligand is well known[3], the extensive chemistry of this ditertiary arsine has only been briefly documented. In 1964 Harris and Livingstone[4] tabulated the metal complexes of diars in a chapter on bidentate chelating agents and Booth[5] surveyed the transition metal complexes of diars in a compilation of the complexes of tertiary phosphines, arsines and stibines. Further syntheses and characterisation of metal complexes since 1964 and the reinvestigation of several metal complexes using modern physical methods in themselves warrant a review of diars chemistry. Furthermore, recent development of the coordination chemistry of other ditertiary arsines requires correlation in order to optimise the value of future work in the field.

In part 5 of this book the preparation and characterisation of metal complexes of several bidentate arsines are surveyed to the end of 1970; more recent work, some of which is unpublished, is also included in several instances. Information discussed in the sections on various physical methods is not a rigorous coverage of the topics but rather, illustrative of the various techniques and the type of information that can be obtained in specific cases.

Material that is relevant to the coordinating ability of ditertiary arsines is also discussed in other parts; for example, the nature of the metal–phosphorus bond is dealt with in part 1 and the metal complexes of monodentate arsines in part 3. Recent reviews concerning the complexes of 2,2'-dipyridyl-1,10-phenanthroline[6], and of ditertiary phosphines[7] are pertinent for comparative purposes.

---

† Abbreviations used: X = Cl, Br, I (or pseudohalides); M = metal; R = alkyl or aryl; Me = methyl; Et = ethyl; Pr = n-propyl; Bu = n-butyl; Ph = phenyl; C_5H_5 = cyclopentadienyl; L = neutral ligand; L—L' = bidentate neutral ligand; en = ethylenediamine; dipy = 2,2'-dipyridyl; o-phen = 1,10-phenanthroline; asq = 8-dimethylarsinoquinoline.

‡ See section 32.1.1(2) for the first uranium (IV) complex.

31 SYNTHESES OF LIGANDS

In contrast to the wide range of metal complexes with monodentate arsines (part 3) relatively few complexes with bidentate arsines are known except for diars[4,5]. The scarcity of metal complexes with ditertiary arsines can be related to the previous lack of suitable preparative methods of the bidentate arsines. For example, the original synthetic method[8] for diars requires four steps with an overall yield of only 20 per cent and this 'energy barrier' apparently restricted the development of the coordination chemistry of diars largely to the research efforts of Nyholm and associates. Several general synthetic routes as well as specific ones are now known[9,10] and wider utilisation of chelating diarsines in coordination chemistry can reasonably be expected.

Synthetic details for some ditertiary arsines of current research interest are outlined in table 33. It is obvious that the most important route is the reaction of an alkali metal derivative of a secondary arsine with an organic dihalide. Quite frequently the alkali metal derivative is Me_2AsNa, prepared *in situ* from Me_2AsI [9-12], or Ph_2AsK, isolable as the 1:2 dioxan adduct[13]. The organic dihalide is normally added to a solution of the R_2AsM reagent in tetrahydrofuran or dioxan; the reaction probably proceeds by nucleophilic substitution of the halogen atoms by the anion R_2As^- [9]. These substitution reactions proceed in better yield with the diphenyl-arsino salts than with the dimethylarsino derivatives. For example, *cis*-1,2-bis (diphenylarsino)-ethylene (vdiars) is produced in 61 per cent yield from Ph_2AsNa and *cis*-1,2-dichloroethylene[14]. The analogous reaction of Me_2AsNa and *cis*-1,2-dichloroethylene yields only 30–40 per cent of the substitution product, and furthermore, only 10 per cent of the product retains the *cis* configuration[9,11,15,16]. The *trans* derivative, which is also obtained, gives bridged polymeric species with transition metal halides of the type (CLXXXVI)[15-18]. The lower yields in the reactions of Me_2AsNa have been attributed to the production of only 50 per cent yields of the compound Me_2AsNa in the first step[9(a)].

(CLXXXVI)

Feltham *et al.* have recently investigated the reaction of Me_2AsNa with *o*-dichloro-benzene[9(a),19] and with *cis*-1,2-dichloroethylene[11,15] in greater detail and identified the products by proton nuclear magnetic resonance (n.m.r.) and mass spectral measurements[9(b)]. In the vigorous reaction between the nucleophilic anion Me_2As^- and *o*-dichlorobenzene the desired diars constituted 40 per cent of the recovered

TABLE 33 SYNTHESIS OF SOME DITERTIARY ARSINES

Ligand name and formula	Ligand abbreviation	Method of synthesis	% Yield	M.p(°C) and/or B.p(°C/mm Hg∥)	Other data	Ref.†
o-Phenylenebis(dimethylarsine) cis-$(Me_2As)_2C_6H_4$	diars	o-$Cl_2C_6H_4$ Me_2AsNa	45–48	100–101/1.0 –12;153–158/20	$n_D^{25} = 1.6156$ m.p of MeI adduct = 226–230°C	8,9*,19, 29,30
1,2-Bis(dimethylarsino)-ethylene cis-$Me_2AsCH=CHAsMe_2$	edas	cis-ClCH=CHCl Me_2AsNa	30–40	90–93/24‡	$n_D^{25} = 1.5567$ m.p of MeI adduct = 260°C	9,11*,15* 16
1,2-Bis(diphenylarsino)-ethylene cis-$Ph_2AsCH=CHAsPh_2$	vdiars	cis-ClCH=CHCl Ph_2AsLi	61	112–113	Dipole moment = 1.37 D	14*
1,2-Bis(diphenylarsino)-ethane $Ph_2AsCH_2CH_2AsPh_2$	dae	$ClCH_2CH_2Cl$ Ph_2AsK . 2D	57	100	m.p of disulphide = 200°C	13*,8, 31,32
1,1-Bis(diphenylarsino)-methane $Ph_2AsCH_2AsPh_2$	dam	$Cl_2AsCH_2AsCl_2$ PhMgBr	72	96–97	HNO_3 gives Ph_2AsO_2H	33*,34
1,2-Bis(methylphenylarsino)-ethane $MePhAsCH_2CH_2AsMePh$	dmpa	$(PhClAsCH_2)_2$ MeMgI	71–74	156–158/0.1§ (racemic) 142/0.05 (meso)	m.p of MeI adduct = 284–285°C	22,23*
1,2-Bis(dimethylarsino)-3,3,4,4-tetrafluorocyclobutene $Me_2AsC(CF_2CF_2)CAsMe_2$	ffars	$ClC(CF_2CF_2)CAsMe_2$ Me_2AsH	52	120/47	Absorption very weak in ν(C=C) region of i.r.	25*,26

† Data from references with asterisk.
‡ Mixture of 10 per cent cis- and 90 per cent trans-product; see reference 15 for separation.
§ Initial mixture has b.p. of 140–155/0.05.
∥ 1 mm Hg = 133.3 N m⁻².

arsenic compounds; in addition they identified the following products: dimethylarsine, trimethylarsine, dimethylphenylarsine, 5,10-dimethyl-5,10-dihydroarsanthren, and methylbis[*o*-(dimethylarsino)phenyl] arsine. The latter triarsine was identified when a fragment with small relative abundance at *m/e* 450 was observed in the mass spectrum of the crude product mixture. It is also formed from diars in boiling diethylene glycol in the presence of a nickel(II) salt[20]. Feltham and Metzger also report an improved synthesis of *cis*-1,2-bis(dimethylarsino)ethylene by hydroboration of bis-(dimethylarsino)acetylene. The latter ditertiary arsine, $[Me_2AsC{\equiv}CAsMe_2]$, obtained in 75 per cent yield, would undoubtedly lead to polymeric structures rather than to chelate compounds with transition metal salts. An analogous ditertiary phosphine, bis(diphenylphosphino)acetylene, stabilises a variety of binuclear and polymeric diphosphine-bridged halide complexes of the later transition elements[21].

The second most important route to ditertiary arsines appears to be the reaction of organohaloarsines with Grignard reagents. Indeed, it is the most convenient method of forming ligands of the type $[RR'As(CH_2)_nAsRR']$. For example, bis(methylphenylarsino)ethane is obtained in 74 per cent yield from the reaction of $(PhClAsCH_2)_2$ with MeMgX in benzene[22,23]. Other methods of preparation are known for specific ligands, as for ditertiary arsines with perfluorocarbon linkages[24-6]; some of the various methods that have been employed are summarised in the following equations (1-10)[†]. Although only alkylene

$$2R_2AsM + X(CH_2)_nX \rightarrow R_2As(CH_2)_nAsR_2 + 2MX \tag{1}$$

$$4RMgX + X_2As(CH_2)_nAsX_2 \rightarrow R_2As(CH_2)_nAsR_2 + 4MgX_2 \tag{2}$$

$$2RMgX + XR'As(CH_2)_nAsR'X \rightarrow RR'As(CH_2)_nAsRR' + 2MgX_2 \tag{3}$$

$$R_2As(CH_2)_nX + Mg \rightarrow R_2As(CH_2)_nMgX \tag{4a}$$

$$R_2As(CH_2)_nMgX + R'_2AsX \rightarrow R_2As(CH_2)_nAsR'_2 + MgX_2 \tag{4b}$$

$$X(CH_2)_nX + 2Mg \rightarrow XMg(CH_2)_2MgX \tag{5a}$$

$$XMg(CH_2)_2MgX + 2R_2AsX \rightarrow R_2As(CH_2)_nAsR_2 + 2MgX_2 \tag{5b}$$

$$Li(CH_2)_nLi + 2R_2AsX \rightarrow R_2As(CH_2)_nAsR_2 + 2LiX \tag{6}$$

$$R_2As(CH_2)X + R'Li \rightarrow R_2As(CH_2)_nLi + R'X \tag{7a}$$

$$R_2As(CH_2)_nLi + R''_2AsX \rightarrow R_2As(CH_2)_nAsR''_2 + LiX \tag{7b}$$

$$2R_2AsH + [-CH_2CH_2O-]_1 \rightarrow R_2As(CH_2)_2AsR_2 + H_2O \tag{8}$$

$$R_2As-AsR_2 + R'_2C{=}CR'_2 \rightarrow R_2As(CR'_2)_2AsR_2 \tag{9}$$

$$R_2As-AsR_2 + R'C{\equiv}CR' \rightarrow R_2AsC(R')C(R')AsR_2 \tag{10}$$

linkages between the arsenic atoms are represented for the sake of clarity, the backbone of the diarsine ligands may be aromatic (for example, diars) or olefinic (for example, edas) or a fluorinated moiety (for example, ffars)[‡]. Whilst all of the reaction schemes represented in equations (1-10) have been successful in specific cases, no generality is implied in the present outline. In fact, there are notable exceptions. The reaction of Ph_2AsK with $BrCH_2CH_2Br$ does not give 1,2-bis(diphenylarsino)-ethane (dae) as expected from equation (1); instead, the products are $CH_2{=}CH_2$ and $Ph_2As-AsPh_2$. The desired product, dae, can be obtained, however, by employing

† Usually R = Me, Ph; M = Na, K; X = Cl, Br.

‡ See table 33 for ligand nomenclature.

$ClCH_2CH_2Cl$ in equation (1)[13]. Such reactions apparently account for the failure to synthesise $[R_2AsCH_2CH_2AsR_2]$ (R = Me, Et) by equation (1), though the analogous diphosphine ligands are known and lead to some interesting metal complexes[7,27]. It is feasible that the bis(dialkylarsino)ethane ligands will be accessible by the facile route recently developed for the syntheses of polytertiary phosphines and arsines[28]. The synthesis of $[Ph_2PCH_2CH_2AsPh_2]$ from Ph_2AsH and $Ph_2PCH=CH_2$ in 65 per cent yield is the basis of this prediction.

References to the synthesis of specific ditertiary arsines may be ascertained from papers referred to in the sections describing their particular metal complexes; references for the most studied ligands are included in table 33. Information concerning the syntheses of ditertiary arsines for which no or only a few metal complexes are known may be obtained from the comprehensive compilation by Dub[10].

32 SURVEY OF METAL COMPLEXES

32.1 o-PHENYLENEBIS(DIMETHYLARSINE)

32.1.1 Group IVA

(1) Titanium

Investigations of the diars complexes of titanium(IV) halides resulted in the first examples of eight coordination for atoms in the first transition series. The 1:2 complexes $[MX_4(diars)_2]$ (M = Ti, X = Cl, Br; M = V, X = Cl) have been shown to be isostructural by X-ray powder photography, and single-crystal diffraction methods show that the coordination polyhedron around the titanium atom in $[TiCl_4(diars)_2]$ is a dodecahedron (figure 30). The two bidentate ligands form an elongated tetrahedron and the four chlorine atoms a flattened tetrahedron with respect to the fourfold inversion axis[35]. Eight complexes of the type $[MX_4(diars)_2]$ (M = Zr, Hf, V, Nb, X = Cl; M = Ti, Zr, Hf, Nb, X = Br) are known to be isomorphous with the eight-coordinate $[TiCl_4(diars)_2]$[35-7]. Although several attempts have been made to prepare similar complexes of titanium tetrachloride with other bidentate ligands[38], success has only been attained with the diars-like ligands of structures (CLXXXVII), (CLXXXVIII) and (CLXXXIX)[39,40]. Even the closely related o-phenylenebis-(diethylarsine) (Et-diars)[39,40], o-phenylenebis(diethylphosphine) (Et-diphos)[41] and

As(CD$_3$)$_2$ AsMe$_2$ PMe$_2$

As(CD$_3$)$_2$ Me   AsMe$_2$ PMe$_2$

(CLXXXVII)   (CLXXXVIII)   (CLXXXIX)

1,2-bis(dimethylarsino) ethylene(edas)[18] gave only the six-coordinate $[TiCl_4(L-L')]$. These studies suggest that the formation of eight-coordinate $[TiCl_4(L-L')_2]$ complexes is independent of the electronic properties of the ligand, crystal packing forces and solvent but is dependent upon the electronic and steric effects of the alkyl groups attached to the donor atoms. It is worth noting that eight coordination for titanium has only been proved for two other cases, namely $[Ti(NO_3)_4]$[42] and $[Ti(S_2CNEt_2)_4]$[43].

The 1:1 complexes $[TiX_4(diars)]$ (X = Cl, Br), readily obtained from inert solvents, are diamagnetic non-electrolytes[35]; the analogous complexes of TiCl$_4$ with the ditertiary arsines of structure (CLXXXVII) and (CLXXXVIII) were also characterised as octahedral complexes of titanium(IV)[39]. With TiF$_4$ and TiI$_4$ diars forms complexes of the type $[(TiF_4)_2(diars)]$ and $[TiI_4(diars)_2]$, respectively. Their i.r. spectra are characteristic of bidentate diars, and other physical properties suggested their formulation as a fluorine-bridged polymer and as the ionic structure $[TiI_2(diars)_2]^{2+} 2I^-$ [37].

No authentic complexes of titanium(III) salts with diars have been prepared, although $[TiX_3(H_2O)_6]$ was reported to react with diars in acetic acid solution to yield the complex $[TiX_3(diars)(H_2O)]$ (X = Cl, Br)[44]. An attempt to repeat the preparation for the chloro case gave the well-characterised $[TiCl_4(diars)_2]$ in about 50 per cent yield[35].

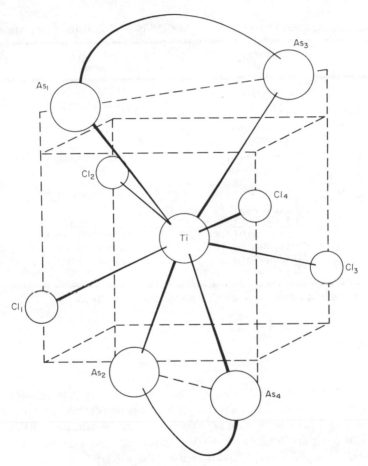

Figure 30 The structure of $[TiCl_4(diars)_2]$

(2) Zirconium and Hafnium

The 1:2 complexes $[MX_4(diars)_2]$ (M = Zr, Hf; X = Cl, Br) obtained from the reactions in acetone solution are all isostructural with the eight-coordinate titanium analogue. It is significant that the 1:1 complexes could not be isolated with these larger atoms. Since $[ZrCl_4(diars)_2]$ is formed at a faster rate in tetrahydrofuran than the corresponding hafnium complex, separation of the elements may be achieved[37].

The lesser affinity of Group IVB metal chlorides for ditertiary arsines is reflected in the failure of silicon and germanium tetrachlorides to form diars complexes[35], while tin tetrachloride forms only the 1:1 complex[35,45]. Also worth mention at this point are the observations that thorium tetrachloride does not complex with diars[35] but that uranium tetrachloride forms the 1:2 adduct of dodecahedral symmetry[46]. The latter complex is the first diars complex of the actinide elements to be characterised. Group IVA metal–diars complexes are listed in table 34.

TABLE 34 DIARSINE COMPLEXES OF GROUP IVA AND GROUP VA METALS

Coordination number	Group IVA complex	Ref.	Group VA complex	Ref.
Coordination 6	$[(TiF_4)_2(diars)]$	37	$[VCl_4(Et-diars)]$	41
	$[TiCl_4(diars)]$ †	35,39,40	$[TaCl_4(diars)]$	46
	$[TiCl_4(Et-diars)]$	39,40	$[TaCl_4(Et-diars)]$	46
	$[TiBr_4(diars)]$	35		
	$[TiI_2(diars)_2]I_2$	37		
Coordination 7	$[V(CO)_4(diars)]_2$	50	$[NbCl_5(diars)]$	51
	$[V(CO)_4(diars)I]$	50	$[NbBr_5(diars)]$	51
			$[TaCl_5(diars)]$	51
			$[TaBr_5(diars)]$	51
Coordination 8	$[TiCl_4(diars)_2]$ †	35,39	$[VCl_4(diars)_2]$	35
	$[TiBr_4(diars)_2]$	35	$[NbCl_4(diars)_2]$	36,46
	$[ZrCl_4(diars)_2]$	37	$[NbCl_4(Et-diars)_2]$	46
	$[ZrBr_4(diars)_2]$	37	$[NbCl_4(4Me-diars)_2]$	46
	$[HfCl_4(diars)_2]$	37	$[TaCl_4(diars)_2]$	46
	$[HfBr_4(diars)_2]$	37		

† Also with diars-like ligands of formulae (CLXXXVI) and (CLXXXVII).

32.1.2 Group VA

(1) Vanadium

The orange complex $[VCl_4(diars)_2]$, isolated from carbon tetrachloride solution, has the expected magnetic moment of 1.74 B.M. X-ray powder photography verified that the geometry was dodecahedral as in the previously described metal(IV) complexes[35]. Eight-coordination is rare in V(IV) complexes, other examples being $[V(S_2CNR_2)_4]$[43], $K_3[V(O_2)_4]$[47] and probably $[V(S_2CR)_4]$[48].

Surprisingly the 1:1 adduct of VCl_4 and diars could not be prepared[35], though six-coordinate complexes are well known[38,49]. However, the analogous Et-diars ligand gives only the *cis* octahedral complex[41]. Presumably the change in the steric and inductive effects between the methyl and ethyl groups on the arsine atoms are responsible for this remarkable contrast of complexing ability between diars and Et-diars.

The only other report of a vanadium complex with diars is $[V(CO)_4(diars)]$, readily prepared from the paramagnetic vanadium hexacarbonyl[50]. The compound is dimeric and diamagnetic in both the solid state and solution, implying a vanadium–vanadium bond. Spectrophotometric titration with iodine indicates that the seven-coordinate $[V(CO)_4(diars)I]$ is formed.

(2) Niobium and tantalum

Niobium(IV) and niobium(V) complexes of diars have been well characterised. The reaction of excess diars with niobium pentahalide, tetrahalide or oxyhalide in sealed tubes leads to the formation of the eight-coordinate complexes $[NbX_4(diars)_2]$ (X = Cl, Br, I). The iodo-complex is not isomorphous with the known dodecahedral compounds of this type (compare titanium) but the same stereochemistry is indicated by its visible spectrum[36]. These compounds have recently been prepared from the tetrahalides in acetonitrile solution and the complexes are free of ferromagnetic impurities by this method[46]. Deutscher and Kepert have also made tentative electronic spectral assignments for the analogous dodecahedral complexes $[NbX_4(Et-diars)_2]$ (X = Cl, Br, I) and $[NbCl_4(4-Me-diars)_2]$. The latter ligand, 4-Me-diars, is represented in formula (CLXXXVIII). Interestingly, the ethyldiarsine ligand fails to give the 1:1 adduct with niobium tetrachloride. This contrast to the behaviour of the tetrachlorides of titanium and vanadium, which form only the six-coordinate adducts with Et-diars, is believed to reflect the greater radius of niobium(IV). The ligand o-phenylenebis(diphenylarsine)(Ph-diars) does not react with $NbCl_4$ in aceto-nitrile[46].

Previous attempts to prepare Ta(IV) complexes of diars failed[36], but $[TaCl_4(L-L')]$ (L–L' = diars and Et-diars) have recently been isolated by precipitation from aceto-nitrile. The 1:2 adducts were not obtained by this method but $[TaCl_4(diars)_2]$ is formed in an impure state from a sealed-tube reaction[46]. Obviously, diars-like ligands emphasise the differences in the chemistry of niobium and tantalum.

Complexes of the type $[MX_5(diars)]$ (M = Nb, Ta; X = Cl, Br) precipitate when diars is added to dry non-hydroxylic solutions of the pentahalides. These compounds are isomorphous, diamagnetic, monomeric and non-conducting, and are therefore seven coordinate. The compounds are very sensitive to oxygen and moisture, and the niobium oxy-compounds formed, $[NbOX_3(diars)]$ and $[NbCl_4(diars)]_2O$, have characteristic Nb=O and Nb—O—Nb stretching frequencies, respectively[51]. The ready formation of oxo-niobium compounds is well established[52] and therefore it is note-worthy that the reaction of $[NbOX_3]$ with excess diars results in the formation of $[NbX_4(diars)_2]$ [36]. The complexes of Group VA metals with diars-like ligands are summarised in table 34.

32.1.3 Group VIA

(1) Chromium

Chromium(III) halides form complexes with diars that have been characterised by conductivity, molecular weight and magnetic measurements. When the proportions of diarsine to chromic salt are equimolecular, the very soluble blue compounds of the type $[CrX_3(diars)(H_2O)]$ are formed preferentially in a number of solvents. The less soluble green compounds of the type $[CrX_2(diars)_2][CrX_4(diars)]$ and $[CrX_2(diars)_2]$-ClO_4 can be prepared readily using a $2:1$ ratio of diars to chromic salt. As expected for the d^3 configuration the magnetic moment is near 3.9 B.M. for all of the octa-hedral complexes[53]. Efforts by Nyholm and Sutton to prepare a $[Cr(diars)_3]^{3+}$ ion were unsuccessful even though ions such as $[Cr(dipy)_3]^{3+}$ are known[53].

The perchlorate salt $[CrCl_2(diars)_2]ClO_4$ has also been prepared more recently from the reaction involving $[CrCl_3(THF)_3]$, diars and perchloric acid in tetrahydro-furan[54]. Feltham and Silverthorn studied the electronic spectra of the $[CrX_2(diars)_2]ClO_4$ ($X = Cl$, Br) complexes and concluded that diars has the same ligand-field strength as ethylenediamine[55]. This relatively low place for diars in the spectro-chemical series has also been deduced from a study of the octahedral anions $[Cr(NCS)_4L_2]^-$ [56]. Aspects of the electronic spectra are discussed in section 33.3.

The action of halogens on $[Cr(CO)_4(diars)]$ results in complete replacement of carbon monoxide and the formation of chromium(III) complexes. These products, $[CrX_3(diars)]$ ($X = Br$, I), are believed to be halogen-bridged, octahedrally co-ordinated dimers. The bromo-derivative very readily forms the known monomeric hydrate[53], whose reflectance spectrum is very similar to that of the dimer[57]. The complete displacement of carbonyl groups in the oxidation of $[Cr(CO)_4(diars)]$ with halogens is in sharp contrast, it should be noted, to the behaviour of the other Group VIA elements.

Feltham and coworkers have studied the complex $[CrCl(NO)(diars)_2]^+$, in which it is possible to consider that Cr(III) is coordinated to $N=O^-$. The yellow per-chlorate derivative has a doublet ground state and the electron spin resonance spectrum shows hyperfine splitting due to both the nitrogen and arsenic nuclei. The diamagnetic orange compound $[CrCl(NO)(diars)_2]$ is readily obtained from the cation by reduction[58]. X-ray powder patterns show that the three compounds $[MCl(NO)(diars)_2]ClO_4$ ($M = Cr$, Fe, Co) are isomorphous and the X-ray crystal structure of the iron compound $[FeBr(NO)(diars)_2]ClO_4$ shows it to be six-coordinate, with the halogen *trans* to the nitrosyl group[58,59].

Complexes of diars with Cr(II) halides could not be prepared[53] but some diars–carbonyl complexes of chromium in which the chromium is formally divalent have been characterised by Nyholm and coworkers[57]. Oxidation of $[Cr(CO)_2(diars)_2]$ at room temperature produces seven-coordinate complexes of chromium(II), $[Cr(CO)_2(diars)_2X]X$ ($X = Br$, I), isomorphous with the corresponding molybdenum(II) and tungsten(II) compounds[60,61]. With excess halogen the metal is not further oxidised and the trihalide anion is produced; this latter behaviour is also found for the analogous molybdenum(II) and tungsten(II) compounds. The properties of halogenodiarsine chromium complexes are outlined in table 35 and those of carbonyldiarsine chromium complexes in table 36.

(2) Molybdenum and tungsten

The predominant feature of the diarsine complexes of these elements is the occurrence of carbonyl derivatives of the divalent metal. However, a variety of complexes have been studied and these will be described first. The orange-brown tetravalent molybdenum compound [MoBr$_4$(diars)] is obtained by the oxidation of [Mo(CO)$_4$(diars)] with an excess of bromine. The low value of the magnetic moment (1.96 B.M.) for two unpaired electrons was attributed to the high value of the spin–orbit coupling constant for molybdenum[60]. The values of the magnetic moment (2.8–2.9 B.M.) for the complexes [MoX$_2$(diars)$_2$] (X = Cl, Br, I) indicate a

TABLE 35 PROPERTIES OF HALOGENODIARSINE–GROUP VIA COMPLEXES

Compound	Colour	μ_{eff} B.M.	Λ_M (ohm^{-1} cm^2 mol^{-1}) 10^{-3}M(PhNO$_2$)	Ref.
[CrCl$_3$(diars)(H$_2$O)]	blue	3.82	0.7	53
[CrBr$_3$(diars)]$_2$	green	3.87	2.4	57
[CrBr$_3$(diars)(H$_2$O)]	blue	3.85	1.5	53,57
		3.87	0.6	
[CrI$_3$(diars)]$_2$	brown	3.82	1.7	57
[CrCl$_2$(diars)$_2$] ClO$_4$	pale green	3.87	28.2	53,54, 55
[CrBr$_2$(diars)$_2$]ClO$_4$	deep green	3.96	26.9	53,55
[CrI$_2$(diars)$_2$] ClO$_4$	pale green	3.94	26.9	53
[CrI$_2$(diars)$_2$] I$_3$	dark green	3.83	26.1	57
[CrCl$_2$(diars)$_2$] [CrCl$_4$(diars)]	bright green	3.88	40.0	53
[CrBr$_2$(diars)$_2$] [CrBr$_4$(diars)]	dark green	3.88	38.1	53
[CrI$_2$(diars)$_2$] [CrI$_4$(diars)]	brown-green	3.89	41.9	53
[MoCl$_2$(diars)$_2$]	bright yellow	2.9	20†	62
[MoBr$_2$(diars)$_2$]	buff	2.85	8†	62
[MoI$_2$(diars)$_2$]	light brown	2.8	23†	62,63
[MoBr$_4$(diars)]	orange-brown	1.96	11.6	60
[WI$_2$(diars)$_2$]	brown	2.70	non-electrolyte	63

† Conductivity in CH$_3$NO$_2$ solution.

low-spin d^4 configuration. These complexes, obtained from [MoCl$_5$(H$_2$O)]$^{2-}$ and [MoCl$_6$]$^{3-}$ in the presence of the appropriate hydrogen halide, are isomorphous with the corresponding iron, technetium and rhenium compounds[62]. The complexes [MI$_2$(diars)$_2$] (M = Mo, W) are also produced by the reaction of diars with MI$_3$ [63]. Under the latter conditions, pyridine gives the adduct [MoI$_3$(py)$_3$], indicating that diars preferentially stabilises the divalent oxidation state of molybdenum.

Complexes of the type [(Mo$_6$Cl$_8$)X$_2$(L–L')$_2$]X$_2$ (X = Cl, I; L–L' = diars, dipy, o-phen) are 2:1 electrolytes and it is suggested that coordination of the bidentate ligand is to a single molybdenum of the Mo$_6$Cl$_8$ cluster[64]. The far-i.r. spectra of these compounds as well as those of analogous tungsten compounds show that two bands above 200 cm^{-1} can be attributed to vibrations of the [M$_6$Cl$_8$]$^{4+}$ core[65].

A single-crystal X-ray study has shown that the configuration around the tungsten atom in [WOCl$_4$(diars)] is approximately pentagonal-bipyramidal, the oxygen atom and one of the chlorine atoms being in the apical positions[66].

TABLE 36 PROPERTIES OF CARBONYLDIARSINE–GROUP VIA COMPLEXES

Compound	Colour	Λ_M (ohm^{-1} cm^2 mol^{-1}) 10^{-3}M(PhMO$_2$)	Carbonyl i.r. bands (cm^{-1})	Ref.
[Cr(CO)$_4$(diars)]	pale yellow	0.8	2012,1922,1898 1845,1771	57,67
[Cr(CO)$_2$(diars)$_2$]	yellow	0.7	1845,1771	57,67
[Cr(CO)$_2$(diars)$_2$Br] Br	pale yellow	22.7	1925,1865	57
[Cr(CO)$_2$(diars)$_2$Br] Br$_3$	yellow	26.2	1925,1865	57
[Cr(CO)$_2$(diars)$_2$I] I	deep orange	25.8	1923,1863	57
[Cr(CO)$_2$(diars)$_2$I] I$_3$	brown	29.5	1923,1863	57
[Mo(CO)$_4$(diars)]	white	1.0	2026,1938,1923 1914	67
[Mo(CO)$_2$(diars)$_2$]	pale yellow	0.9	1859,1786	67
[Mo(CO)$_2$(diars)$_2$Br] Br	pale yellow	23.5	1959,1888	60
[Mo(CO)$_2$(diars)$_2$Br] Br$_3$	deep yellow	20.5	1959,1890	60
[Mo(CO)$_2$(diars)$_2$I] I	pale yellow	27.4	1960,1888	60
[Mo(CO)$_2$(diars)I$_2$]	deep orange	1.5	1887,1942	61
[Mo(CO)$_2$(diars)I$_3$]	dark brown	1.6	1905,1960	61
[Mo(CO)$_3$Br$_2$(diars)]	deep orange	2.2	2023,1971,1921	60
[Mo(CO)$_3$I$_2$(diars)]	golden yellow	1.2	2053,1982,1925	60
[Mo(CO)$_2$(diars)$_2$Cl] BPh$_4$	cream	93.5†	1949,1895	68
[W(CO)$_4$(diars)]	pale yellow	0.5	2016,1923,1905 1885	67
[W(CO)$_2$(diars)$_2$]	bright yellow	0.7	1850,1774	67
[W(CO)$_2$(diars)$_2$Cl] BPh$_4$	cream	92.3†	1941,1882	68
[W(CO)$_2$(diars)$_2$Br] Br	yellow	24.3	1927,1853	61
[W(CO)$_2$(diars)$_2$Br] Br$_3$	deep yellow	28.1	1927,1853	61
[W(CO)$_2$(diars)$_2$I] I	deep yellow	26.4	1925,1852	61
[W(CO)$_2$(diars)$_2$I] I$_3$	deep orange	27.2	1925,1852	61
[W(CO)$_3$(diars)Br$_2$]	yellow	2.1	2030,1942,1905	61
[W(CO)$_3$(diars)Br$_2$]Br	yellow-green	24.0	2041,1950,1915	61
[W(CO)$_4$(diars)I] I	orange	27.3	2040,2005,1960 2082	61
[W(CO)$_4$(diars)I] I$_3$	deep orange	25.0	2040,2000,1960 2080	61

† Conductivity in acetone solution.

Feltham *et al.* have studied the rates of isomerisation of the green diamagnetic nitrosyl complexes, *cis,cis*-[MoCl$_2$(NO)$_2$(L–L′)] (L–L′ = diars and ethylenediamine) to the mononitrosyl hyponitrito complexes {[MoCl(NO)(L–L′)]$_2$N$_2$O$_2$}Cl$_2$ [58].

Many of the carbonyldiarsine derivatives of the Group VIA elements are obtained by halogen oxidation of the compounds [M(CO)$_4$(diars)] and [M(CO)$_2$(diars)$_2$]. These precursors are formed by heating the metal hexacarbonyls with an excess of diars in sealed tubes for several hours. Higher temperatures (about 200–240°C) and removal of carbon monoxide are necessary to effect the second stage of substitution, particularly with chromium[67].

Treatment of the diamagnetic monomeric non-electrolyte [Mo(CO)$_4$(diars)] with bromine or iodine in an approximately 1:1 molar ratio gives the seven-coordinate molybdenum(II) complexes [Mo(CO)$_3$(diars)X$_2$][60]. Under more vigorous conditions, in boiling carbon tetrachloride and in boiling chloroform, the reaction with iodine yields the paramagnetic compounds, [Mo(CO)$_2$(diars)I$_2$], μ_{eff} = 1.98 B.M., and [Mo(CO)$_2$(diars)I$_3$], μ_{eff} = 1.40 B.M., respectively[61]. The seven-coordinate bivalent molybdenum complexes [Mo(CO)$_2$(diars)$_2$X]X result from the reaction of [Mo(CO)$_2$(diars)$_2$] with two equivalents of iodine or bromine. The same bisdiarsine reactant with excess bromine gives the product [Mo(CO)$_2$(diars)$_2$Br]Br$_3$, which is also a 1:1 electrolyte[60]. The compounds [M(CO)$_2$(diars)$_2$Cl]BPh$_4$ have been obtained for both molybdenum and tungsten by the reaction of diars with [M(CO)$_3$(hmb)Cl]BPh$_4$, where hmb is hexamethylbenzene[68]. The 'nine-orbital rule' was invoked to explain the occurrence of the above seven-coordinate molybdenum(II) carbonyl compounds[60].

Irradiation of mixtures of [C$_5$H$_5$Mo(CO)$_3$I] and diars for thirty and sixty hours yields products whose physical properties indicate formulation as [C$_5$H$_5$Mo(CO)$_2$(diars)]I and [C$_5$H$_5$Mo(CO)(diars)$_2$I] respectively[69].

The complex [W(CO)$_4$(diars)] is oxidised by iodine in any proportion to the triiodide [W(CO)$_4$(diars)I]I$_3$; reduction of this complex with sulphur dioxide leads to the more soluble monoiodide, which is also a diamagnetic 1:1 electrolyte. With bromine and [W(CO)$_4$(diars)] in equimolar proportions the seven-coordinate tungsten(II) complex [W(CO)$_3$(diars)Br$_2$] is obtained. Excess bromine does not displace all of the carbon monoxide (as with Cr and Mo) and the tungsten(III) complex [W(CO)$_3$(diars)Br$_2$]Br is formed. The magnetic moment, μ_{eff} = 1.54 B.M., of this compound is compatible with the formulation as a d^3 spin-paired complex. Halogenation of the complex [W(CO)$_2$(diars)$_2$] gives the seven-coordinate tungsten(II) cations [W(CO)$_2$(diars)$_2$X]$^+$ as the mono- and trihalides; these complexes are analogous to the molybdenum(II) species[61]. Some properties of the carbonyl–diarsine complexes of molybdenum and tungsten are given in table 36.

32.1.4 Group VIIA

(1) Manganese

Complexes of the type [MnX$_2$(diars)$_2$] (X = Cl, Br, I) were obtained as white crystalline compounds using dioxan as the solvent for the manganese(II) halides. Attempts to prepare complexes with other anions or using other solvents failed. The divalent complexes are monomeric and non-electrolytes and the magnetic moments, μ_{eff} = 6.0 B.M., indicate that spin pairing does not occur as it does for other metal complexes of diars. Nyholm and Sutton were also successful in isolating the five-coordinate cation [MnCl$_2$(diars)(H$_2$O)]$^+$, which is a d^4 spin-free complex, though manganese(III) salts are normally easily reduced by arsines and phosphines[70]. The complex [MnCl$_3$(diars)] is reported to be formed from the reaction of [Mn(CO)$_3$(diars)] with chlorine in chloroform[71].

Dimanganese decacarbonyl and diars form $[Mn(CO)_3(diars)]_2$ at 130°C in a
Carius tube. On further heating or by recrystallisation from chloroform the
monomer is obtained. Oxidation with iodine and bromine yields the complexes
$[Mn(CO)_3(diars)I]$ and $[Mn(CO)_2(diars)Br_2]$ [71,72]. The orange diamagnetic complexes $[Mn(CO)_3(diars)X]$ (X = I, Br) may also be obtained by heating diars with
$[Mn(CO)_5X]$ [73].

Irradiation of a cyclohexane solution of diars and either $[C_5H_5Mn(CO)_3]$ or
$[(MeC_5H_4)Mn(CO)_3]$ yields the yellow complexes $\{[C_5H_5Mn(CO)_2]_2$ diars$\}$ and
$\{[(MeC_5H_4)Mn(CO)_2]_2$ diars$\}$, in which diars acts as a bridging group between the
manganese atoms [74]. This unique rôle of diars is confirmed by the single-crystal

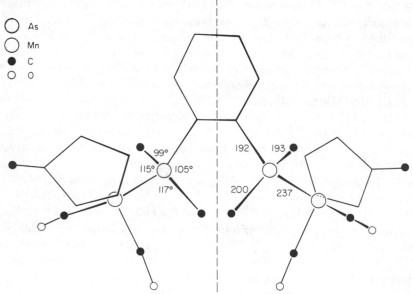

Figure 31   The structure of $[(MeC_5H_4)Mn(CO)_2]_2$ (diars)

X-ray structural determination of the methylcyclopentadienyl derivative [75]. A
representation of this structure is given in figure 31.

(2) Technetium and rhenium

Fergusson and Nyholm have prepared six-coordinate Tc(II) and Tc(III), and
eight-coordinate Tc(V) complexes with diars. All of the compounds are paramagnetic
and it is noteworthy that the compounds $[TcX_2(diars)_2]$ (X = Cl, Br, I) are spin
paired, in contrast to their manganese analogues. The Tc(III) complexes
$[TcX_2(diars)_2]X$ may be reversibly reduced to the neutral Tc(II) complexes or oxidised
to the uni-univalent Tc(V) electrolytes, $[TcCl_4(diars)_2]Y$ (Y = Cl, ClO_4) [76].

Rhenium also forms divalent $[ReX_2(diars)_2]$ complexes, trivalent $[ReX_2(diars)_2]ClO_4$
complexes and the pentavalent $[ReCl_4(diars)_2]ClO_4$ and $[ReBr_4(diars)_2]Br_3$
complexes that are analogous to the technetium compounds. The low values

of the magnetic moments (1.8–2.1 B.M.) for the low-spin Re(III) d^4 complexes were explained in terms of Kotani theory for octahedral complexes[77,78]. Similar metal(III) $[MX_2(diars)_2]^+$ cations are known for chromium(III) and several Group VIII elements (sections 32.1.3 and 32.1.5–7).

The pink-red complex obtained by reaction of the trinuclear chloride $[Re_3Cl_9]$ with diars in ethanol at room temperature is formulated as $[Re_3Cl_8(diars)_2]Cl$. In contrast to other neutral ligands diars will also break down the triangular $[Re_3Cl_9]$ structure and the yellow complex $[ReCl_2(diars)_2Cl]$ has been isolated from ethanol and dimethylformamide solutions[79].

The rhenium(I) complexes $[Re(CO)_3(diars)X]$, obtained from $[Re(CO)_5X]$ in a similar manner to the manganese analogues[73], are precursors of a variety of products[80]. Oxidation with chlorine or bromine yields the complexes $[ReX_4(diars)]$ and $[ReOCl_3(diars)]$ in which all of the carbon monoxide has been displaced. Heating $[Re(CO)_3(diars)X]$ with diars in sealed tubes at 210°C yields $[Re(CO)(diars)_2X]$, from which the seven-coordinate rhenium(III) species $[Re(CO)(diars)_2X_2]Y$ (Y = Br_3, I_3, ClO_4) have been obtained with the appropriate halogen.

32.1.5 Group VIIIA

(1) Iron

Complexes containing diars have been characterised, in which the iron has formal oxidation states from zero to four.

The yellow diamagnetic Fe(0) derivatives, $[Fe(CO)_3(diars)]$ and $[Fe(CO)(diars)_2]$, are obtained by u.v. irradiation or by sealed tube reactions of diars with iron pentacarbonyl[81,82]. Reinvestigation of this reaction led to an improved synthesis of $[Fe(CO)_3(diars)]$ and isolation of $[Fe_2(CO)_8(diars)]$ in low yield[83]. In addition Cullen and Harbourne collected spectroscopic data on two products in which fragmentation of diars appears to have occurred. The structure shown in (CXC) was

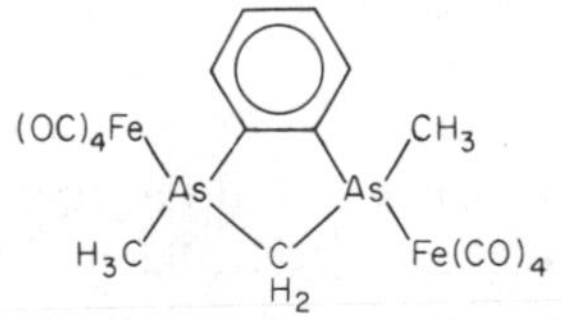

(CXC)

consistent with the evidence obtained for a yellow oil produced in the reaction of diars with $[Fe(CO)_5]$. No structure was suggested for the paramagnetic halogenocarbonyl derivative that formed when $[Fe(CO)_3(diars)]$ was added to chloroform, though the Mössbauer data and the magnetic susceptibility indicated that Fe(IV) could be present[83].

Treatment of $[Fe(CO)_3(diars)]$ with iodine and bromine yields the complexes $[Fe(CO)_2(diars)I]$ and $[Fe(CO)_2(diars)X_2]$. Oxidation of $[Fe(CO)(diars)_2]$ with iodine in ether gave diamagnetic $[FeI_2(diars)_2]$ [81]. The diffuse reflectance spectra of the spin-paired tetragonal d^6 complexes $[FeX_2(diars)_2]$ [81,84,85] have been recorded by Feltham and Silverthorn[55]. Evaluation of the ligand field parameters for the orange octahedral complex $[Fe(diars)_3](ClO_4)_2$ indicates strong σ-bonding but no π-bonding between the iron and the diarsine[55].

The spin-paired iron(III) complexes $[FeX_2(diars)_2]Y$ ($X = Cl$, $Y = ClO_4$, $FeCl_4$; $X = Br$, $Y = Br$, $FeBr_4$) were amongst the first metal complexes of diars studied by Nyholm[84]. The recent characterisation by Feltham and associates of these *trans*-$[FeX_2(diars)_2]^+$ species using a number of physical techniques indicates that there is a small tetragonal splitting of the octahedral ground state, $^2T_{2g}$. Feltham and co-workers have also formulated the six-coordinate monomeric cations $[FeX(NO)(diars)_2]^+$ in which the halide is *trans* to the nitrosyl group, as Fe(III) complexes that contain the nitrosyl group bonded as $(N=O)^-$ [59]. Variable temperature magnetic susceptibility and Mössbauer data are compatible with a ligand-field model with a rhombically distorted ^2E ground term for the nitrosyl complexes. In this excellent study of iron nitrosyl compounds the dark blue diperchlorate $[Fe(NO)(diars)_2](ClO_4)_2$ served as starting material for preparation of the other iron(III) complexes[59].

The magnetic and spectral properties of several spin-paired complexes of iron(III), $[Fe(L-L')_3](ClO_4)_3$, have been examined by Feltham and Silverthorn[86]. The values of the ligand-field parameters, Dq, B, and C are closely similar for all of the chelating arsenic and nitrogen ligands studied (diars, edas, asq, *o*-phen, dipy, en)[†]. The iron(III) complexes were prepared from $[Fe(L-L')_3]^{2+}$ by oxidation in concentrated nitric acid, the only solvent in which they were stable towards reduction.

The first cationic iron(IV) complexes, $[FeCl_2(diars)_2]Y_2$ ($Y = FeCl_4ClO_4$), were obtained by the oxidation of the compound $[FeCl_2(diars)_2][FeCl_4]$ with concentrated nitric acid[87(a)]. The same authors have more recently confirmed the assignment of oxidation state IV from a study of the magnetic and spectral properties of the octahedral complexes $[FeX_2(diars)_2]Y_2$ ($Y = BF_4$, ReO_4)[87(b)]. The variable temperature magnetic susceptibility data indicate a large tetragonal distortion in the spin-paired d^4 cations. Conductivity and far-i.r. spectral data (see section 33.1) were also consistent with the stated formulation[87b].

(2) Ruthenium and osmium

Both elements form the diamagnetic divalent complexes $[MX_2(diars)_2]$ and the paramagnetic trivalent complexes $[MX_2(diars)_2]^+$ [88,89]. As with the analogous iron(III) species the latter complexes may be oxidised by nitric acid to the tetravalent state in the case of osmium; no ruthenium(IV) complexes were isolated. The low values of the magnetic moments of the $[OsX_2(diars)_2]^{2+}$ complexes were attributed to the large spin–orbit coupling constant of osmium. The six-coordinate anions, $[RuX_4(diars)]^-$ ($X = Br$, Cl), were formed when excess halogen was used with the appropriate $[RuX_2(diars)_2]$ [88,89]. The band in the absorption spectrum of $[RuCl_2(diars)_2]$ at $22\,400\ cm^{-1}$ ($\epsilon = 60$) is assigned to the $^1E_g \rightarrow {}^1A_{1g}$ transition expected for the *trans* tetragonally distorted diamagnetic complex; this places *o*-phenylenebis(dimethylarsine) just under the ditertiary phosphine ligands, $[R_2PCH_2CH_2PR_2]$ ($R = Me$, Ph), in the spectrochemical series[90].

† See table 33 and footnote on page 311 for ligand nomenclature.

Hydrido and alkyl derivatives of $[RuCl_2(diars)_2]$ have been reported by Chatt and coworkers from reduction with lithium aluminium hydride[91(a)] and reaction with trimethylaluminium respectively[91(b)]. The complex *trans*-$[RuHCl(diars)_2]$ has $\nu(Ru-H)$ at 1804 cm^{-1} and the complex $[RuMeCl(diars)_2]$ has a melting point greater than 350°C; the hydride derivative is unstable in air relative to similar complexes with aliphatic ditertiary phosphines[91].

Ruthenium nitrosyl trichloride reacts readily with diars in ethanol to give the non-ionic, diamagnetic orange compound $[Ru(NO)Cl_3(diars)]$ [92(a)]. The nitrosyl stretching frequency in similar ruthenium compounds of the type $[Ru(NO)X_3L_2]$ was found to increase in more polar solvents[92(b)]. Feltham and coworkers have recently communicated some reactions of *trans*-$[Ru(NO)Cl(diars)_2]Cl_2$ that are relevant to the fixation of dinitrogen. With hydrazine this complex gives the azido-complex, *trans*-$[Ru(N_3)Cl(diars)_2]$, which with an excess of $NOPF_6$ in methanol yields both *trans*-$[Ru(N_2)Cl(diars)_2]PF_6$ and *trans*-$[Ru(NO)Cl(diars)_2](PF_6)_2$. Infrared data using ^{15}N-enriched samples of $[Ru(NO)Cl(diars)_2]Cl_2$ are being studied to elucidate the mechanism of formation of the dinitrogen complex[93].

Osmium tricarbonyl diiodide reacts with diars in refluxing benzene to give the white complex $[Os(CO)_2I_2(diars)]$. The occurrence of two CO bands in the i.r. spectra (2043 cm^{-1}, 1972 cm^{-1}) was compared with other derivatives of the general type $[Os(CO)_2X_2L_2]$, suggesting that each pair of ligands is also *cis* in those derivatives[94]. It is reported that $[Os_3(CO)_{12}]$ reacts with diars to give $[Os(CO)_2(diars)]_3$ [95].

32.1.6 Group VIIIB

(1) Cobalt

Nyholm studied all of the major types of cobalt complexes with diars in the early 1950s[85,96] but considerable structural data using various physical methods have been reported in recent years.

The yellow to brown products obtained with cobalt(II) salts were originally formulated as the square-planar complexes $[Co(diars)_2]X_2$ (X = Cl, Br, I, CNS) [96] and the subsequent observation that $[Co(diars)_2](ClO_4)_2$ is a 2:1 electrolyte in nitromethane and nitrobenzene[97] supports this conclusion for the structure, at least in ionising solvents. However, the work of Rodley and coworkers indicates that the complexes of the type $[Co(diars)_2X_2]$ are best viewed as tetragonally distorted octahedral compounds in the solid state. For example, the X-ray powder pattern of the chloride is similar to that of the analogous $[Ni(diars)_2X_2]$ complexes (section 32.1.7 (1))[98]. Infrared data for the complexes $[M(diars)_2]Y_2$ (M = Co, Ni; Y = ClO$_4$, NO$_3$) suggest association of the oxy-anion with the metal and an X-ray single-crystal diffraction analysis showed that the perchlorate ions in the cobalt complex are weakly associated in *trans* positions[99]. It is worth noting that the magnetic moments of the $[Co(diars)_2X_2]$ complexes (μ_{eff} = 2.0–2.4 B.M.) are in the range expected for five-coordinate Co(II) [100] and that square-pyramidal geometry is well established for the Ni(II) complexes $[Ni(diars)_2X]Y$ (Y = X, ClO$_4$) [101].

Nyholm found that the cobalt(II) complexes of diars readily undergo aerial oxidation to give tervalent octahedral complexes $[CoX_2(diars)_2]Y$ $(Y = X, ClO_4, CoCl_4,$ $Co(CNS)_4)$ [96]. In the case of cobalt(II) acetate as the starting material, he isolated three complexes: $[Co(diars)_3](ClO_4)_2$, $[Co(diars)_3](ClO_4)_3$ and $[CoX_2(diars)_2]ClO_4$ $(X = acetate)$ [85]. The electronic spectrum of $[Co(diars)_3](ClO_4)_3$ is similar to that of another diamagnetic d^6 complex, $[Fe(diars)_3](ClO_4)_2$, and evaluation of the ligand-field parameters indicates that π-bonding between diars and the metal cannot be important [55].

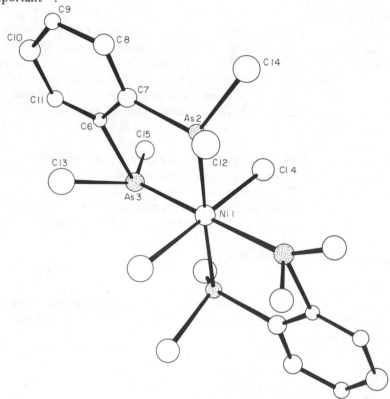

Figure 32 The structure of $[NiCl_2(diars)_2]Cl$

Other recent studies of the tervalent complexes primarily concerns the characterisation of *cis* and *trans* isomers for the complexes $[CoX_2(diars)_2]Y$. Electronic spectra of the complexes $[CoX_2(diars)_2]ClO_4$ $(X = Cl, Br)$ indicate that the purple forms have the *cis* configuration and the green forms have the *trans* configuration [102,10]. The diffuse reflectance spectra of the tetragonal *trans*-$[CoX_2(diars)_2]ClO_4$ complexes have been compared with those of *trans*-$[FeCl_2(diars)_2]$ complexes, which are also spin paired and diamagnetic [55]. Far-i.r. spectroscopy has established that a *trans* configuration is normally found for $[MX_2(diars)_2]$ and $[MX_2(diars)_2]^+$ complexes [98]. Confirmatory X-ray structural data is known for the green cobalt(III) complexes, $[CoCl_2(diars)_2]Cl$ [98,104] and $[CoCl_2(diars)_2]ClO_4$ [98,105]. The structure is virtually identical to that of $[NiCl_2(diars)_2]Cl$ shown in figure 32 [104]. Differentiation between *cis* and *trans* geometry on the basis of proton magnetic resonance and i.r. spectra has been recently reported for several $[CoX_2(diars)_2]^+$ complexes[106], and the method will undoubtedly find wide applicability to diars metal complexes of this type. Baylis and Bailar also reported the *cis*-dinitrato, -carbonato, -oxalato, and -diacetato cobalt(III) derivatives by the reaction of silver salts of the 'hard' bases with the *trans* starting material[106]. Peloso and coworkers have studied the kinetics of isomerisation and of substitution of some $[CoX_2(diars)_2]^+$ complexes[107].

Six-coordinate mononitrosyl complexes of cobalt of the type $[CoX(NO)(diars)_2]^+$ (X = Cl, Br, I, NCS) have been characterised using several physical methods[108], and the presence of *cis* and *trans* isomers was demonstrated. All of the experimental evidence for the red complex $[Co(NO)(diars)_2](ClO_4)_2$ indicated five-coordination.

(2) Rhodium and iridium

Diars reacts with rhodium(III) halides to form the salts $[RhX_2(diars)_2]X$ (X = Cl, Br, I), analogous to the cobalt(III) complexes[109]. Similarly, far-i.r. and electronic spectral data indicate that the corresponding iridium(III) complexes $[IrX_2(diars)_2]Y$ have the same structures, with both *cis* and *trans* configurations being possible for the diamagnetic six-coordinate metal[110].

Diars reacts with $[Ir(CO)_3Br]$ to form the white uni-univalent electrolyte $[Ir(CO)(diars)_2]Br$. This compound reversibly dissociates to form the yellow-orange solid $[Ir(diars)_2]Br$. This latter complex forms oxygen and hydrogen adducts but the corresponding adducts with the chelating phosphine, 1,2-bis(diphenylphosphino)-ethane, are better characterised[111]. These reactions of the apparently square-planar complex $[Ir(diars)_2]Br$ are analogous to the behaviour of Vaska's compound $[IrCl(CO)(PPh_3)_2]$, and thus have relevance to the chemisorption of hydrogen and oxygen on metals and oxidative addition reactions[112].

32.1.7 Group VIIIC

(1) Nickel

The versatility of diars in stabilising a wide range of oxidation states and co-ordination numbers for nickel was demonstrated by Nyholm in part of his original survey of the complexing ability of the ditertiary arsine. Confirmation of his structural assignments using modern physical techniques has been the principal aim of the more recent research on nickel complexes of diars.

Four-coordinate nickel(0) complexes of the types $[Ni(diars)_2]$ [113,114] and $[Ni(CO)_2(diars)]$ [115] are also known for other chelating ligands[113]. Oxidation of the latter complex with halogens provides a synthetic route to the *cis*-planar nickel(II) complexes $[NiX_2(diars)]$ (X = Br, I) [115] since nickel(II) halides form only 1:2 complexes with diars[116]. Nyholm and coworkers have recently developed an improved synthesis of the *cis*-dihalides $[Ni(diars)X_2]$ (X = Cl, Br, I) by the action of gaseous HX on the $[Ni(CO)_2(diars)]$ complex[117].

Red to brown, diamagnetic nickel(II) complexes of the type $[Ni(diars)_2X]X$ (X = Cl, Br, I, NCS, NO_3) and $[Ni(diars)_2X]ClO_4$ (X = Cl, Br, I, CNS, CN) have been formulated as square pyramidal, at least in solution[101,116,118]. Though the compounds are uni-univalent electrolytes in nitromethane or nitrobenzene, as are the corresponding palladium(II) and platinum(II) complexes[118,119], all of the $[M(diars)_2X_2]$ complexes may be considered as tetragonally distorted octahedral complexes in the solid state on the basis of the crystal structure determinations of the diiodides for Pt, Pd and Ni respectively[120-2]. Preer and Gray presented electronic spectral criteria which provide a differentiation of square-pyramidal and trigonal-bipyramidal geometries and assigned the ligand-field bands in the d^8 low-spin electronic configuration[101]. Both the $[Ni(diars)_2X]X$ and $[Ni(diars)_2X]ClO_4$ complexes exhibit three bands in the visible region in solution at 300 K and 77 K, as does the complex $[Ni(diars)(triars)](ClO_4)_2$ {triars = methyl-bis[o-(dimethylarsino)phenyl] arsine}, which is known from an X-ray structural study to be square pyramidal[20].

The conductivity, magnetism and absorption spectrum of $[Ni(diars)_2](ClO_4)_2$ are in agreement with the assignment of square-planar geometry to the cation[101,116,118]. The ready addition of one equivalent of halide or of thiourea to $[Ni(diars)_2]^{2+}$ has been demonstrated[101,116].

The compound originally formulated as $[Ni(diars)_3](ClO_4)_2$ [116] has recently been shown to be $[Ni(diars)(triars)](ClO_4)_2$, in which the nickel(II) atom has nearly a regular square pyramid of As atoms about itself[20]. The anomalous diamagnetism of the complex (since all octahedral Ni(II) complexes are paramagnetic) had previously been rationalised in terms of a large trigonal (D_3) field[123]. It is interesting that nickel(II) is the only transition metal ion that catalyses the disproportionation of the diarsine to triars, though fragmentation of the diars also occurs in its reaction with $Fe(CO)_5$ [83]. Spectral evidence that the d-level ordering is $xy < xz, yz < z^2 \ll x^2 - y^2$ in $[Ni(diars)(triars)](ClO_4)_2$ and the other square-pyramidal $[Ni(diars)_2X]Y$ (Y = X, ClO_4) complexes was taken as support for the presence of nickel($d\pi$) → diarsine(π) back-bonding[101].

Nyholm formulated the complexes of the type $[NiX_2(diars)_2]Y$ (X = Y = Cl, Br, CNS; X = Cl, Y = ClO_4, $PtCl_6$) as octahedral tervalent nickel compounds on the basis of conductivity, potentiometric and magnetic measurements[116]. The brownish-yellow trichloride, whose magnetic moment of 1.89 B.M. is in agreement with the presence of one unpaired electron, was obtained by aerial oxidation of $[NiCl_2(diars)_2]$ in the presence of hydrochloric acid. The other Ni(III) complexes were isolated by metathesis reactions involving the appropriate salt. The magnetic susceptibility of the trichloride complex has recently been measured to 4 K; it obeys the Curie–Weiss law with a Weiss constant of -10 K [124].

The X-ray structural determination of the spin-doublet complex $[Ni(diars)_2Cl_2]Cl$ has verified that the four arsenic atoms are in a square-planar arrangement about the nickel atom with the two chlorides completing a slightly distorted octahedron (figure 32)[104,125]. The longer Ni—Cl bond distance (243 pm) compared with the Co—Cl distance (226 pm) in the corresponding cobalt complex of identical structure is in agreement with the ^2A_g ground state proposed as the result of analysis of the electron spin resonance spectrum of the nickel monocation[104,124]. Electron spin resonance parameters have recently been measured for the complexes $[Ni(diars)_2Cl_2]Y$

$(Y = Cl, ClO_4)$ as powders, in solutions and frozen solutions $(77 \ K)^{124-7}$, and as single crystals doped into a diamagnetic host crystal[124]. Manoharan and Rogers also studied the e.s.r. spectra of the complexes $[Ni(diars)_2X_2]X$ $(X = Br, NCS)$ and $[Co(diars)_2X_2]$ $(X = Cl, Br, NCS)^{126}$. Although a metal-stabilised 'σ-radical' structure for $[Ni(diars)_2Cl_2]^+$ was proposed by Gray *et al.* in a preliminary communication[125], the more detailed work is in agreement with a model based on d^7 Ni(III) with a ^2A_g ground state in which the unpaired electron is strongly delocalised over the metal and all of the ligand atoms[124,126]. Details of the e.s.r. spectral data are outlined further in section 33.5.

An unstable complex formulated as $[NiBr_3(diars)]$, with a magnetic moment of 2.4 B.M. and possibly a polymeric octahedral structure, was reported by Nyholm[115]. The only Ni(III) complex in addition to $[NiCl_2(diars)_2]Cl$, whose structure has been confirmed by X-ray structural data, is the five-coordinate complex $[NiBr_3(PMe_2Ph)_2]^{128}$.

The only complex in which nickel has the formal oxidation state of IV is in the octahedral cation $[NiCl_2(diars)_2]^{2+}$ [129]. Nyholm *et al.* has isolated the analogous Pd(IV) and Pt(IV) complexes by oxidation of lower valent diars complexes with concentrated nitric acid and precipitation as the diperchlorate[118].

(2) Palladium and platinum

The 1:1 and 1:2 adducts of palladium(II) chloride with diars were amongst the first known complexes of ditertiary arsines[1]. Since that time, many M(II) and M(IV) (M = Pd, Pt) complexes have been characterised; although most are analogous to the nickel species already described, there are significant differences.

Both metals form the square-planar cations $[M(diars)_2]^{2+}$, isolated as the perchlorate or tetrachlorometallate(II) salts, which are precursors to five- and six-coordinate cations[1,118,119]. The neutral *cis*-square-planar complexes of palladium $[PdX_2(diars)]$ $(X = Cl, Br)$ are also known[1,130] and treatment of the dibromide with methyl lithium yields the colourless dimethyl derivative[131]. A convenient method of synthesis of the complexes $[PdX_2(diars)]$ $(X = Cl, Br, I, CNS)$ using $[PdX_2(PhCN)_2]$ as starting material was recently reported[117], but this method only gives the chloride derivative with platinum. The 1:1 palladium–diars complexes have also been obtained from $[PdX_2en]$ (en = ethylenediamine) by a ligand-displacement reaction[132].

Several complexes of the types $[MX_2(diars)_2]$ and $[MX(diars)_2]ClO_4$ have been synthesised, mainly from the dichlorides and diperchlorates[1,118,119]. Although the first type are uni-univalent electrolytes in nitromethane or nitrobenzene and presumably contain five-coordinate palladium or platinum[118,119], the tetragonally distorted octahedral geometry can be assumed in the solid state on the basis of the X-ray crystal structure of the diiodides $[MI_2(diars)_2]^{120,121}$. However, the X-ray structure of $[Pt(diars)_2Cl_2]$ indicates that the chlorine atoms are at ionic distances; the platinum atom is virtually square-planar[120(b)], as schematically represented in figure 33[120(b)]. Additional evidence for the existence of five-coordinate $[M(diars)_2X]^+$ cations in solution has been provided by Peloso and coworkers, who have determined formation constants and thermodynamic parameters for a wide range of such species[133-6]. Such thermodynamic studies support the conclusions based on the synthetic work[118,119] that the stability of the five-coordinate cations is greater for Ni(II) than for Pd(II) and Pt(II)[137].

In a recent study of palladium and platinum complexes of the type [M(L–L')$_2$L]ClO$_4$ (L–L' = diars, o-phen, asq; L = 4-methylpyridine) n.m.r. spectroscopy indicated that the monodentate ligand was not coordinated in the case of diars[138].

As indicated for Ni(II) the tetravalent complexes [MX$_2$(diars)$_2$](ClO$_4$)$_2$, which are uni-divalent electrolytes in nitromethane, are obtained by nitric acid oxidation of the octahedral metal(II) complexes [MX$_2$(diars)$_2$] [118]. Barclay *et al.* have also characterised the analogous tetravalent palladium and platinum complexes

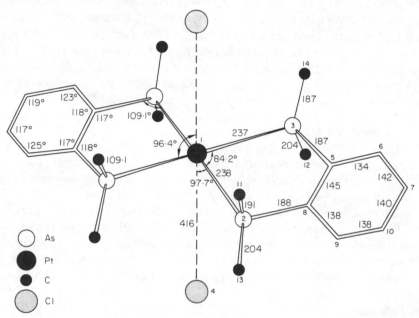

Figure 33 The structure of [Pt(diars)$_2$]Cl$_2$

[MCl$_2$(asq)$_2$](ClO$_4$)$_2$] (asq)[138]. Peloso and coworkers have also studied the rates of reduction of the platinum(IV) cations [PtX$_2$(diars)$_2$]$^{2+}$ [139–41].

No Pt(0) complexes of diars are known but Pd(0) is found in [Pd(diars)$_2$] and [Pd(diars)(Et-diphos)] (Et-diphos = o-C$_6$H$_4$(PEt$_2$)$_2$) [113].

32.1.8 Group IB

(1) Copper

Although attempts have been made to isolate Cu(II) complexes of diars[15,142], only Cu(I) derivatives are formed. This is also the case for complexes with other arsines and phosphines.

Two types of complexes are formed: [Cu(diars)$_2$]X (X = Br, I, ClO$_4$) and

[Cu(diars)$_2$][CuX$_2$] (X = Cl, Br, I). All are white diamagnetic solids and are uni-univalent electrolytes in nitrobenzene. The geometry of the cation is not known but is assumed to be tetrahedral[142,143].

(2) Silver and gold

As with copper, the known complexes of diars with silver are of the type [Ag(diars)$_2$]X and [Ag(diars)$_2$][AgX$_2$]; their physical properties support these formulations except that the [Ag(diars)$_2$][AgI$_2$] complex has a tendency to exist as a non-ionic dimer in solution[144].

In addition to forming complexes of the types [Au(diars)$_2$]X and [Au(diars)$_2$]Y (Y = AuI$_2$, CuI$_2$) gold also forms four-, five- and six-coordinate Au(III) complexes[145,146]. The tervalent gold complexes can be prepared directly from sodium tetrachloroaurate(III) and the diarsine in alcohol, and by the oxidation of the univalent complexes. The colourless diamagnetic triperchlorate [Au(diars)$_2$](ClO$_4$)$_3$ presumably has square-planar geometry, whilst conductivity and visible absorption spectra for nitrobenzene solutions of [Au(diars)$_2$I](ClO$_4$)$_2$ and [Au(diars)$_2$NO$_3$](NO$_3$)$_2$ suggest five-coordinate stereochemistry for these cations. Harris and Nyholm demonstrated that the former cation will become [Au(diars)$_2$I$_2$]$^+$ in nitrobenzene with excess iodide ion and that the latter cation becomes [Au(diars)$_2$]$^{2+}$ in aqueous solution[146]. That Au(III) can adopt five-coordination has been confirmed by X-ray structural analysis of [AuCl$_3$(biq)] (biq = 2,2'-biquinolyl)[147]. Stephenson has also confirmed that [AuI$_2$(diars)$_2$]I has a six-coordinate Au(III) cation[148,149] in the solid state and thus the compounds [AuX$_2$(diars)$_2$](ClO$_4$) (X = Br, I) also described earlier[145,146] can be assigned the same tetragonal symmetry with the diarsine ligands in a square plane. The structure is thus similar to that of [NiCl$_2$(diars)$_2$]Cl shown in figure 32. The occurrence of coordination numbers greater than four for Au(III) and other d^8 ions has been discussed by Harris and Livingstone[150].

32.1.9 Group IIB

(1) Zinc

The four-coordinate white adducts [Zn(diars)X$_2$] (X = Cl, Br, I) are non-electrolytes in solution. With Zn(NO$_3$)$_2$ the complex [Zn(diars)$_2$](ClO$_4$)$_2$, which is a bi-univalent electrolyte in nitromethane, can be precipitated by adding HClO$_4$[145,151]. In these zinc complexes and in other diars complexes with metals having the d^{10} configuration, the possibility of π-bonding is considered to be negligible[145].

Complexes of the type [R$_2$Zn(diars)] (R = Bu, Ph, C$_6$F$_5$) have been prepared as part of a study of complex formation of organozinc compounds. The stability of the Zn–As bonds appears to be enhanced with the more electronegative organo-groups[152].

(2) Cadmium and mercury

Both of the elements give the complexes [MX$_2$(diars)] (X = Cl, Br, I) and

[M(diars)$_2$](ClO$_4$)$_2$ that are analogous to the Zn(II) species. However, in these cases the non-electrolytic type show significant formation of salts of the type [M(diars)$_2$][MX$_4$] especially for the mercury(II) derivative. The much greater resistance of [MX$_2$(diars)] to aquation when M = Hg than Zn or Cd has been related to the greater polarising power of the heavier element[145]. Recent assignments of the i.r. spectral bonds of most of the known Group IIB complexes with diars indicate tetrahedral geometry for the complexes [MX$_2$(diars)] (M = Zn, Hg). However, the ν(Cd—Cl) stretching mode of the corresponding [CdCl$_2$(diars)] complex is in the range found for cadmium complexes that are believed to have polymeric octahedral structures[153].

32.1.10 Group IIIB

Diamagnetic colourless complexes of the type [M(diars)$_2$X$_2$][MX$_4$] (X = Cl, Br, I) are usually formed when the tervalent d^{10} metals gallium(III) and indium(III) are treated with the stoichiometric amount of diarsine[154]. However, with GaCl$_3$ and InI$_3$, the products are formulated as [Ga(diars)$_2$Cl$_2$][Ga$_3$Cl$_{10}$] and [In(diars)$_2$I$_2$][In(diars)I$_4$] respectively. The compound [In(diars)$_2$Br$_2$]ClO$_4$ was isolated and also shown to be a 1:1 electrolyte in nitrobenzene. The tervalent metal is six-coordinate in all of the diars complexes that were prepared. No complexes of univalent gallium, indium or thallium could be isolated[154]. Nyholm and Ulm were also unsuccessful in isolating any diars complexes of thallium(III), although complexes of the type [Ti(L—L')$_2$X$_2$][TlX$_4$] [155] and [Tl(C$_6$F$_5$)$_2$(L—L')X] [156] have been characterised with the more basic bidentate ligands, 2,2'-bipyridyl and *o*-phenanthroline.

32.1.11 Group VB

Sutton has shown that the trihalides of these elements form 1:1 adducts with diars[157]. Since these [MX$_3$(diars)] complexes are monomeric non-electrolytes in nitrobenzene, they are probably five-coordinate. With the arsenic and phosphorus complexes the iodo-derivatives have some tendency to form [MI$_2$(diars)]$^+$I$^-$ in solution [157(a)]. Bismuth also forms the six-coordinate complex [Bi(diars)$_3$](ClO$_4$)$_3$ [157(b)].

32.2 1,2-BIS(DIMETHYLARSINO)ETHYLENE

Relatively few complexes of 1,2-bis(dimethylarsino)ethylene, hereafter abbreviated edas, are known because the olefinic arsine was only prepared in 1968[11,15,16]. Though the ligand preparation gives a mixture of the *cis* and *trans* isomers in a 1:9 molar ratio, the pure *cis* isomer may be isolated by formation of the iron complex [FeCl$_2$(*cis*-edas)$_2$][FeCl$_4$] and subsequent regeneration as needed[15]. While *trans*-edas can only act as a bridging ligand (see later in this section), *cis*-edas is expected to form chelate complexes that are comparable to those formed by *o*-phenylenebis-(dimethylarsine).

32.2.1 First transition series

The only complexes of edas with early transition metals reported to date are the 1:1 adducts with $TiBr_4$ [18]. In an apparent attempt to form an eight-coordinate analogue of $[TiBr_4(diars)_2]$ [35] Clark *et al.* added at least a seven-fold excess of the ligand as the *cis–trans* mixture to $TiBr_4$ in cyclohexane. The physical properties (monomer in benzene, non-electrolyte in nitrobenzene, strong i.r. absorption at 1560 cm^{-1}, 328 cm^{-1} and 278 cm^{-1}) indicated a *cis*-octahedral structure for the bright red crystalline material that was formed. When more $TiBr_4$ was used in the reaction with edas, a deep red insoluble product was formed that was formulated as in (CLXXXVI), with the *trans*-ligand bridging *cis*-octahedral positions of the titanium. Interestingly, both adducts can be sublimed without change.

Feltham and coworkers have characterised several transition metal derivatives of *cis*-edas; those of the first transition series are $[Cr(CO)_4(cis\text{-edas})]$, $[Ni(CO)_2(cis\text{-edas})]$ [158], $[Fe(cis\text{-edas})_3](ClO_4)_2$, $[Fe(cis\text{-edas})_3](ClO_4)_3$, *trans*-$[FeCl_2(cis\text{-edas})_2][FeCl_4]$, *trans*-$[CoBr_2(cis\text{-edas})_2]Br$ and $[Cu(cis\text{-edas})_2]ClO_4$ [15]. Comparison with the corresponding diars complexes as outlined below provides a valuable assessment of the coordinating ability of edas.

The white complex $[Ni(CO)_2(cis\text{-edas})]$ could be prepared under mild conditions (pentane, 0°C) and the initial rate of reaction of nickel carbonyl with edas was the same as with diars. However, the product loses carbon monoxide on standing, suggesting that the olefinic double bond must also be capable of coordinating to the nickel. Apparently 1,2-bis(dimethylarsino)tetrafluorocyclobutene (ffars) behaves in a similar manner with nickel carbonyl [159] and the coordinating ability of the olefinic double bond in ffars has been verified by an X-ray structure determination of an iron derivative [160]. While the nitrosyl complexes, $[Co(CO)(NO)(diars)]$ and $[Fe(NO)_2(diars)]$, could be isolated, the corresponding *cis*-edas derivatives were too unstable to permit detection. The mass spectra of some of these *cis*-edas complexes will be discussed in section 33.6 [158].

Assignment of the electronic spectra of the d^6 complexes of *cis*-edas (Fe(II) and Co(III) complexes listed above) reveals that the ligand-field parameters are virtually identical to those of the analogous octahedral spin-paired diars complexes. X-ray structural investigations have shown that $[CoCl_2(diars)_2]^+$ has *trans*-octahedral geometry [104,105]. Spectral data for the octahedral spin-paired iron(III) compounds $[Fe(edas)_3]^{3+}$, $[Fe(diars)_3]^{3+}$, $[Fe(asq)_3]^{3+}$, $[Fe(phen)_3]^{3+}$, $[Fe(en)_3]^{3+}$ and $[Fe(dipy)_3]^{3+}$ also showed that the ligand-field parameters (Dq, B and C) are very similar for all of the bidentate ligands. Accordingly, the degree of participation by the d orbitals of the arsenic in π-bonding with the t_{2g} orbitals of iron(III) appears to be negligible for both edas and diars [86]. This apparent lack of π-bonding was also noted previously for the complexes of Cr(III), Co(III) and Fe(II) with diars [55]. It is noteworthy that $[Fe(cis\text{-edas})_3](ClO_4)_3$ is particularly stable relative to $[Fe(diars)_3](ClO_4)_3$, which decomposes on standing.

A white insoluble copper(I) complex, formulated as $[Cu(cis\text{-edas})_2]ClO_4$, has been prepared, but, as with diars, no copper(II) derivatives could be isolated [15].

Bennett *et al.* [16] mentioned that both *cis*-edas and *trans*-edas complexes of nickel(II) and cobalt(II) could be isolated and used as sources of the pure *cis* and *trans* isomers of edas; however, this work has not yet been published[†]. Obviously many other complexes of *cis*-edas with the elements of the first transition series could be prepared and comparison of their magnetic and spectral properties with the analogous diars complexes would help elucidate the nature of the transition metal–arsenic bond.

32.2.2 Other metal complexes

In one of the first studies of the complexing properties of edas, Bennett and coworkers prepared several platinum complexes containing either *cis*-edas or *trans*-edas[16]. Reaction of the mixture of ligand isomers with [K_2PtCl_4] gives an insoluble white complex [$PtCl_2(edas)$] with equimolar quantities and a water-soluble yellow complex [$Pt_2(edas)_4$]$Cl_4 . 3H_2O$ with two equivalents of the ligand. Infrared and n.m.r. spectra suggest that *trans*-edas is present in both instances. The complex [$Pt_2(\textit{trans}\text{-edas})_4$]$Cl_4$, which reverts to the insoluble complex [$PtCl_2(edas)$] very readily, also forms derivatives of formula [$Pt_2(\textit{trans}\text{-edas})_4$]$Y_4$ (Y = PF_6, ClO_4, $\frac{1}{2}PtCl_6$), which are 4:1 electrolytes in nitromethane. Dimeric structures in which the *trans*-olefinic ditertiary arsine bridges two square-planar platinum atoms were suggested; the lower conductivity of the tetrachloride derivative indicated a higher coordination number for the platinum atoms in that case. Derivatives of formula [$Pt(\textit{cis}\text{-edas})_2$]$X_2$ (X = Cl, ClO_4, PF_6, $\frac{1}{2}PtCl_6$) are obtained by u.v. irradiation of solutions of the complex [$Pt_2Cl_4(\textit{trans}\text{-edas})_4$]. The i.r. spectra of these complexes show bands at about 1570 cm^{-1}, 1090 cm^{-1} and 700 cm^{-1} not observed in the complexes of *trans*-edas and thus isomerisation appears to occur. The solid-state structures of the complexes [$Pt(\textit{cis}\text{-edas})_2$]$X_2$ are thought[16] probably to contain the planar [$Pt(\textit{cis}\text{-edas})_2$]$^{2+}$ cation by analogy with the known geometry of [$Pt(diars)_2$]Cl_2 (figure 33) where the Pt–Cl distance of 416 pm shows the interaction between the cation and chloride ions is electrostatic only[120(b)].

Erskine has prepared palladium(II) complexes of the types [$Pd(edas)X_2$] (X = Cl, Br) and [$Pd(edas)_2X_2$]$_n . 3H_2O$ (X = Cl, Br, ClO_4, PF_6), where $n = 2$ when the ligand is in the *trans* form[17]. These complexes were formulated as described above for the analogous platinum(II) complexes, except that in [$Pd_2(\textit{trans}\text{-edas})_2Cl_4$] the bridging ligands occupy *cis* positions about the square-planar palladium atoms. Infrared evidence suggested that the palladium may be five-coordinate in the anhydrous form of [$Pd_2(\textit{trans}\text{-edas})_4$]$Cl_4 . 3H_2O$; that is, Cl^- ions have replaced the water in the first coordination sphere. Erskine also demonstrated that u.v. irradiation of non-aqueous solutions of [$Pd_2(\textit{trans}\text{-edas})_4$]$Cl_4 . 3H_2O$ produces a *trans*-to-*cis* conversion of the ligand[17].

The brown complex [$Pd(\textit{cis}\text{-edas})_2Cl$]$ClO_4$, isolated from the reaction of [K_2PdCl_4] with a mixture of the *cis* and *trans* isomers of edas in aqueous ethanol, appears to be five-coordinate but no physical evidence was given to support this possibility[15]. The square-pyramidal cations [$M(diars)_2X$]$^+$ (M = Ni, Pd, Pt) are well characterised[101,118,119] and a study of the corresponding *cis*-edas complexes would
† See additional references 220, 221.

help elucidate the factors promoting five-coordination. In this regard McAuliffe and coworkers[161] have isolated square-pyramidal complexes $[Ni(VAs_2MePh)_2X]^+$ with the new ditertiary arsine, dimethylarsinodiphenylarsinoethylene, shown in structure (CXCI).

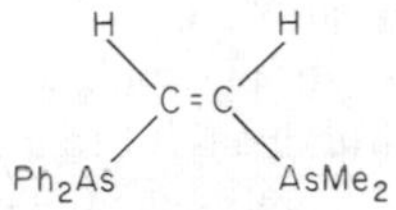

(CXCI)

32.3 1,2-BIS(DIPHENYLARSINO)ETHYLENE

The coordinating ability of this ditertiary arsine, abbreviated vdiars, has been investigated primarily by Mague and coworkers since the report of its synthesis in 1968[14]. An alternative preparative route, involving the reaction of diphenylarsine with diphenylarsinoacetylene, has been reported by some Russian workers[162].

32.3.1 Metal complexes
The reaction of 1,2-bis(diphenylarsino)ethylene, also called *cis*-vinylenebis(diphenylarsine), with rhodium dicarbonyl chloride dimer in methanol followed by the addition of $NaBF_4$ produced the orange monomeric complex $[Rh(vdiars)_2]BF_4$ as the methanol solvate[14]. By contrast the same reaction with the *trans* isomer of vdiars yields an insoluble yellow complex of stoichiometry $[Rh(CO)Cl(vdiars)]$, for which the polymeric structure shown in (CXCII) was suggested [14,163]. Since this is the

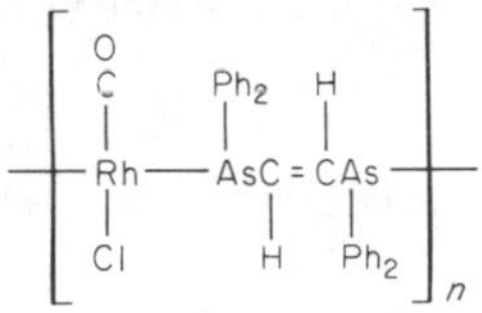

(CXCII)

only report of a *trans*-vdiars complex, the abbreviation vdiars is hereafter understood to mean the *cis* isomer of the ligand.

The square-planar Rh(I) cation $[Rh(vdiars)_2]^+$, which is more conveniently studied as the chloride because the solubility is greater than that of the tetrafluoroborate salt, readily reacts with many molecules to produce addition and oxidative-addition products. All of these complexes are 1:1 electrolytes in nitromethane[164]. The cation readily and reversibly activates molecular hydrogen but the *cis*-$[RhH_2(vdiars)_2]Cl$ complex does not act as a homogeneous hydrogenation catalyst for olefins because, unlike the dihydride formed with the Wilkinson catalyst $[RhCl(PPh_3)_3]$, no site is available for olefin coordination.

Nuclear magnetic resonance and i.r. spectra indicate that the complexes $[RhHX(vdiars)_2]X$ (X = Cl, Br, I) probably have *trans* configurations. With the exception of $[Rh(CO)(vdiars)_2]BPh_4$ and $[RhCl_2(vdiars)_2]Cl$, the five- and six-coordinate adducts of $[Rh(vdiars)_2]^+$ with species such as halogens, allyl chloride, methyl iodide, sulphur dioxide, nitric oxide and trifluorophosphine have not been fully characterised[164]. It is interesting that the complex $[Rh(dppe)_2]Cl$ [dppe = 1,2-bis(diphenylphosphino)ethane] is inactive towards molecular hydrogen and carbon monoxide but the less electronegative ditertiary phosphine 1,2-bis(dimethylphosphino)ethane (dmpe) enables the complex $[Rh(dmpe)_2]Cl$ to undergo the same oxidative additions as $[Rh(vdiars)_2]Cl$ [165]. The latter complex may be more conveniently prepared by the reaction of stoichiometric quantities of $[Rh(COD)Cl]_2$ (COD = 1,5-cyclooctadiene) and vdiars in acetonitrile. The square-planar rhodium(I) complex $[Rh(diars)_2]Cl$, which also undergoes a variety of oxidative-addition reactions, may be prepared in the same manner[166].

Under mild conditions with $[Rh(CO)_2Cl_2]_2$ vdiars gives orange crystalline complexes, which can be formulated as $[Rh(CO)(vdiars)_2][Rh(CO)_2X_2]$ (X = Cl, Br) on the basis of their conductivity in acetonitrile[163] and the fact that the five-coordinate cation can be isolated as the tetraphenylborate[164]. In their study of phosphine and arsine complexes of Rh(I), Mague and Mitchener also isolated some analogous complexes with the ditertiary phosphine, 1,2-bis(diphenylphosphino)ethylene[163].

A recent attempt to prepare complexes of the type $[Pd(vdiars)_2]^{2+}$ or $[Pd(vdiars)_2X]^+$ was unsuccessful but McAuliffe and coworkers isolated the square-planar non-electrolytes $[Pd(vdiars)X_2]$ [167].

Just as further comparison of metal complexes of diars and edas would permit an evaluation of the importance of the aromatic ring in stabilising so many metal-diars complexes, an extension of the known complexes of vdiars would provide useful comparisons with both edas and the saturated analogue, 1,2-bis(diphenylarsino)-ethane. Such studies have been reported for the corresponding ditertiary phosphines[7].

32.4 1,2-BIS(DIPHENYLARSINO)ETHANE

Although the synthesis of 1,2-bis(diphenylarsino)ethane, abbreviated as dae, and its palladium complexes was reported by Chatt and Mann at the same time as for diars[1], relatively few metal complexes of dae are known. By comparison with its ditertiary phosphine analogue, 1,2-bis(diphenylphosphino)ethane (dpe), of which many more complexes have been studied[7], dae might be expected to act as a bridging ligand as well as a chelating one.

32.4.1 Metal complexes

Westland *et al.* have studied the complexes of dae with Group IVA metal halides[168,169]. Titanium tetrachloride forms only a 2:3 complex $[(TiCl_4)_2(dae)_3]$ although dpe also forms 1:1 and 3:2 adducts. Both ligands form the 1:1 complexes $[MX_4(L—L')]$ (M = Zr, X = Cl, Br; M = Hf, X = Cl), suggestive of six-coordination for these heavier elements. The insolubility and the complexity of the far-i.r. spectrum of

$[(TiCl_4)_2(dae)_3]$ does not allow its structure to be assigned[168] but it is feasible that dae acts as both a bridging ligand and a monodentate ligand.

With the exception of the complexes of diars and Et-diars discussed in section 32.1.2 (1) vanadium complexes with ditertiary arsines are unknown. An attempt to prepare complexes of the type $[VOX_2(dae)]$ (X = Cl, Br) gave only green impure products although the analogous diphosphine dpe apparently gave complexes of the same formulation[170].

Zingales and coworkers have prepared Group VIA carbonyl derivatives of the types $[M(CO)_4(dae)]$, $[M(CO)_2(dae)_2]$ and $[M_2(CO)_6(dae)_3]$ (M = Cr, Mo, W), depending upon the mode of preparation[171(a)]. The corresponding dpe complexes were also studied by i.r. spectroscopy[171(b),(c)]. Complexes of the first two types are analogous to the bidentate derivatives of diars[67] discussed in section 32.1.3. The binuclear complexes $[M_2(CO)_6(dae)_3]$ (M = Mo, W), in which the i.r. spectra suggest the presence of *cis*-CO groups, might have either one or three bridging dae ligands, assuming the metals are six-coordinate. The oxidative halogenation of these Group VIA complexes of dae would be of interest for comparison with the known results for the diars compounds[57,60,61]. In this regard the seven-coordinate carbonyl complexes of molybdenum and tungsten, $[M(CO)_3(dae)I_2]$, are known from the reaction of the diarsine with $[Bu_4N][M(CO)_3I_3]$[172].

Dobson and coworkers have investigated the kinetics and mechanism of the reaction of triethyl phosphite (L) with $[Cr(CO)_4(dae)]$ in mesitylene[173]. The products of the reaction are *trans*- and *cis*-$[Cr(CO)_4L_2]$ and *trans*-$[Cr(CO)_3L_3]$, in contrast to the reaction of Lewis bases with $[Cr(CO)_4(dpe)]$, in which CO is replaced and not the bidentate phosphine[174]. The mechanism is believed to involve initial reversible dissociation of one end of the bidentate dae and is compatible with other kinetic data that suggest that, in general, M—As bonds are weaker than M—P bonds. Since the i.r. spectra of $[Cr(CO)_4(dae)]$ and $[Cr(CO)_4(dpe)]$ are nearly identical in the carbonyl stretching region, Dobson concludes that the kinetic results illustrate the non-correlation of reactivity and carbonyl stretching frequencies in metal carbonyl complexes[173].

The zerovalent chromium compound $[Cr(dae)_3]$, which is a monomeric non-electrolyte, was obtained in the reaction of $K_6[Cr(CN)_6]$ with dae in liquid ammonia at room temperature[175]. The isolation of several products of the type $[Cr(L—L')_3]$ (L—L' = dpe, o-phen, dipy) by this method suggests that CN^- in $K_6[Cr(CN)_6]$ is more susceptible to substitution than is CO in $Cr(CO)_6$ because the latter gives only $[Cr(CO)_4(dpe)]$ and $[Cr(CO)_2(dpe)_2]$ as reaction products with dpe[176].

The observance of two carbonyl stretching frequencies for the compounds $\{[LMn(CO)_2]_2dae\}$ (L = C_5H_5 and MeC_5H_4) suggested a binuclear structure with bridging dae. This structure has been confirmed for the diars analogue (figure 31)[74,75]. The ditertiary arsine, 1,2-bis(diphenylarsino)butane (dab), also appears to act as a bridging ligand although the corresponding ditertiary phosphine, 1,2-bis(diphenyl-phosphino)butane (dpb), forms only the chelate complex $[LMn(CO)(dpb)]$ when $[LMn(CO)_3]$ and the ligand are subjected to u.v. irradiation[177].

The reaction of dae with $[C_8H_8Fe(CO)_3]$ produces the five-coordinate complex

[Fe(CO)$_3$(dae)] [179]. The displacement of two terminal carbonyl groups in the dimer [LFe(CO)$_2$]$_2$ (L = C$_5$H$_5$, MeC$_5$H$_4$) by donor ligands such as dae and dpe gave compounds of type {[LFe(CO)]$_2$(L—L')}. Infrared and n.m.r. spectral data indicated the presence of bridging ligand groups[180]. The proposed structure is shown in figure 34, where M = As, P, and R' = CH$_2$, C$_2$H$_4$. Cobalt and iron complexes of dae have been postulated as intermediates in patents concerning organometallic catalysts for preparing 1,4-dienes[178].

Yellow crystalline complexes of formulation {RhCl$_2$(L—L')[(*o*-tolyl)$_3$P] }(L—L' = dpe, dae, dipy) have been well characterised by Bennett and Longstaff in a study

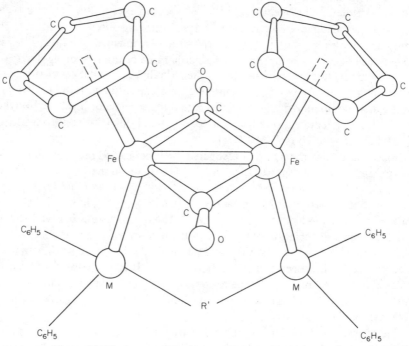

Figure 34 The proposed structure of {[(C$_5$H$_5$)Fe(CO)]$_2$dae}

of homogeneous rhodium catalysts. These complexes are monomeric non-electrolytes and their structure, which is compatible with n.m.r. spectra, is represented in formula (CXCIII), where R = *o*-tolyl [181]. The insoluble yellow powders [Rh(CO)X(dae)] are probably dimeric[163] (see section 32.5.1 and figure 35).

(CXCIII)

The complex [Ni(CO)$_2$(dae)] is analogous to [Ni(CO)$_2$(diars)], and the i.r. spectrum and dipole moment (5.07 D) evidence are compatible with tetrahedral geometry[182]. Doubtless, halogen oxidation would yield the nickel(II) complexes of the type [NiX$_2$(dae)]. Sacconi *et al.* have characterised the complexes [NiX$_2$(dae)] (X = Br, I), obtained by direct reaction of the nickel salt and dae in butanol, as square planar on the basis of conductivity, molecular weight, magnetic susceptibility and electronic spectral data. By contrast, when the ditertiary arsine has four methylene groups in its backbone rather than two as in dae, the resultant complex [NiI$_2$(dab)] is tetrahedral [183].

The zerovalent complex [Ni(dae)$_2$] was prepared in liquid ammonia from K$_4$[Ni(CN)$_4$], and Behrens and Mueller showed that CN$^-$ was easily replaced by a variety of ligands to yield monomeric zerovalent derivatives[175].

The square-planar complex [PdCl$_2$(dae)] was the first metal complex of dae to be prepared, and Chatt and Mann used the usual [PdCl$_4$]$^{2-}$ anion as the starting material[1]. The series [PdX$_2$(dae)] was also obtained by Watt and Layton by the ligand displacement of ethylenediamine in [PdX$_2$(en)] and identified by X-ray diffraction patterns[184]. The same *cis*-square-planar series was also isolated from the reaction of potassium halides with the complex [Pd(dae)$_2$](ClO$_4$)$_2$.

The 1:1 adducts of copper(I) halides with dae and dab (two and four methylene groups as 'backbones', respectively) have negligible solubility and polymeric tetrahedral structures were proposed. The molar conductivity in nitrobenzene of the copper(I) chloride complexes with dae and its phosphine analogue dpe suggested the formulation [Cu(L—L')$_2$][CuCl$_2$] in that solvent[185].

The reaction of dae with UCl$_4$ in tetrahydrofuran solution in air yielded only a green complex of the arsine dioxide although the diphosphine dpe gave the complex [(UCl$_4$)$_2$(dpe)] [186]. Similarly Westland has noted the failure to obtain the desired [AuX(dae)], whilst triphenylphosphine yields the complex [AuX(Ph$_3$P)$_2$], which may be three-coordinate in nitrobenzene solution[187].

32.5 BIS(DIPHENYLARSINO)METHANE

The ditertiary arsine, bis(diphenylarsino)methane (dam), might not be expected to be a good chelating ligand and probably explains the fact that the coordinating ability of dam has been little investigated. The recent comparison by Colton and coworkers of the complexes formed by halocarbonyls of molybdenum and tungsten with dam, its diphosphine analogue dpm and dpe has demonstrated the types of bonding employed by dam. These results will only be briefly discussed since a review by Colton has recently appeared[188].

32.5.1 Metal complexes

The molybdenum(II) and tungsten(II) halo-tetracarbonyls [M(CO)$_4$X$_2$] (X = Br, Cl) react with dam at room temperature to give the complexes [M(CO)$_3$(dam)$_2$X$_2$], which on warming in solution are converted to [M(CO)$_2$(dam)$_2$X$_2$]; the latter complex will readily reform the tricarbonyl by the action of carbon monoxide on its solutions. Both types of complexes are non-electrolytes and their n.m.r. spectra (methylene protons of dam) show that both dam ligands are monodentate in the tricarbonyls and that one is monodentate and the other bidentate in the dicarbonyl derivatives[189]. X-ray crystallography has confirmed the seven-coordinate geometry in $[W(CO)_3(dam)_2Br_2]$ and $[Mo(CO)_2(dam)_2Br_2]$; each metal atom has a distorted

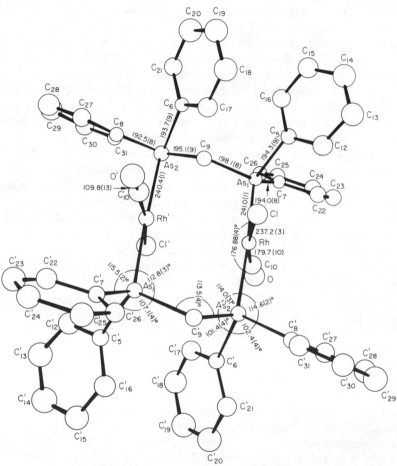

Figure 35 The structure of $[Rh(CO)Cl(dam)]_2$

capped octahedral environment[190]. The weak chelating character of dam relative to dpm and dpe may be attributed to the inability of one methylene group to bridge the much larger arsenic atoms in the chelated complex[188]. Alternatively, the difference in reactivity of dam may be simply a difference in strength in M—P and M—As bonds, arsenic being a poorer σ-donor and π-acceptor than phosphorus[190].

The reactions of dam with the diiodotetracarbonyls of molybdenum and tungsten are more complex than those with the other halocarbonyls[191]. Colton has established by i.r. and n.m.r. spectroscopic techniques that the following equilibria (equation 11) exist in solution

$$M(CO)_3(dam)I_2 + dam \rightleftarrows M(CO)_3(dam)_2I_2 \rightleftarrows M(CO)_2(dam)_2I_2 + CO \qquad (11)$$

The rapid interchange between these complexes confirms the facile cleavage of the chelated dam ligand by either excess ligand or carbon monoxide. The complex $[M(CO)_3(dam)I_2]$, in which the ligand is chelated, is analogous to dpe and dpm derivatives, and could be considered as a reactive intermediate in the other halo-dam systems[188].

In a search for rhodium(I) complexes that would undergo oxidative-addition reactions, Mague and Mitchener prepared the complexes $[Rh(CO)Cl(L-L')]$ (L—L' = dam and dpm). Molecular weight studies suggested a dimeric formulation for the isomorphous complexes[163] and this has been confirmed by a single-crystal X-ray structure determination[192]. Both ligands bridge the rhodium atoms in $[Rh(CO)Cl(dam)_2Rh(CO)Cl]$, holding each rhodium atom in the apical position of a square-pyramidal geometry of the other; however, the Rh—Rh distance of 339.6 pm does not signify a metal–metal interaction. Each rhodium atom is bonded to terminal carbonyl and chlorine groups in *trans* positions, and to two arsenic atoms, one from each of the two bridging dam molecules, in a *trans* configuration. This structure is shown in figure 35. The same dimeric structure is probably found in $[Rh(CO)X(dae)]$ [163].

The cyclopentadienyldicarbonyliron dimer reacts with dam (also dae, dpm, dpe) under u.v. radiation to give the binuclear complex $\{[(C_5H_5)Fe(CO)]_2dam\}$. In addition to indicating that the diarsine bridges the iron atoms, the i.r. and n.m.r. evidence suggests that both carbonyl groups are also bridging[181]. The proposed structure is represented in figure 34 (see section 32.4.1).

Iridium complexes of the type $[Ir(CO)X(dam)_2]$ have been patented as hydrogenation catalysts[193].

In a study of the coordination complexes of bis(pentafluorophenyl)mercury, Canty and Deacon have shown that the complexes $\{[(C_6F_5)_2Hg]_2(L-L')\}$ (L—L' = dam, dpm) are isomorphous. Preliminary crystal structure data for the dam derivative shows that each arsenic atom is coordinated to a mercury atom, presumably in a dimeric structure[194]. It appears that dam will favour a bridging rôle in its metal complexes.

32.6 1,2-BIS(ALKYLPHENYLARSINO)ETHANE

Although one ditertiary arsine of this type, 1,2-bis(butylphenylarsino)ethane (dbpa), was originally synthesised by Chatt and Mann in 1939[8], their investigation of the palladium chloride complex[1] was the only study of metal complexes of these ligands until 1970. The interesting feature of ligands of the type $[RR'AsCH_2CH_2AsRR']$ is that both racemic and meso forms might be expected, and, indeed, $[PdCl_2(dbpa)]$ was isolated mechanically into two forms that were believed to correspond to the two conformations of the ligand. However, only recently the isomers of an analogous diarsine 1,2-bis(methylphenylarsino)ethane (dmpa) have been separated and their structures established by resolution of the racemic form[23].

Bosnich and Wild separated the yellow complex $[PdCl_2(dmpa)]$ into two fractions by chromatography on silica gel. The pure meso and racemic diarsines are obtained by decomposing these respective complexes, which have distinctly different melting points and i.r. and n.m.r. spectra, and distilling out the isomers at about 150°C *in vacuo*; interconversion does not occur since the dichloropalladium isomeric complexes may be reprepared[23].

Further metal complexes of dmpa, and the corresponding ligand with a propylene backbone, $[MePhAsCH_2CH_2CH_2AsPhMe]$, are currently being studied by McAuliffe and coworkers[195].

The preparation of 1,2-bis(ethylphenylarsino)ethane has been reported[196] but no metal complexes are known. Further studies of metal complexes of this type of ditertiary arsine would be of great interest.

32.7 1,2-BIS(DIMETHYLARSINO)-3,3,4,4-TETRAFLUOROCYCLOBUTENE

Since 1966 Cullen and coworkers have studied metal complexes of a series of fluoroalicyclic-bridged arsines and phosphines of the type represented in (CXCIII), where n = 2, 3, 4 and L, L′ are Me_2As- and/or Ph_2P- groups. The presence of the unsaturated bridging linkage as well as the electronegative nature of the bridging groups make these ligands of considerable interest for comparison of their coordinating ability with the other ditertiary arsines of current research activity. Since the preparation and coordination complexes of these ligands are expertly discussed by Cullen in a current review on the fluoroalicyclic derivatives of metals and metalloids[26], only the complexes of 1,2-bis(dimethylarsino)3,3,4,4-tetra-fluorocyclobutene (ffars)† will be discussed here. The variety of products formed by ffars with metal halides and metal carbonyls are quite representative of the whole series of ligands and serve as the most comparable complexes to those of diars and edas.

32.7.1 Metal complexes

Preliminary attempts to form complexes of ffars with first row transition metal halides gave no isolable products (at least with Ni(II), Fe(II) or Fe(III) salts). Furthermore, although some complexes of ffars with heavier elements are known, namely $[HgCl_2(ffars)]$, $[PdCl_2(ffars)]$ and $[RhCl_3(ffars)_2]$, their stability is low[160]. This greater reluctance of ffars to form metal halide complexes as compared to diars and edas appears to be related to both the inductive effect and the geometric factors associated with the tetrafluorocyclobutene ring. It is not possible at present to suggest a more specific reason for the failure of ffars to chelate readily. For example,

† ffars: n = 2, L=L′ = $AsMe_2$ of formula (CXCIII); Cullen[26] denotes the ligands for n = 2, 3 and 4 as $f_4$fars, $f_6$fars and $f_8$fars, respectively.

Cullen has pointed out that the L to L′ distance or the bite of the ligands of the type represented in (CXCIV) can depend upon the bonding situation, and also that ligands such as those illustrated in (CXCV) with electronegative substituents can still form chelate complexes with transition metal elements[26]. The idea that the tetra-fluorobutene backbone makes ffars a better π-acceptor ligand and a poorer σ-donor ligand is supported by the fact that a wide range of metal carbonyl complexes are known. The various types of bonding found for ffars in these compounds will now be outlined in the order of the Groups.

$$(CF_2)_n \begin{array}{c} C \diagup L \\ \| \\ C \diagdown L' \end{array}$$

(CXCIV)

(CXCV a)   (CXCV b)

With Group VIA hexacarbonyls ffars gives the chelate complexes [M(CO)$_4$(ffars)] (M = Cr, Mo) after brief refluxing in tetrahydrofuran; after 36 hours ffars displaces all of the carbonyl from [Mo(CO)$_6$] but the resultant pink solid was not characterised. Though the three carbonyl bands in the i.r. spectrum of [Mo(CO)$_4$(ffars)] are little changed from those in [Mo(CO)$_4$(diars)], suggesting that the π-acceptor ability of ffars is the same as that of diars, the greater reactivity of ffars and other fluorinated ditertiary ligands is quite evident in their easier displacement of carbonyl groups[160]. Dobson *et al.* noted the lack of correlation of carbonyl stretching frequencies with the reactivity of other [M(CO)$_4$(L—L′)] complexes[173].

Bridged derivatives {[M$_2$(CO)$_8$](ffars)} are obtained readily in the reaction of ffars with [M$_2$(CO)$_{10}$] (M = Mn, Re)[197,198], although other bidentate ligands apparently form chelate complexes[199,200]. X-ray crystallography has shown the structure to be that in figure 36, where the As—As distance is 402 pm in the bridged ligand[201]. Iodine cleaves the Mn—Mn bond in {[Mn$_2$(CO)$_8$](ffars)} and the X-ray structure confirms that ffars bridges two [Mn(CO)$_4$I] moieties, each iodine atom being *cis* to the separate Me$_2$As— groups[197,198,201]. When the complexes {[M$_2$(CO)$_8$](ffars)} (M = Mn, Re) are refluxed in xylene solution, the decomposition products of the same formulation have a structure in which the metal atoms are bonded directly to the fluorocyclobutene ring[198,201,202]. Thus, the structure is similar to the decomposition 'isomer' of [Fe$_3$(CO)$_9$(ffars)] (see below), whose structure has also been determined by X-ray methods. Heating of the bridged binuclear species {[M(CO)$_4$I]$_2$(ffars)} appears to give the chelated mononuclear complexes [M(CO)$_3$I(ffars)][197] in spite of the usual reluctance of ffars to chelate.

The yellow complex [Fe(CO)$_4$(ffars)], obtained by u.v. irradiation of a petroleum ether solution of [Fe(CO)$_5$] and ffars, has an i.r. spectrum (showing three carbonyl bands) and an n.m.r. spectrum (showing both free and coordinated Me$_2$As groups) indicative of apical substitution of the monodentate ligand in the trigonal-bipyramidal geometry about the iron atom[26,203]. This structure has been confirmed by an X-ray structural determination for the ligand of type (CXCIV) in which the coordinated donor group is Ph$_2$P (that is, $n = 2$, L = AsMe$_2$, L' = PPh$_2$)[204]. By contrast diars forms the chelate compound [Fe(CO)$_3$(diars)][81-3] in which an X-ray study has shown the existence of apical–equatorial substitution[205]. When ligands of the type

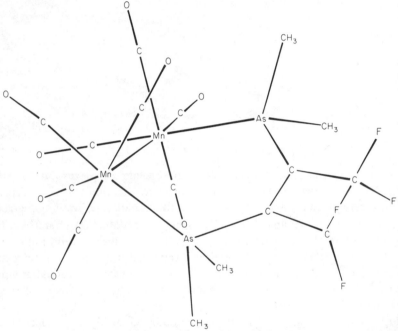

Figure 36 The structure of {[Mn$_2$(CO)$_8$] (ffars)}

(CXCIV) contain at least one Ph$_2$P group, reactions with iron pentacarbonyl usually give chelate complexes of formula [Fe(CO)$_3$(L–L')]. In solution these complexes are stereochemically non-rigid on the n.m.r. time-scale[26].

With iron pentacarbonyl ffars also forms the complexes [Fe$_2$(CO)$_8$(ffars)] and [Fe$_2$(CO)$_6$(ffars)][203], though the latter is better obtained from [Fe$_3$(CO)$_{12}$][206]. Spectroscopic measurements indicate that the former complex contains a ligand bridge between two Fe(CO)$_4$ moieties[203]. This structure is analogous to that found for {[Mn(CO)$_2$(MeC$_5$H$_4$)] diars} (figure 31[75]; the only known example of diars acts as a bridging ligand. In the novel complex [Fe$_2$(CO)$_6$(ffars)] ffars acts as a tridentate ligand, insofar as both arsenic atoms coordinate to one Fe(CO)$_3$ group while the double bond of the cyclobutene ring coordinates to the other Fe(CO)$_3$ moiety[160].

The lower symmetry of this second iron atom, which also has the first iron atom in one of its coordination sites, as shown in figure 37, is reflected in the Mössbauer spectrum by a higher quadrupole splitting and a greater isomer shift[206]. The spectroscopic properties[207,208] of other complexes of the type $[Fe_2(CO)_6(L-L')]$, where $L-L'$ is a fluorocarbon-bridged ligand as represented in structural formulae (CXCIV) and (CXCVb), indicate the same structure as proved by X-ray crystallography for $[Fe_2(CO)_6(ffars)]$.

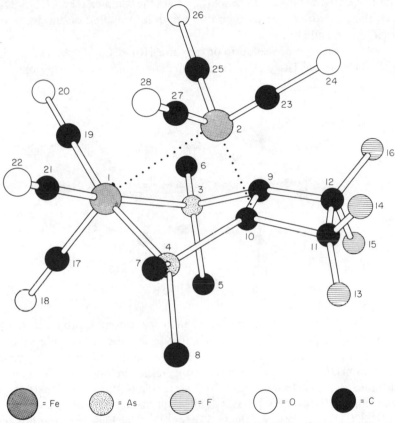

Figure 37 The structure of $\{[Fe_2(CO)_6](ffars)\}$

The reaction of ffars with $[Fe_3(CO)_{12}]$ also yields $[Fe_3(CO)_{10}(ffars)]$ as a by-product[209]. The structure deduced by Cullen and coworkers on the basis of spectroscopic properties (i.r., ^1H and ^{19}F n.m.r., Mössbauer) has been verified by X-ray investigation[210]. As shown in figure 38, the ffars ligand bridges two iron atoms, and the basic skeleton of $[Fe_3(CO)_{12}]$ is little changed. Heating of this compound in cyclohexane solution results in fragmentation of the ligand, though the chemical analysis of the product $[Fe_3(CO)_9(ffars)]$ does not indicate that cleavage of a Me_2As- group has occurred since it remains in the derivative as a bridging moiety. The X-ray structure confirms spectroscopic data that suggests little symmetry in the molecule; that is, all three iron atoms have different environments and make up a distorted square plane with the displaced arsenic atom; also, one iron atom forms a weak bond with the double bond of the cyclobutene ring[209,211]. The structure is shown in figure 39.

The reluctance of ffars to chelate is apparently reflected in its reaction with iron dicarbonyl dinitrosyl since only one carbonyl group is displaced; other ligands of the type (CXCIV) give the chelated dinitrosyl iron derivatives in good yield[212]. Refluxing ffars with $[Fe(CO)_3 SCH_3]_2$ in xylene solution gives the complex $\{[Fe(CO)_2 SCH_3]_2(ffars)\}$ for which the structure shown in figure 40 has been proposed on the basis of Mössbauer, n.m.r. and i.r. spectral data[213].

Ultraviolet irradiation of a solution of ffars and $[Ru_3(CO)_{12}]$ gave two products, $[Ru_3(CO)_{10}(ffars)]$ and $[Ru_3(CO)_8(ffars)_2]$, depending upon the conditions[214]. The

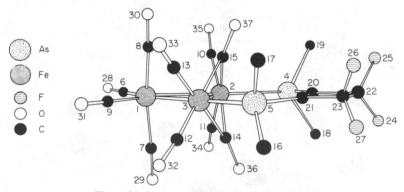

Figure 38 The structure of $[Fe_3(CO)_{10}]$ (ffars)]

high symmetry of the products suggested by their spectroscopic properties has been verified by X-ray single-crystal diffraction methods, and the ligand bridges the ruthenium triangle in equatorial positions. The replacement of carbonyl groups by the arsenic atoms, which are poorer π-acceptors, results in longer Ru—Ru bonds for the ruthenium atoms bridged by ffars[215,216]. Reaction of ffars with $[Ru_3(CO)_{12}]$ also gives the complex $\{[Ru_2(CO)_6] ffars\}$[214], in which the ligand is tridentate as in the $\{[Fe_2(CO)_6](L-L')\}$ complexes (figure 37)[160,206-8]. The high temperatures needed to cause reaction of ffars with the trinuclear osmium carbonyl $[Os_3(CO)_{12}]$ apparently cause decomposition and no complexes were isolated[217].

The complex of formula $[Co_2(CO)_6(ffars)]$ has an i.r. spectrum in the carbonyl region indicative of displacement of two terminal carbonyl groups from the bridging form of $[Co_2(CO)_8]$ [218]. This structure, in which each cobalt atom is bonded to two terminal and two bridging carbonyl groups and to an arsenic atom, has been recently confirmed by X-ray crystallography[219]. Other complexes of the type $\{[Co_2(CO)_6](L-L')\}$ can be assumed to have the same structure[218].

The reaction between $[CH_3 CCo_3(CO)_9]$ and ffars yields a red air-stable solid whose stoichiometry $[CH_3 CCo_3(CO)_7 ffars]$ was confirmed by its mass spectrum.

The structure postulated for this complex in solution on the basis of the observance of both terminal and bridging carbonyl bands in its i.r. spectrum[218] has been confirmed in the solid state by an X-ray investigation[220]. Similarly ffars acts as a chelating ligand, bridging a tetrahedron of cobalt atoms, in the structure of $[Co_4(CO)_8(ffars)_2]$. This black complex, obtained by refluxing ffars with $[CF_3CCo_3(CO)_9]$ in hexane, has no bridging carbonyl groups in the solid state but they are present in solution[218,221].

The unstable product isolated by mild heating of $\{[Co_2(CO)_6](ffars)\}$ has been characterised by Crow and Cullen by mass spectral and i.r. evidence[218]. This reaction intermediate' in the decomposition of $\{[Co_2(CO)_6](ffars)\}$ contains one less carbonyl

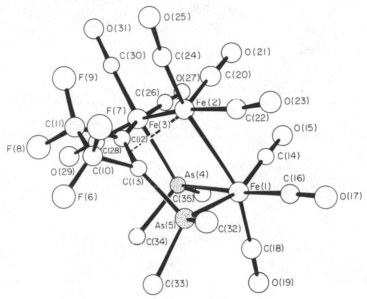

Figure 39 The structure of $[Fe_3(CO)_9(ffars)]$

group and is believed to have a structure analogous to that of $\{[Fe_2(CO)_6](ffars)\}$ [160], that is, tridentate ligand and no bridging carbonyl groups (figure 37). Another product of empirical formula $[Co_4(CO)_9(ffars)_2]$, obtained in low yield from the thermal decomposition of $\{[Co_2(CO)_6(ffars)\}$, contains the ligand in a considerably transformed condition[26]. The crystal structure shows that two cleaved Me_2As- groups bridge separate pairs of cobalt atoms and the remaining fragments have combined to form a bicyclobutyl system $[Me_2AsC(CF_2CF_2)C-C(CF_2CF_2)CAsMe_2]$; this latter system acts as a tetradentate ligand towards the four cobalt atoms[222]. Although ffars seems most susceptible to fragmentation, rearrangement of diars has occurred with nickel salts[20] and iron carbonyls[83], and thus the possibility should be considered for ditertiary arsines.

The red complex formed by treating $[Rh(CO)_2Cl]_2$ with ffars appears to have the same bridging structure[223] as that confirmed by X-ray crystallography for

[Rh(CO)Cl(dam)]$_2$ (figure 35)[163,192]. By contrast other ligands of the type represented in (CXCIV) (for example, L=L' = PPh$_2$) chelate to one of the rhodium atoms[197] rather than bridge the two square-planar metal atoms as in [Rh(CO)Cl(ffars)].

All of the carbonyl groups are displaced from Ni(CO)$_4$ by reaction with ffars at room temperature, in contrast to the forcing conditions needed to give Ni(diars)$_2$ [114]. The insoluble product was not characterised but Cullen *et al.* suggested that ffars could be acting as a tridentate group[160]. Although the yellow complexes [MCl$_2$(ffars)](M = Pd, Pt) appear to be square planar, the reaction of ffars with Pt(II) salts is complex[160] and Cullen suggests an analogy with the bridged complexes of *trans*-edas with Pt(II)[16,26] (section 32.2.2).

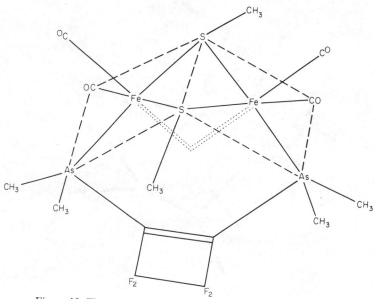

Figure 40 The proposed structure of {[Fe(CO)$_2$(SCH$_3$)]$_2$(ffars)}

32.8 MISCELLANEOUS DITERTIARY ARSINES

There are several ditertiary arsines for which only a few metal complexes have been reported; some examples are discussed in this section.

A communication by Nyholm and coworkers has summarised the preparation and coordinating ability of the tetrafluoro analogue of diars, called F-diars, shown in formula (CXCVa)[224]. Although most of the complexes reported are analogous to those of diars already described in sections 32.1.3 and 32.1.5-7, there are significant differences as a result of fluorination of the aromatic ring. The preparations of [Mo(CO)$_4$(F-diars)] and [Mo(CO)$_2$(F-diars)$_2$] are effected under milder conditions than the diars compounds; the higher carbonyl stretching frequencies in the F-diars derivatives are compatible with the expected greater amount of molybdenum–arsenic double bonding. The ease of oxidation of the Co(II) and Fe(II) complexes is much greater in the F-diars compounds. The same tetragonally-distorted octahedral geometry can be assumed for the $[FeCl_2(F\text{-diars})_2]^{n+}$ (n = 0, 1, 2) complexes as was surmised for the diars complexes[84–6]. Although the five-coordinate nickel(II) cations $[Ni(F\text{-diars})_2X]^+$ could be isolated as the perchlorate salts, the green paramagnetic solids of formula $[Ni(F\text{-diars})_2X_2]$ (X = Cl, Br) exist as non-electrolytes in benzene solution and only partially dissociate in nitromethane. In contrast the brown diamagnetic $[Ni(F\text{-diars})_2I_2]$ complex behaves like the diars derivatives in being a uni-univalent electrolyte. With palladium and platinum dihalides the 1:1 adducts are readily isolated, while diars favours the formation of 1:2 complexes[117–19]. These results suggest that F-diars has reduced σ-donor capacity relative to diars. It would be interesting to compare the ligand-field parameters of F-diars with those calculated for diars and other bidentate ligands[55,86].

Mann and coworkers have prepared other *o*-phenylenebis(dialkylarsines) and *o*-phenylenebis(diphenylarsine) and demonstrated their coordinating ability by the synthesis of a metal derivative[8,113,225,226]. The steric effect of these diars analogues, such as *o*-phenylenebis(diethylarsine)(Et-diars), has been shown to prevent the attainment of higher coordination numbers of Group IVA and VA metal atoms[38–41,46] (section 32.1.1, 32.1.2). Although Et-diars and other analogues of diars would doubtless form a wide variety of metal complexes, there is no advantage in using these ligands save that of steric hindrance. On the other hand Kepert and associates[39] have shown that 4-methyl-diars (structural formula (CLXXXVIII)) has the same coordinating ability as diars and it is feasible that optical isomers and *cis–trans* isomers might be formed in certain instances (for example, four-coordinate complexes) by employing the unsymmetrical ditertiary arsine. The 4-methyl derivative of Et-diars is also known[227].

Mann and coworkers have prepared the unusual ditertiary arsines with formulae (CXCVI) (R = Me, Et)[228] and investigated the complexes of one of these ligands,

| As | As |
| R_2 | R_2 |

(CXCVI)

| As | As |
| Me_2 | Me_2 |

(CXCVII)

2,2'-biphenylenebis(diethylarsine), with palladium(II) bromide and nickel(II) halides. Although a stable 1:1 adduct was readily formed by mixing hot ethanolic solutions of the diarsine and potassium tetrabromopalladate(II), no nickel(II) complexes could be isolated; in contrast the analogous diphosphine gave several four-coordinate complexes[229].

Another interesting ligand with an aromatic backbone is 1,8-naphthalenebis-(dimethylarsine)(nas), prepared from Me_2AsNa and 1,8-dichloronaphthalene in tetrahydrofuran[230]. To date only d^8 metal complexes have been characterised for this ditertiary arsine, shown in (CXCVII). The bite (As—As distance) in nas is esti-

mated to be 249 pm as compared to 321 pm in diars[231]. Comparison of the electronic spectra of $[Ni(diars)_2]^{2+}$ and $[Ni(nas)_2]^{2+}$ shows that nas is significantly lower
in the spectrochemical series. Di Sipio *et al.* found both orange and blue forms of
this four-coordinate cation and suggested that conformational isomers were possible
depending upon the arrangement of the nas ligands in the complex. Although bidentate ligands apparently do tend to form square-pyramidal $[Ni(L-L')_2X]^+$
complexes[27], the assumption of this geometry for the complexes $[Ni(nas)_2X]X$
$(X = Cl, Br, I, CNS)^{231}$ is questionable. Application of Gray's electronic spectral
criteria for distinguishing between square-pyramidal and trigonal-bipyramidal
geometries[101] indicates that the latter geometry is just as likely. In that instance
static distortion of the blue $[Ni(nas)_2X]^+$ complexes would explain the assymetry
(about 15 000 cm^{-1} and 18 000 cm^{-1} components) of the main band that is diagnostic
of trigonal-bipyramidal geometry. Observation of the temperature dependence of
the ligand-field bands would give more conclusive evidence for the choice of five-
coordinate geometry[101].

Ros and Tondello have recently reported the physical properties of palladium(II)
and platinum(II) complexes of the type $[M(nas)_2X_2]$ for a variety of anions[232].
Conductivity measurements clearly indicate that the planar $[M(nas)_2]^{2+}$ species
readily form five-coordinate $[M(nas)_2X]^{2+}$ complexes; however, the electronic
spectra can be interpreted in terms of trigonal-bipyramidal geometry although the
authors assumed square-pyramidal geometry, as is reasonable for $[M(diars)_2X]^+$
complexes[118,119].

Cullen *et al.* have recently initiated n.m.r. spectroscopic studies to evaluate the
conformations of chelate rings in solutions of metal complexes[233]. Several new
ditertiary arsines of the type $[Me_2AsC(R)HCH_2AsMe_2]$ (R = Me$_3$Si, Cl$_3$Si, F, Cl,
CN) were prepared from the sealed tube reaction of tetramethyldiarsine $[Me_2AsAsMe_2]$
with the appropriate vinyl derivative[233(b)]. Detailed analyses of the n.m.r. spectra of
some five-membered chelate complexes of chromium carbonyl indicate that some
derivatives have a definite conformational preference, though others exist in solution
in several different conformations. For example, in the compounds
$[Cr(CO)_4(Me_2AsC(R)HCH_2AsMe_2)]$ the favoured rotamer about the C—C bond
for R = Me$_3$Si and F corresponds with that depicted in the projection (CXCVIIIa) and
(CXCVIIIb) respectively. This implies that the five-membered chelate ring places the
trimethylsilyl substituent in an 'equatorial' and the fluorine substituent in an 'axial'
orientation relative to the chromium atom, in the respective compounds. It is apparent
that coupling-constant information from the n.m.r. spectra could provide an important
means of determining conformations in other chelate rings in addition to those of
ditertiary arsines[233].

(CXCVIII a)   (CXCVIII b)

Cullen and associates[207] have also recently studied the iron carbonyl complexes of fluorocarbon-bridged ditertiary arsines related to ffars and to those described in the previous paragraph. In complexes of the type $[Fe_3(CO)_{10}(L-L')]$ Mössbauer data suggests that $[Me_2AsCH_2CF_2AsMe_2]$ and $[Me_2AsC(CF_3)FCF_2AsMe_2]$ chelate to one iron atom but that $[Me_2AsCHFCF_2AsMe_2]$ bridges two equivalent iron atoms as does ffars (section 32.7) (figure 38). The ligand $[Me_2AsC(CF_3)=C(CF_3)AsMe_2]$ (CXCVb) does not form a $Fe_3(CO)_{10}$ derivative but does give a chelated $Fe(CO)_3$ complex and a $Fe_2(CO)_6$ derivative; the spectroscopic data indicate the latter structure is similar to that of $[Fe_2(CO)_6ffars]$, in which the ligand is tridentate (figure 37). Formation of the latter complex is surprising since the removal of ring strain (as in ffars) should reduce the tendency of $[Me_2AsC(CF_3)=C(CF_3)AsMe_2]$ to coordinate through its olefinic linkage[207]. The isolation of other transition metal complexes of $[Me_2AsC(CF_3)=C(CF_3)AsMe_2]$ would provide an interesting comparison with those of the less electronegative ditertiary arsine $[Me_2AsCH=CHAsMe_2]$ (edas, section 32.2).

The preparation and characterisation of the potentially chelating diarsine $[o\text{-}B_{10}H_{10}C_2(AsMe_2)_2]$(bicars), which incorporates the carborane framework as the backbone, has been reported[234,235]. Smith indicated that a green unstable solid could be isolated with nickel chloride; also, the phenyl analogue of bicars gave palladium complexes with the correct analyses for the formulae $[B_{10}H_{10}C_2(AsPh_2)_2PdX_2]$ [234]. Zaborowski and Cohn have studied the complexes formed by bicars with a few metal carbonyls[235]. The isomer shift in the Mössbauer spectrum of the yellow iron complex $[Fe(CO)_3(bicars)]$ suggests that the unusual ligand has electron-donating properties similar to those of Ph_3P. Infrared band complexity in the carbonyl region of this complex and of the molybdenum derivative $[Mo(CO)_4(bicars)]$ indicate lower symmetry than that in the analogous diars complexes. Although the significance of shifts in carbonyl-stretching frequencies has been questioned[236,237], it was suggested that the higher values of the two carbonyl bands for $[Ni(CO)_2(bicars)]$ as compared to $[Ni(CO)_2(diars)]$ indicated better π-acceptor ability by the carborane ligand[235].

Abel *et al.* have recently investigated the reaction of nickel, chromium and molybdenum carbonyls with the unconventional chelating ditertiary arsines $[Me_2Si(AsMe_2)_2]$ and $[Me_2Sn(AsMe_2)_2]$ [238]. In complexes such as $[Me_2Si(AsMe_2)_2Cr(CO)_4]$ and $[Me_2Sn(AsMe_2)_2Ni(CO)_2]$ the arsine atoms apparently bridge the metal atoms, though the organometallic complexes are too insoluble for n.m.r. measurements. The lowered i.r. frequencies of the metal carbonyl stretching modes, assuming the complexes to be monomeric, suggest that these unusual ligands are better electron donors toward metal carbonyls than ditertiary arsines with carbon backbones (for example, diars, dae).

There are literature reports of polydentate arsenic-containing ligands acting only as bidentate chelating agents; three such instances are now mentioned. The palladium(II) complexes of the type $[PdX_2(OAS)]$ (X = Cl, Br, I, CNS) are square planar and the potentially tetradentate ligand $[(o\text{-}XC_6H_4YCH_2)_2]$ (X = Ph$_2$As, Y = O) appears to be bonding only through the two arsenic atoms, occupying *trans* positions. By contrast the ligand SAS (X = Ph_2As, Y = S) apparently bonds to palladium via adjacent arsenic and sulphur atoms[239]. Ellermann and Dorn found that the tetradentate arsine $C[CH_2AsPh_2]_4$ reacted with various metal carbonyls and/or nitrosyls to give complexes in which all four arsenic atoms were coordinated and the ligand bridged two metal moieties, as shown in structural formula (CXCIXa). However, with tungsten hexacarbonyl the derivative contains the ligand acting only as a

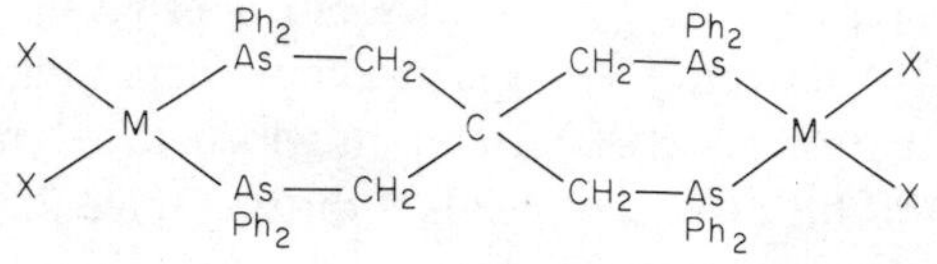

(X = CO , NO ; M = Fe , Co , Ni)

(CXCIX a)

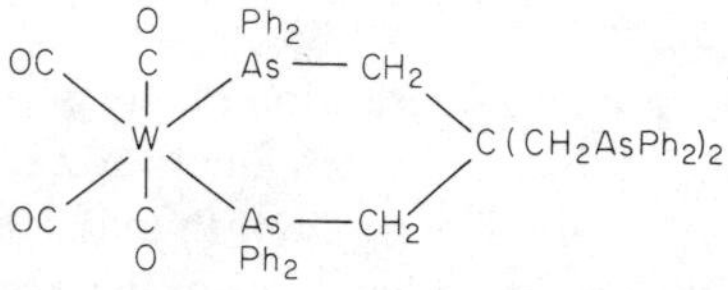

(CXCIX b)

bidentate, as shown in (CXCIXb)[240]. Structural assignments have been made on the basis of considerable physical evidence for the variety of palladium(II) complexes that can be prepared with the new polydentate ligands shown in (CC). In non-polar solvents the monomeric square-planar palladium(II) complexes of the ligands ($n = 2, 3$) appear to have only the arsenic atoms of the ligand coordinated[241].

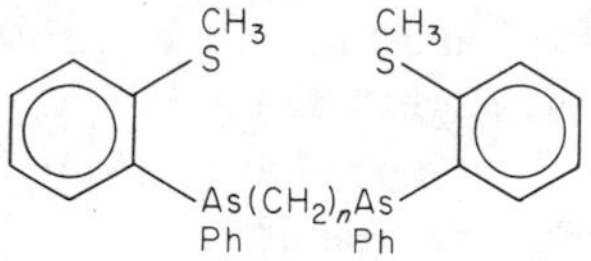

(CC)

33 PHYSICAL METHODS

The purpose of this chapter is to illustrate the usefulness of several common physical techniques in the study of metal complexes of ditertiary arsines. Applications of these physical methods have been mentioned previously in the appropriate sections but some representative data is discussed at present in order to demonstrate the detailed information that might be obtained from a particular method.

33.1 INFRARED SPECTROSCOPY

The i.r. spectra of free and coordinated ditertiary arsines are usually very similar, so that, in general, no structural information is obtained by recording the vibrational spectrum of a metal complex. In addition assignments of metal–arsenic stretching frequencies are scarce, though they are expected to occur below 300 cm^{-1} [98]. It is possible, however, to deduce the stereochemistry of metal–ditertiary arsine complexes by observing the metal–halogen stretching frequencies in their far-i.r. spectra[98,153]. Furthermore, for certain ditertiary arsines, as described below for diars and edas, there are i.r. bands that are characteristic of the coordinated ligand.

The vibrational frequencies in the range 4000–200 cm^{-1} have been assigned for diars by comparison with o-dibromobenzene, related arsines and the deuterated derivative $[o\text{-}C_6H_4(As(CD_3)_2)_2]$ [242]. Though most of the bands remain unchanged when diars is coordinated to a metal, it has been noted that the methyl rocking modes are sensitive to the stereochemistry of the complex[35,98]. In a recent study Baylis and Bailar have shown that observance of the methyl rocking vibrations in the i.r. (in the region of 920–840 cm^{-1}) allows cis and $trans$ isomers of $[Co(diars)_2X_2]^+$

TABLE 37 INFRARED SPECTRA OF BIS(DIARSINE)COBALT(III) COMPLEXES[†]
IN THE REGION 900–730 cm^{-1}

Assignment	Diars	$Trans$ complexes	Cis complexes
b_1 fundamental	744 s[‡]	752–784 ms	753–766 ms
b_1 combination band	785 w	813 vw	
b_1 combination band	826 vw	833 w	824–833 w
			846–864 w
CH_3 rocking mode	846 s	846–860 sh	864–873 s
		864–873 s	874–889 ms
a_2 fundamental	865 sh	898–902 m	900–901 m
CH_3 rocking mode	885 s	910–919 ms	909–921 ms

[†] From reference 106.
[‡] Intensities are denoted as follows: w, weak; m, medium; s, strong; vw, very weak; sh, shoulder.

complexes to be distinguished[106]. If the *cis* isomer is free of *trans* impurities, then it will have more bands than the *trans* isomer in this region; also, a band between 874 cm^{-1} and 889 cm^{-1} is unique to the *cis* form. The positions of the relevant bands for the free ligand[242] and for the coordinated ligand in the *cis-* and *trans-*cobalt(III) complexes[106] are given in table 37 for the region 920–730 cm^{-1}. Geometrical isomers of octahedral $[M(diars)_2X_2]^+$ complexes have previously been differentiated on the basis of far-i.r. spectra (that is, $\nu(M-X)$ frequencies)[98] and electronic spectra[102,103], but the new method appears to be more readily applicable. It should be possible to extrapolate the criterion to isomeric metal complexes of other ditertiary arsines containing Me_2As- groups (for example, edas).

As shown in table 37 the free diars i.r. bands at 885, 846 and 744 cm^{-1} move about 25 cm^{-1} to higher frequencies in the bis(diarsine) cobalt(III) complexes. These frequency shifts are characteristic of diars behaving as a chelate because for the monoquaternary salt $[o\text{-}Me_2AsC_6H_4AsMe_3^+Cl^-]$ bands occur in both regions (for example, 762 cm^{-1} and 744 cm^{-1} for the C–H deformation mode). Thus, bands near 900, 860 and 760 cm^{-1} were considered evidence for the presence of chelating diars in the Group IVA and VA complexes $[MX_4(diars)_2]$ [35]; subsequently, X-ray crystallography verified that each metal atom is eight-coordinate (figure 30). Deacon and Green reported similar band shifts in diars complexes of Group IIB elements[153].

The i.r. spectra of some platinum complexes with 1,2-bis(dimethylarsino)-ethylene (edas) indicate that the presence of *trans*-edas may be easily distinguished from that of *cis*-edas. Identification of the isomers of edas is especially important because the ligand as prepared (chapter 31) is a mixture of both forms, and metal salts may selectively react with either isomer in the mixture[15–18]. The *trans* isomer has a strong band at about 970 cm^{-1} that can be assigned to the olefinic C–H deformation mode, but no band in the region near 1600 cm^{-1} that can be assigned to the C=C stretching vibration. However, the latter vibration is i.r.-active for a *cis*-disubstituted ethylenic derivative, and a band at 1570 cm^{-1} occurs in those complexes believed to contain *cis*-edas. In addition *cis*-edas has a strong i.r. band near 700 cm^{-1} that is characteristic of the out-of-plane C–H deformation vibration of a *cis*-disubstituted ethylenic derivative[16]. Of the two red complexes isolated from the reaction of titanium tetrabromide with the ligand mixture, the one with the i.r. absorption at 1560 cm^{-1} is believed to contain *cis*-edas[18].

33.2 MAGNETIC MEASUREMENTS

The value of the magnetic moment of a transition metal complex is often characteristic of the stereochemistry and oxidation state of the central metal atom. Thus, a survey of the magnetic properties of transition metal ions[243] shows that formal oxidation states in metal–diars complexes have often been assigned on the basis of bulk magnetic susceptibility measurements. For example, the diamagnetism of the complexes $[M(CO)_2(diars)_2X]X$ (M = Cr, Mo, W) is compatible with the Group VIA metal having a spin-paired d^4 configuration in the seven-coordinate metal(II) complexes[57,60,61]. Similarly, the octahedral complexes $[MoX_2(diars)_2]$ have magnetic moments (μ_{eff} = 2.8–2.9 B.M.) that signify a low-spin $(t_{2g})^4$ configuration for Mo(II) [60]. In fact, where a choice between high-spin and low-spin configurations exist (d^{4-7} configurations for octahedral complexes), it is usually found that diars causes spin-pairing. Other ditertiary arsines are also 'strong-field' ligands and all of their known metal complexes are low-spin.

An exception to spin-pairing occurs in some diars complexes of manganese. The monomeric non-electrolytes $[MnX_2(diars)_2]$ have magnetic moments ($\mu_{eff} \approx 6.0$ B.M.) indicative of five unpaired electrons. Since low-spin complexes of manganese(II) are known with other ligands that are high in the spectrochemical series[243], it is apparent that no significant covalent bonding occurs in the octahedral $[MnX_2(diars)_2]$ complexes, because this might be expected to reduce interelectronic repulsions sufficiently to induce spin-pairing. Obviously, it would be of interest to prepare other manganese(II) complexes of ditertiary arsines and investigate their magnetic properties.

Variable-temperature magnetic susceptibility data can yield the magnitude of splitting of ground states in favourable circumstances but the method has been little employed for metal complexes of ditertiary arsines[243]. Feltham and coworkers have recently interpreted magnetic susceptibility data in the range 77–300 K for the iron(III) complexes, *trans*-$[FeX_2(diars)_2]^+$ and *trans*-$[FeX(NO)(diars)_2]^+$, in terms of tetra-gonal and rhombic splitting of the $^2T_{2g}$ ground state, respectively. The change in sign of the splitting of the 'octahedral' ground state is attributed to the strong axial field of the nitrosyl group[59(b)].

33.3 ELECTRONIC SPECTROSCOPY

Although the absorption spectra of various metal–diars complexes have been recorded, until recently only those of Group VIIIB had been considered in any detail[102,103]. In particular Yamada showed that the electronic absorption spectra distinguished between the *cis* and *trans* configurations in the complexes $[CoX_2(diars)_2]^+$. The green *trans* isomers have ligand-field bands at lower frequencies than the violet *cis* isomers; in addition their band intensities are lower than for the *cis* complexes. It should be noted that the ligand-field bands of all metal–diars complexes have high intensities relative to the corresponding ethylenediamine complexes and this is attributed to the nearness of strong charge-transfer bands[103].

Deutscher and Kepert[46] have recently investigated the diffuse reflectance and solution spectra for several eight-coordinate d^1 complexes of the type $[MX_4(L-L')_2]$ (M = Nb, Ta; L–L' = diars and related ligands). These pale green complexes have similar absorption spectra in the visible region, and, since $[NbCl_4(diars)_2]$ is iso-morphous with $[TiCl_4(diars)_2]$ (figure 30)[35,36], dodecahedral geometry can be assumed. Thus, the three main bands in the visible spectra (at about 10 000 cm^{-1},

13 000 cm^{-1} and 15–18 000 cm^{-1}) are assigned to the three d–d transitions expected for dodecahedral geometry[244]. With [VCl$_4$(diars)$_2$] only a broad band with a maximum at 13 250 cm^{-1} was observed in earlier work[35]. In these spectra of eight-coordinate complexes the intense bands above 20 000 cm^{-1} are assigned to arsenic-metal charge-transfer transitions. Similarly, intense charge-transfer and intraligand transitions have been recorded for other metal–diars complexes (for example, Ti [39], Ni [101]) but specific assignments have not been attempted; there is no discernible correlation of these band positions with the geometry of the metal complexes.

The electronic absorption spectrum of diars itself[245] is featureless at 300 K, showing only rising absorption above 35 000 cm^{-1}; however, at 77 K five intense bands (ϵ_M = 1300–9200) are resolved between 35 000 cm^{-1} and 41 000 cm^{-1} as shoulders on the main absorption at 41 000 cm^{-1}. The shoulders are assigned as the vibronic structure of the $\pi \to \pi^*$ transition localised on the benzene ring whereas the main charge-transfer transition involves excitation of an essentially non-bonding As electron to the π^* level of the benzene ring. Gray and associates have also estimated the ionisation potential of diars as 7.90 eV (1.27 aJ) from the lowest energy charge-transfer band in the spectrum of the complex of diars with tetracyanoethylene (TCNE). The electronic absorption spectrum of the [diars . TCNE] complex (bands at 17 200 cm^{-1} and 23 400 cm^{-1}) shows that the two highest occupied levels of diarsine are in an arsenic non-bonding orbital and a π molecular orbital confined to the aromatic ring[245].

Feltham and associates have interpreted the visible spectra of octahedral chromium(III), iron(III), iron(II) and cobalt(III) complexes of diars on a ligand-field model[15,55,86]. The value of Dq for these complexes falls in the order: Co(III) > Fe(II) > Cr(III) > Fe(III) and is approximately the same as that of ethylenediamine, being greater only for the d^6 complexes. Furthermore, for the spin-paired complexes [Fe(L–L')$_3$](ClO$_4$)$_3$ the positions of various bidentate ligands (diars, edas, asq, *o*-phen, dipy, and en; see footnotes on page 314 for ligand nomenclature) in the spectrochemical series is virtually identical. Other work places diars just under the ditertiary phosphines [R$_2$PCH$_2$CH$_2$PR$_2$] (R = Me, Ph) in ligand-field strength[90] and the study of complexes of the type [Cr(NCS)$_4$L$_2$]$^-$ gives the position of diars relative to many common ligands[56].

In the above metal complexes of diars, where the oxidation number of the metal is high, Feltham concluded that back π-bonding was not important, though the interelectron repulsion parameters B and C are slightly reduced from the free ion values[15,55,86]. In contrast Preer and Gray invoked metal–diars π-bonding to rationalise their electronic spectral data, which places the nickel d$_{xy}$ orbitals lower than d$_{xz}$ and d$_{yz}$ orbitals in the square-pyramidal [Ni(diars)$_2$X]$^+$ complexes[101]. Whether π-bonding generally plays a significant rôle in complexes of metals in their normal oxidation states with diars and other ditertiary arsines must remain an open question at this time. Although such back-bonding was often considered in the past to explain the properties of metal–arsenic and metal–phosphorus bonds[5], the importance of π-bonding has been questioned recently[246] and is the subject of current research (see part 1 for further discussion).

33.4 NUCLEAR MAGNETIC RESONANCE SPECTROSCOPY

The n.m.r. spectra of most of the common ditertiary arsines have been reported and are relatively simple, even though the ^{75}As isotope has a nuclear spin of 3/2. No spin–spin coupling occurs with adjacent protons because the arsenic nucleus has a large quadrupole moment, which also results in narrow lines for the observed proton resonances. In metal complexes the protons in the ditertiary arsines are deshielded and the ^{1}H resonance peaks move downfield, thus providing a criterion of complex formation. Examples are now discussed.

The methyl proton resonance for diars occurs at τ 8.80 (relative to τ 10.00 ppm for tetramethylsilane) and the phenyl proton resonance at τ 2.67 (as an $A_2 B_2$ multiplet)[9(b)]. In carbon disulphide solution the trigonal-bipyramidal complex [Fe(CO)$_3$(diars)] has the methyl and phenyl resonances at τ 8.28 and τ 2.23, respectively[203]. Since the Me$_2$As— groups occupy apical and equatorial sites[205], inequivalence could occur, but the observance of only one ^{1}H n.m.r. peak for the methyl protons indicates that the five-coordinate complex is stereochemically non-rigid on the n.m.r. time-scale[247]. By contrast the two methyl resonances found for [Fe(CO)$_4$(ffars)] at τ 8.56 and τ 8.10 (in a 1:1 ratio) correspond to a free and a coordinated Me$_2$As— group, respectively[203]. The green form of [CoCl$_2$(diars)$_2$]Cl, which has the *trans* conformation according to electronic spectral data[102,103], has only one methyl resonance peak (at τ 7.53 in D$_2$O solution) as expected for equivalent methyl groups (see below)[106]. Similarly only one methyl resonance might be anticipated for [Co(diars)$_3$]$^{3+}$, in which octahedral symmetry is indicated by the electronic spectrum[55]. However in D$_2$O solution two sharp and equal methyl resonances are observed at τ 7.63 and τ 7.93 [123]. Since [Co(diars)$_3$](ClO$_4$)$_3$ is known to decompose to *trans*-[CoCl$_2$(diars)$_2$]$^+$ on standing[55], a possible explanation is that the τ 7.63 peak is actually due to this product. An alternative and more plausible explanation for the occurrence of two equal methyl resonances for the [Co(diars)$_3$]$^{3+}$ cation is that the symmetry is actually less than O$_h$ when the orientations of the methyl groups are taken into account, in which case equal proportions of two types of methyl environment are expected.

The racemic and meso forms of 1,2-bis(methylphenylarsino)ethane (dmpa) are readily distinguished by ^{1}H n.m.r. spectroscopy. In racemic-[PdCl$_2$(dmpa)] the methyl resonance has shifted downfield to τ 8.00 from τ 8.92 in the free ligand, whilst in meso-[PdCl$_2$(dmpa)] the methyl resonance has moved to τ 7.95 from τ 8.86 [23]. Significant differences are also evident in the methylene resonances of the two forms of dmpa. In the same way the methylene resonance peak at τ 7.42 in the ^{1}H n.m.r. spectrum of {RhCl$_2$(dae)[(o-tolyl)$_3$P]} (section 32.4.1) is characteristic of coordinated 1,2-bis(diphenylarsino)-ethane(dae)[181]. Colton and coworkers have also used the methylene proton resonance of 1,1-bis(diphenylarsino)-methane(dam) to distinguish between monodentate and bidentate behaviour in Group VIA halocarbonyl complexes. For example, the resonances at τ 4.35 and τ 7.35 are assigned to chelated dam in [W(CO)$_3$(dam)I$_2$] and to free dam, respectively, when [W(CO)$_3$(dam)$_2$I$_2$] is dissolved in CDCl$_3$ solution; the latter complex is also seven-coordinate and an absorption at τ 6.05 is characteristic of monodentate dam[189,191]. These n.m.r. assignments have been verified by X-ray crystallography, in that the geometries of $[W(CO)_3(dam)_2Br_2]$ and $[Mo(CO)_2(dam)_2Br_2]$ are seven-coordinate as predicted[190].

Several workers have reported n.m.r. spectra data for 1,2-bis(dimethylarsino)-ethylene (edas) and its metal complexes; some representative data is given in table 38.

Baylis and Bailar have demonstrated that *trans* and *cis* isomers of the type $[CoX_2(diars)_2]^+$ can be readily differentiated by their n.m.r. spectra, even if the samples are not completely pure[106]. Since all of the methyl groups in the *trans*

TABLE 38 NUCLEAR MAGNETIC RESONANCE SPECTRAL DATA[†]
FOR 1,2-BIS(DIMETHYLARSINO)-ETHYLENE COMPLEXES

Complex	Vinyl protons (τ)	Methyl protons (τ)	Solvent	Ref.
cis-edas	3.00	8.95	n.i.[‡]	9(b)
	3.06	9.02	$(CD_3)_2SO$	17
trans-edas	3.35	8.95	n.i.	9(b)
	3.35	n.i.	$CDCl_3$	16
cis and *trans* mixture	3.22	8.98	n.i.	11
	3.23	9.01	$(CD_3)_2SO$	17
$[Cr(CO)_4(cis\text{-}edas)]$	2.6	8.5	$CDCl_3$	158
$[Mo(CO)_4(cis\text{-}edas)]$	2.5	8.5	$CDCl_3$	158
$[W(CO)_4(cis\text{-}edas)]$	2.8	8.4	$CDCl_3$	158
$[PdCl_2(cis\text{-}edas)]$	2.59	8.23	$(CD_3)_2SO$	17
$[PdCl_2(cis\text{-}edas)_2]$	2.36	8.10	$(CD_3)_2SO$	17
$[PtCl_2(cis\text{-}edas)_2]$	2.75	7.95	D_2O	16
$[CoBr_2(cis\text{-}edas)_2]Br$	2.6	8.4	$(CD_3)_2SO$	158
$[Pd_2(trans\text{-}edas)_4Cl_4]$	2.45	8.05	D_2O	17
$[Pt(trans\text{-}edas)_2Cl_2](1.5H_2O)$	2.55	8.05	D_2O	16

[†] Resonance absorptions are quoted in ppm downfield from tetramethylsilane = 10.
[‡] Not indicated.

isomers are in equivalent environments, only one methyl resonance is expected, and only one was observed for the samples of the diiodo-, diisothiocyanato- and *trans*-dichloro- complexes made by Nyholm's method[96]. The four methyl groups on each diars molecule in the *cis* isomers are in four different environments, and thus the four expected methyl resonances were observed for the purple *cis*-dichloro- complex, as well as for other derivatives prepared using silver salts. In addition to the expected diars methyl resonances the *trans* and *cis* forms of the diacetato- complex also show one acetato–methyl resonance in each case. As for the other ditertiary arsines, the diars methyl resonances are centred about one ppm downfield from the position of uncoordinated diars; this is compatible with deshielding of the methyl groups in the complexes.

33.5 ELECTRON SPIN RESONANCE SPECTROSCOPY

Since the nuclear spin of arsenic ($I = 3/2$) can lead to hyperfine splitting in the electron spin resonance (e.s.r.) spectrum of transition metal–arsine complexes, this method has great potential in evaluating the degree of delocalisation of unpaired electron density onto any arsenic ligand. However, the e.s.r. spectra of only a few metal–ditertiary arsine complexes have been examined[58,124-7]; also, u.v. photolysis of diars and dae in frozen solutions gives paramagnetic products with complicated e.s.r. spectra[245].

The strong signal at $g = 2.075$ with seven hyperfine lines in the spectrum of u.v.-irradiated diarsine is assigned to (diars)$^+$. Similarly, after u.v.-irradiation at low

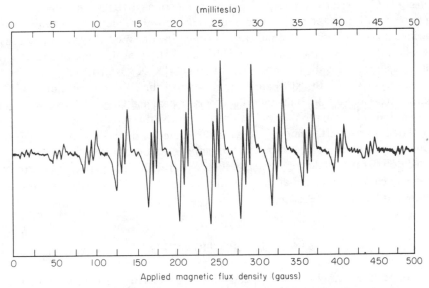

Figure 41 The electron spin resonance spectrum of [CrCl(NO)(diars)$_2$] ClO$_4$

temperatures dae and Ph$_3$As give molecular cations that show e.s.r. spectra that are almost identical to each other. In all cases the hyperfine splitting parameters indicate that the unpaired electron is delocalised over the ligand π orbital network[245].

Feltham and coworkers have observed arsenic hyperfine splitting in the yellow complex [CrCl(NO)(diars)$_2$]ClO$_4$ [58]. The chromium(III) complex, which is isomorphous with the corresponding *trans* iron and cobalt complexes, has one unpaired electron per chromium ($\mu_{eff} = 1.70$ B.M.). The e.s.r. spectrum of the octahedral complex in dichloromethane solution is shown in figure 41. The thirteen hyperfine lines arise from the interaction of the unpaired electron with the arsenic nuclei, in which the total spin of the four coordinated arsenic atoms is $I = 6$. The splitting of each of these $2I + 1 = 13$ lines into a triplet can be attributed to interaction with the nitrogen nucleus, which has $I = 1$. Estimation of the spin density on the arsenic atoms in $[CrCl(NO)(diars)_2]^+$ indicates only a small amount of delocalisation (less than two per cent on each arsenic atom) of the unpaired electron onto the diars ligand. Recalling that the ligand-field parameters of the analogous $[CrX_2(diars)_2]ClO_4$ complexes indicated that π-bonding was negligible[55], it might be concluded that the electron density delocalisation in the nitrosyl complex is via a σ-mechanism; however, the question again arises as to when π-bonding is 'significant' (section 33.3). Neither the e.s.r. solution spectrum nor the electronic spectrum of the nitrosyl complex allows a choice between the two possible ground states, $^2B_2$ or ^2E. Therefore, the magnetic anisotropy of single crystals of $[CrCl(NO)(diars)_2]ClO_4$ diluted in a diamagnetic host lattice is under investigation[58,248].

Independent investigations reported in 1968 of the e.s.r. spectrum of the $[Ni(diars)_2Cl_2]^+$ cation led to different ground states being proposed[125,127]. Kreisman *et al.*[125] rejected the $^2A_g(d_{z^2})$ ground state of a Ni(III) d^7 formulation as being inconsistent with the anisotropic three g-value powder spectrum. They proposed an unconventional 'σ-radical' ground state in which the unpaired electron is delocalised over the ligand atoms via the d_{xy} orbital. Subsequent study of the powder and solution e.s.r. spectra of the complexes $[Ni(diars)_2X_2]X$ (X = Br, NCS) and $[Co(diars)_2X_2]$ (X = Cl, Br, NCS) appeared to support the Ni(III) d^7 formulation[126], as initially suggested by the Italian workers[127]. Bernstein and Gray[124] have now obtained detailed information from e.s.r. studies on $[Ni(diars)_2Cl_2]^+$ doped into diamagnetic host crystals. Their e.s.r. data is interpreted in terms of a ^2A_g ground state for the diluted complex, whereas the rhombic g tensor observed for the concentrated powder is explained by d_{xz} as well as d_{z^2} orbital participation in the ground state. In any case the unpaired electron in the a_g molecular orbital is delocalised over the entire $[NiAs_4Cl_2]$ system.

33.6 MASS SPECTROSCOPY

There have been relatively few mass spectral investigations of ditertiary arsines and their complexes[9(b),83,158,249,250] but the method has enabled reaction products to be identified and modes of decomposition to be established.

Feltham and Metzger[9(b)] obtained the mass spectra of the crude products of the reaction of Me_2AsNa with o-dichlorobenzene and identified two fragments as $[AsMe_3]^+$ and $[MeAs(C_6H_4AsMe_2)_2]^+$; neither parent compound has yet been isolated from the reaction mixture. These results indicate that methyl groups are readily transferred from one arsenic atom to another and this loss of methyl groups is common to the proposed mass spectral fragmentation scheme of all dimethylarsino-compounds. With *cis*- and *trans*-edas, which have identical mass spectra, the subsequent formation of As_2 fragments was also inferred. The same authors have shown that the mass spectra of the Group VIA complexes $[M(CO)_4(cis\text{-edas})]$ (M = Cr, Mo, W) show the same features, that is, loss of methyl groups and formation of ion fragments containing two arsenic atoms. In each case the parent ion and ions corresponding to the loss of each carbonyl group have high abundance[158].

Colton and coworkers[249,250] have analysed the fragmentation patterns of the

ligands $[Ph_2M(CH_2)_nMPh_2]$ (M = P, As; n = 1, 2) as well as of their Group VIA
carbonyl derivatives. Both dam and dae fragment in the same way as triphenylarsine,
thus indicating migration of phenyl groups as discussed for methylarsino compounds.
The electron-impact induced rearrangements of the arsenic ligands are similar to
those of the phosphorus ligands (dpm and dpe), though ions involving multiple
bonds to arsenic are much less abundant. In compounds of the type
$[M(CO)_2X_2(L-L')_2]$, in which n.m.r. and X-ray crystallographic studies indicate
that one L—L' is monodentate and the second is bidentate[189-91], the mass spectra
does not differentiate the non-equivalent ligands; that is, only the mass spectra of
the ligands are obtained, indicating that initial loss of ligand is the major mode of
decomposition.

33.7 X-RAY STRUCTURAL DETERMINATIONS

At the time of previous reviews on metal complexes of ditertiary arsines in 1964[4,5],
crystal structures had only been obtained for two diars complexes. X-ray single-
crystal diffraction data is now available for a number of metal–ditertiary arsine
complexes and this structural information is summarised in table 39. As indicated in
section 32.1, several other diars complexes have been shown by X-ray powder data
to be isomorphous with certain of the listed complexes.

    Most of the structural determinations of the metal complexes containing diars
are for the type $[MX_2(diars)_2]^{n+}$ (n = 0, 1); all of these complexes have *trans*-
octahedral geometry. Several workers have discussed the factors that influence the
specific bond angles and bond lengths found in these cases[104,105,122,149]. Two of the
most important factors are methyl–methyl interactions and methyl–axial X group
interactions; non-bonded intermolecular interactions have also been considered.
Such factors explain such general observations as the M$\widehat{As}$Me bond angles being
significantly greater than the tetrahedral angle of 109.5° and the axial groups being
'cushioned' on the methyl groups of the four Me_2As- groups, thereby giving long
M—X bond lengths. The magnitude of the dihedral angle between the metal–arsenic
and benzene planes and the extent of distortion of the axial X groups from the perpendicular to the metal–arsenic plane have also been related to these factors[104]. The
much longer Ni—Cl bond length in $[NiCl_2(diars)_2]Cl$ than the Co—Cl distance in
the otherwise identical $[CoCl_2(diars)_2]Cl$ has been attributed to the presence of the
unpaired electron in the d_{z^2} orbital in the nickel case[104,124].

    The structure of $\{[(MeC_5H_4)Mn(CO)_2]_2(diars)\}$, shown in figure 31, is unique in
that this is the only instance of diars acting as a bridging ligand. However, as indicated
by the known crystal structures of dam and ffars complexes, the tendency for ditertiary arsines other than diars to bridge metal atoms is much greater. The additional
tendency of ffars to fragment and yield decomposition 'isomers' of the same stoichiometry makes the determination of crystal structures by X-ray diffraction techniques
especially important in characterising metal complexes of ffars[26]. The fact that most
of the novel geometries of metal-ffars compounds were predicted by a combination
of spectroscopic techniques lends confidence to the predictive ability of the various
physical methods for analogous metal complexes.

TABLE 39 X-RAY CRYSTAL STRUCTURE DETERMINATIONS

Complex	Coordination number	Structure	Remarks	Ref.
[TiCl$_4$(diars)$_2$]	8	dodecahedral	figure 30; several other [MX$_4$(diars)$_2$] are isostructural	35
[WOCl$_4$(diars)]	7	pentagonal bipyramid	one Cl and O are *trans*-axial	66
{[(MeC$_5$H$_4$)Mn(CO)$_2$]$_2$(diars)}	4 or 6	similar to [(MeC$_5$H$_4$)Mn(CO)$_3$]	figure 31; diars bridges the two Mn atoms	75
[Fe(CO)$_3$(diars)]	5	trigonal bipyramidal	apical–equatorial substitution	205
[FeCl(NO)(diars)$_2$]ClO$_4$	6	*trans*-octahedral	Cr and Co analogues are isomorphous	58,59
[Co(diars)$_2$](ClO$_4$)$_2$	4 or 6	square planar or *trans*-octahedral	ClO$_4$ groups are weakly associated in axial positions	99
[CoCl$_2$(diars)$_2$]Cl	6	*trans*-octahedral	Co—Cl = 226 pm	98,104
[CoCl$_2$(diars)$_2$]ClO$_4$	6	*trans*-octahedral	green form; also *trans* by i.r., n.m.r. and u.v.–visible evidence	98,105
[NiCl$_2$(diars)$_2$]Cl	6	*trans*-octahedral	figure 32 Ni—Cl = 243 pm	104,105
[NiI$_2$(diars)$_2$]	6	*trans*-octahedral	1:1 electrolyte in PhNO$_2$	122
[Ni(diars)(triars)(ClO$_4$)$_2$]	5	square pyramidal	originally formulated as [Ni(diars)$_3$](ClO$_4$)$_2$	20
[PdI$_2$(diars)$_2$]	6	*trans*-octahedral	five-coordinate in solution	121
[PtI$_2$(diars)$_2$]	6	*trans*-octahedral	five-coordinate in solution	120(a)
[Pt(diars)$_2$]Cl$_2$	4	square planar	figure 33 Pt—Cl = 416 pm	120(b)
[AuI$_2$(diars)$_2$]I	6	*trans*-octahedral	Au—I = 335 pm	149
[W(CO)$_3$(dam)$_2$Br$_2$]	7	capped octahedral	both dam ligands are monodentate	190
[Mo(CO)$_2$(dam)$_2$Br$_2$]	7	capped octahedral	one dam is monodentate, the other, bidentate	190
[RhCl(CO)(dam)]$_2$	4	square planar	figure 35 Rh—Rh = 339.6 pm bridging dam ligands	192
[(C$_6$F$_5$)$_2$Hg(dam)]	4	polymeric tetrahedral	bridging dam ligands?	194
{[Mn$_2$(CO)$_8$](ffars)}	6	dimer; Mn—Mn bond	figure 36; ffars bridges Mn(CO)$_4$ moieties	201
{[Mn(CO)$_4$I]$_2$(ffars)}	6	dimer; no Mn—Mn bond	ffars bridges Mn(CO)$_4$I groups	201

Complex	Coordination number	Structure	Remarks	Ref.
{[$Mn_2(CO)_8$] (ffars)}	6	unsymmetrical dimer with Me_2As bridge	as figure 39; ffars is cleaved	201,202
{[$Fe_2(CO)_6$] (ffars)}	5 or 6	dimer Fe—Fe bond	figure 37; ffars chelates to one Fe; double bond coordinates to second Fe	160
[$Fe_3(CO)_{10}$ (ffars)]	6 or 7	similar to $Fe_3(CO)_{12}$	figure 38; ffars bridges two Fe atoms	210
[$Fe_3(CO)_9$ (ffars)]	6	unsymmetrical trimer with Me_2As bridge	figure 39; ffars is cleaved	211
[$Ru_3(CO)_{10}$ (ffars)]	6	similar to $Ru_3(CO)_{12}$	ffars bridges two Ru atoms	215
[$Ru_3(CO)_8$ (ffars)$_2$]	6	trinuclear	ffars bridges equatorially	216
[$Co_2(CO)_6$ (ffars)]	6	similar to $Co_2(CO)_8$	ffars bridges two Co atoms	219
[$CH_3CCo_3(CO)_7$ (ffars)]	7	trinuclear Co_3 bridged by CO and CH_3C groups	ffars bridges two Co atoms	220
[$Co_4(CO)_8$ (ffars)$_2$]	6	tetranuclear Co_4; no CO bridges	ffars bridges pairs of Co atoms	221
[$Co_4(CO)_9$ (ffars)$_2$]	5	see text	ffars is rearranged	222

34 CONCLUSIONS

This part of the book has reviewed the metal complexes of several ditertiary arsines. The most extensively studied of the bidentate ligands, diars, forms a wide variety of metal complexes, many of which contain the metal atom in an unusual oxidation state or with an uncommon coordination number. The versatility of diars as a chelating ligand has not yet been demonstrated by other ditertiary arsines but further studies of their metal complexes are needed in order to understand the 'universal' coordinating ability of diars. An extension of the recently initiated investigations of *cis*-edas complexes would be particularly worthwhile in evaluating the nature of the metal–arsenic bond. Other suggestions for future work have been made throughout part 5 and it is also hoped that the chapter on ligand synthesis provides impetus for further research in this area of coordination chemistry.

ACKNOWLEDGMENTS

In Part 5 tribute is paid to the memory of the late Sir Ronald Nyholm, whose pioneering research on the chemistry of o-phenylenebis(dimethylarsine) will continue to stimulate research efforts in this field. The author also wishes to express his gratitude to Professors W. R. Cullen, R. D. Feltham, H. B. Gray, J. T. Mague, and C. A. McAuliffe for communicating the unpublished results cited in this chapter.

The author is grateful to the following for permission to reproduce illustrations: The Chemical Society (figures 30, 31, 36, 37); The American Chemical Society (figures 32, 35, 38, 39, 41); The Canadian Chemical Society (figure 40); *Acta Crystallographica* (figure 33); *The Journal of Organometallic Chemistry* (figure 34).

REFERENCES

1. Chatt, J., and Mann, F. G., *J. chem. Soc.* (1939), 1622.
2. Nyholm, R. S., and Tobe, M. L., *Adv. inorg. Chem. Radiochem.*, **5** (1963), 1.
3. Cotton, F. A., and Wilkinson, G., *Advanced Inorganic Chemistry,* 2nd edn, Interscience – Wiley, New York (1966).
4. Harris, C. M., and Livingstone, S. E., in *Chelating Agents and Metal Chelates,* (eds F. Dwyer and D. Mellor), Academic Press, New York (1964), chapter 3, pp. 129–31.
5. Booth, G., *Adv. inorg. Chem. Radiochem* **6** (1964), 1.
6. McWhinnie, W. R., and Miller, J. D., *Adv. inorg. Chem. Radiochem.* **12** (1969), 135.
7. Levason, W., and McAuliffe, C. A., *Adv. inorg. Chem. Radiochem* **14** (1972) 173.
8. Chatt, J., and Mann, F. G., *J. chem. Soc.* (1939), 610.
9. (a) Feltham, R. D., Kasenally, A., Nyholm, R. S., *J. organometal. Chem.* **7** (1967), 285;
   (b) Feltham, R. D., and Metzger, H. G., *J. organometal. Chem.,* personal communication.
10. *Organometallic Compounds,* Vol. III, (ed. M. Dub, Springer, New York (1968), pp. 79–104.
11. Phillips, J. R., and Vis, J. H., *Can. J. Chem.,* **45** (1967), 675.
12. Millar, I. T., Heaney, H., Heinekey, D., and Fernelius, W. C., *Inorg. Synth.* **6** (1960), 116.
13. Tzschach, A., and Lange, W., *Chem. Ber.* **95** (1962), 1360.
14. Aguiar, A. G., Mague, J. T., Aguiar, H. J., Archibald, T. G., and Prejean, B., *J. org. Chem.* **33** (1968), 1681.
15. Feltham, R. D., Metzger, H. G., and Silverthorn, W., *Inorg. Chem.,* **7** (1968), 2003.
16. Bennett, M. A., Erskine, G. J., and Wild, J. D., *Inorg. chim. Acta,* **2** (1968), 379.
17. Erskine, G. J., *Can. J. Chem.,* **47** (1969), 2699.
18. Clark, R. J. H., and Negrotti, R. H. U., *Chem. Ind.* (1968), 154.
19. Feltham, R. D., and Silverthorn, W., *Inorg. Synth.,* **10** (1967), 159.
20. Bosnich, B., Nyholm. R. S., Pauling, P. J., and Tobe, M. L., *J. Am. chem. Soc.,* **90** (1968), 4741.
21. Carty, A. J., Ng, T. W., Efraty, A. E., and Birchall, T., *Inorg. Chem.,* **9** (1970), 1263.
22. Jones, E. R. H., and Mann, F. G., *J. chem. Soc.* (1955), 401.
23. Bosnich, B., and Wild, S. B., *J. Am. chem. Soc.,* **92** (1970), 459.
24. Cullen, W. R., and Hota, N. K., *Can. J. Chem.,* **42** (1964), 1123.
25. Cullen, W. R., Dhaliwal, P. S., and Styan, G. E., *J. organometal. Chem.,* **6** (1966), 364.
26. Cullen, W. R., *Adv. inorg. Chem. Radiochem.,* in press.
27. Alyea, E. C., and Meek, D. W., *Inorg. Chem.,* **11** (1972), 1029.
28. King, R. B., and Kapoor, P. N., *J. Am. chem. Soc.,* **91** (1969), 5191.
29. Eberly, K. C., and Smith, G. E. P., Jr, *J. org. Chem.,* **22** (1957), 1710.
30. Jones, E. R. H., and Mann, F. G., *J. chem. Soc.* (1955), 4472.
31. Hewertson, W., and Watson, H. R., *J. chem. Soc.* (1962), 1490.
32. Braz, G. I., Berlin, A. Ya., and Markova, Yu. V., *J. gen. Chem. USSR,* **18** (1948), 316; *Chem. Abstr.,* **42** (1948), 6764c.
33. Titov, A. I., and Levin, B. B., *Sb. Stat. Obshch. Khim.,* **2** (1953), 1478; *Chem. Abstr.,* **49** (1955), 4504g.
34. Cookson, R. C., and Mann, F. G., *J. chem. Soc.* (1949), 2895.
35. (a) Clark, R. J. H., Lewis, J., Nyholm, R. S., Pauling, P., and Robertson, G. B., *Nature, Lond.,* **192** (1961), 222;
   (b) Clark, R. J. H., Lewis, J., and Nyholm, R. S., *J. chem. Soc.* (1962), 2460.
36. Clark, R. J. H., Kepert, D. L., Lewis, J., and Nyholm, R. S., *J. chem. Soc.* (1965), 2865.
37. Clark, R. J. H., Errington, W., Lewis, J., and Nyholm, R. S., *J. chem. Soc. (A)* (1966), 989.
38. Clark, R. J. H., *The Chemistry of Titanium and Vanadium,* Elsevier, Amsterdam (1968).
39. Crisp, W. P., Deutscher, R. L., and Kepert, D. L., *J. chem. Soc. (A)* (1970), 2199.
40. Clark, R. J. H., Negrotti, R. H. U., and Nyholm, R. S., *Chem. Commun.* (1966), 486.

41. Clark, R. J. H., *J. chem. Soc.* (1965), 5699.
42. Garner, C. D., and Wallwork, S. C., *J. chem. Soc. (A)* (1966), 1496.
43. Colapietro, M., Vaciago, A., Bradley, D. C., Hursthouse, M. B., and Rendall, I. F., *Chem. Commun.* (1970), 743.
44. Sutton, G. J., *Aust. J. Chem.* **12** (1959), 122.
45. Allison, J. A. C., and Mann, F. G., *J. chem. Soc.* (1949), 2915.
46. Deutscher, R. L., and Kepert, D. L., *Inorg. Chem.,* **9** (1970), 2305.
47. Fergusson, J. E., Wilkins, C. J., and Young, J. F., *J. chem. Soc.* (1962), 2136.
48. Piovesana, O., and Furlani, C., *Chem. Commun.* (1971), 256.
49. Bridgland, B. E., Fowles, G. W. A., and Walton, R. A., *J. inorg. nucl. Chem.,* **27** (1965), 383.
50. Kasenally, A. S., Nyholm, R. S., O'Brien, R. J., and Stiddard, M. H. B., *Nature, Lond.,* **204** (1964), 871.
51. Clark, R. J. H., Kepert, D. L., and Nyholm, R. S., *J. chem. Soc.* (1965), 2877.
52. Fairbrother, F., *The Chemistry of Niobium and Tantalum,* Elsevier, Amsterdam (1967).
53. Nyholm, R. S., and Sutton, G. J., *J. chem. Soc.* (1958), 560.
54. Gustav, K., and Moerke, H., *Z. Chemie, Lpz.,* **9** (1969), 32.
55. Feltham, R. D., and Silverthorn, W., *Inorg. Chem.,* **7** (1968), 1154.
56. Bennett, M. A., Clark, R. J. H., and Goodwin, A. D. J., *Inorg. Chem.,* **6** (1967), 1625.
57. Lewis, J., Nyholm, R. S., Pande, C. S., Sandhu, S. S., and Stiddard, M. H. B., *J. chem. Soc.* (1964), 3009.
58. Feltham, R. D., Silverthorn, W., and McPherson, G., *Inorg. Chem.,* **8** (1969), 344.
59. (a) Silverthorn, W., and Feltham, R. D., *Inorg. Chem.,* **6** (1967), 1622;
    (b) Feltham, R. D., Silverthorn, W., Wickman, H., and Wesolowski, W., *Inorg. Chem.,* in press, 1971.
60. Nigam, H. L., Nyholm, R. S., and Stiddard, M. H. B., *J. chem. Soc.* (1960), 1806.
61. Lewis, J., Nyholm, R. S., Pande, C. S., and Stiddard, M. H. B., *J. chem. Soc.* (1963), 3600.
62. Lewis, J., Nyholm, R. S., and Smith, P W., *J. chem. Soc.* (1962), 2592.
63. Djordjevic, C., Nyholm, R. S., Pande, C. S., and Stiddard, M. H. B., *J. chem. Soc. (A)* (1966), 16.
64. Fergusson, J. E., Robinson, B. H., Wilkins, C. J., *J. chem. Soc. (A)* (1967), 486.
65. Clark, R. J. H., Kepert, D. L., Nyholm, R. S., and Rodley, G. A., *Spectrochim. Acta,* **22** (1966), 1697.
66. Drew, M. G. B., and Mandyczewsky, R., *Chem. Commun.* (1970), 292.
67. Nigam, H. L., Nyholm, R. S., and Stiddard, M. H. B., *J. chem. Soc.* (1960), 1803.
68. Stiddard, M. H. B., and Townsend, R. E., *J. chem. Soc. (A)* (1969), 2355.
69. Haines, R. J., Nyholm, R. S., and Stiddard, M. H. B., *J. chem. Soc. (A)* (1967), 94.
70. Nyholm, R. S., and Sutton, G. J., *J. chem. Soc.* (1958), 564.
71. Ethyl Corp. and Nyholm, R. S., *US Pat.* 3 037 037 (1960); *Chem. Abstr.,* **57** (1962), 13806.
72. Nyholm, R. S., and Ramana Rao, D. V., *Proc. chem. Soc.* (1959), 130.
73. Osborne, A. G., and Stiddard, M. H. B., *J. chem. Soc.* (1962), 4715.
74. Nyholm, R. S., Sandhu, S. S., and Stiddard, M. H. B., *J. chem. Soc.* (1963), 5916.
75. Bennett, M. J., and Mason, R., *Proc. chem. Soc.* (1964), 395.
76. (a) Fergusson, J. E., and Nyholm, R. S., *Nature, Lond.,* **183** (1959), 1039;
    (b) Fergusson, J. E., and Nyholm, R. S., *Chem. Ind.* (1960), 347.
77. Curtis, N. F., Fergusson, J. E., and Nyholm, R. S., *Chem. Ind.* (1958), 625.
78. Fergusson, J. E. and Nyholm, R. S., *Chem. Ind.* (1958), 1555.
79. Fergusson, J. E., and Hickford, J. H., *Inorg. chim. Acta,* **2** (1968), 475.
80. Kirkham, W. J., Osborne, A. G., Nyholm, R. S., and Stiddard, M. H. B., *J. chem. Soc.* (1965), 550.
81. Nigam, H. L., Nyholm, R. S., and Ramano Rao, D. V., *J. chem. Soc.* (1959), 1397.
82. Lewis, J., Nyholm, R. S., Sandhu, S. S., and Stiddard, M. H. B., *J. chem. Soc.* (1964), 2825.
83. Cullen, W. R., and Harbourne, D. A., *Can. J. Chem.,* **47** (1969), 3371.
84. Nyholm, R. S., *J. chem. Soc.* (1950), 851.
85. Burstall, F. H., and Nyholm, R. S., *J. chem. Soc.* (1952), 3570.

86. Feltham, R. D., and Silverthorn, W., *Inorg. Chem.,* **9** (1970), 1207.
87. (a) Nyholm, R. S., and Parish, R. V., *Chem. Ind.* (1956), 470;
    (b) Hazeldean, G. S. F., Nyholm, R. S., and Parish, R. V., *J. chem. Soc. (A)* (1966), 162.
88. Nyholm, R. S., and Sutton, G. J., *J. chem. Soc.* (1958), 567.
89. Nyholm, R. S., and Sutton, G. J., *J. chem. Soc.* (1958), 572.
90. Klassen, D. M., and Crosby, G. A., *J. molec. Spectrosc.,* **25** (1968), 398.
91. (a) Chatt, J., and Hayter, R. G., *Proc. chem. Soc.* (1959), 153;
    (b) *J. chem. Soc.* (1963), 6017.
92. (a) Fairey, M. B., and Irving, R. J., *J. chem. Soc. (A)* (1966), 475;
    (b) *Spectrochim. Acta,* **20** (1964), 1757.
93. (a) Douglas, P. G., Feltham, R. D., and Metzger, H. G., *Chem. Commun.* (1970), 889;
    (b) Feltham, R. D., unpublished observations, 1971.
94. Hales, L. A. W., and Irving R. J., *J. chem. Soc. (A)* (1967), 1932.
95. Bradford, C. W., *Platin. Metals Rev.* **11** (1967), 104.
96. Nyholm, R. S., *J. chem. Soc.* (1950), 2071.
97. Gill, N. S., and Nyholm, R. S., *J. chem. Soc.* (1959), 3997.
98. Lewis, J., Nyholm, R. S., and Rodley, G. A., *J. chem. Soc.* (1965), 1483.
99. (a) Rodley, G. A., and Smith, P. W., *J. chem. Soc. (A)* (1967), 1580;
    (b) Einstein, F. W. B., and Rodley, G. A., *J. inorg. nucl. Chem.,* **29** (1967), 347.
100. Dyer, G., and Meek, D. W., *J. Am. chem. Soc.,* **89** (1967), 3983.
101. Preer, J. R., and Gray, H. B., *J. Am. chem. Soc.,* **92** (1970), 7306.
102. Dunn, T. M., Nyholm, R. S., and Yamada, S., *J. chem. Soc.* (1962), 1564.
103. Yamada, S., *Coord. Chem. Rev.* **2** (1967), 89.
104. Bernstein, P., Rodley, G. A., Marsh, R., and Gray, H. B., *Inorg. Chem.,* personal communication.
105. Pauling, P. J., Porter, D. W., and Robertson, G. B., *J. chem. Soc. (A)* (1970), 2728.
106. Baylis, B. K. W., and Bailar, J. C., Jr, *Inorg. Chem.,* **9** (1970), 641.
107. (a) Peloso, A., and Tobe, M. L., *J. chem. Soc.* (1964), 5063;
    (b) Dolcetti, G., and Peloso, A., *Gazz. chim. ital.,* **97** (1967), 230;
    (c) Peloso, A., and Dolcetti, G., *J. chem. Soc. (A)* (1969), 1506.
108. Feltham, R. D., and Nyholm, R. S., *Inorg. Chem.,* **4** (1965), 1334.
109. Nyholm, R. S., *J. chem. Soc.* (1950), 857.
110. Volponi, L., Zarli, B., and Columbini, C., *Gazz. chim. ital.,* **98** (1968), 413.
111. Canziani, F., Sartorelli, U., and Zingales, F., *Rc. Ist. lomb. Sci. Lett.,* **A101** (1967), 227; *Chem. Abstr.,* **68** (1968), 83947s.
112. (a) Vaska, L., and Diluzio, J. W., *J. Am. chem. Soc.,* **84** (1962), 679;
    (b) Vaska, L., *Acct chem. Res.,* **1** (1968), 335.
113. Chatt, J., Hart, F. A., and Watson, H. R., *J. chem. Soc.* (1962), 2537.
114. Nyholm, R. S., and Ramana Rao, D. V., unpublished observations, cited in refs. 113, 118.
115. Nyholm, R. S., *J. chem. Soc.* (1952), 2906.
116. Nyholm, R. S., *J. chem. Soc.* (1952), 2061.
117. Hudson, M. J., Nyholm, R. S., and Stiddard, M. H. B., *J. chem. Soc. (A)* (1968), 40.
118. Harris, C. M., Nyholm, R. S., and Phillips, D. J., *J. chem. Soc.* (1960), 4379.
119. Harris, C. M., and Nyholm, R. S., *J. chem. Soc.* (1956), 4375.
120. (a) Stephenson, N. C., *J. inorg. nucl. Chem.,* **24** (1962), 791;
    (b) *Acta. crystallogr.,* **17** (1964), 1517.
121. Stephenson, N. C., *J. inorg. nucl. Chem.,* **24** (1962), 797.
122. Stephenson, N. C., *Acta. crystallogr.,* **17** (1964), 592.
123. Bosnich, B., Bramley, R., Nyholm, R. S., and Tobe, M. L., *J. Am. chem. Soc.,* **88** (1966), 3926.
124. Bernstein, P. K., and Gray, H. B., *Inorg. Chem.,* personal communication.
125. Kreisman, P., Marsh, R., Preer, J. R., and Gray, H. B., *J. Am. chem. Soc.,* **90** (1968), 1067.
126. Manoharan, P. T., and Rogers, M. T., *J. chem. Phys.,* **53** (1970), 1682.
127. Corvaja, C., and Nordio, P. L., *Ricerca scient.,* **38** (1968), 44; *Chem. Abstr.,* **69** (1968), 14760j.

128. Meek, D. W., Alyea, E. C., Stalick, J. K., and Ibers, J. A., *J. Am. chem. Soc.,* **91** (1969), 4920.
129. Nyholm, R. S., *J. chem. Soc.* (1951), 2902.
130. Heaney, H., Mann, F. G., and Millar, I. T., *J. chem. Soc.* (1957), 3930.
131. Calvin, G., and Coates, G. E., *J. chem. Soc.* (1960), 2008.
132. Watt, G. W., and Layton, R., *Inorg. Chem.,* **1** (1962), 496.
133. Dolcetti, G., Peloso, A., and Sindellari, L., *Gazz. chim. ital.,* **96** (1966), 1648.
134. Ettore, R., Peloso, A., and Dolcetti, G., *Gazz. chim. ital.,* **97** (1967), 968.
135. Peloso, A., and Ettore, R., *J. chem. Soc. (A)* (1968), 2253.
136. Peloso, A., Corain, B., and Bressan, M., *Gazz. chim. ital.,* **99** (1969), 111.
137. Ettore, R., Dolcetti, G., and Peloso, A., *Gazz. chim. ital.,* **97** (1967), 1681.
138. Barclay, G. A., Collard, M. A., Harris, C. M., and Kingston, J. V., *J. chem. Soc. (A)* (1969), 830.
139. Dolcetti, G., Peloso, A., and Tobe, M. L., *J. chem. Soc.* (1965), 5196.
140. Dolcetti, G., and Peloso, A., *Gazz. chim. ital.,* **97** (1967), 1540.
141. Peloso, A., *Gazz. chim. ital.,* **99** (1969), 1025.
142. Kabesh, A., and Nyholm, R. S., *J. chem. Soc.* (1951), 38.
143. Burkin, A. R., *J. chem. Soc.* (1954), 71.
144. Lewis, J., Nyholm, R. S., and Phillips, D. J., *J. chem. Soc.* (1962), 2177.
145. Harris, C. M., Nyholm, R. S., and Stephenson, N. C., *Recl Trav. chim. Pays-Bas Belg.,* **75** (1956), 687.
146. Harris, C. M., and Nyholm, R. S., *J. chem. Soc.* (1957), 63.
147. Charlton, R. J., Harris, C. M., Patil, H., and Stephenson, N. C., *Inorg. nucl. Chem. Lett.,* **2** (1966), 409.
148. Duckworth, V. F., and Stephenson, N. C., *Inorg. Chem.,* **8** (1969), 1661.
149. Duckworth, V. F., Harris, C. M., and Stephenson, N. C., *Inorg. nucl. Chem. Lett.,* **4** (1968), 419.
150. Harris, C. M., and Livingstone, S. E., *Rev. pure appl. Chem.,* **12** (1962), 16.
151. Sutton, G. J., *Aust. J. Chem.,* **14** (1961), 545.
152. Noltes, J. G., and van den Hurk, J. W. G., *J. organometal. Chem.,* **1** (1964), 377.
153. Deacon, G. B., and Green, J. H. S., *Spectrochim. Acta.,* **A24** (1968), 959.
154. Nyholm, R. S., and Ulm, K., *J. chem. Soc.* (1965), 4199.
155. Sutton, G. J., *Aust. J. Chem.,* **11** (1958), 120.
156. Deacon, G. B., Green, J. H. S., and Nyholm, R. S., *J. chem. Soc.* (1965), 3411.
157. (a) Sutton, G. J., *Aust. J. Chem.,* **11** (1958), 420;
     (b) *Aust. J. Chem.* **11** (1958), 415.
158. Metzger, H. G., and Feltham, R. D., *Inorg. Chem.,* **10** (1971), 951.
159. Cullen, W. R., Dhaliwal, P. S., and Stewart, C. J., *Inorg. Chem.,* **6** (1967), 2256.
160. Einstein, F. W. B., and Trotter, J., *J. chem. Soc. (A)* (1967), 824.
161. McAuliffe, C. A., Chow, K. K., and Halfpenny, M. T., unpublished observations, 1971.
162. Nesmeyanov, A. N., Borisov, A. E., and Kudryavtseva, L. V., *Izv. Akad. Nauk SSSR, Ser. Khim.* (1969), 1973; *Chem. Abstr.,* **72** (1970), 21754a.
163. Mague, J. T., and Mitchener, J. P., *Inorg. Chem.,* **8** (1969), 119.
164. Mague, J. T., and Mitchener, J. P., *Chem. Commun.* (1968), 911.
165. Chatt, J., and Butler, S. A., *Chem. Commun.* (1967), 501.
166. Mague, J. T., unpublished observations, 1971.
167. Chow, K. K., Halfpenny, M. T., and McAuliffe, C. A., unpublished observations, 1971.
168. Westland, A. D., and Westland, L., *Can. J. Chem.,* **43** (1965), 426.
169. Ray, T. C., and Westland, A. D., *Inorg. Chem.,* **4** (1965), 1501.
170. Selbin, J., and Vigee, G., *J. inorg. nucl. Chem.,* **30** (1968), 1644.
171. (a) Zingales, F., Canziani, F., and Ugo, R., *Gazz. chim. ital.,* **92** (1962), 761;
     (b) Zingales, F., and Canziani, F., *Gazz. chim. ital.,* **92** (1962), 343;
     (c) Canziani, F., Zingales, F., and Sartorelli, N., *Gazz. chim. ital.,* **94** (1964), 841.
172. Tsang, W. S., Meek, D. W., and Wojcicki, A., *Inorg. Chem.,* **7** (1968), 1263.
173. Powers, D. R., Faber, G. C., and Dobson, G. R., *J. inorg. nucl. Chem.,* **31** (1969), 2970.
174. Dobson, G. R., and Houk, L. W., *Inorg. chim. Acta,* **1** (1967), 287.
175. Behrens, H., and Mueller, A., *Z. anorg. allg. Chem.,* **341** (1965), 124.
176. Chatt, J., and Watson, H. R., *J. chem. Soc.* (1961), 4980.

177. Sandhu, S. S., and Mehta, A. K., *Inorg. nucl. Chem. Lett.,* 7 (1971), 891.
178. Du Pont de Nemours, E. I. and Co. and Sarafidis, C., *US Pat.* 3 407 244 and 3 407 245 (1968); *Chem. Abstr.,* 70 (1969), 11097h, j.
179. Zingales, F., Canziani, F., and Ugo, R., *Chimica Ind.,* Milano, 44 (1962), 1394; *Chem. Abstr.,* 61 (1964), 3910e.
180. Haines, R. J., and Du Preez, A. L., *J. organometal. Chem.,* 21 (1970), 181.
181. Bennett, M. A., and Longstaff, P. A., *J. Am. chem. Soc.,* 91 (1969), 6266.
182. Chatt, J., and Hart, F. A., *J. chem. Soc.* (1960), 1378.
183. Sacconi, L., Bertini, I., and Mani, F., *Inorg. Chem.,* 7 (1968), 1417.
184. Watt, G. W., and Layton, R., *Inorg. Chem.,* 1 (1962), 496.
185. Sandhu, S. S., and Sandhu, R. S., *Indian J. Chem.,* 8 (1970), 189.
186. Selbin, J., and Ortego, J. D., *J. inorg. nucl. Chem.,* 29 (1967), 1449.
187. Westland, A. D., *Can. J. Chem.,* 47 (1969), 4135.
188. Colton, R., *Coord. Chem. Rev.,* 6 (1971), 269.
189. Anker, M. W., Colton, R., and Tomkins, I. B., *Aust. J. Chem.,* 21 (1968), 1159.
190. Drew, M. G. B., Johans, A. W., Wolters, A. P., and Tomkins, I. B., *Chem. Commun.* (1971), 819.
191. Colton, R., and Rix, C. J., *Aust. J. Chem.,* 23 (1970), 441.
192. Mague, J., *Inorg. Chem.,* 8 (1969), 1975.
193. I.C.I. Ltd, *Neth. Pat. Appl.* 6 605 627 (1966); *Chem. Abstr.,* 66 (1966), 55581h.
194. Canty, A. J., and Deacon, G. B., *Aust. J. Chem.,* 24 (1971), 489.
195. McAuliffe, C. A., Baracco, L., and Levason, W., unpublished observations, 1971.
196. Kamai, G., and Gatilov, Yu.F., *Zh. Obshch. Khim.,* 33 (1963), 1189; *Chem. Abstr.,* 59 (1963), 10115.
197. Crow, J. P., Cullen, W. R., and Hou, F. L., *Inorg. Chem.,* in press, 1971.
198. Cullen, W. R., and Hou, F. L., unpublished observations, 1971.
199. Hieber, W., and Freyer, W., *Chem. Ber.,* 93 (1960), 462.
200. Sacco, A., and Ugo, R., *J. chem. Soc.* (1964), 3274.
201. Crow, J. P., Cullen, W. R., Hou, F. L., Chan, L. Y. Y., and Einstein, F. W. B., *Chem. Commun.* (1971), 1229.
202. Einstein, F. W. B., Jones, R. D. G., MacGregor, A. C., and Cullen, W. R., personal communication.
203. Cullen, W. R., Harbourne, D. A., Liengme, B. V., and Sams, J. R., *Inorg. Chem.,* 8 (1969), 1464.
204. Einstein, F. W. B., and Jones, R. D. G., cited in ref. 26.
205. Brown, D. A., and Bushnell, G. W., *Acta. crystallogr.,* 22 (1967), 292.
206. Cullen, W. R., Harbourne, D. A., Liengme, B. V., and Sams, J. R., *Inorg. Chem.,* 8 (1969), 95.
207. Crow, J. P., Cullen, W. R., Sams, J. R., and Ward, J. E. H., *J. organometal. Chem.,* 22 (1970), C29.
208. Cullen, W. R., and Chia, L. S., unpublished observations, 1971.
209. Cullen, W. R., Harbourne, D. A., Liengme, B. V., and Sams, J. R., *Inorg. Chem.,* 9 (1970), 702.
210. Roberts, P. J., Penfold, B. R., and Trotter, J., *Inorg. Chem.,* 9 (1970), 2137.
211. Einstein, F. W. B., Pilotte, A. M., and Restivo, R., *Inorg. Chem.,* 10 (1971), 1947.
212. Crow, J. P., Cullen, W. R., Herring, F. G., Sams, J. R., and Tapping, R. L., *Inorg. Chem.,* 10 (1971), 1616.
213. Crow, J. P., and Cullen, W. R., *Can. J. Chem.,* 48 (1971), 2948.
214. Cullen, W. R., and Harbourne, D. A., *Inorg. Chem.,* 9 (1970), 1839.
215. Roberts, P. J., and Trotter, J., *J. chem. Soc. (A)* (1971), 1479.
216. Roberts, P. J., and Trotter, J., *J. chem. Soc. (A)* (1970), 3246.
217. Crow, J. P., and Cullen, W. R., *Inorg. Chem.,* 10 (1971), 1529.
218. Crow, J. P., and Cullen, W. R., *Inorg. Chem.,* 10 (1971), 2165.
219. Harrison, W., and Trotter, J., *J. chem. Soc. (A)* (1971), 1607.
220. Einstein, F. W. B., and Jones, R. D. G., personal communication.
221. Einstein, F. W. B., and Jones, R. D. G., personal communication.
222. Einstein, F. W. B., Jones, R. D. G., MacGregor, A. C., and Cullen, W. R., personal communication.

223. Cullen, W. R., and Thompson, J. A. J., *Can. J. Chem.,* **48** (1970), 1730.
224. Duffy, N. V., Layton, A. J., Nyholm, R. S., Powell, D., and Tobe, M. L., *Nature, Lond.,* **221** (1966), 177.
225. Cochran, W., Hart, F. A., and Mann, F. G., *J. chem. Soc.* (1957), 2816.
226. Chatt, J., and Hart, F. A., *J. chem. Soc.* (1960), 1378.
227. Hart, F. A., and Mann, F. G., *J. chem. Soc.* (1960), 3939.
228. (a) Heaney, H., Heinekey, D. M., Mann, F. G., and Millar, I. T., *J. chem. Soc.* (1958), 3838;
     (b) Forbes, M. H., Heinekey, D. M., Mann, F. G., and Millar, I. T., *J. chem. Soc.* (1961), 2762.
229. Allan, D. W., Millar, I. T., Mann, F. G., Canadine, R. M., and Walker, J., *J. chem. Soc. (A)* (1969), 1097.
230. Sindellari, L., and Deganello, G., *Ricerca scient.,* **35 (II, A)** (1965), 744.
231. Di Sipio, L., Sindellari, L., Tondello, E., DeMichelis, G., and Oleari, L., *Coord. Chem. Rev.,* **2** (1971), 129.
232. Ros, R., and Tondello, E., *J. inorg. nucl. Chem.,* **33** (1971), 245.
233. (a) Cullen, W. R., Hall, L. D., and Ward, J. E. H., *Chem. Commun.* (1970), 625;
     (b) Cullen, W. R., Hall, L. D., and Ward, J. E. H., unpublished observations, 1971.
234. Smith, H. D., Jr, *Inorg. Chem.,* **8** (1969), 676.
235. Zaborowski, R., and Cohn, K., *Inorg. Chem.,* **8** (1969), 678.
236. Dobson, G. R., Stolz, I. W., and Sheline, R. K., *Adv. inorg. Chem. Radiochem.,* **8** (1966), 1.
237. Haines, L. M., and Stiddard, M. H. B., *Adv. inorg. Chem. Radiochem.,* **12** (1969), 53.
238. Abel, E. W., Crow, J. P., and Illingworth, S. M., *J. chem. Soc. (A)* (1969), 1631.
239. Cannon, R. D., Chiswell, B., and Venanzi, L. M., *J. chem. Soc. (A)* (1967), 1277.
240. Ellermann, J., and Dorn, K., *Chem. Ber.,* **101** (1968), 643.
241. Dutta, R. L., Meek, D. W., and Busch, D. H., *Inorg. Chem.,* **9** (1970), 1215.
242. Green, J. H. S., Kynaston, W., and Rodley, *Spectrochim. Acta.,* **24**A (1968), 853.
243. Figgis, B. N., and Lewis, J., *Prog. inorg. Chem.,* **6** (1964), 37.
244. Parish, R. V., and Perkins, P. G., *J. chem. Soc. (A)* (1967), 345.
245. Preer, J. R., Tsay, F.-D., Gray, H. B., *J. Am. chem. Soc.,* personal communication.
246. Venzani, C. M., *Chem. Br.* (1968), 162
247. Muetterties, E. L., and Schunn, R. A., *Q. Rev. chem. Soc.,* **20** (1966), 245.
248. Feltham, R. D., unpublished observations, 1971.
249. Colton, R., and Porter, Q. N., *Aust. J. Chem.,* **21** (1968), 2215.
250. Anker, M. W., Colton, R., and Tomkins, I. B., *Rev. pure appl. Chem.,* **18** (1968), 23.

APPENDIX

Additional References
PART 2
MISCELLANEOUS

1. Leigh, G. J., *Prep. inorg. React.* **7** (1971), 165. A review of metal–dinitrogen complexes.

2. Beck, W., *Organometal. chem. Rev.,* **7** (1971), 159. A review of metal fulminate complexes.

3. Mann, B. E., Shaw, B. L., and Stainbank, R. E., *Chem. Commun.* (1972), 151. The use of ^{13}C n.m.r. spectra to determine the structure of transition-metal phosphine complexes.

4. Dessy, R. E., Rheingold, A. L., and Howard, G. H., *J. Am. chem. Soc.,* **94** (1972), 746. An n.m.r. spectral study of some phosphido-bridged complexes.

5. Leigh, G. J., and Bremser, W., *J. chem. Soc. (Dalton)* (1972), 1216. The X-ray photoelectron spectra of some heavy transition-metal tertiary phosphine complexes.

V, Nb, Ta

6. Davison, A., and Ellis, J. E., *J. organometal. Chem.,* **31** (1971), 239. The preparation of $[NEt_4][M(CO)_5PPh_3]$ (M = Nb, Ta), and an improved method for preparing $[V(CO)_5L]^-$ (L = PPh_3, PBu^n_3).

7. Henrici-Olivé, S., and Olivé, S., *J. Am. chem. Soc.,* **91** (1971), 4154. Some e.s.r. evidence for the existence of a series of complexes $VOCl_2(PR_3)_2$ (PR_3 = various phosphines).

Cr, Mo, W

8. King, R. B., and Korenowski, T. F., *Inorg. Chem.,* **10** (1971), 1188. Reaction of $(Me_2N)_3P$ with $M(CO)_6$ (M = Cr, Mo, W) yielding $LM(CO)_5$ and $L_2M(CO)_4$, and with $Fe(CO)_5$ yielding $LFe(CO)_4$.

9. Ehrl, W., and Vahrenkamp, H., *Chem. Ber.,* **104** (1971), 3261. The reaction of $NaM(CO)_5$ (M = Mn, Re) with $[(OC)_5M'(PMe_2Cl)]$ (M' = Cr, Mo) to give $(OC)_5M-PMe_2-M'(CO)_5$.

10. Huttner, G., and Schelle, O., *J. Crystallogr. molec. Struct.,* **1** (1971), 69. The X-ray structure of *cis*-$Cr(PH_3)_4(CO)_2$.

11. Bowden, J. A., and Colton, R., *Aust. J. Chem.,* **24** (1971), 2471. The reaction of *o*-, *m*- and *p*-tolylphosphines with $M(CO)_6$ (M = Cr, Mo, W) to give $M(CO)_5L$ and *cis*- and *trans*-$M(CO)_4L_2$. The formation of (π-tri-*o*-tolylphosphine)$M(CO)_3$ complexes is described.

12. Bowden, J. A., and Colton, R., *Aust. J. Chem.,* **25** (1972), 17. The reaction of $M(CO)_4X_2$ (M = Mo, W; X = Cl, Br, I) with *o*-, *m*- and *p*-tolylphosphines. A reinvestigation of the reaction of PPh_3 with $M(CO)_4I_2$, and the oxidation of *trans*-$M(CO)_4(PPh_3)_2$ with halogens.

13. Atkinson, L. K., and Smith, D. C., *J. organometal. Chem.,* **33** (1971), 189. The preparation of $LMo(CO)_5$ and *trans*-$L_2Mo(CO)_4$ (L = R_2NPPh_2; R = Me, Et, Pr^n, Bu^n).

14. Chatt, J., Leigh, G. J. and Thankarajan, N., *J. organometal. Chem.,* **29** (1971), 105. Improved synthetic methods for a range of phosphine-substituted carbonyl complexes of Cr, Mo, W. A number of new complexes of W are also reported.

15. Friedel, H. F., Renk, I. W., and Dieck, H. T., *J. organometal. Chem.,* **26** (1971), 247. The reaction of π-allyldicarbonylmolybdenum complexes with PPh_3 or PBu_3 to give $Mo(CO)_2(PR_3)_2L_2$.

16. Brown, R. A., and Dobson, G. R., *Inorg. Chim. Acta,* **6** (1972), 65. The preparation of $W(CO)_5(PR_3)$ (R = Ph, Bu^n) and their i.r. spectra.

17. Nixon, J. F., and Swain, J. R., *J. chem. Soc. (Dalton)* (1972), 1038. The preparation and n.m.r. spectra of cis-$ML_2(CO)_4$ $\{L = (F_3C)_2PX, F_3CPY_2; M = Cr, Y = H; M = Mo, Y = Cl, Br, H; X = Cl, Br, I, H\}$ and evaluation of the phosphorus–phosphorus coupling constants.

18. Nassimbeni, L. R., *Inorg. nucl. Chem. Lett.,* **7** (1971), 909. The structure of $[M(CO)_4PMe_2]_2$.

19. Keiter, R. L., and Shah, D. P., *Inorg. Chem.,* **11** (1972), 191. The preparation of tungsten carbonyl complexes of monoquaternised diphosphine ligands.

20. Mawby, A., and Pringle, G. E., *J. Inorg. nucl. Chem.,* **34** (1972), 517. The structure of $Mo(CO)_2(PPhMe_2)_3Cl_2$. MeOH, showing the Mo to be seven-coordinate.

21. Bishop, J. K. B., Cullen, W. R., and Gerry, M. C. L., *Can. J. Chem.,* **49** (1971), 3910. The ^{35}Cl n.q.r. spectra of fac-$Mo(CO)_3(PCl_3)_3$ and $Ni(PCl_3)_4$.

22. Davis, R., Johnson, B. F. G., and Al-Obaidi, K. H., *J. chem. Soc. (Dalton)* (1972), 508. The reaction of $[Mo(NO)Cl_3]_n$ with PPh_3 to give $[Mo(NO)(OPPh_3)_2Cl_3]$; this, and the W analogue, were also obtained from NOCl and various phosphine-substituted carbonyls of Mo, W.

23. Hidai, M., Tominari, K., and Uchida, Y., *J. Am. chem. Soc.,* **94** (1972), 110. The preparation of $Mo(N_2)(PPh_3)_2 . C_6H_5CH_3$ from $Mo(acac)_3$, AlR_3, PPh_3 and N_2. No dinitrogen complexes were isolated with PBu^n_3, $P(p\text{-tolyl})_3$, and PPh_2Et.

24. Darensbourg, D. J., *Inorg. nucl. Chem. Lett.,* **8** (1972), 529. Photochemical reaction of dinitrogen complexes of Mo, Re, Fe and Os with carbon monoxide, resulting in elimination of the dinitrogen.

25. Butcher, A. V., Chatt, J., Leigh, G. J., and Richards, P. L., *J. chem. Soc. (Dalton)* (1972), 1064. The synthesis of $WCl_4(PR_3)_2$ ($R_3 = Ph_3$, Ph_2Me, Ph_2Et, Ph_2Pr^n, Ph_2Bu^n, $PhMe_2$, $PhEt_2$, $PhPr^n_2$, $PhBu^n_2$, Et_3), $WCl_4(RCN)(PPhMe_2)_2$, $WOCl_2(PR_3)_3$ ($R_3 = Ph_2Me$, $PhMe_2$, $PhEt_2$) and $WOX_2(PPhMe_2)_3$ (X = NCO, NCS).

26. Bright, D. G., Kepert, D. L., Mandyczewsky, R., and Trigwell, K. R., *J. chem. Soc. (Dalton)* (1972), 313. A brief report of $WCl_4(PPh_3)_2$ and $WCl_4(PrCN)(PPh_3)$, and of the failure to obtain W(III) complexes of PMe_3, PPh_3.

27. Manojlović-Muir, L., and Muir, K. W., *J. chem. Soc. (Dalton)* (1972), 686. The structure of cis-mer-$[MoOCl_2(PPhEt_2)_3]$.

28. Millar, J. R., and Myers, D. H., *Inorg. Chim. Acta,* **5** (1971), 215. A reinvestigation of the reaction of $Mn_2(CO)_{10}$ with PPh_3; the isolation of $Mn_2(CO)_8(PPh_3)_2$ and confirmation of the presence of a paramagnetic species in the products. The formation of trans-$HMn(CO)_4(PPh_3)$ is also proposed.

Mn, Tc, Re (see also refs. 8, 24)

29. Fischer, E. O., and Herrmann, W. A., *Chem. Ber.,* **105** (1972), 286. The preparation of $Mn_2(CO)_9PH_3$, in which phosphine occupies an equatorial position.

30. Bennett, R. L., Bruce, M. I., and Stone, F. G. A., *J. organometal. Chem.,* **38** (1972), 325. The preparation of $(PhCH_2)PMe_2$ complexes of Mn, Fe, Ru, Rh, Ni, Pd, Pt. Among the complexes prepared are $[Mn(CO)_3L_2]_2$, $Mn(CO)_3L_2Br$, $Fe(CO)_4L$, $Ru_3(CO)_9L_3$, $Ru(CO)_2LCl_2$, $Ru(CO)L_3Cl_2$, $RuCl_2L_4$, $Rh(CO)L_2Cl$, PdL_3Cl_2, PdL_2Cl_2, PtL_2Cl_2 (cis and trans).

31. Yasafuku, K., and Yamazaki, H., *J. organometal. Chem.,* **28** (1971), 415. Mixed-metal (Ni–Fe, Mn–Fe, Co–Fe, Fe–Ni) phosphido-bridged complexes.

32. Schumann, H., and Knoth, H. J., *J. organometal. Chem.*, **32** (1971), C47. The reaction of $(Me_3M)_3P$ (M = Ge, Sn) with $Mn(CO)_5Br$ to give $(OC)_4Mn\{(Me_3M)_2P\}_2Mn(CO)_4$.

33. Curtis, M., *Inorg. Chem.*, **11** (1972), 802. The reaction of silyl, germyl and stannyl halides with some phosphine-substituted carbonyls of Mn and Co.

34. Cariati, F., Sgamellotti, A., Morazzoni, F., and Valenti, V., *Inorg. Chim. Acta*, **5** (1971), 531. The electronic spectra and magnetic properties of $[PPh_3H][ReX_5(PPh_3)]$ and $[ReX_4(PPh_3)_2]$ (X = Cl, Br).

35. Chatt, J., Crabtree, R. H., and Richards, R. L., *Chem. Commun.* (1972), 534. The relative base strengths of *trans*-$[ReCl(N_2)(PPhMe_2)_4]$, *trans*-$[ReCl(CO)(PPhMe_2)_4]$ and *mer*-$[OsCl_2(N_2)(PPhEt_2)_3]$ towards Al_2Me_6.

Fe, Ru, Os

36. Dobbie, R. C., Hopkinson, M. J., and Whittaker, D., *J. chem. Soc. (Dalton)* (1972), 1030. The reaction of $(F_3C)_2PH$ with $Fe(CO)_5$, $Fe_3(CO)_{12}$ to give $Fe_2(CO)_6\{P(CF_3)_2\}_2$ and $H_2Fe_2(CO)_6\{P(CF_3)_2\}_2$, with $Fe_2(CO)_9$ to give $Fe(CO)_4\{P(CF_3)_2H\}$ and with $Fe(NO)_2(CO)_2$ to give $Fe_2(NO)_4\{P(CF_3)_2\}_2$.

37. Raper, G., and McDonald, W. S., *J. chem. Soc. (A)* (1971), 3430. The structure of $[Fe_3(CO)_9(PPhMe_2)_3]$.

38. Birck, J. L., Le Cars, Y., Baffier, N., Legendre, J. J. and Huber, M., *C.r. hebd. Séanc. Acad. Sci., Paris*, **C273** (1971), 880. The structure of *cis*-$Fe(CO)_2(PH_3)_2I_2$, showing *cis*-I, *cis*-CO and *trans*-P groups.

39. De Beer, J. A., Haines, R. J., Greatrex, R., and Greenwood, N. N., *J. chem. Soc. (A)* (1971), 3271. The reaction of $[Fe(CO)_3SMe]_2$ with PPh_3 and PEt_3; i.r. and Mössbauer spectra of the products.

40. De Beer, J. A., and Haines, R. J., *J. organometal. Chem.*, **36** (1972), 297. The reaction of $[Fe(CO)_3SPh]_2$ with tertiary phosphines to give mono-, di-, or tri-substitution products.

41. De Beer, J. A., and Haines, R. J., *J. organometal. Chem.*, **37** (1972), 173. The reaction of $[Fe(CO)_3SBu^n]_2$ with tertiary phosphines.

42. Mann, B. E., *Chem. Commun.* (1971), 1173. The ^{57}Fe-^{31}P coupling constants in iron carbonyl–phosphine complexes.

43. Johnson, B. F. G., and Segal, J. A., *J. chem. Soc. (Dalton)* (1972), 1268. The preparation of $[Fe(CO)_2(NO)(PR_3)_2]PF_6$ and $[Fe(CO)(NO)(PR_3)_3]^+$.

44. Crow, J. R., Cullen, W. R., Herring, F. G., Sams, J. R., and Tapping, R. L., *Inorg. Chem.*, **10** (1971), 1616. Mössbauer data on $Fe(NO)_2L_2$ (L = PPh_3, PPh_2Me) and $Fe(NO)_2(PPh_3)X$ (X = Br, I).

45. Bianco, V. D., Doronzo, S., and Rossi, M., *J. organometal. Chem.*, **35** (1972), 337. The insertion of CO_2 into Fe—H bonds in $FeH_4(PPh_2Et)_3$ and $FeH_2(N_2)(PPh_2Et)_3$ to give the diformate complex $Fe(OOCH)_2(PPh_2Et)_2$.

46. Aresta, M., Giannoccarro, P., Rossi, M., and Sacco, A., *Inorg. Chim. Acta*, **5** (1971), 203. The preparation and properties of $FeH_2(N_2)(PR_3)_3$ (R_3 = Ph_2Me, Ph_2Et, Ph_2Bu), and the formation of $FeH_2(CO)(PPh_2Et)_3$ from $FeH_2(N_2)(PPh_2Et)_3$ and CO is described.

47. Bradford, C. W., Nyholm, R. S., Gainsford, G. J., Guss, J. M., Ireland, P. R., and Mason, R., *Chem. Commun.* (1971), 87. Reaction of $Os_3(CO)_{12}$ with PPh_3 in 1:2 ratio in xylene solution, producing $Os_3(CO)_{12-n}(PPh_3)_n$, three hydrido complexes, and three complexes containing phosphine moieties which have lost phenyl groups. X-ray structures of $HOs_3(CO)_9(PPh_3)(Ph_2PC_6H_4)_7$, $Os_3(CO)_7(PPh_2)_2(C_6H_4)$ and $Os_3(CO)_8(PPh_2)(Ph)(PPhC_6H_4)$.

48. Cavit, B. E., Grundy, K. R., and Roper, W. R., *Chem. Commun.* (1972), 60. The preparation of $M(CO)_2(PPh_3)_2$ (M = Ru, Os) and of $Os(O_2)(CO)_2(PPh_3)_2$.

49. Deeming, A. J., Nyholm, R. S., and Underhill, M., *Chem. Commun.* (1972), 224. The formation of $Os_3(CO)_{12-n}(PPhMe_2)_n$ (n = 1–3) and their conversion into $Os_3H(C_6H_4)(PMe_2)(CO)_9$ and $Os_3(C_6H_4)(PMe_2)_2(CO)_7$.

50. Piacenti, F., Bianchi, M., Frediani, P., and Benedetti, E., *Inorg. Chem.*, **10** (1971), 2759. The reaction of $H_4Ru_4(CO)_{12}$ with PBu^n_3 and PPh_3, producing $H_4Ru_4(CO)_{12-n}(PR_3)n$($n$ = 1–4).

51. Moers, F. G., and Langhout, J. P., *Recl Trav. chim. Pays-Bas Belg.*, **91** (1972), 591. The preparation of PCy_3 complexes of $MH(CO)(PCy_3)_2Cl$ (M = Ru, Os) from the reaction of $(NH_4)_2OsCl_6$ or $RuCl_3 . 3H_2O$ in 2-methoxyethanol with PCy_3.

52. Christian, D. F., and Roper, W. R., *Chem. Commun.* (1971), 1271. Preparation of $Ru(CO)_2(CNR)(PPh_3)_2$, $Ru(CO)(CNR)(PPh_3)_2$ and $Ru(O_2)(CO)(CNR)(PPh_3)_2$.

53. Christian, D. F., Clark, G. R., Roper, W. R., Waters, J. M., and Whittle, K. R., *Chem. Commun.* (1972), 458. The reaction of $Ru(O_2)(CO)(CNR)(PPh_3)_2$ with ethanol to give the novel $Ru(CO)(HCNR)(CH_3COO)(PPh_3)_2$.

54. Clark, G. R., Grundy, K. R., Roper, W. R., Waters, J. M. and Whittle, K. R., *Chem. Commun.* (1972), 119. The preparation of $[Os(NO)(CO)_2(PPh_3)_2]Y$ (Y = ClO_4, PF_6, BF_4) from $[OsH(CO)(NO)(PPh_3)_2]$, CO and HY. The cation is trigonal bipyramidal with axial phosphines and a linear NO group.

55. Poddar, R. K., and Agarivala, U., *Indian J. Chem.*, **9** (1971), 477. The reaction of $RuCl_3 . 3H_2O$ with various alcohols in the presence of PPh_3, leading to $Ru(PPh_3)_2Cl_2$; and Ru(II) carbonyl complexes.

56. Robinson, S. D. and Uttley, M. F., *J. chem. Soc. (Dalton)* (1972), 1. A convenient synthesis for some complexes of the type $[M(NO)(PR_3)_2X_3]$ (M = Ru, Os; X = Cl, Br, I; R = alkyl, aryl, mixed alkyl–aryl).

57. Graham, B. W., Laing, K. R., O'Connor, C. J., and Roper, W. R., *J. chem. Soc. (Dalton)* (1972), 1237. The oxidation of PPh_3 to $OPPh_3$ using $Ru(NCS)(NO)(PPh_3)_2L$ (L = CO, O_2) catalysts.

58. Pierpont, C. G., and Eisenberg, R., *Inorg. Chem.*, **11** (1972), 1088. The X-ray structure of $[Ru(NO)_2(PPh_3)_2Cl]PF_6 . C_6H_6$, showing the cation to be square pyramidal with an apical bent NO group and a basal linear NO group.

59. Pierpont, C. G., and Eisenberg, R., *Inorg. Chem.*, **11** (1972), 1094. The X-ray structure of $RuH(NO)(PPh_3)_3$ shows it to be trigonal bipyramidal with axial H and an axial linear NO group.

60. Prater, B. E., *J. organometal. Chem.*, **34** (1972), 379. The preparation of $Ru(EtCN)_2(PPh_3)_2X_2$ (X = Cl, Br), and an indication that *cis* and *trans* nitriles are present.

61. Cenini, S., Fusi, A., and Capparella, G., *Inorg. nucl. Chem. Lett.*, **8** (1972), 127. The reactivity of $Ru(PPh_3)_3Cl_2$ towards CO, NO, SO_2, O_2 and olefins.

62. Ewing, D. F., Hudson, B., Webster, D. E. and Wells, P. B., *J. chem. Soc. (Dalton)* (1972), 1287. Olefin isomerisation catalysed by $RuH(PPh_3)_3Cl$.

63. Switkes, E. S., Ruiz-Ramirez, L., Stephenson, T. A., and Sinclair, J., *Inorg. nucl. Chem. Lett.*, **8** (1972), 593. The preparation of some new mixed-ligand Ru(II) and Ru(III) phosphine complexes.

64. Noth, W. H., *J. Am. chem. Soc.*, **94** (1972), 104. The reaction of $RuHCl(PPh_3)_3$ with $AlEt_3$ and N_2 to give $Ru(N_2)H_2(PPh_3)_3$. The reported $RuH_2(PPh_3)_3$ is suggested to be $RuH_4(PPh_3)_3$.

65. Ahmad, N., Robinson, S. D., and Uttley, M. F., *J. chem. Soc. (Dalton)* (1972), 843. Improved syntheses of some PPh_3 complexes of Ru, Os, Rh, Ir.

66. Sanders, J. R., *J. chem. Soc. (A)* (1971), 2991. The preparation of diphenylphosphine complexes: *cis*- and *trans*-$[M(PPh_2H)_4Cl_2]$ (M = Ru, Os), *cis*-$[Re(PPh_2)(PPh_2H)_3Cl_2]$ and salts of $[Rh(PPh_2H)_4]^+$, $[Ir(CO)(PPh_2H)_4]^+$, $[RuH(PPh_2H)_5]^+$, *trans*-$[RuH(CO)(PPh_2H)_4]^+$ and *cis*-$[Os(PPh_2H)_4H_2]$.

Appendix

Co, Rh, Ir (see also refs. 30, 31, 33, 65, 66)

67. Pregaglia, G. F., Andretta, A., Ferrari, G. F., Montrasi, G., and Ugo, R., *J. organometal. Chem.*, **33** (1971), 73. The preparation and catalytic activity of a series of compounds of type $[Co(CO)_2PR_3]_3$ ($R_3 = Ph_3$, $PhBu^n{}_2$, $Bu^n{}_3$).

68. Newman, J., and Manning, A. R., *J. chem. Soc. (Dalton)* (1972), 241. The preparation of $Hg[Co(CO)_2(PR_3)_2]_2$ from $Hg[Co(CO)_4]_2$ and a range of tertiary phosphines.

69. Bercaw, J., Gaustalla, G., and Halpern, J., *Chem. Commun.* (1971), 1594. The reaction of phosphines with $[Co(CO)_2(CN)_3]^-$ affording $[Co(CN)(CO)_2(PR_3)_2]$, $[Co(CN)_2(CO)_2(PR_3)]^-$ $[Co(CN)_2(CO)(PR_3)_2]^-$.

70. Ogino, K., and Brown, T. L., *Inorg. Chem.*, **10** (1971), 517. The reaction of $Cl_3MCo(CO)_4$ (M = Si, Ge, Sn) with $PBu^n{}_3$ and PPh_3.

71. Ward, D. L., Caughlin, C. N., Voecks, G. E., and Jennings, P. W., *Acta crystallogr.*, **B28** (1972), 1949. The crystal structure of $[Co(CO)_2(NO)(PPh_3)]$.

72. Klein, H. F., *Angew. Chem. int. Edn*, **10** (1971), 343. The preparation of $Co(PMe_3)_4$ and $Co(NO)(PMe_3)_3$.

73. Matheson, T. W., Robinson, B. H., and Tham, W. S., *J. chem. Soc. (A)* (1971), 1437. The preparation of mono-, di-, and trisubstituted tertiary phosphine complexes of methinyltricobaltenneacarbonyls.

74. Abbano, V. G., Bellon, P. L., and Ciani, G., *J. organometal Chem.*, **38** (1972), 155. The structures of $Co(CO)_2(NO)(PPh_3)$ and $Co(CO)(NO)(PPh_3)_2$, both containing cobalt in a distorted tetrahedral environment.

75. Tayim, H. A., Thabet, S. K., and Karkarawi, M. U., *Inorg. nucl. Chem. Lett.*, **8** (1972), 235. Facile solution syntheses of $Cu_2Cl_2(PPh_3)_3$ and $CoCl_2(PPh_3)_2$.

76. Dammann, C. B., Singh, P., and Hodgson, D. J., *Chem. Commun.* (1972), 586. The structure of $[(PPh_3)_2Rh(CO)]_2 \cdot 2CH_2Cl_2$, showing the presence of bridging carbonyls and a Rh—Rh bond (263 pm).

77. Gallay, J., De Montauzon, D., and Poilblanc, R., *J. organometal. Chem.*, **38** (1972), 179. A detailed study of the systems $[Rh(CO)_2Cl]_2–PR_3\{PR_3 = PPh_3, PPhMe_2, P(NMe_2)_3\}$, in which evidence was found for $Rh(CO)_{3-n}Cl(PR_3)_n$ ($n = 1$–3) and $Rh_2Cl_2(CO)_{4-n}(PR_3)_n$ ($n = 0$–4). Some reactions of the dinuclear complexes are reported.

78. Intille, G. M., *Inorg. Chem.*, **11** (1972), 695. A comprehensive study of phosphine complexes of rhodium, including the preparation of thirty-eight new complexes of types $RhLCl_3$, RhL_3Cl, RhL_3HCl_2, RhL_3H_2Cl, RhL_2HCl, $Rh(CO)L_2Cl$, $Rh(CO)L_2Cl_3$, $Rh(CO)L_2HCl_2$, and the dimeric $[Rh_2Cl_6L_4]$ (L = wide range of tertiary phosphines).

79. Vaska, L., Millar, W. V., and Flynn, B. R., *Chem. Commun.* (1971), 1615. Synthesis of *trans*-$[M(CO)(PR_3)_2L]$ (M = Rh, Ir; R = Ph, Cy; L = BH_3CN^-, $Bh_4{}^-$).

80. Intille, G. M., and Braithwaite, M. J., *J. chem. Soc. (Dalton)* (1972), 645. Preparation of Rh(I) and Ir(I) organomercury complexes containing PPh_3.

81. Hall, D. I. and Nyholm, R. S., *J. chem. Soc. (Dalton)* (1972), 804. The preparation of Rh(I) complexs of tris(*o*-vinylphenyl)phosphine and bis(*o*-vinylphenyl)phenylphosphine of type RhLX.

82. Arai, H., and Halpern, J., *Chem. Commun.* (1971), 1571. A spectrophotometric study of the dissociation of $Rh(PPh_3)_3Cl$ in solution.

83. Meakin, P., Jesson, J. P., and Tolman, C. A., *J. Am. chem. Soc.*, **94** (1972), 3240. Further studies of the nature of $Rh(PPh_3)_3Cl$ in solution and its reaction with H_2 to give $RhH_2Cl(PPh_3)_3$.

84. Masters, C., and Shaw, B. L., *J. chem. Soc. (A)* (1971), 3679. The preparation of para-magnetic Rh(II) complexes, *trans*-$[Rh(PBu^t{}_2R)_2Cl_2]$; rhodium(III) hydridocomplexes, $RhHCl_2(PBu^tR_2)_2$, $RhHCl_2(PBu^t{}_2R)_2$, $RhH_2Cl(PBu^t{}_2Me)_2$; Rh(I) carbonyl complexes are also described.

85. Mitchell, R. W., Ruddick, J. D., and Wilkinson, G., *J. chem. Soc. (A)* (1971), 3224. The reaction of Rh_2^{4+} with PPh_3, alkali metal carboxylates and dithiocarbamates.

86. Coulton, K. G., and Cotton, F. A., *J. Am. chem. Soc.*, 93 (1971), 1915. The X-ray structure of $[Rh(PPh_3)(DMG)]_2$ (DMG = monoanion of dimethylglyoxime).

87. Mann, B. E., Masters, C., and Shaw, B. L., *J. chem. Soc. (Dalton)* (1972), 704. The ^{31}P n.m.r. spectra of a large range of complexes of types *mer*-$Rh(PR_3)_3Cl_3$, *mer*-$Ir(PR_3)_3X_3$ and $Ir(PR_3)_3Cl_2X$ (X = Cl, Br, I).

88. Malatesta, L., Angoletta, M., and Conti, F., *J. organometal. Chem.*, 33 (1971), C43. The synthesis of $Ir_2(CO)_6(PPh_3)_2$, $HIr(CO)_3(PPh_3)$ and $H_3Ir(CO)_2(PPh_3)$.

89. Shaw, B. L., and Stainbank, R. E., *J. chem. Soc. (A)* (1971), 3716. A study of the importance of steric factors in the oxidative addition of acids to *trans*-$Ir(CO)(PBu^tR_2)_2Cl$ and *trans*-$Ir(CO)(PBu^t_2R)_2$. New Ir(III) hydrido complexes are also described.

90. Shaw, B. L., and Stainbank, R. E., *J. chem. Soc. (Dalton)* (1972), 233. A study of the effect of bulky *t*-butyl and di-*t*-butylphosphines (L) upon oxidative addition reactions of *trans*-$Ir(CO)L_2Cl$. The preparation of $[Ir(CO)_3L_2]^+$.

91. Wickman, H. H., and Silverthorn, W. E., *Inorg. Chem.*, 10 (1971), 2333. A Mössbauer study of some adducts of *trans*-$Ir(CO)(PPh_3)_2Cl$.

92. Cash, D. N., and Harris, R. O., *Can. J. Chem.*, 49 (1971), 3821. The reaction of *trans*-$Ir(CO)(PPh_3)_2Cl$ with $AgNO_3$ to give $[Ir(CO)(PPh_3)_2NO_3]$ and $[Ir(CO)(PPh_3)_2(NO_3)_2Ag]$, and some reactions of the products.

93. Colburn, C. B., Hill, W. E., and Sharp, D. W. A., *Inorg. nucl. Chem. Lett.*, 8 (1972), 625. The reaction of $Ir(CO)(PPh_3)_2Cl$ and $Ir(CO)(PPh_2Me)Cl$ with PF_2X (X = O, S).

94. Vaska, L., *Inorg. Chim. Acta*, 5 (1971), 295. A discussion of the rôle of the metal in some reversible addition reactions of small molecules and d^8 metal ions.

95. Chen, J. Y., and Halpern, J., *J. Am. chem. Soc.*, 91 (1971), 4939. The addition of small amounts of $PPhMe_2$ to $Ir(CO)(PPhMe_2)Cl$ increases the rate of reaction with H_2; the intermediate $[Ir(CO)(PPhMe_2)_3]^+$ is suggested.

96. Masters, C., Shaw, B. L., and Stainbank, R. E., *J. chem. Soc. (Dalton)* (1972), 664. Iridium complexes of *t*-butyl- and di-*t*-butylphosphines, including the $[IrHCl_2(PBu^t_2R)_2]$ complexes which have τ values of about 6.0; and the preparation of $[PBu^t_2Pr^nH][Ir_2Cl_7(PBu^t_2Pr^n)_2]$.

97. Mann, B. E., Masters, C., and Shaw, B. L., *J. chem. Soc. (Dalton)* (1972), 48. A ^{31}P INDOR study of $Ir(PPhMe_2)(Ph_2PCH_2CH_2PPh_2)Cl_3$ and some related complexes.

98. Reed, C. A., and Roper, W. R., *J. chem. Soc. (Dalton)* (1972), 1243. The reaction of $[Ir(NO)(PPh_3)_2Cl_3]^+$ with alcohols to give alkyl nitrite complexes of Ir(III). New routes to $[Ir(NO)_2L_2]^+$ cations.

99. Mays, M. J., and Stephanini, F. P., *J. chem. Soc. (A)* (1971), 2747. The preparation of some five-coordinate thiocarbonyl complexes of iridium.

100. Brookes, P. R., Masters, C., and Shaw, B. L., *J. chem. Soc. (A)* (1971), 3756. An important paper describing the photochemical isomerisation of Ir(III) complexes of type *mer*-$Ir(PR_3)_3X_3$, $Ir(PR_3)_3HX_2$ and $Ir(CO)(PR_3)_2X_3$. Especially interesting is the conversion of *mer*-$Ir(PR_3)_3X_3$ into *fac*-$Ir(PR_3)_3X_3$ in high yield. *Fac* isomers are difficult to prepare by other routes.

101. Smith, S. A., Blake, D. M., and Kubota, M., *Inorg. Chem.*, 11 (1972), 660. The preparation of some iridium carboxylate complexes.

Ni, Pd, Pt (see also refs. 21, 30)

102. Kang, D. K., and Burg, A. B., *Inorg. Chem.*, 11 (1972), 802. The preparation of $LNi(CO)_3$ (L = $P(CF_3)_3$, $P(CF_3)_2Me$, $P(CF_3)_2Et$, $P(CF_3)_2Pr^i$, and $P(CF_3)Et_2$).

103. Pánkowski, M., and Bigorgne, M., *J. organometal. Chem.*, **35** (1971), 397. The preparation of the first carbonyl halide of nickel, $Ni(CO)(PMe_3)_2I_2$, which is probably trigonal bipyramidal with axial phosphines.

104. Jolly, P. W., Jonas, K., Krüger, C., and Tsay, Y-H, *J. organometal. Chem.*, **33** (1971), 109. The preparation and structure of $\{Ni(PCy_3)_2\}_2(N_2)$. The dinitrogen is readily displaced to give $Ni(PCy_3)_2$, some reactions of which are described.

105. Bachman, D. F., Stevens, E. D., Lane, T. A., and Yoke, J. T., *Inorg. Chem.*, **11** (1972), 109. The methanolysis of $Ni(PCl_3)_4$ and a study of the reaction products.

106. Weston, C. W., Bailey, G. W., Nelson, J. H., and Jonassen, H. B., *J. inorg. nucl. Chem.*, **34** (1972), 1755. The characterisation of $Ni(PPh_2H)_4$ and $Pd(PPh_2H)_4$, confirming that '$Ni(PPh_2)_2(PPh_2H)_2$' is, in fact, the former. Thermal analysis data are reported.

107. Uhlig, E., and Walther, H., *Z. Chem.*, **11**(1971), 26. The preparation of NiL_4 ($L = PPh_2Bu^n$, $PPhBu^n_2$, PBu^n_3) from nickelocene.

108. Tolman, C. A., Seidel, W. C., and Gerlach, D. H., *J. Am. chem. Soc.*, **94** (1972), 2669. A study of zerovalent phosphine complexes of Ni, Pd and Pt in solution. The $M(PR_3)_4$ complexes are substantially dissociated at ambient temperatures into PR_3 and $M(PR_3)_3$.

109. Merle, A., Obier, M. F., Dartinguenave, M., and Dartinguenave, Y., *C.r. hebd. Séanc. Acad. Sci.*, Paris, **C272** (1971), 1956. The preparation of $Ni(PMe_3)_3(NCS)_2$.

110. Merle, A., Dartinguenave, M., and Dartinguenave, Y., *C.r. hebd. Séanc. Acad. Sci.*, Paris, **C272** (1971), 2046. The preparation of $Ni(PMe_3)_3(NO_2)_2$.

111. Merle, A., Dartinguenave, M., and Dartinguenave, Y., *Bull. Soc. chim. Fr.* (1972), 87. Electronic spectra of $Ni(PMe_3)_3X_2$ ($X = NCS, NO_2$).

112. Powell, H. M., and Chiu, K. M., *Chem. Commun.* (1971), 1037. The structures of ML_3X_2 ($M = Ni, Pd, Pt; L = $ 5-alkyl-5H-dibenzophosphole; $X = Br, CN$). Powell, H. M., Watkins, D. J., and Wilford, J. B., *J. chem. Soc. (A)* (1971), 1803. The crystal and molecular structures of two forms of dicyano-(5-methyl-5H-dibenzophosphole)-nickel(II) and of dicyano-(5-ethyl-5H-dibenzophosphole)nickel(II). Tetragonal-pyramidal and trigonal-bipyramidal molecules.

113. Rigo, P., and Bressan, M., *Inorg. Chem.*, **11** (1972), 1314. The preparation of diamagnetic four- and five coordinate Ni(II) complexes of diethylphosphine, $[Ni(PHEt_2)_3X]^+$ ($X = Cl, Br, I, NCS$), $[Ni(PHEt_2)_3X_2]$ ($X = Br, I$) and $[Ni(PHEt_2)_4X]^+$ ($X = Cl, Br, I, NCS$). The previously reported six-coordinate complexes, $[Ni(PHEt_2)_4X_2]$ are reformulated $[Ni(PHEt_2)_4X]X$.

114. Kudo, K., Hidai, M., and Uchida, Y., *J. organometal. Chem.*, **33** (1971), 393. The synthesis of $Pd(CO)(PPh_3)_3$ from $Pd(acac)_2$ and $AlEt_3$, or from $Pd(PPh_3)_2Cl$ and $NaBH_4$, in the presence of CO and PPh_3. The preparation of $Pd_3(CO)_3(PPh_3)_3$ and $Pd_3(CO)_3(PPh_3)_4$ are also reported.

115. Cenini, S., Ugo, R., La Monica, G., and Robinson, S. D., *Inorg. chim. Acta*, **6** (1972), 182. The reaction of $Pt(PPh_3)_4$ and $Pt(PPh_3)_3$ with nitric oxide to give the hyponitrite derivative $Pt(PPh_3)_2(N_2O_2)$.

116. Cheng, P. T., Cook, C. D., Nyburg, S. C., and Wan, K. Y., *Can. J. Chem.*, **49** (1972), 3772. The X-ray crystal structure of $(Ph_3P)_2Pt(O_2) \cdot 2CHCl_3$, with O—O = 150.5 pm.

117. Fink, W., and Wenger, A., *Helv. chim. Acta*, **54** (1971), 2186. The preparation of Pt–Si compounds from $Pt(PPh_3)_4$ and $MeHSiCl_2$.

118. Durkin, T. R., and Schramm, E. P., *Inorg. Chem.*, **11** (1972), 1054. The reaction of $Pt(PPh_3)_4$ and $Pt(PPh_3)_3$ with Al_2Me_6 and BCl_3.

119. Durkin, T. R., and Schramm, E. P., *Inorg. Chem.*, **11** (1972), 1048. The formation of $Pt(PPh_3)_2SiF_4$ from $Pt(PPh_3)_4$ and SiF_4.

120. Mason, R., Zubieta, J., Hsieh, A. T. T., Knight, J., and Mays, M. J., *Chem. Commun.* (1972), 200. The structure of $[(Ph_3P)(OC)Fe_2Pt(CO)_8]$, one of the products from the reaction of $Pt(PPh_3)_4$ and $Fe_3(CO)_{12}$.

121. Bruce, M. I., Shaw, G., and Stone, F. G. A., *J. chem. Soc. (Dalton)* (1972), 1082. The preparation of $Fe_2Pt(CO)_9(PR_3)$ and $Fe_2Pt(CO)_8(PR_3)_2$ (PR_3 = various phosphines) from $Fe_2(CO)_9$ and $Pt(PR_3)_4$.

122. Nakamura, A., Tatsuno, Y., Yamamoto, M., and Otsuka, S., *J. Am. chem. Soc.*, 93 (1971), 6052. An ^{18}O isotopic i.r. study of $Pt(O_2)(PPh_3)_2$ and $[Rh(O_2)(PPh_3)_2(Bu^tCN)X]$.

123. Albano, V. G., Bellon, P. L., and Manassero, M., *J. organometal. Chem.*, 35 (1972), 423. The structure of $Pt(CO)_2(PPh_2Et)_2$ is approximately tetrahedral.

124. Otsuka, S., Tatsuno, Y., and Ataka, K., *J. Am. chem. Soc.*, 93 (1971), 6705. A report of a palladium(I) complex, $[Pd(PPh_3)(Bu^tCN)I]_2$.

125. Clark, H. C., and Kurosawa, H., *J. organometal. Chem.*, 36 (1972), 399. The synthesis of *trans*-$Pt(H)(PR_3)_2X$ by borohydride reduction of *cis*-$Pt(PR_3)_2X_2$ (R_3 = Ph_2Me, $PhMe_2$; X = Cl, Br, I, CN). Cationic hydrido complexes were also isolated.

126. Adlard, M. W., and Socrates, G., *J. chem. Soc. (Dalton)* (1972), 797. The preparation of *trans*-$PtHX(PR_3)_2$ (X = NCO, NCS, NCSe, CN). Only the thiocyanate complexes exhibit linkage isomerism.

127. Adlard, M. W., and Socrates, G., *Chem. Commun.* (1972), 17. Ligand exchange in solutions of *trans*-$PtH(PR_3)_2X$.

128. Beck, W., and Werner, K. V., *Chem. Ber.*, 104 (1971), 2901. The reaction of $(Ph_3P)_2M(NCO)_2$ (M = Pd, Pt) and $[(Ph_3P)_2PtX_2Pt(PPh_3)_2](BF_4)_2$ (X = N_3, NCO) with CO.

129. Cheney, A. J., and Shaw, B. L., *J. chem. Soc. (Dalton)* (1972), 754. Internal metallation of platinum complexes of $PBu^t_2(o\text{-}tolyl)$ and $PBu^t(o\text{-}tolyl)_2$.

130. Cheney, A. J., and Shaw, B. L., *J. chem. Soe. (Dalton)* (1972) 860. Internal metallation of *trans*-PdL_2Cl_2 {L = $PBu^t(o\text{-}tolyl)_2$ and $PBu^t_2(o\text{-}toly)$}

131. Mann, B. E., Shaw, B. L., and Slade, R. M., *J. chem. Soc.(A)* (1971), 2976. The preparation of $[Pd(PBu^t_2R)_2X_2]$, $[Pd(PBu^tR_2)_2X_2]$ and $[Pd_2(PBu^t_2R)_2X_4]$, and a study of the ^1H and ^{31}P n.m.r. spectra of these and other Pd(II) phosphine complexes.

132. Cheney, A. J., Mann, B. E., Shaw, B. L., and Slade, R. M., *J. chem. Soc. (A)* (1971), 3833. Internal metallation reactions of Pt(II) complexes of bulky phosphines, *trans*-$Pt(PR_3)_2X_2$ (R_3 = Bu^tR_2, Bu^t_2R, $Ph(o\text{-}tolyl)_2$, etc.).

133. Graziani, R., *Inorg. nucl. Chem. Lett.*, 8 (1972), 701. The X-ray crystal structure of *trans*-$Pt(PEt_3)_2(NO_2)_2$ shows the presence of N-bonded nitro groups.

134. Haszeldine, R. N., Lunt, R. J., and Parish, R. V., *J. chem. Soc. (A)* (1971), 3705. The preparation of Pd(II) and Pt(II) complexes of $CH_2=CH(CH_2)_nPPh_2$ (n = 1–3), and the reaction of some of these with nucleophiles.

135. Kumar, G., Blackburn, J. R., Albridge, R. G., Moddenan, W. E., and Jones, M. M., *Inorg. Chem.*, 11 (1972), 296. Photoelectron spectroscopy of $Pd(PPh_3)_2X_2$ (X = Cl, Br, I, CN).

136. Payne, D. H., and Frye, H., *Inorg. nucl. Chem. Lett.*, 8 (1972), 73. The preparation of some Pt(II) complexes of $Ph_2P(o\text{-}C_6H_4CN)$.

137. Haake, P., and Martin, S. M., *J. Am. chem. Soc.*, 93 (1971), 6823. The mass spectra of *cis*- and *trans*-$Pt(PR_3)_2X_2$ (X = Cl, Br, I; R = Ph, Et).

138. Treichel, M., Knebel, W. J., and Hess, R. W., *J. Am. chem. Soc.*, 93 (1971), 5424. The preparation of Pt(II) isocyanides of types $[Pt(MeCN)_2(PR_3)_2X]Y$ and $[Pt(MeCN)_2(PPh_3)_2]Y_2$, and some reactions.

139. Druce, P. M., Lappert, M. F., and Riley, P. N. K., *J. chem. Soc. (Dalton)* (1972), 438. The preparation of halogen-bridged dinuclear cations from the reaction of *cis*-$Pt(PBu^n_3)_2X_2$ with boron halides.

140. Cherwinski, W. J., and Clark, H. C., *Inorg. Chem.*, 10 (1971), 2263. The preparation of $Pt(PR_3)_2(X)Cl$ (X = anionic ligand; R_3 = Ph_3, Et_3, $PhMe_2$) from $[Pt_2Cl_2(PR_3)_4](BF_4)_2$, and their conversion into $[Pt(PR_3)_2(X)(CO)]BF_4$ and *trans*-$PtHX(PR_3)_2$.

141. Kikukawa, K., Yamane, T., Takagi, M., and Matsuda, T., *Chem. Commun.* (1972), 695. The arylation of olefins by $Pd(OAc)_2(PAr_3)$ (Ar = p-tolyl). The fate of dearylated phosphine is unknown.

142. Powell, J., and Jack, T., *Inorg. Chem.*, 11 (1972), Conformational studies of some carboxylate-bridged complexes of Pd and Pt, containing $PPhMe_2$ ligands.

Cu, Ag, Au

143. Zelonka, R. A., and Baird, M. C., *Can. J. Chem.*, 50 (1972), 1269. The preparation of a series of copper(II) complexes, $Cu(hfac)_2(PR_3)_2$ (hfac = hexafluoroacetylacetone).

144. Lippard, S. J., and Weckler, P. S., *Inorg. Chem.*, 11 (1972), 6. The preparation and properties of $CuL(PPh_3)_3$ and $AgL(PPh_3)_3$ (L = cyanotrihydroborate).

145. Lippard, S. J., and Mayerle, J. J., *Inorg. Chem.*, 11 (1972), 753. The equilibria in chloroform solutions of some Cu(I) complexes of PPh_3, PPh_2Me and $PPhMe_2$, of types $Cu(PR_3)_nX$ (n = 3, 2, $1\frac{1}{2}$; X = Cl, Br, I) are described. New complexes of these ligands are also prepared.

146. Albano, V. G., Bellon, P. L., Ciani, G., and Manassero, M., *J. chem. Soc. (Dalton)* (1972), 171. The X-ray crystal structure of $Cu_2Cl_2(PPh_3)_3$.

147. Mathew, M., Palenik, G. J., and Carty, A. J., *Can. J. Chem.*, 49 (1972), 4119. The crystal structure of $Cu(PPh_2Me)_3NO_3$. The nitrate group is monodentate, and there are three different Cu—P distances.

148. Ziolo, R. F., Thich, J. A., and Dori, Z., *Inorg. Chem.*, 11 (1972), 626. The preparation and some reactions of $M(PR_3)_nX$ {n = 1, M = Au; n = 2, M = Ag, Cu; R_3 = Ph_3, Ph_2Me, $Ph_2(o$-tolyl); X = N_3, NCS}.

149. Casey, M., and Manning, A. R., *J. chem. Soc. (A)* (1971), 2989. New complexes containing Au—Fe bonds: $(R_3P)AuFe(CO)_3(NO)$ and $(Ph_3P)AuFe(CO)_2(NO)(PPh_3)$.

150. Bellon, P. F., Cariati, F., Manassero, M., Naldini, L., and Sansoni, M., *Chem. Commun.* (1971), 1423. The preparation of $[Au_9L_8]X_3$ (L = tris-p-substituted phenylphosphines; X = NO_3, PF_6, picrate), and the crystal structure of $[Au_9\{P(p\text{-}MeC_6H_4)_3\}_8](PF_6)_3$.

151. Cariati, F., and Naldini, L., *Inorg. Chim. Acta*, 5 (1971), 172. The preparation of $Au_{11}(PR_3)_9X_3$ (R = p-ClC_6H_4, p-FC_6H_4, p-MeC_6H_4; X = I, SCN, CN) by reaction of $Au(PR_3)X$ with $NaBH_4$. The structure of $Au_{11}\{P(p\text{-}ClC_6H_4)_3\}_9I_3$ is described.

Zn, Cd, Hg

152. De Simone, R. E., and Stucky, G. D., *Inorg. Chem.*, 10 (1971), 1808. The X-ray crystal structure of $[4\text{-}MeC_5H_4NH][Zn(PPh_3)Br_3]$; the coordination about the zinc is distorted tetrahedral.

153. Forbes-Cameron, A., Forrest, K. P., and Ferguson, G., *J. chem. Soc. (A)* (1971), 1286. The structure of $Cd(PPh_3)_2Cl_2$, showing the tetrahedral environment about the cadmium.

154. Mann, B. E., *Inorg. nucl. Chem. Lett.*, 7 (1971), 595. The ^{111}Cd and ^{113}Cd cadmium–phosphorus coupling constants in $Cd(PR_3)_2I_2$.

155. Goggin, P. L., Goodfellow, R. J., Haddock, S. R., and Eary, J. G., *J. chem. Soc. (Dalton)* (1972), 647. Vibrational and ^1H n.m.r. spectra of $[Hg(PMe_3)_2X]^+$ (X = Cl, Br, I, CN, Me) and $[Hg(PMe_3)_2]^{2+}$.

PART 3

156. Mann, B. E., Shaw, B. L., and Stainbank, R. E., *Chem. Commun.* (1972), 151. The use of
^{13}C n.m.r. spectra to assign structures to tertiary arsine complexes.

157. Adlard, M. W., and Socrates, G., *Chem. Commun.* (1972), 17. Ligand exchange in *trans*-
Pt(AsEt$_3$)$_2$HX (X = NCO, NCS) in solution.

158. Hall, D. I., and Nyholm, R. S., *J. chem. Soc. (Dalton)* (1972), 804. The preparation of
Rh(I) complexes of tris(*o*-vinylphenyl)arsine.

159. Adlard, M. W., and Socrates, G., *J. Chem. Soc. (Dalton)* (1972), 797. Of the *trans*-
Pt(AsEt$_3$)$_2$HX (X = NCO, NCS, CN, NCSe) only the thiocyanate complex exhibits linkage
isomerism.

160. Butcher, A. V., Chatt, J., Leigh, G. J., and Richards, P. L. *J. chem. Soc. (Dalton)* (1972),
1064. The preparation of WCl$_4$L$_2$ (L = AsPhMe$_2$, AsPh$_2$Me).

161. Bruce, M. I., Shaw, G., and Stone, F. G. A., *J. chem. Soc. (Dalton)* (1972), 1082. The
reaction of Pt(AsPh$_3$)$_4$ with Fe$_2$(CO)$_9$ giving Fe$_2$Pt(CO)$_9$(AsPh$_3$).

162. Johnson, B. F. G., and Segal, J. A., *J. chem. Soc. (Dalton)* (1972), 1268. The preparation
of [Fe(CO)$_2$NO(AsPh$_3$)$_2$]PF$_6$.

163. Brown, R. A., and Dobson, G. R., *J. inorg. nucl. Chem.*, **33** (1971), 892; *Inorg. Chim. Acta*,
6 (1972), 65. The preparation and i.r. spectra of [M(CO)$_5$L] (M = Cr, W; L = AsPh$_3$, SbPh$_3$,
BiPh$_3$, SbBu$_3$).

164. Newman, J., and Manning, A. R., *J. chem. Soc. (Dalton)* (1972), 241. The preparation of
Hg[Co(CO)$_3$(AsPh$_2$Me)]$_2$ and (MePh$_2$As)CoHgBr . Me$_2$CO.

165. Robinson, S. D., and Uttley, M. F., *J. chem. Soc. (Dalton)* (1972), 1. A convenient
synthesis of M(NO)(AsPh$_3$)$_2$X$_3$ (M = Ru, Os; X = Cl, Br, I) using N-methyl-N-nitrosotoluene-*p*-
sulphonamide or pentyl nitrite.

166. Khan, M. M. T., Andal, R. K., and Manoharan, P. T., *Chem. Commun.* (1971), 561. The
activation of O$_2$, H$_2$, CO and olefins by Ru(AsPh$_3$)$_3$Cl$_2$ in benzene solution.

167. de Beer, J. A., Haines, R. J., Greatrex, R., and Greenwood, N. N., *J. chem. Soc. (A)* (1971),
3271. The reaction of [Fe(CO)$_3$SMe]$_2$ with AsPh$_3$ and SbPh$_3$ giving mono-, di-, and trisubstituted products.

168. Brooks, P. R., Masters, C., and Shaw, B. L., *J. chem. Soc. (A)* (1971), 3756. The photochemical reaction of *mer*-Ir(AsPhMe$_2$)$_3$Cl$_3$ with PPhMe$_2$ to give Ir(AsPhMe$_2$)$_2$(PPhMe$_2$)Cl$_3$.
The Ir(CO)(AsR$_3$)$_2$Cl$_3$ (R$_3$ = PhMe$_2$, Et$_3$) complexes are also described.

169. Gunz, H. P., and Leigh, G. J., *J. chem. Soc. (A)* (1971), 2229. The preparation and
properties of *mer*-Re(AsR$_3$)$_3$X$_3$ (R$_3$ = PhMe$_2$, PhEt$_2$; X = Cl, Br) and *trans*-Os(AsPhMe$_2$)$_2$Cl$_4$.

170. Shaw, B. L., and Slade, R. M., *J. chem. Soc. (A)* (1971), 1184. The preparation of
Ir(AsPhMe$_2$)$_3$X$_2$(NO$_3$) (X = halogen). The nitrate is readily displaced by neutral ligands to give
cationic complexes.

171. Zelonka, R. A., and Baird, M. C., *Can. J. Chem.*, **50** (1972), 1269. The preparation of
Cu(hfac)$_2$(AsPh$_3$)$_n$ (hfac = hexafluoroacetylacetone; n = 1, 2).

172. Tayim, H. A., Thabet, S. K., and Karkanawi, M. U., *Inorg. nucl. Chem. Lett.*, **8** (1972),
235. The synthesis of Cu(AsPh$_3$)$_3$Cl.

173. Switkes, E. S., Ruiz-Ramirez, L., Stephenson, T. A., and Sinclair, J., *Inorg. nucl. Chem.
Lett.*, **8** (1972), 593. The preparation of Ru(AsPh$_3$)$_2$(S)X$_3$ (X = halogen) and its reactions, in
which the solvent molecule (S) is displaced to give Ru(III) complexes, Ru(AsPh$_3$)$_2$X$_3$L, and in
which both (S) and one AsPh$_3$ are replaced to give Ru(AsPh$_3$)X$_3$L$_2$.

174. Bowen, L. H., Garrou, P. E., and Long, G. C., *Inorg. Chem.*, 11 (1972), 182. The [127]Sb and [57]Fe Mössbauer spectra of $(Ph_3Sb)Fe(CO)_4$ and $(Ph_3Sb)_2Fe(CO)_3$. The [57]Fe Mössbauer data is also reported for the $AsPh_3$ analogues.

175. Lippard, S. J., and Mayerle, J. J., *Inorg. Chem.*, 11 (1972), 753. The equilibria in chloroform solutions of $[Cu(AsPh_{3-n}Me_n)_mX]$ (n = 0, 1, 2; m = 3, 2, 1.5; X = halogen) are reported. The trisarsine complexes are less stable than the phosphine analogues.

176. Powell, J. and Jack, T., *Inorg. Chem.*, 11 (1972), 1034. Carboxylate-bridged complexes of Pd and Pt containing $AsPhMe_2$.

177. Vahrenkamp, H., *Chem. Ber.*, 105 (1972), 1486. The structure of $(OC)_5Cr-AsMe_2-Mn(CO)_5$.

178. Chatt, J., Leigh, G. J., and Thankajan, N., *J. organometal. Chem.*, 29 (1971), 105. Improved syntheses of some $AsPh_3$, $AsEt_3$ and $SbPh_3$ (L) complexes derived from $M(CO)_6$ (M = Cr, Mo, W). New complexes, *cis*-$W(CO)_4L$, are also described.

179. Grobe, J., and Kober, F., *J. organometal. Chem.*, 29 (1971), 295. The reactions of $Mn_2(CO)_{10}$ with Me_2AsX leading to $Mn_2(CO)_8AsMe_2X$ (X = I) and to polymeric products when X = Cl, Br. New complexes containing $As(CF_3)_2$, $P(CF_3)_2$, $AsMe_2$ and PMe_2 bridges are described; as are substitution reactions of $Mn_2(CO)_8AsMe_2I$.

180. Bennett, M. A., Hoskins, K., Kneen, W. R., Nyholm, R. S., Hitchcock, P. B., Mason, R., Robertson, G. B., and Towl, A. D. C., *J. Am. chem. Soc.*, 93 (1971), 4593. The bromination of

$Pt(o\text{-}Me_2AsC_6H_4CH{=}CH_2)_2Br_2$ giving $\overline{PtBr_3[o\text{-}Me_2AsC_6H_4CH(CH_2Br)]}[o\text{-}Me_2AsC_6H_4CH{=}CH_2]$.

181. De Stefano, N. J., and Burmeister, J. L., *Inorg. Chem.*, 10 (1971), 998. The bonding of pseudohalide ions in Rh(I), Ir(I), Au(I) and Au(III) complexes containing triphenylarsine.

182. King, R. B., and Korenowski, T. F., *Inorg. Chem.*, 10 (1971), 1188. Complexes of tris-(dimethylamino)arsine with the carbonyls of Cr, Mo, W and Fe.

183. Ehrl, W., and Vahrenkamp, H., *Chem. Ber.*, 104 (1971), 3261. The synthesis of $(OC)_5M-AsMe_2-M'(CO)_5$ (M = Cr, Mo, W; M' = Mn, Re) from $(OC)_5MAsMe_2Cl$ and $NaM'(CO)_5$.

184. Fischer, E. O., Bathelt, W., and Müller, J., *Chem. Ber.*, 104 (1971), 986. The preparation, ^1H n.m.r., i.r. and mass spectra of $M(CO)_5(SbH_3)$ (M = Cr, Mo, W).

185. Davison, A., and Ellis, J. E., *J. organometal. Chem.*, 31 (1971), 239. The preparation of $[NEt_4][V(CO)_5L]$ (L = $AsPh_3$, $SbPh_3$).

186. Prater, B. E., *J. organometal. Chem.*, 34 (1972), 379. The preparation of $Ru(EtCN)_2L_2X_2$ (L = $AsPh_3$, $SbPh_3$; X = Cl, Br).

187. Thornhill, D. J., and Manning, A. R., *J. organometal. Chem.*, 37 (1972), C41. The tautomerism of $[(R_3M)Co(CO)_3]_2$ (R = alkyl; M = As, Sb). The presence of species containing bridging carbonyls, and those without bridging carbonyls in solution.

188. Bennett, R. L., Bruce, M. I., and Stone, F. G. A., *J. organometal. Chem.*, 38 (1972), 325. Benzyldimethylarsine complexes of Mn, Fe, Ru, Rh, Pd and Pt.

189. Mann, B. E., Masters, C., and Shaw, B. L., *J. inorg. nucl. Chem.*, 33 (1971), 2195. The preparation of $IrH_3(PR_3)_2L$ (L = $AsPhMe_2$, $SbPh_3$).

190. Goggin, P. L., Knight, R. J., Sindellari, L., and Venanzi, L. M., *Inorg. Chim. Acta*, 5 (1971), 62. A study of the electronic spectra of a large series of complexes of types $[M(AsMe_3)X_3]^-$, $[M(SbMe_3)X_3]^-$, *cis*-$[M(AsMe_3)_2Cl_2]$, *trans*-$[M(AsMe_3)_2I_2]$, *cis*-$[M(SbMe_3)_2Cl_2]$, *trans*-$[M(SbMe_3)_2I_2]$ (M = Pd, Pt; X = Cl, Br) and some R_3P analogues; and a discussion of the position of antimony ligands in the spectrochemical series.

191. Reichle, W. T., *Inorg. Chim. Acta*, 5 (1971), 325. The preparation, physical properties and reactions of $Cu(MPh_3)_3Cl$ (M = As, Sb).

PART 4

192. Cunninghame, R. G., Nyholm, R. S., and Tobe, M. L., *J. chem. Soc. (Dalton)* (1972), 229. The preparation of M(ttas)X$_2$ (ttas = bis-(*o*-dimethylarsinophenyl)-methylarsine; M = Ni; X = Cl, Br, I, SCN; M = Pd, Pt; X = Cl, Br, I). The Pd and Pt complexes are generally planar, 1:1 electrolytes, while the Ni complexes are five coordinate. In hydroxylic solvents [Ni(ttas)$_2$]$^{2+}$ is formed. The structures of M(ttas)$_2$Y$_2$ (M = Ni; Y = ClO$_4$, NO$_3$, I; M = Pd, Pt; Y = ClO$_4$) have not been definitely elucidated.

193. Sacconi, L., and Midollini, S., *J. chem. Soc. (Dalton)* (1972), 1213. The terdentate ligand 1,1,1-tris(diphenylphosphinomethyl)ethane forms cobalt(I) and nickel(I) complexes of formula MLX (X = Cl, Br, I). All the complexes are non-electrolytes, isomorphous with the analogous Cu(I) complexes and probably have a pseudotetrahedral structure.

194. Mealli, C., Orioli, P. L., and Sacconi, L., *J. chem. Soc. (Dalton)* (1972), 2691. The crystal structure of the low-spin [CoLI] I (L = tris(2-diphenylphosphinoethyl)amine) shows that in the cation the cobalt atom exists in a distorted square-pyramidal configuration with one phosphorus atom at the apex of the pyramid, the two others, together with the nitrogen and the iodine atoms, forming the base.

195. Dapporto, P., and Fallani, G., *J. chem. Soc. (Dalton)* (1972), 1498. The crystal structure of chloro{bis-[2-(diethylamino)ethyl]-2-(diphenylphosphino)ethylamine}cobalt(II)perchlorate shows the cobalt to exist in a remarkably distorted trigonal-bipyramidal N$_3$PCl donor set.

196. King, R. B., Kapoor, P. N., and Kapoor, R. N., *Inorg. Chem.,* **10** (1971), 1841. An important paper describing the reaction of a tritertiary phosphine bis(2-diphenylphosphino-ethyl)phenylphosphine with Cr, Mo, W, Mn, Fe, Co, Rh, Ir, Ni, Pd and Pt compounds.

197. King, R. B., Kapoor, R. N., Saran, M. S., and Kapoor, P. N., *Inorg. Chem.,* **10** (1971), 1851. An important paper describing the reactions of two isomeric tetratertiary phosphines, tris(2-diphenylphosphinoethyl)phosphine and hexaphenyl-1,4,7,10-tetraphosphadecane, with Cr, Mo, W, Mn, Fe, Co, Rh, Ir, Ni, Pd and Pt compounds.

198. King, R. B., and Saran, M. S., *Inorg. Chem.,* **10** (1971), 1861. The complexes formed between a hexatertiary phosphine 1,1,4,4-tetrakis(2-diphenylphosphinoethyl)-1,4-diphospha-butane with Cr, Mo, W, Mn, Rh, Ir, Ni, Pd and Pt compounds.

199. Siegl, W. O., Lapporte, S. J., and Collman, J. P., *Inorg. Chem.,* **10** (1971), 2158. Rhodium(I) and iridium(I) complexes of the tripodal ligand 1,1,1-tris-(diphenylphosphinomethyl)ethane, ML(CO)Cl. The rhodium complex is a mixture of four- and five-coordinate species; the iridium complex is five coordinate in the solid state.

200. DuBois, T. D., *Inorg. Chem.,* **11** (1972), 718. The tetradentate ligand 2,3-butanedionebis-(2-diphenylphosphinoethylimine) forms a four-coordinate, planar [NiL](ClO$_4$)$_2$ complex, and the five-coordinate square-pyramidal [NiLX]ClO$_4$ (X = Cl, Br, I) complexes.

201. Nappier, T. E., and Meek, D. W., *J. Am. chem. Soc.,* **94** (1972), 306. Five-coordinate rhodium(I) complexes containing small molecules and a chelating triphosphine ligand C$_6$H$_5$P{CH$_2$CH$_2$CH$_2$P(C$_6$H$_5$)$_2$}$_2$.

202. Dawson, J. W., Gray, H. B., Hix, J. E., Preer, J. R., and Venanzi, L. M., *J. Am. chem. Soc.,* **94** (1972), 2979. The solution and solid-state electronic spectra of some low-spin trigonal-bi-pyramidal complexes containing nickel(II), palladium(II) and platinum(II). The temperature dependence of the lowest energy ligand-field band is discussed.

203. Berglund, D., and Meek, D. W., *Inorg. Chem.,* **11** (1972), 1493. The preparation of [Ni(tpp)NO]X and [Ni(tep)NO]X {X = Cl, Br, I, BF$_4$, BPh$_4$; tpp = CH$_3$C(CH$_2$PPh$_2$)$_3$, tep = CH$_3$C(CH$_2$PEt$_2$)$_3$}. An X-ray structural determination of [Ni(tep)NO]BF$_4$ has shown

that the nickel atom is surrounded in a pseudotetrahedral fashion by three P atoms and a linearly bound nitrosyl group.

204. King, R. B., and Kapoor, P. N., *Inorg. Chem.,* **11** (1972), 1524. Some zerovalent platinum complexes containing a range of polytertiary phosphines and arsines.

205. King, R. B., *Accs chem. Res.,* **5** (1972), 177. An account of the recent studies on poly-tertiary phosphines and their metal complexes in King's laboratory.

206. Baracco, L., Halfpenny, M. T., and McAuliffe, C. A., *Chem. Commun.* (1971), 1502. The bis[tris-*o*-dimethylarsinophenyl)stibine]nickel(II) cation; a five-coordinate complex containing the novel [Ni(Sb$_2$As$_3$)]$^{2+}$ donor set.

207. Venanzi, L. M., Spagna, R., and Zambonelli, L., *Chem. Commun.* (1971), 1570. The preparation and structure of an unusual trigonal-bipyramidal complex [Ir(PPh$_3$)(QP)]BPh$_4$ (QP = tris-(*o*-diphenylphosphinophenyl)phosphine).

208. Dalton, J., Levason, W., and McAuliffe, C. A., *Inorg. nucl. Chem. Lett.,* **8** (1972), 797. Dimethyl(*o*-dimethylarsinophenyl)stibinebis(3-dimethylarsinopropyl)methylarsinenickel(II) bromide: a five-coordinate [Ni(bidentate)(tridentate)]$^{2+}$ cation containing the novel [Ni(SbAs$_4$)]$^{2+}$ donor set.

209. Morgan, T. D. B., and Tobe, M. L., *Inorg. Chim. Acta,* **5** (1971), 563. The kinetics of the reactions [M(qas)X]$^+$ + Y$^-$ → [M(qas)Y]$^+$ + X$^-$ (qas = tris(*o*-dimethylarsinophenyl)arsine) have been studied in methanol. A mechanism is proposed in which substitution occurs in planar four-coordinate intermediates, in which one or two of the four arsenics of the tetradentate ligand are temporarily displaced.

210. Mathew, M., Palenik, G. J., Dyer, G., and Meek, D. W., *Chem. Commun.* (1972), 379. The first structural study of a low-spin trigonal-bipyramidal nickel(II) complex containing an unsymmetric trigonal field: [NiLBr]ClO$_4$ (L = As(*o*-C$_6$H$_4$AsPh$_2$)$_2$(*o*-C$_6$H$_4$SMe)).

211. Bianchi, A., Ghilardi, C. A., Mealli, C., and Sacconi, L., *Chem. Commun.* (1972), 651. X-ray structural analysis of two five-coordinate complexes of cobalt(II) and nickel(II) having the MN$_2$P$_2$Br chromophore and the same ligands shows that axial elongation in square-pyramidal low-spin complexes is more favourable for nickel than cobalt.

PART 5

212. Feltham, R. D., and Metzger, H. G., *J. organometal. Chem.*, **33** (1971), 347. The synthesis of vicinal bis(dimethylarsino) compounds and their proton n.m.r. and mass spectra; see main reference 9(b).

213. Preer, J. R., Tsay, Fun-Dow, and Gray, H. B., *J. Am. chem. Soc.*, **94** (1972), 1875. Electronic spectra of diars and the e.s.r. spectra of u.v.-irradiated diars and dae; see main reference 245.

214. Blight, D. G., Kepert, D. L., and Mandyczewsky, R., *J. chem. Soc. (Dalton)* (1972), 313. Synthesis and characterisation of $[WCl_5(diars)]$, $[WOCl_4(diars)]$, $[MoOCl_3(diars)]$, $[Mo_2OCl_8(diars)_2(dioxan)]$, $[MoCl_4(diars)]$ and $[WCl_4(diars)]$. Both metals (Mo, W) adopt seven- rather than eight-coordinate geometries in these diars complexes.

215. Douglas, P. G., Feltham, R. D., and Metzger, H. G., *J. Am. chem. Soc.*, **93** (1971), 84. The reactions of *trans*-$[RuCl(NO)(diars)_2]Cl_2$ with nitrogen bases is investigated and the conversion of *trans*-$[RuN_3Cl(diars)_2]$ to dinitrogen and nitro derivatives is studied using ^{15}N-substituted complexes; see main reference 93.

216. Finn, P., and Jolly, W. L., *Inorg. Chem.*, **11** (1972), 893. The nitrogen 1s binding energies (*ca.* 400 eV) of several transition-metal nitrosyls that also contain diars are reported.

217. Cheney, A. J., and Shaw, B. L., *J. chem. Soc. (A)* (1971), 3545. The oxidative addition reactions of $[PtMe_2(diars)]$ with various alkyl, acyl and allylic halides to give either *cis*- or *trans*-platinum(IV) adducts are studied using i.r. and n.m.r. spectroscopy.

218. McCleverty, J. A., and Orchard, D. G., *J. chem. Soc.* (A) (1971), 3784. The preparation and voltammetric behaviour of several complexes of the type $[ML(S-S)_2]^{n-}$ ($n = 0, 1$) are reported. In these complexes M may be Fe, Co or Ni; L is diars or dae; S—S is various dithiolene ligands.

219. Braterman, P. S., Wilson, V. A., and Joshi, K. K., *J. organometal. Chem.*, **31** (1971), 123. The complexes $[(Et-diars)M(SR)_2]$ (M = Pd, Pt; R = Me, Ph) act as chelating disulphide ligands toward Group VIA carbonyls. There is no spectroscopic evidence of metal–metal bond formation in the products (for example $[(Et-diars)Pd(SMe)_2Mo(CO)_4]$).

220. Bennett, M. A., and Wild, J. D., *J. chem. Soc.* (A) (1971), 545. The preparation and characterisation of Co(II) and Co(III) complexes of edas are described. The Co(II) complexes, $[CoX_2(cis\text{-}edas)_2]$ (X = Cl, Br, I, NO_3), are readily oxidised to Co(III) complexes of the type $[CoX_2(cis\text{-}edas)_2]Y$ (X = Cl, Br; Y = Cl, Br, PF_6; X = NCS; Y = PF_6, $Co(SCN)_4$). One complex of *trans*-edas was isolated, $[CoI_2(trans\text{-}edas)_2]$, and the physical measurements indicate that the structure is polymeric with bridging *trans*-edas groups.

221. Bennett, M. A., and Wild, J. D., *J. chem. Soc.* (A) (1971), 536. The preparation and characterisation of Ni(II) and Ni(III) complexes of edas are described. The diamagnetic $[NiX_2(cis\text{-}edas)_2]$ (X = Cl, Br, I, NCS, NO_3) and $[Ni(cis\text{-}edas)_2X]PF_6$ (X = Cl, Br, NCS) complexes are probably five-coordinate in solution and tetragonally-distorted octahedral in the solid state. The six-coordinate nickel(III) complexes $[NiX_2(cis\text{-}edas)_2]Y$ (X = Cl, Br; Y = Cl, Br, PF_6) have one unpaired electron. Octahedral nickel(II) complexes of *trans*-edas appear to contain bridging ligand groups. Other nickel(II) complexes of *cis*-edas, as well as an unstable nickel(IV) complex, are also described in this extensive study.

222. Anker, M. W., and Cotton, R., *Aust. J. Chem.*, (1971), 2223. Preparation and characterisation of $[Mo(CO)_3(dae)X_2]$ and $[Mo(CO)_2(dae)_{1.5}X_2]_2$ (X = Cl, Br or I); the latter species, which contain a bridging dae group, react with carbon monoxide to reform the tricarbonyl complex, and with excess dae to give $[Mo(CO)_2(dae)_2X_2]$.

223. Mann, B. E., Masters, C., and Shaw, B. L., *J. chem. Soc. (Dalton)* (1972), 276. A ^{31}P-INDOR investigation of $[IrCl_3(PMe_2Ph)(dpe)]$ and related complexes, such as $[IrCl_3(M'Me_2Ph)(dae)]$ (M' = As, P).

224. Watt, G. W., and Cuddeback, J. E., *J. inorg. nucl. Chem.*, 33 (1971), 259. Synthesis of $[Pt(dae)X_2]$ (X = Cl, Br or I) and $[Pt(dae)en]Y_2$ (Y = Cl, Br) and comparison with the analogous dpe complexes.

225. Newman, J., and Manning, A. R., *J. chem. Soc. (Dalton)* (1972), 241. Reactions of bis-(tetracarbonylcobaltio)mercury with dae, dpe and monodentate phosphine ligands. Structures for the products of the types $\{[(dae)_2Co(CO)_2]_2Hg\}$, $\{[(dae)Co(CO)_3]_2Hg\}$ and $\{[Co(CO)_3HgBr]_2(dae)\}$ are proposed from a study of the carbonyl region of their i.r. spectra.

226. Drew, M. G. B., *J. chem. Soc. (Dalton)* (1972), 626. A report of the crystal structure of $[MoBr_2(CO)_2(dam)_2]$, wherein the seven-coordinate molybdenum(II) atom has a distorted capped octahedron geometry and one dam ligand is unidentate. The preliminary X-ray data was communicated in main reference 190.

227. Canty, A. J., and Gatehouse, B. M., *J. chem. Soc. (Dalton)* (1972), 511. The crystal structure of $\{[(C_6F_5)_2Hg]_2(dam)\}$ shows a T-shaped distribution of two C_6F_5 groups and an As atom about each Hg atom. The bridging dam ligand is weakly bonded to each mercury atom, as was suggested by the properties of the adduct, described in main reference 194.

228. Cheney, A. J., and Shaw, B. L., *J. chem. Soc. (A)* (1971), 3549. The complexes $[MX_2(dmpa)]$ (M = Pd or Pt; X = Cl, Br, I) are separated into the rac- and meso-forms. Oxidative addition reactions of both forms of $[PtMe_2(dmpa)]$ with alkyl, acetyl and allylic halides to give either *cis*- or *trans*-platinum(IV) adducts are studied using i.r. and n.m.r. spectroscopy.

229. Einstein, F. W. B., and Jones, R. D. G., *J. chem. Soc. (A)* (1971), 3359. The crystal structure of $[Co_4(CO)_8(ffars)_2]$ reveals that the cobalt atoms form a tetrahedral cluster and each ffars ligand forms a bridge between two Co atoms; see main reference 221.

230. Roberts, P. J., and Trotter, J., *J. chem. Soc. (A)* (1971), 1501. The crystal structure of $[Mo(CO)_4(Me_2AsCF(CF_3)CF_2AsMe_2)]$ is described. The five-membered chelate ring is non-planar, and the staggered conformation places the CF_3 group in an equatorial position relative to the molybdenum atom.

231. Nowell, I. W., and Trotter, J., *J. chem. Soc. (A)* (1971), 2922. The crystal structure of $[Mo(CO)_4(Me_2AsCHFCF_2AsMe_2)]$ is described. The non-planar chelate ring has the hydrogen atom in an equatorial position.

232. Carlton, T. R., and Cook, C. D., *Inorg. Chem.*, 10 (1971), 2628. Synthesis and n.m.r. data for o-phenylenediarsine, o-phenylene-bis(methylarsine) and o-phenylene-As, As, As'-trimethyl-arsine, as well as related primary and secondary phosphines.

233. Bishop, J. J. and Davidson, A., *Inorg. Chem.*, 10 (1971), 826. Group VIA carbonyl complexes of the type $[LM(CO)_4]$ (M = Cr, Mo, W) where L is the new ditertiary arsine, ferrocene-1,1'-bis(dimethylarsine)(fdma), or ferrocene-1,1'-bis(diphenylarsine)(fdpa). The former ligand acts as a bridging group in μ-fdma-$[(fdma)Mo(CO)_3]_2$ and as both a monodentate and a bidentate in $[(fdma)_2Mo(CO)_3]$.

234. Bishop, J. J., and Davidson, A., *Inorg. Chem.*, 10 (1971), 832. Complexes of Group VIII salts and fdma or fdpa. The complexes $[M(fdma)X_2]$ (M = Pd; X = Cl, Br; M = Pt; X = Cl, Br, I) react with an excess of fdma to give $[M(fdma)_2X]^+$ and $[M(fdma)_2]^{2+}$ salts but the corresponding fdpa complexes do not add a second fdpa ligand. Whilst the analogous nickel(II) complexes could not be prepared, the interesting carbonyl complexes of nickel(II), $[(fdma)Ni(CO)I_2]$ and $[(fdpa)Ni(CO)I_2]$ were characterised.

235. Pierpont, C. G., and Eisenberg, R., *Inorg. Chem.*, 11 (1972), 828. The crystal structure of the nickel(II) carbonyl complex of fdma, isolated as described in reference 234, shows that $[(fdma)Ni(CO)I_2]$ has a nearly regular trigonal-bipyramidal geometry. The fdma ligand, which has a 'stepped' conformation of the cyclopentadienyl rings, occupies an axial and an equatorial position, and the carbonyl group fills the other axial position.

236. Feltham, R. D., Silverthorn, W., Wickman, H., and Wesolowski, W., *Inorg. Chem.*, 11 (1972), 676. The magnetic susceptibilities, electron spin resonance, and Mössbauer spectra of the complexes *trans*-[FeX(NO)(das)$_2$]$^+$ and *trans*-[FeX$_2$(das)$_2$]$^+$ have been obtained. See main reference 59(b).

AUTHOR INDEX

Numbers in bold type refer to the reference number; numbers in parentheses refer to the pages in the text containing the particular reference. The final number refers to the page containing the full reference citation.

398 *Author Index*

196; **2.1030** (143) 197; **2.1294** (169, 170)
203; **3.96** (214, 215) 257; **3.142** (219)
258; **3.145** (220) 258; **3.146** (219, 220)
258; **3.152** (220) 258; **3.262** (231) 260;
5.3 (311) 368; **A.86,** 379
Cotton, J. D.   **2.408** (73, 74) 185
Coulson, D. R.   **2.1154** (154) 200
Coulton, K. G.   **A.86,** 379
Coussmaker, C. R. C.   **2.971** (135, 136, 138, 139) 196
Crabtree, R. H.   **A.35,** 376
Craddock, J. H.   **3.352** (238) 262
Cradock, S.   **2.1159** (155) 200
Craig, D. P.   **1.4** (3) 29; **3.560** (252) 266
Cramer, R. D.   **2.1166** (155) 200
Cresswell, P. J.   **1.27** (8, 15) 29
Crisp, W. P.   **5.39** (316, 318, 351, 358) 368
Crociani, B.   **2.619** (102, 103) 190; **2.1341** (173) 204; **3.473** (246) 265; **3.474** (246) 265; **3.493** (247, 248) 265
Crooks, G. R.   **2.369** (70) 184; **2.439** (77) 185; **2.717** (104, 123) 191; **3.251** (230) 260; **3.372** (237, 239, 241, 243) 262
Crosby, G. A.   **5.90** (326, 358) 370
Cross, R. J.   **2.1145** (154) 200; **2.1163** (155) 200
Crow, J. R.   **2.918** (130) 195; **3.423** (244) 264; **3.424** (244) 264; **5.197** (345, 350) 372; **5.201** (345, 364, 365) 372; **5.207** (347, 348, 353) 372; **5.212** (348) 372; **5.213** (348) 372; **5.217** (348) 372; **5.218** (348, 349) 372; **5.238** (353) 373; **A.44,** 376
Cuddebank, J. E.   **A.224,** 388
Cule-Davis, W.   **2.1215** (159) 201
Cullen, W. R.   **3.192** (225) 259; **3.196** (225) 259; **5.24** (314) 368; **5.25** (313, 314) 368; **5.26** (313, 314, 344, 345, 346, 349, 350, 363) 368; **5.83** (323, 330, 346, 349, 362) 369; **5.159** (335) 371; **5.197** (345, 356) 372; **5.201** (345, 364, 365) 372; **5.202** (345, 365) 372; **5.203** (346, 359) 372; **5.206** (346, 347, 348) 372; **5.207** (347, 348, 353) 372; **5.208** (347, 348) 372; **5.209** (347, 348) 372; **5.212** (348) 372; **5.213** (348) 372; **5.214** (348) 372; **5.217** (348) 372; **5.218** (348, 349) 372; **5.222** (349, 365) 372; **5.223** (349) 373; **5.233** (352) 373; **A.21,** 375; **A.44,** 376
Cundy, C. S.   **2.1026** (143) 197
Cunninghame, R. G.   **4.30** (275) 305; **4.31** (275) 305; **A.192,** 385
Cuppers, H. G. A. V.   **2.896** (128) 195
Curran, C.   **2.1348** (174) 204
Curtis, N. F.   **2.281** (64) 182; **4.4** (271) 305; **5.77** (325) 369; **A.33,** 376

Dahl, L.   **2.531** (89) 187
Dahl, L. F.   **2.673** (101) 190; **2.929** (132) 915; **2.1045** (145) 198; **4.86** (286, 294) 306

Dahl, L. W.   **2.120** (45) 179
Dahl, O.   **2.970** (135, 137, 138, 140, 141, 142) 196; **2.1307** (170, 171) 203
Dahlhoff, W. V.   **2.611** (96, 137) 189; **4.54** (279) 306
Dahn, D. J.   **2.325** (67) 183
Dalton, J.   **3.73** (213) 256; **3.84** (213) 256; **3.111** (216) 257; **A.208,** 386
Daly, J. J.   **2.494** (85) 186; **2.495** (85) 186
Damman, C. B.   **A.76,** 378
Dapporto, P.   **4.95** (288, 294) 306; **4.105** (289, 292, 293) 307; **4.110** (291) 307
Darensbourg, D. J.   **1.63** (17, 19) 30; **1.72** (19) 30; **3.75** (213) 256
Dartinguenave, M.   **A.109,** 380; **A.110,** 380
Dartinguenave, Y.   **A.109,** 380; **A.109,** 380
Davidson, J. M.   **2.1144** (154, 165) 200; **2.1273** (168) 202; **3.543** (250) 266
Davies, G. R.   **2.337** (68) 183
Davies, N. R.   **1.5** (3) 29; **3.3** (207) 255
Davies, R. F. B.   **2.625** (95, 135, 136, 137, 138) 189
Davis, A. R.   **2.1359** (175) 204
Davis, B. R.   **2.266** (63) 182; **2.571** (93) 188; **2.572** (93) 188
Davis R.   **4.15** (274) 305; **4.16** (274) 305; **A.22,** 375
Davison, A.   **2.35** (38) 177; **2.345** (69) 184; **2.514** (88) 187; **3.184** (223, 224) 258; **A.6,** 374; **A.185,** 384; **A.233,** 388; **A.234,** 388
Dawson, J. W.   **4.69** (283, 287, 294, 295, 296, 297) 306; **4.78** (284) 306; **4.128** (304) 307; **A.202,** 385
Day, J. P.   **2.13** (36) 177
Deacon, G. B.   **2.1345** (174) 204; **2.1346** (174) 204; **2.1347** (174) 204; **2.1351** (174) 204; **2.1354** (174) 204; **2.1357** (175) 204; **5.153** (334, 355, 365) 371; **5.156** (334) 371; **5.194** (343, 364) 372
Dean, R. R.   **2.1127** (152) 199
De Beer, J. A.   **A.29,** 376; **A.40,** 376; **A.41,** 376; **A.167,** 383
De Boer, J. L.   **1.99** (21) 31; **2.765** (110) 192
De Charentenay, F.   **2.769** (110) 192
Deeming, A. J.   **2.401** (73, 74) 185; **2.404** (73, 74, 75) 185; **2.415** (73) 185; **2.676** (101) 192; **2.809** (115, 117, 122) 193; **2.811** (115) 193; **2.835** (119) 193; **2.840** (120, 122) 193; **2.842** (120) 194; **3.353** (238) 262; **3.396** (241, 242) 263; **3.397** (242) 263; **3.402** (243) 263; **3.403** (243) 263; **3.406** (241, 243) 263; **A.49,** 377
Degischer, C.   **4.60** (280, 291) 306
Degonello, G.   **2.690** (102) 190; **2.691** (102, 103) 190; **2.1168** (155) 200; **2.1169** (155) 200; **3.519** (249) 266; **5.230** (351) 373
Dehn, W. M.   **3.2** (207) 255
Delbeke, F. T.   **2.55** (42, 43) 178; **2.90** (43) 178; **2.317** (63) 183

418   *Author Index*

Rossetti, R.   **2.527** (89) 187; **3.191** (225)
    229; **3.281** (233) 260
Rossi, M.   **2.376** (70, 71) 184; **2.563** (92)
    188; **2.565** (92) 188; **2.566** (92, 93) 188;
    2.567 (92) 188; **2.581** (93) 188; **2.587**
    (94) 188; **2.589** (94) 188; **2.590** (94) 188;
    2.722 (104) 191; **3.289** (233) 261; **A.45,** 376
Roth, J. F.   **3.337** (238) 262; **3.351** (238) 262;
    3.352 (238) 262
Roundhill, D. M.   **2.1116** (151, 152) 199;
    2.1123 (152) 199
Rouschias, G.   **2.293** (65) 182; **2.296** (65)
    183; **3.147** (220) 258; **3.148** (219, 220)
    258
Rowe, G. A.   **2.257** (62) 182; **2.258** (62, 63,
    64, 65, 66) 182; **2.259** (62, 63, 65) 182;
    2.292 (65) 182; **2.295** (65, 66) 182;
    2.1068 (146) 198; **3.149** (219, 220) 258
Rowe, J. M.   **1.10** (4) 29; **2.623** (97) 189;
    2.1124 (152) 919; **2.1255** (163) 202
Rubessa, F.   **2.1287** (168, 169) 203
Rubtrova, N. D.   **2.741** (107) 192
Ruddick, J. D.   **1.37** (11) 29; **2.1148** (154,
    156) 200; **A.85,** 379
Rule, L.   **2.1288** (168, 169) 203
Rüsch, L.   **2.919** (130) 195
Rycheck, M.   **2.1055** (145) 198

Saalfield, F. E.   **1.106** (23) 31
Sabatini, A.   **3.466** (246) 265
Sabherwal, I. H.   **2.183** (53, 57) 180; **2.556**
    (91) 188; **2.557** (91) 188; **2.921** (130)
    195; **2.1023** (143, 167, 170) 197
Sacco, A.   **2.223** (56) 181; **2.375** (70, 71) 184;
    2.376 (70, 71) 184; **2.511** (88) 187;
    2.512 (88, 89) 187; **2.518** (88) 187; **2.529**
    (89, 91) 187; **2.565** (92) 188; **2.566** (92,
    93) 188; **2.567** (92) 188; **2.581** (93)
    188; **2.587** (94) 188; **2.589** (94) 188;
    2.590 (94) 188; **2.597** (95, 96) 189; **2.722**
    (104) 191; **2.730** (105) 191; **2.731** (105,
    106, 108) 191. **2.933** (132) 195; **3.101**
    (216) 257; **3.289** (233) 261; **5.200** (345)
    372; **A.46,** 376
Sacconi, L.   **2.1006** (137) 197; **4.11** (273,
    296, 297, 298) 305; **4.47** (277) 306;
    4.48 (277) 306; **4.49** (278) 306; **4.50**
    (278, 279, 280, 281) 306; **4.55** (279, 280,
    289) 306; **4.62** (280) 306; **4.63** (280)
    306; **4.93** (288, 289) 306; **4.94** (288)
    306; **4.95** (288, 294) 306; **4.96** (288)
    306; **4.97** (288) 306; **4.98** (288, 289)
    307; **4.102** (289) 307; **4.103** (289, 290,
    292) 307; **4.104** (289, 290, 292) 307;
    4.105 (289, 292, 293) 307; **4.107** (290)
    307; **4.109** (290) 307; **4.110** (291) 307;
    4.116 (292, 298) 307; **4.120** (296, 298)
    307; **5.183** (341) 372; **5.193** (385) 385:
    5.194 (385) 385; **A.211,** 386

Saffer, B. A.   **3.430** (244) 264; **3.435** (244) 264
Saito, T.   **2.569** (93) 188; **2.580** (93) 188;
    2.959 (134) 196; **2.960** (134) 196; **2.961**
    (134) 196; **2.963** (134, 151) 196
Salthouse, J. A.   **1.93** (21, 22) 31
Sams, J. R.   **3.192** (225) 259; **5.203** (346,
    357) 372; **5.206** (346, 347, 348) 372;
    5.207 (347, 348, 353) 372; **5.209**
    (347, 348) 372; **5.212** (348) 372; **A.44,**
    376
Sanders, J. R.   **2.72** (63) 182; **A.66,** 377
Sandhu, R. S.   **5.185** (341) 372
Sandhu, S. S.   **2.172** (52, 53) 180; **2.307** (67)
    183; **2.310** (67) 183; **3.175** (223, 224)
    258; **3.177** (223) 258; **3.558** (252) 266;
    5.57 (320, 321, 322, 357, 359) 369;
    5.74 (324, 339) 369; **5.82** (325, 346) 369;
    5.177 (339) 372; **5.185** (341) 372
Sansoni, M.   **1.110** (23) 31; **2.861** (23, 24)
    194; **2.1051** (145) 198; **2.1318** (171) 203;
    2.1338 (173) 204; **A.150,** 382
Sarafidis, C.   **5.178** (340) 372
Saran, M. S.   **A.197,** 385; **A.198,** 385
Satorelli, N.   **2.58** (42) 178; **2.235** (57) 181;
    3.376 (239) 262; **3.394** (241, 242, 243)
    263. **5.111** (329) 370; **5.171** (339) 371
Sauerborn, H.   **2.103** (41) 179
Saunders, V. R.   **1.82** (21, 22) 30; **1.84** (21,
    22) 30; **1.85** (21, 22) 31; **1.87** (21) 31;
    1.88 (21, 22) 31; **1.90** (21) 31
Savage, C. A.   **4.24** (276, 284) 305; **4.25** (276,
    284) 305
Sbrana, G.   **2.394** (72, 73) 184
Scaife, D. E.   **1.27** (8, 15) 29
Scatturin, V.   **1.108** (23) 31; **2.774** (113) 192;
    2.975 (135, 137, 138) 196; **2.976** (135,
    137, 138, 144) 196; **2.978** (138, 143) 196;
    2.994 (139) 197; **2.995** (139) 197; **2.1000**
    (139) 197; **2.1035** (144) 197; **2.1050** (145)
    198; **2.1080** (147) 198
Schaal, W.   **2.1022** (143) 197
Schaechl, H.   **2.1136** (152) 200; **2.1189** (156)
    201
Schelle, O.   **A.10,** 374
Schindler, H.   **2.140** (48) 179
Schiurer, E.   **2.973** (138, 156, 157) 196;
    2.1204 (157) 201; **2.1206** (157) 201;
    2.1313 (171, 175) 203
Schlegel, H.   **2.1269** (167, 170, 172) 202
Schmelz, A.   **2.1348** (174) 204
Schmid, D. D.   **2.618** (97) 189
Schmid, G.   **2.187** (54) 180; **2.1170** (155) 200;
    2.1244 (163, 164) 202; **3.273** (231) 260
Schmidbauer, H.   **2.942** (132) 196; **3.587**
    (254) 267
Schmitz-Dumont, O.   **2.226** (56) 181; **2.1022**
    (143) 197
Schmutzler, R.   **2.937** (132) 195
Schneider, R. J. J.   **2.36** (38) 177

SUBJECT INDEX